COLLEGE MATHEMATICS
for Business and the Social Sciences

COLLEGE MATHEMATICS
for Business and the Social Sciences

Arthur Lieberman *Cleveland State University*

Brooks/Cole Publishing Company
Monterey, California

Consulting Editor: Robert J. Wisner

Brooks/Cole Publishing Company
A Division of Wadsworth, Inc.

© 1982 by Wadsworth, Inc., Belmont, California 94002. All rights reserved. No part of this book may be reproduced, stored in a retrieval system, or transcribed, in any form or by any means—electronic, mechanical, photocopying, recording, or otherwise—without the prior written permission of the publisher, Brooks/Cole Publishing Company, Monterey, California 93940, a division of Wadsworth, Inc.

Printed in the United States of America

10 9 8 7 6 5 4 3 2 1

Library of Congress Cataloging in Publication Data

Lieberman, Arthur.
    College mathematics for business and the social sciences.
    Includes index.
    1. Mathematics—1961–   2. Business mathematics. 3. Social sciences—Mathematics. I. Title.
QA39.2.L53   510   81-17976
ISBN 0-8185-0474-9   AACR2

Subject Editor: Craig Barth
Manuscript Editor: Margaret E. Hill
Production Editor: Cece Munson, Joan Marsh
Interior Design: Stan Rice
Cover Design: Stan Rice, Graphics Plus
Illustrations: Scientific Illustrators
Typesetting: Science Typographers

Preface

It is commonly agreed that mathematics is useful in engineering and in the physical sciences. But all too few students are aware of the many uses for mathematics in other aspects of everyday life. The purpose of this textbook is to reveal some areas of applied mathematics to students who need tools for problems of the real world of business and the social sciences.

The approach of this book is to present the concepts and techniques of mathematics in a manner that is easily understood by anyone who makes a reasonable effort. Each important point is illustrated by examples that demonstrate how mathematics can be used by decision makers in industry, business, government, and various other organizations in our society. Except for the first, every chapter contains applications and applied exercises in almost every section. Problems that apply especially to business and social science are marked accordingly. The topics covered include algebra, linear algebra, linear programming, mathematics of finance, probability, and calculus.

Since examples provide the easiest and most natural vehicle for learning, numerous examples of graduated complexity are included throughout the exposition. These are very carefully detailed in order to demonstrate the myriad techniques for problem solving. In order to keep things "concrete," the numerous exercises are nearly all computational. Exercises preceded by an asterisk (*) go beyond the material in the text and are intended for advanced students or for those students who have special interests. All the other exercises are intended for every student to do.

Several features of this book are intended to facilitate learning and retention of the important ideas that students will need for further study. For example, although some formulas are derived by algebraic manipulation, many are merely stated, their utility being justified by use and example. There are no phony proofs, loose statements of mathematical principles, or invalid reasoning. Logarithms are no longer needed for numerical computation since the advent of inexpensive pocket calculators, and so they are not used for such purposes; however, common and natural logarithmic functions are included, along with their applications. The concept of area is approached by a method of exhaustion, a method that is more intuitive and easier to understand than others. Students thus find themselves more comfortable with the concept of area and the corresponding use of the definite integral. The basic ideas of probability are presented before introducing the technical procedures and formulas for counting techniques, such as permutations and combinations. Binomial probability is covered, but the binomial theorem—which many students find difficult—is not.

Class testing with a preliminary edition of this text was conducted at Cleveland State University. Over 3000 students participated during the year-long test period, and they were judged to have used the material successfully.

The text is organized to allow for a great deal of flexibility to meet the needs of course and students. It is important to note that Chapters 1 and 2 cover algebra at the level required later in the book: they are thus prerequisite to the other chapters. Chapter 3 is necessary for Chapter 4, but taken together, the two are independent of the rest of the book. Chapter 5 is also an independent chapter. Chapter 6 provides the basis for Chapter 7, Chapter 8, and Section 13.3; but Chapters 7 and 8 are independent of one another. Chapters 9 through 13 (except for Section 13.3) form a completely separate sequence, wherein functions and graphs are covered at the level required for calculus. In its entirety, this book is suitable for a full year course of study.

A calculator is recommended but not required for using this text. The sort of calculator used is not important, just so the student knows its uses and limitations. But it is important to have the calculator capability to compute e^x, $\ln x$, and y^x, however those keys are labeled. Some of the exercises are designed especially for calculators, and they are marked with ◈.

All numerical answers for examples and exercises were computed by means of a hand calculator; thus, they are sometimes not exact answers. For instance, a statement such as $\sqrt{2} = 1.414$ means that 1.414 is a "good" approximation for $\sqrt{2}$ in the sense that neither 1.413 nor 1.415 is a better approximation (notice that no special symbol is used for approximate equality).

This book started as a rough manuscript in 1977 and has gone through several revisions since. I wish to thank Craig Barth of Brooks/Cole for having the persistence to guide and develop this project and Ann Melville for doing all the typing and retyping.

Valuable suggestions have been made by Robert J. Wisner, Consulting Editor, and the following reviewers: Wilson Banks, Illinois State University (Normal, IL); Leonard Bruening, Cleveland State University (Cleveland, OH); Allen Emerson, Montgomery College (Rockville, ND); Gary Grimes, Mount Hood Community College (Gresham, OR); Shirley Lilge, Cleveland State University (Cleveland, OH); Laurence Maher, Jr., North Texas State University (Denton, Texas); Roger Marty, Cleveland State University (Cleveland, OH); Mike Murphy, University of Houston—Downtown Campus (Houston, TX); Wesley Sanders, Sam Houston State University (Huntsville, TX); Sherwood D. Silliman, Cleveland State University (Cleveland, OH); Richard M. Smith, Bryant College (Smithfield, RI).

My colleagues, Professors David Bittker, Richard H. Black, Leonard F. Bruening, Allen W. Brunson, S.H. Chang, John Chao, Pratibha Ghatage, Paula C. Gnepp, Rasul A. Khan, Frank W. Lozier, Roger H. Marty, Charles Reno, Stewart M. Robinson, Brian M. Scott, Allan J. Silberger, Sherwood D. Silliman, and Bhushan Wadhwa, of Cleveland State University, have taught from a preliminary edition of this book and have been very helpful.

A.L.

Contents

1 Sets, Numbers, and Algebraic Manipulation 1
- 1-1 Sets 1
- 1-2 Numbers and the Number Line 6
- 1-3 Exponents 11
- 1-4 Polynomials 22
- 1-5 Factoring 32
- 1-6 Rational Expressions 36
- 1-7 Algebraic Expressions 45
- Chapter 1 Review 50

2 Equations and Inequalities 52
- 2-1 Linear Equations in One Variable 52
- 2-2 Quadratic Equations 59
- 2-3 More General Equations 66
- 2-4 Inequalities 73
- 2-5 Word Problems 80
- 2-6 Coordinates and Graphs 88
- Chapter 2 Review 93

3 Linear Algebra 95
- 3-1 Matrices 95
- 3-2 Matrix Multiplication 104
- 3-3 Two Linear Equations in Two Unknowns 119
- 3-4 Systems of Three Linear Equations in Three Unknowns 129
- 3-5 Gauss–Jordan Elimination 135
- 3-6 Inverse of a Matrix 144
- 3-7 Input-Output Analysis 154
- Chapter 3 Review 157

4 Linear Programming 159

4-1 Systems of Linear Inequalities 159
4-2 The Graphical Method 168
4-3 Applications 175
4-4 The Simplex Method 183
4-5 The Dual Problem 198
Chapter 4 Review 210

5 Mathematics of Finance 212

5-1 Interest 212
5-2 Annuities 220
5-3 Present Value and Amortization 229
5-4 Amortization Tables 236
Chapter 5 Review 242

6 Probability 244

6-1 An Introduction to Probability 244
6-2 Cartesian Products 251
6-3 Set Operations and Counting 255
6-4 Complements, Unions, and Intersections 262
6-5 Independent Events 268
6-6 Conditional Probability 273
6-7 Bayes' Theorem 281
Chapter 6 Review 287

7 More Topics in Probability 290

7-1 Permutations 290
7-2 Combinations and Hypergeometric Probability 295
7-3 Binomial Probability 301
7-4 More Examples 306
Chapter 7 Review 310

8 Applications of Probability to Decision Making 312

8-1 Expected Value and Odds 312
8-2 Decision Theory 319
8-3 Game Theory 323
8-4 Games with Mixed Strategies 328
Chapter 8 Review 335

9 Functions and Graphs 337

9-1 The Concept of Function 337
9-2 Lines and Linear Functions 346

9-3 Quadratic Functions 355
9-4 Arithmetic and Composition of Functions 361
9-5 Inverse of a Function 365
9-6 The Exponential Function 370
9-7 The Natural Logarithm Function 376
9-8 The Common Logarithm Function 383
Chapter 9 Review 387

10 Differential Calculus 390
10-1 Limits and Continuity 390
10-2 The Derivative 396
10-3 Formulas for Derivatives 404
10-4 Rates of Change 407
10-5 Marginal Analysis 411
10-6 The Product and Quotient Rules 415
10-7 The Chain Rule 420
10-8 Higher Order Derivatives 426
Chapter 10 Review 428

11 Applications of Derivatives 431
11-1 Extreme Values 431
11-2 The First and Second Derivative Tests 444
11-3 Applications to Optimization 449
11-4 Partial Derivatives 459
11-5 Extreme Values of Functions of Two Variables 466
11-6 Lagrange Multipliers 472
11-7 Implicit Differentiation and Related Rates 476
Chapter 11 Review 480

12 Integral Calculus 482
12-1 The Antiderivative 482
12-2 Area 489
12-3 The Fundamental Theorem of Calculus 495
12-4 The Area Between Curves 503
12-5 Computation of More Antiderivatives 511
Chapter 12 Review 519

13 Applications of Integration 521
13-1 Applications to Economics 521
13-2 Separable Differential Equations 528
13-3 Probability Density Functions 535

13-4 Integration by Parts 543
13-5 The Trapezoidal Rule 547
    Chapter 13 Review 550

Appendix: Tables for Mathematics of Finance 553
Answers to Odd-Numbered Exercises and Review Exercises 566
Index 596

1

Sets, Numbers, and Algebraic Manipulation

In this book, we study several different branches of mathematics and their application to real-world situations. The first chapter is about sets, numbers, and algebraic manipulation. You need to understand and be able to use this material in order to master topics central to this course, such as linear algebra, probability, and calculus.

Section 1-1 Sets

Mathematicians speak in the language of sets. *Set* is an undefined term, just as *point* and *line* are undefined in geometry. A set is understood to be a collection of objects. A set may be described by listing its members between the symbols { and }, which are called *set braces*. Thus, the expression $\{5, 6, 7, 8\}$ is read: the set 5, 6, 7, and 8. The *members* of the set are $5, 6, 7,$ and 8. The symbol for set membership is $\in$. Thus $5 \in \{5, 6, 7, 8\}$ is read: 5 is a member (or *element*) of $\{5, 6, 7, 8\}$. The symbol $\notin$ is the negation of $\in$. Thus, $4 \notin \{5, 6, 7, 8\}$ is read: 4 is not a member of $\{5, 6, 7, 8\}$.

A set can also be described by a condition. One can speak of {United States senators} or {states in the United States} or {first five letters of the alphabet}.

Another procedure for describing a set is *set-builder notation*: $\{x|x$ has property $P\}$. This is read: the set of all x such that x has property P. Here P is some property that each element x either satisfies or does not satisfy. The symbol x is a variable, and any other symbol could be used instead of x, so that $\{x|x$ has property $P\}$ is precisely the same as $\{y|y$ has property $P\}$. If $A = \{2, 3, 5, 8, 10, 12, 16\}$, then $\{s|s \in A$ and s is even$\}$ is $\{2, 8, 10, 12, 16\}$ because 2, 8, 10, 12, and 16 are the even members of A.

Examples Let $A = \{a, b, c, d, e, f, g\}$, $B = \{c, d, e, f, g, h, i, j, k\}$ and $C = \{f, g, h, i\}$.

1. $c \in A$, $c \in B$, but $c \notin C$
2. $5 \notin A$, $5 \notin B$, and $5 \notin C$
3. $f \in A$, $f \in B$, and $f \in C$
4. $\{x|x \in A$ and $x \notin B\} = \{a, b\}$ because a and b are in A but not in B. Every other member of A is in B.
5. $\{y|y \in B$ and $y \in A\} = \{c, d, e, f, g\}$

■

Two sets are equal precisely when they have the same members, so that $\{2, 3, 4\} = \{4, 2, 3\} = \{2, 2, 2, 3, 4\} = \{2, 2, 4, 3, 3, 4, 2, 4\}$, since each of these sets comprises exactly the numbers 2, 3, and 4.

There is exactly one set with no members. This set is called the *empty set*, or the *null set*, and is denoted by the symbol $\emptyset$. Then $\emptyset = \{$odd integers divisible by 2$\} = \{$three-sided squares$\}$.

A set A is a *subset* of a set B (written $A \subseteq B$) if every member of A is a member of B. If A is not a subset of B, we write $A \not\subseteq B$. Notice that $B \subseteq B$, since every member of B is a member of B. It is also true that $\emptyset \subseteq B$.

Examples

6. $\{a, b, c\} \subseteq \{a, b, c, d, e\}$
7. $\{a, b, c\} \subseteq \{a, c, q, r, r, c, b\}$
8. $\{a, b, c\} \not\subseteq \{a, b, d, e, f, g\}$ because $c \notin \{a, b, d, e, f, g\}$
9. $\{a, a, c, d, d, c, b, a, b, c\} = \{a, b, c, d\}$ because these sets have the same members

■

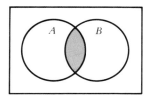
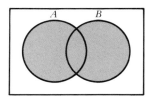

Diagram 1.
$A \cap B$ is shaded.

Diagram 2.
$A \cup B$ is shaded.

Now let's define certain operations that involve two sets.

Definition. Let A and B be sets. The *intersection* of A and B (written $A \cap B$) is the set of elements that belong to both A and B.
$$A \cap B = \{x \mid x \in A \text{ and } x \in B\}$$

Definition. Let A and B be sets. The *union* of A and B (written $A \cup B$) is the set of elements that belong either to A or to B (or to both).
$$A \cup B = \{x \mid x \in A \text{ or } x \in B\}$$

The intersection and the union of two sets are illustrated in Diagrams 1 and 2.

Examples 10. Let $A = \{2, 4, 6, 7\}$ and $B = \{4, 5, 6, 7, 8\}$. Then $A \cap B = \{4, 6, 7\}$ and $A \cup B = \{2, 4, 5, 6, 7, 8\}$.

11. $\{a, b, c, d, e\} \cap \{b, d, q, x, y, z\} = \{b, d\}$
$\{a, b, c, d, e\} \cup \{b, d, q, x, y, z\} = \{a, b, c, d, e, q, x, y, z\}$
■

We can define the intersection and the union of any number of sets. The *intersection* $C_1 \cap C_2 \cap \cdots \cap C_n$ of the sets $C_1, C_2, \ldots, C_n$ is the set of elements common to all of these sets. The *union* $C_1 \cup C_2 \cup \cdots \cup C_n$ of the sets $C_1, C_2, \ldots, C_n$ is the set of elements that belong to at least one of these sets.

Examples 12. Let $C_1 = \{a, b, c\}$, $C_2 = \{b, c, d, e\}$, $C_3 = \{c, 3, 5, 7\}$, and $C_4 = \{a, b, c, 5\}$. $C_1 \cap C_2 \cap C_3 \cap C_4 = \{c\}$ and $C_1 \cup C_2 \cup C_3 \cup C_4 = \{a, b, c, d, e, 3, 5, 7\}$.

13. Let $D = \{4, 6, 8, 10\}$, $E = \{4, 5, 8, 9\}$, and $F = \{4, 5, 7, 8, 10, 11\}$. $D \cap E \cap F = \{4, 8\}$ and $D \cup E \cup F = \{4, 5, 6, 7, 8, 9, 10, 11\}$.

∎

Now let $A = \{1, 2, 3\}$, $B = \{3, 4, 5\}$, and $C = \{3, 4, 6\}$. To compute $A \cup (B \cap C)$, first compute $B \cap C$. Now $B \cap C = \{3, 4\}$. Then

$$A \cup (B \cap C) = A \cup \{3, 4\} = \{1, 2, 3, 4\}$$

Now compute $(A \cup B) \cap C$. First, $A \cup B = \{1, 2, 3, 4, 5\}$. Then

$$(A \cup B) \cap C = \{1, 2, 3, 4, 5\} \cap C = \{3, 4\}$$

Note that $A \cup (B \cap C) \neq (A \cup B) \cap C$. Because of this, the expression $A \cup B \cap C$ is not defined. In attempting to compute $A \cup B \cap C$, one would obtain different answers depending on whether a union or an intersection was taken first.

Now consider a special case of intersection—when the intersection of two sets is empty. For example, the set of all dogs and the set of all cats have empty intersection.

Definition. Two sets A and B are *disjoint* if $A \cap B = \varnothing$.

Examples 14. $\{3, 4, 5\}$ and $\{1, 2, 6, 7\}$ are disjoint, because they have no element in common—their intersection is empty.

15. $\{3, 4, 5\}$ and $\{2, 4, 6\}$ are not disjoint, because they have the element 4 in common—their intersection is not empty.

Definition. A collection of two or more sets is *disjoint* if the intersection of every two sets in the collection is empty.

Examples 16. $\{1, 2, 3\}$, $\{a, b, c\}$, $\{d, q, r, 5, 7\}$, and $\{e, f, 9\}$ is a disjoint collection of sets, because the intersection of any two of these sets is empty.

17. $\{1, 5, 8\}$, $\{a, b, c\}$, $\{d, q, r, 5, 7\}$, and $\{e, f, g\}$ is not a disjoint collection of sets because $\{1, 5, 8\} \cap \{d, q, r, 5, 7\} = \{5\} \neq \varnothing$.

∎

Usually, only certain objects are under discussion at one time. The *universe* or *universal set* is the set of all objects under discussion. For example, suppose the manager of the Los Angeles Dodgers is making up the starting line-up for tonight's game. Then the universal set consists of all players on the team's roster. The set of those players who do not start is called the *complement* of the set of starters.

Definition. If U is the universal set and A is a set, the *complement* of A (written A') is $\{x \mid x \in U \text{ and } x \notin A\}$.

The complement of A is the set of those elements in the universal set that are not members of A. It is always true that $(A')' = A$, $U' = \varnothing$, and $\varnothing' = U$.

Examples 18. If $U = \{0, 1, 2, 3, 4, 5, 6, 7, 8, 9\}$, then $\{2, 4, 6, 8\}' = \{0, 1, 3, 5, 7, 9\}$, $\{1, 2, 3, 4, 5\}' = \{0, 6, 7, 8, 9\}$, and $\varnothing' = \{0, 1, 2, 3, 4, 5, 6, 7, 8, 9\} = U$.

19. If $U = \{$Arnold, Beth, Carl, Duncan, Elsie, Francine$\}$, then $\{$Beth, Duncan$\}' = \{$Arnold, Carl, Elsie, Francine$\}$.

Exercises 1-1

Fill in the blank with $\in$ or $\notin$ to make the statement true.

1. 4 ____ $\{2, 4, 6, 8\}$
2. 3 ____ $\{2, 4, 6, 8\}$
3. $\varnothing$ ____ $\{p, q, r, s, t\}$
4. r ____ $\{p, q, s, t\}$
5. 15 ____ $\{$odd integers$\}$
6. 1.4 ____ $\{$even integers$\}$

Fill in the blank with $=$ or $\neq$ to make the statement true.

7. $\{1, 3, 5\}$ ____ $\{3, 1, 5\}$
8. $\{3, 4, 7, 8\}$ ____ $\{3, 4, 7, 4, 4, 8\}$
9. $\{p, r, w, t\}$ ____ $\{p, q, r, w, t\}$
10. $\{a, b, c, d\}$ ____ $\{d, c, c, b, c\}$
11. $\{2, 3, 4, 8\}$ ____ $\{2, 3, 4, 9\}$
12. $\{9, 7, 3, 5\}$ ____ $\{3, 5, 5, 7, 9\}$

Fill in the blank with $\subseteq$ or $\not\subseteq$ to make the statement true.

13. $\{2, 4, 6, 8\}$ ____ $\{2, 4, 6\}$
14. $\{2, 4, 6\}$ ____ $\{2, 4, 6, 8\}$
15. $\{a, b, d, j\}$ ____ $\{d, a, j, b\}$
16. $\{5\}$ ____ $\{1, 2, 3, 4, 5, 6\}$
17. $\varnothing$ ____ $\{0, 1, 2\}$
18. $\varnothing$ ____ $\{a, b, c\}$

Let $U = \{a, b, c, d, e, f, g, h\}$, $A = \{b, c, d\}$, $B = \{c, f, g, h\}$, $C = \{d, b\}$, and $D = \{a, b, f, g, h\}$. Determine each of the following.

19. $A \cap B$
20. $A \cup B$
21. C'
22. D'

6 CHAPTER 1 SETS, NUMBERS, AND ALGEBRAIC MANIPULATION

23. $A \cap B \cap C$
25. $(C \cup D)'$
27. $A \cap (C \cup D)$
29. $(A \cap C) \cup D$
31. $D \cup U$
33. $A \cap C \cap D$

24. $A \cup B \cup C$
26. $(C \cap D)'$
28. $(A \cap C) \cup (A \cap D)$
30. $D \cap U$
32. $\emptyset \cup D$
34. $A \cup B \cup C \cup D$

Suppose the universal set U is the set of all employees of the Fifth Federal Savings and Loan Association. Let M be the set of all male employees of Fifth Federal. Let E be the set of all exempt employees of Fifth Federal. (In business, exempt employees are employees who are exempt from the Fair Labor Standards Act.) Let S be the set of all employees of Fifth Federal who have at least 20 years seniority. Verbally describe each of the following sets as simply as possible.

35. M'
37. $E \cap M$
39. $M \cap E \cap S$
41. $M' \cap E' \cap S$
43. $E \cap S$

36. E'
38. $E' \cap M'$
40. $M' \cup M$
42. $E \cup S$
44. $E' \cap S'$

*45. List all subsets of $\{a, b, c\}$.
*46. List all subsets of $\{a, b, c, d\}$.

Section 1-2 Numbers and the Number Line

We now come to numbers. Here we define several different kinds of numbers and review their basic properties. Let's start with the most familiar numbers. The *counting numbers* are 1, 2, 3, 4, ... The set of counting numbers is $\{1, 2, 3, 4, \ldots\}$.

The *integers* are the counting numbers, the negatives of the counting numbers, and 0. The set of integers is $\{\ldots, -3, -2, -1, 0, 1, 2, 3, \ldots\}$.

A *rational number* is a fraction—that is, a number that can be written as a quotient of two integers. The set of rational numbers is

$$\{x \mid x = p/q, \quad p \text{ and } q \text{ integers}, q \neq 0\}.$$

The condition $q \neq 0$ is needed because division by 0 is not defined.

Now look at the number line. Draw a line, choose any point, and mark it 0. This point is called the *origin*. Choose any point to the right of the origin and mark this second point 1. Choose the point for 2 by requiring that the distance from 0 to 1 be the same as the distance from 1 to 2. In this way, a point can be found for each integer. To find a point for 1/3, take the point between 0 and 1 that is twice as far from 1 as from 0. The point 2.5 is halfway between 2 and 3. In this way, a point can be found for each rational number.

The *real numbers* correspond to the set of all points on the number line. A rigorous definition of real number is beyond the scope of this course. An *irrational number* is a real number that is not a rational number. Examples of irrational numbers are $\sqrt{2}$ and π. The following list contains properties of the arithmetic of real numbers.

1. $a + (b + c) = (a + b) + c \quad (a \cdot b) \cdot c = a \cdot (b \cdot c)$ Associative laws.
2. $a + b = b + a \quad\quad a \cdot b = b \cdot a$ Commutative laws.
3. $a + 0 = 0 + a = a \quad\quad b \cdot 1 = 1 \cdot b = b$ Identity laws.
4. $a + (-a) = (-a) + a = 0$
   $b \cdot (1/b) = (1/b)b = 1 \quad \text{for } b \neq 0$ Inverse laws.
5. If $a + b = a + c$, then $b = c$.
   If $a \cdot b = a \cdot c$ and $a \neq 0$, then $b = c$.   Cancellation laws.
6. If $ab = 0$, then $a = 0$ or $b = 0$.   Zero product law.
7. $-(-a) = a \quad\quad (-a) \cdot b = a \cdot (-b) = -ab$
   $(-a)(-b) = ab \quad\quad (-a) + (-b) = -(a + b)$
8. $\dfrac{-c}{d} = \dfrac{c}{-d} = -\dfrac{c}{d}$ and $\dfrac{-c}{-d} = \dfrac{c}{d}$ if $d \neq 0$
9. $a(b + c) = ab + ac$ Distributive law.

The next list shows how to do arithmetic with fractions. In this list, $b \neq 0$ and $d \neq 0$.

1. $\dfrac{a}{b} = \dfrac{c}{d}$ if and only if $ad = bc$
2. $\dfrac{a}{b} + \dfrac{c}{d} = \dfrac{ad}{bd} + \dfrac{bc}{bd} = \dfrac{ad + bc}{bd}$
3. $\dfrac{a}{b} - \dfrac{c}{d} = \dfrac{ad}{bd} - \dfrac{bc}{bd} = \dfrac{ad - bc}{bd}$
4. $\dfrac{a}{b} \cdot \dfrac{c}{d} = \dfrac{ac}{bd}$
5. $\dfrac{a}{b} \div \dfrac{c}{d} = \dfrac{a}{b} \cdot \dfrac{d}{c} = \dfrac{ad}{bc}$ where $c \neq 0$

Examples 1. $\dfrac{2}{3} = \dfrac{8}{12}$ since $2 \cdot 12 = 3 \cdot 8$

2. $\dfrac{3}{5} + \dfrac{8}{5} = \dfrac{11}{5}$

3. $\dfrac{2}{3} - \dfrac{4}{11} = \dfrac{2 \cdot 11 - 3 \cdot 4}{3 \cdot 11} = \dfrac{10}{33}$

4. $\dfrac{2}{3} \cdot \dfrac{5}{7} = \dfrac{10}{21}$

5. $\dfrac{2}{3} \div \dfrac{5}{11} = \dfrac{2}{3} \cdot \dfrac{11}{5} = \dfrac{22}{15}$

Inequality

Frequently two numbers are compared in order to determine which one is larger. Examples are: "this job will cost over $1500" and "our price is lower than our competitor's price." To put such statements in a mathematical form, we need some definitions.

On the number line, the numbers that correspond to points to the right of the origin are *positive*. The numbers that correspond to points to the left of the origin are *negative*.

Definitions. Let a and b denote real numbers.

$a > b$ (a is *greater than* b) means $a - b$ is positive.

$a \geq b$ (a is *greater than or equal to* b) means $a > b$ or $a = b$.

$a < b$ (a is *less than* b) means $a - b$ is negative.

$a \leq b$ (a is *less than or equal to* b) means $a < b$ or $a = b$.

Notice that a is positive precisely when $a > 0$ and a is negative precisely when $a < 0$. The condition $a \geq 0$ is described by saying that a is *nonnegative*.

Many statements that occur in business can be written in mathematical form as inequalities. If we let the variable x denote sales for this year, the statement, "sales this year will exceed $5,000,000," can be written as $x > 5,000,000$.

The properties of inequality are summarized in the following list.

1. If $a < b$ and $b < c$, then $a < c$.

2. If $a < b$, then $a + c < b + c$.
3. If $a < b$ and $c > 0$, then $ac < bc$ and $a/c < b/c$. Note that the inequality sign has the same direction.
4. If $a < b$ and $c < 0$, then $ac > bc$ and $a/c > b/c$. Note that the inequality sign has changed direction. When an inequality is multiplied or divided by a negative number, the inequality sign changes direction.

Example 6 $-2 < 4$
$-2 + 6 < 4 + 6$ Add 6 to both sides.
$4 < 10$

Example 7 $-2 < 4$
$-2 \cdot 5 < 4 \cdot 5$ Multiply by 5. Since $5 > 0$, the inequality sign keeps the same direction.
$-10 < 20$

Example 8 $-2 < 4$
$(-2) \cdot (-8) > 4(-8)$ Multiply by -8. Since $-8 < 0$, the inequality sign changes direction.
$16 > -32$

∎

Sometimes it is important to know how big a number is, but it is not important whether the number is positive or negative. For example, suppose the specifications for a spindle call for a diameter of 3.50 cm accurate to within 0.01 cm. When an actual spindle is measured, what matters is how big the difference between its diameter and 3.50 cm is—no one cares about the sign of this difference.

> **Definition.** If b is a real number, then the *absolute value* of b (written $|b|$) is b if $b \geq 0$ and $-b$ if $b < 0$.

The geometric interpretation of absolute value is that $|b|$ is the distance between the point corresponding to b and the origin.

Example 9 $|7| = 7$, $|-7| = 7$, $|0| = 0$, $|-2\frac{2}{3}| = 2\frac{2}{3}$, $|1.4326| = 1.4326$,
$|-3.14159| = 3.14159$, $|-8| + |-3| \cdot |7| = 8 + 3 \cdot 7 = 8 + 21 = 29$

Exercises 1-2

Perform the indicated arithmetic.

1. $\dfrac{1}{2} + \dfrac{2}{3}$
2. $\dfrac{3}{4} + \dfrac{1}{5}$
3. $\dfrac{3}{5} - \dfrac{6}{5}$
4. $\dfrac{4}{7} - \dfrac{9}{7}$
5. $\dfrac{8}{3} - \dfrac{4}{5}$
6. $\dfrac{3}{7} - \dfrac{4}{11}$
7. $\dfrac{2}{5} + \dfrac{4}{7}$
8. $\dfrac{4}{5} + \dfrac{2}{9}$
9. $-\dfrac{3}{4} - \dfrac{1}{4}$
10. $-\dfrac{6}{7} - \dfrac{8}{7}$
11. $\dfrac{9}{2} - \dfrac{9}{5}$
12. $\dfrac{8}{3} + \dfrac{8}{7}$
13. $\left(\dfrac{2}{3}\right) \cdot \left(\dfrac{4}{7}\right)$
14. $\left(\dfrac{5}{9}\right) \cdot \left(-\dfrac{2}{11}\right)$
15. $\left(\dfrac{3}{11}\right) \cdot \left(-\dfrac{15}{4}\right)$
16. $\left(-\dfrac{2}{11}\right) \cdot \left(-\dfrac{5}{7}\right)$
17. $\left(-\dfrac{3}{8}\right) \cdot \left(-\dfrac{5}{7}\right)$
18. $-\left(-\dfrac{2}{3}\right) \cdot \left(-\dfrac{4}{5}\right)$
19. $\left(\dfrac{1}{2}\right) \div \left(\dfrac{1}{3}\right)$
20. $\left(\dfrac{2}{7}\right) \div \left(-\dfrac{11}{5}\right)$
21. $\left(-\dfrac{3}{11}\right) \div \left(\dfrac{2}{5}\right)$
22. $\left(\dfrac{4}{17}\right) \div \left(\dfrac{3}{7}\right)$
23. $\left(-\dfrac{2}{3}\right) \div \left(\dfrac{5}{2}\right)$
24. $\left(\dfrac{11}{3}\right) \div \left(\dfrac{13}{2}\right)$

Write each expression without using absolute value notation.

25. $|25|$
26. $|3| \cdot |-5|$
27. $|4| - 5|-2|$
28. $3|-1| + |5| \cdot |0|$
29. $|-25| - |25|$
30. $|-8| - 8|-2|$

Use inequality signs to rewrite each of the following statements. Use x as the variable each time.

31. Profits will exceed last year's profits of $43,000,000.
32. Profits will increase by at least $70,000 from last year's level of $200,000.

33. Each of our salespeople earned at least $7500 in commission last year. Our top one earned $18,462.

34. Flight time is a minimum of two hours.

35. Specifications call for ball bearings that are one-half centimeter in diameter to be off by no more than 0.0001 cm in diameter. Gramco ball bearings meet these specifications.

36. I estimate that repairing your car will require 14 hours of labor. This estimate is off by less than one hour.

Suppose that v and w are rational numbers and that x and y are irrational. Decide whether each of the following must be rational, must be irrational, or may be either rational or irrational.

*37. $v + w$ *38. $x + y$

*39. $v + x$ *40. vw

*41. vx *42. xy

Section 1-3 Exponents

Throughout mathematics, symbols are used to represent numbers. Much of mathematics involves the manipulation of these symbols according to carefully defined rules. In the remainder of this chapter, we will study the manipulation of symbols according to the rules of algebra.

A *variable* is a symbol that stands for a member of some set. In this book, the set is usually a set of real numbers. The symbol x is used usually to denote a variable, but other symbols can be used as well.

When x is used as a factor several times, the products can be written: x, $x \cdot x$, $x \cdot x \cdot x$, $x \cdot x \cdot x \cdot x$, $x \cdot x \cdot x \cdot x \cdot x$, and so on. Instead of writing a lot of x's, it would be easier to write one x together with the number of x's. The following definition provides the way.

> **Definition.** If n is a counting number, then
> $$x^n = \underbrace{(x)(x) \cdots (x)}_{(n \text{ factors})}$$

Thus $x^1 = x$, $x^2 = x \cdot x$, $x^3 = x \cdot x \cdot x$, and so on. The number n in the definition is called an *exponent*. Exponents have many useful properties. For example, if m and n are counting numbers, then

$$\boxed{x^m \cdot x^n = x^{m+n} \quad \text{and} \quad (x^m)^n = x^{mn}}$$

To verify the first formula, note that there are $m + n$ factors in the product $x^m \cdot x^n$

$$\underbrace{(x)(x)\cdots(x)}_{(m\text{ factors})} \cdot \underbrace{(x)(x)\cdots(x)}_{(n\text{ factors})}$$

To verify the second formula, note that there are mn factors in $(x^m)^n$

$$\underbrace{(x)(x)\cdots(x)}_{(m\text{ factors})} \cdot \underbrace{(x)(x)\cdots(x)}_{(m\text{ factors})} \cdot \cdots \cdot \underbrace{(x)(x)\cdots(x)}_{(m\text{ factors})}$$

Examples

1. $2^3 = 2 \cdot 2 \cdot 2 = 8$

2. $3^4 = 3 \cdot 3 \cdot 3 \cdot 3 = 81$

3. $2^3 \cdot 2^4 = 8 \cdot 16 = 128 = 2^7$

4. $(2^2)^3 = 4^3 = 64 = 2^6 = 2^{2 \cdot 3}$

∎

Two other important formulas are

$$\boxed{(ab)^n = a^n b^n \quad \text{and} \quad \left(\frac{a}{b}\right)^n = \frac{a^n}{b^n}}$$

Examples

5. $(2 \cdot 3)^3 = 6^3 = 216 = 8 \cdot 27 = 2^3 \cdot 3^3$

6. $(4x)^3 = 4^3 x^3 = 64 x^3$

7. $(-2b)^4 = (-2)^4 b^4 = 16 b^4$

8. $(-3c)^5 = (-3)^5 c^5 = -243c^5$

9. $\left(\dfrac{c}{2}\right)^4 = \dfrac{c^4}{2^4} = \dfrac{c^4}{16}$

10. $\left(\dfrac{x}{r}\right)^7 = \dfrac{x^7}{r^7}$

∎

Parentheses determine the order in which operations are performed. To compute $(ab)^n$, first form the product ab and then form the nth power of this product. The expressions $(ab)^n$ and ab^n are not equal. To compute ab^n, compute b^n and multiply this number by a.

For example, let $a = 4$ and $b = 5$. Then $(ab)^3 = (4 \cdot 5)^3 = 20^3 = 8000$. However, $ab^n = 4 \cdot 5^3 = 4 \cdot 125 = 500$.

Note that $(-1)^n$ equals -1 if n is odd and 1 if n is even. Thus, if $b < 0$, b^n is negative if n is odd and positive if n is even.

Now we define other integer exponents. If n is a counting number and $x \neq 0$,

$$\boxed{x^{-n} = \left(\dfrac{1}{x}\right)^n}$$

If $x = 0$, x^{-n} is undefined because division by 0 is undefined. If $x \neq 0$, then $x^0 = 1$. However 0^0 is not defined.

Examples

11. $2^{-3} = \left(\dfrac{1}{2}\right)^3 = \dfrac{1}{8}$

12. $c^{-1} = \dfrac{1}{c}$

13. $3^{-4} = \left(\dfrac{1}{3}\right)^4 = \dfrac{1}{81}$

14. $8^0 = 1$

15. $(-23)^0 = 1$

∎

Integer exponents satisfy the same algebraic rules as positive integer exponents. However, one must be careful not to divide by 0. If m and n are integers, then

$$x^m \cdot x^n = x^{m+n} \quad \text{and} \quad \frac{x^m}{x^n} = x^{m-n}$$

$$(x^m)^n = x^{mn}$$

$$(xy)^n = x^n y^n \quad \text{and} \quad \left(\frac{x}{y}\right)^n = \frac{x^n}{y^n}$$

Examples 16. $(2x)^{-5} = \dfrac{1}{(2x)^5} = \dfrac{1}{2^5 x^5} = \dfrac{1}{32x^5}$

17. $2^{-4} \cdot 2^7 = 2^3 = 8$

18. $3^{-9} \cdot 3^6 = 3^{-3} = \dfrac{1}{3^3} = \dfrac{1}{27}$

19. $(4a)^7 \cdot (4a)^{-5} = (4a)^2 = 4^2 a^2 = 16a^2$

20. $(2q)^{-5}(2q)^{-3} = (2q)^{-8} = \dfrac{1}{(2q)^8} = \dfrac{1}{2^8 q^8} = \dfrac{1}{256 q^8}$

21. $(q^3)^{-4} = q^{-12} = \dfrac{1}{q^{12}}$

22. $[(-3x)^2]^{-3} = (-3x)^{-6} = \dfrac{1}{(-3x)^6} = \dfrac{1}{(-3)^6 x^6} = \dfrac{1}{729 x^6}$

23. $\left(\dfrac{1}{2}\right)^{-5} = (2^{-1})^{-5} = 2^5 = 32$

24. $\left(\dfrac{3}{4}\right)^{-2} = \left[\left(\dfrac{3}{4}\right)^2\right]^{-1} = \left(\dfrac{3^2}{4^2}\right)^{-1} = \left(\dfrac{9}{16}\right)^{-1} = \dfrac{16}{9}$

25. $(x^2 y^{-3})^4 = (x^2)^4 (y^{-3})^4 = x^8 y^{-12} = \dfrac{x^8}{y^{12}}$

26. $[(x+2)^{-3}]^2 = (x+2)^{-6} = \dfrac{1}{(x+2)^6}$

■

Many problems involving negative exponents can be worked in several ways. Sometimes one way is clearly best since it is much easier than any other way. Example 18 demonstrates this. Example 18 can also be done as follows:

$$3^{-9} \cdot 3^6 = \left(\frac{1}{3}\right)^9 \cdot 3^6 = \frac{729}{19{,}683} = \frac{1}{27}$$

Notice that this method involves a lot of arithmetic. Sometimes several ways of working a problem are of similar difficulty and are equally good. Example 24 demonstrates this. Example 24 can also be worked as follows:

$$\left(\frac{3}{4}\right)^{-2} = \left[\left(\frac{3}{4}\right)^{-1}\right]^2 = \left(\frac{4}{3}\right)^2 = \frac{4^2}{3^2} = \frac{16}{9}$$

Neither method is better than the other.

You may wonder whether the numbers $(x + y)^n$ and $x^n + y^n$ are equal or not. In general, these numbers are unequal. For example, if $x = 5$, $y = 8$, and $n = 2$; then $(x + y)^n = (5 + 8)^2 = 13^2 = 169$ but $x^n + y^n = 5^2 + 8^2 = 25 + 64 = 89 \neq 169$.

Scientific notation

Now look at the powers of 10. If n is a counting number, 10^n is written as a 1 followed by n zeros.

$$10^1 = 10 \qquad 10^2 = 100 \qquad 10^3 = 1000 \qquad 10^4 = 10{,}000 \qquad 10^5 = 100{,}000$$

If n is a counting number, 10^{-n} is written as a zero followed by a decimal point followed by $n - 1$ zeros followed by a 1.

$$10^{-1} = 0.1 \qquad 10^{-2} = 0.01 \qquad 10^{-3} = 0.001 \qquad 10^{-4} = 0.0001 \qquad 10^{-5} = 0.00001$$

Any positive number can be written in the form $b \times 10^n$, where $1 \leq b < 10$ and n is an integer. Any negative number can be written in the same form with a minus sign before the b. A number written this way is said to be in *scientific notation*.

Examples Express each number in scientific notation.

Example 27 23.472

Solution Recall that to multiply a number by 10, you move the decimal point one place to the right. Now 2.3472 and 23.472 have the same sequence of digits and $1 \leq 2.3472 < 10$. Since the decimal point in 23.472 is one place to the right of the decimal point in 2.3472, $23.472 = 2.3472 \times 10^1$.

Examples 28. $123{,}567 = 1.23567 \times 10^5$

29. $-27{,}000 = -2.7 \times 10^4$

30. $100 = 1 \times 10^2$

Example 31 0.0035

Solution To divide a number by 10, you move the decimal point one place to the left. Now 3.5 and 0.0035 have the same sequence of digits and $1 \leq 3.5 < 10$. Since the decimal point in 0.0035 is three places to the left of the decimal point in 3.5, 3.5 must be divided by 10^3 (or multiplied by 10^{-3}) to obtain 0.0035. Therefore $0.0035 = 3.5 \times 10^{-3}$.

Examples 32. $0.1462 = 1.462 \times 10^{-1}$

33. $-0.00326 = -3.26 \times 10^{-3}$

34. $970{,}000{,}000{,}000{,}000 = 9.7 \times 10^{14}$

35. $0.000{,}000{,}000{,}000{,}000{,}341 = 3.41 \times 10^{-16}$
■

Scientific notation (or something similar) occurs frequently in computer output. In addition, calculators use scientific notation to express numbers, as in the last two examples, which would otherwise require too many digits to be displayed.

Roots and rational exponents

So far, our exponents have all been integers. We will now generalize this by allowing rational numbers to be exponents. To do so, we must first consider exponents that are rational numbers of the form $1/n$, where n is a counting number.

Definition. If n is odd, the *nth root of c*, written $c^{1/n}$, is the unique real number whose nth power is c.

If n is even, and $c \geq 0$, $c^{1/n}$ is the unique nonnegative real number whose nth power is c.

If n is odd, the equation $x^n = c$ always has exactly one solution. If n is even, $x^n = (-x)^n$, and the equation $x^n = c$ has two solutions if $c > 0$. For example, $3^2 = 9$ and $(-3)^2 = 9$. However, $9^{1/2} = 3$, since the square root of 9 is 3. And $9^{1/2} \neq -3$, because we require this root to be nonnegative. If n is even, $x^n \geq 0$ for all real numbers x, so that the nth root of a negative number is undefined.

Examples
36. $8^{1/3} = 2$ because $2^3 = 8$
37. $(-8)^{1/3} = -2$ because $(-2)^3 = -8$
38. $0^{1/n} = 0$ because $0^n = 0$
39. $1^{1/n} = 1$ because $1^n = 1$ (and $1 \geq 0$)
40. $16^{1/4} = 2$ because $2^4 = 16$ and $2 \geq 0$
41. $(-25)^{1/2}$ is undefined because -25 is negative and 2 is even
42. $\left(\dfrac{4}{9}\right)^{1/2} = \dfrac{2}{3}$ because $\left(\dfrac{2}{3}\right)^2 = \dfrac{4}{9}$ and $\dfrac{2}{3} \geq 0$
43. $\left(\dfrac{125}{64}\right)^{1/3} = \dfrac{5}{4}$ because $\left(\dfrac{5}{4}\right)^3 = \dfrac{125}{64}$

Roots are sometimes written using *radical signs*, so that $c^{1/n}$ is written as $\sqrt[n]{c}$. Since square roots are the most commonly used roots, $\sqrt[2]{c}$ is usually written $\sqrt{c}$. An interesting fact about roots is that if m is an integer and $m^{1/n}$ is defined, then $m^{1/n}$ is either an integer or an irrational number. Thus, $\sqrt{2}$, $\sqrt{3}$, $\sqrt{5}$, $\sqrt[3]{2}$, $\sqrt[3]{3}$, and $\sqrt[4]{5}$ are all irrational numbers.

Now let's define rational exponents. If q is a counting number, p is an integer, and $b^{1/q}$ is defined, then

$$b^{p/q} = (b^{1/q})^p \qquad \text{Rational exponent}$$

Examples 44. $8^{2/3} = (8^{1/3})^2 = 2^2 = 4$

45. $32^{3/5} = (32^{1/5})^3 = 2^3 = 8$

46. $64^{-2/3} = (64^{1/3})^{-2} = 4^{-2} = \dfrac{1}{4^2} = \dfrac{1}{16}$

47. $9^{5/2} = (9^{1/2})^5 = 3^5 = 243$

48. $(-9)^{5/2} = [(-9)^{1/2}]^5$, which is undefined

49. $\left(\dfrac{4}{9}\right)^{3/2} = \left[\left(\dfrac{4}{9}\right)^{1/2}\right]^3 = \left(\dfrac{2}{3}\right)^3 = \dfrac{2^3}{3^3} = \dfrac{8}{27}$

∎

If r is any rational number, then $1^r = 1$. If $r > 0$, then $0^r = 0$. If $r \le 0$, then 0^r is undefined.

You may wonder if $b^{p/q} = (b^p)^{1/q}$. If $b^{p/q}$ is defined, then this equation is true. However, it is usually easier to compute $(b^{1/q})^p$ than $(b^p)^{1/q}$. For example, let $b = 16$, $p = 5$, and $q = 4$. Then

$$(b^{1/q})^p = (16^{1/4})^5 = 2^5 = 32 \quad \text{but}$$

$$(b^p)^{1/q} = (16^5)^{1/4} = (1{,}048{,}576)^{1/4} = 32$$

You can do the first computation easily in your head. The second computation requires a calculator or a lot of time.

The following list shows properties of rational exponents. In these formulas, x and y are positive numbers and r and s are any rational numbers.

1. $x^r x^s = x^{r+s}$
2. $\dfrac{x^r}{x^s} = x^{r-s}$
3. $(xy)^r = x^r y^r$
4. $\left(\dfrac{x}{y}\right)^r = \dfrac{x^r}{y^r}$
5. $x^r = x^s$ if $r = s$
6. $x^{-r} = \dfrac{1}{x^r}$
7. $(x^r)^s = (x^s)^r = x^{rs}$

1-3 EXPONENTS

If we drop the requirement that $x > 0$ and $y > 0$, the equations remain valid provided all terms in the equation are defined. Note especially that in the last equation all three terms must be defined. For example, if $b < 0$, $(b^2)^{1/2} = |b| = -b$. However, $(b^{1/2})^2$ is undefined now.

Examples 50. $\dfrac{16^{9/2}}{16^{5/2}} = 16^{(9/2)-(5/2)} = 16^{4/2} = 16^2 = 256$

51. $\left(\dfrac{1}{2}\right)^5 \cdot 2^9 = 2^{-5} \cdot 2^9 = 2^4 = 16$

52. $(9^{9/2})^{1/3} = 9^{(9/2)\cdot(1/3)} = 9^{3/2} = (9^{1/2})^3 = 3^3 = 27$

53. $\left(\dfrac{25}{16}\right)^{3/2} = \dfrac{25^{3/2}}{16^{3/2}} = \dfrac{5^3}{4^3} = \dfrac{125}{64}$

54. $25^{50/100} = 25^{1/2} = 5$

55. $49^{-3/2} = (49^{1/2})^{-3} = 7^{-3} = \dfrac{1}{7^3} = \dfrac{1}{343}$

56. $(-8)^{5/3} = [(-8)^{1/3}]^5 = (-2)^5 = -32$

Examples In the following examples, $b > 0$ and $c > 0$.

57. $b^{2/3} b^{16/3} = b^{18/3} = b^6$

58. $\dfrac{c^{7/3}}{c^{5/3}} = c^{(7/3)-(5/3)} = c^{2/3}$

59. $\left(\dfrac{b^2}{4}\right)^{5/2} = \dfrac{(b^2)^{5/2}}{4^{5/2}} = \dfrac{b^5}{2^5} = \dfrac{b^5}{32}$

60. $(4b)^{3/2} = 4^{3/2} b^{3/2} = 2^3 b^{3/2} = 8 b^{3/2}$

61. $\dfrac{1}{c^{-4/3}} = \dfrac{1}{\left(\dfrac{1}{c^{4/3}}\right)} = c^{4/3}$

62. $\left(\dfrac{16b^8}{c^{16}}\right)^{3/4} = \dfrac{16^{3/4}(b^8)^{3/4}}{(c^{16})^{3/4}} = \dfrac{2^3 b^{24/4}}{c^{48/4}} = \dfrac{8b^6}{c^{12}}$

63. $(-32b^{10})^{2/5} = (-32)^{2/5}(b^{10})^{2/5} = (-2)^2 b^{20/5} = 4b^4$

64. $\left(\dfrac{b^3}{8}\right)^{-2/3} = \left(\dfrac{8}{b^3}\right)^{2/3} = \dfrac{8^{2/3}}{(b^3)^{2/3}} = \dfrac{2^2}{b^2} = \dfrac{4}{b^2}$

65. $(b^{10/7})^{7/10} = b^1 = b$

Exercises 1-3

Evaluate each expression.

1. 5^3
2. 6^2
3. 7^{-2}
4. 2^{-6}
5. $(-3)^0$
6. $(-3)^{-2}$
7. $(-2)^6$
8. $(-3)^{-5}$
9. $\left(\dfrac{2}{3}\right)^4$
10. $\left(\dfrac{3}{4}\right)^3$
11. $\left(\dfrac{5}{2}\right)^{-3}$
12. $\left(-\dfrac{4}{5}\right)^{-3}$
13. $\left(-\dfrac{1}{3}\right)^5$
14. $\left(-\dfrac{5}{2}\right)^{-4}$
15. $\left(-\dfrac{2}{5}\right)^{-3}$
16. $\left(\dfrac{3}{2}\right)^5$

Simplify each expression.

17. $(5x)^4$
18. $(3x)^{-2}$
19. $\left(-\dfrac{x}{4}\right)^3$
20. $\left(-\dfrac{2x}{3}\right)^4$
21. $(xy)^7$
22. $(-bc)^5$
23. $(x^2 y)^3$
24. $(-2xy^3)^{-2}$
25. $(3c)^{-4}(3c)^7$
26. $\dfrac{(4b)^5}{(4b)^4}$
27. $(-5b^3)^3$
28. $(-4c^{-3})^2$
29. $\left(-\dfrac{y}{2}\right)^{-3}$
30. $\left(\dfrac{4z}{3}\right)^{-2}$

31. $\dfrac{(-4b)^4}{(-4b)}$
32. $\left(\dfrac{x^5}{x^2}\right)^4$

Express each number in scientific notation.
33. 12.3
34. 1254
35. -0.00005
36. 2.41
37. 473,000,000
38. -0.00000000871
39. 1.414
40. 0.325

Express each number in standard notation.
41. 2.36×10^4
42. 3.59×10^0
43. -1.37×10^{-3}
44. -4.18×10^5
45. 2.1×10^{12}
46. 3.61×10^{-6}
47. -3.46×10^{-6}
48. 9.75×10^2

Compute each of the following.
49. $64^{1/3}$
50. $(-81)^{1/4}$
51. $121^{1/2}$
52. $(125)^{1/3}$
53. $(-36)^{1/2}$
54. $\left(-\dfrac{1}{8}\right)^{1/3}$
55. $\left(\dfrac{16}{25}\right)^{1/2}$
56. $\left(-\dfrac{125}{216}\right)^{1/3}$
57. $\left(\dfrac{25}{4}\right)^{3/2}$
58. $\left(\dfrac{36}{49}\right)^{-3/2}$
59. $\left(-\dfrac{8}{27}\right)^{2/3}$
60. $(-125)^{4/3}$
61. $\left(\dfrac{1}{36}\right)^{-3/2}$
62. $\left(\dfrac{49}{4}\right)^{-1/2}$
63. $\left(\dfrac{27}{125}\right)^{2/3}$
64. $\left(-\dfrac{1}{32}\right)^{-7/5}$

Simplify each of the following. Assume $x > 0$ and $y > 0$.
65. $x^{2/3} \cdot x^4$
66. $y^{1/3} \cdot y^{1/6}$
67. $\dfrac{x^{12/5}}{x^{2/5}}$
68. $(x^3 y^6)^{1/3}$

69. $(x^{3/2})^{4/3}$

70. $(x^{-3})^{-2/3}$

71. $\dfrac{(xy)^{7/3}}{(xy)^{1/3}}$

72. $\dfrac{x^{7/3} \cdot x^{5/3}}{x}$

73. $\left(\dfrac{x^2}{y^2}\right)^{1/2}$

74. $(4x^4 y^2)^{1/2}$

75. $(-x^3 y^3)^{1/3}$

76. $(-x^4 y^4)^{1/4}$

77. $\dfrac{x^{1/3}}{x^{10/3}}$

78. $(x+y)^{3/4} \cdot (x+y)^{5/4}$

79. $(x^4)^{3/2}$

80. $(x^5 y)^{2/5}$

81. $(-x^2)^{3/2}$

82. $\left(\dfrac{y^4}{x^8}\right)^{1/2}$

83. $\left(\dfrac{x^3}{y^6}\right)^{-2/3}$

84. $(-x^3)^{5/3}$

Section 1-4 Polynomials

A *monomial* is an expression of the form b or bx^n, where b is a real number, n is a counting number, and x is a variable. A *polynomial* is a monomial or a sum of monomials. Examples of polynomials are

1. $5x^3 + (-4x^2) + (-17x) + 23$, which simplifies to $5x^3 - 4x^2 - 17x + 23$
2. $10x^4$
3. $8x^2 - 5x + 3$
4. $4x + 2$

A *term* is any of the monomials that are added to obtain the polynomial. The first polynomial has four terms: $5x^3$, $-4x^2$, $-17x$, and 23. The second polynomial is a monomial. The *degree* of the monomial bx^n is n if $b \neq 0$ and the *coefficient* is b.

A polynomial is in *standard form* if different terms have different degrees, and if the terms are written in order of decreasing degree. In order to write a polynomial in standard form, first rearrange the terms so that no term is followed by a term with higher degree. Then combine like terms. *Like terms* are terms with the same degree, that is, with the same power of x.

Thus, in the polynomial

$$8x^3 + 4x^2 + 11x^2 - 2x + 9$$

$4x^2$ and $11x^2$ are like terms since they are both of degree 2. These like terms may be combined using the distributive law:

$$4x^2 + 11x^2 = (4 + 11)x^2 = 15x^2$$

Then the original polynomial, in standard form, is

$$8x^3 + 15x^2 - 2x + 9$$

Examples Write each polynomial in standard form.

Example 1 $4 + 3x + 8 + 10x^2 - 2x^2$

Solution $4 + 3x + 8 + 10x^2 - 2x^2$
$= 10x^2 - 2x^2 + 3x + 4 + 8$ Rearrange terms.
$= 8x^2 + 3x + 12$ Combine like terms.

Example 2 $5x^3 - 10x + 7x^2 + 8 - 4x$

Solution $5x^3 - 10x + 7x^2 + 8 - 4x$
$= 5x^3 + 7x^2 - 10x - 4x + 8$ Rearrange terms.
$= 5x^3 + 7x^2 - 14x + 8$ Combine like terms.

Example 3 $4 + 8x^3 - 7x^3 + 3x^2 - 14x + x^3 - 2x^3$

Solution $4 + 8x^3 - 7x^3 + 3x^2 - 14x + x^3 - 2x^3$
$= 8x^3 - 7x^3 + x^3 - 2x^3 + 3x^2 - 14x + 4$ Rearrange terms.
$= 3x^2 - 14x + 4$ Combine like terms.
∎

If a polynomial is in standard form, its *degree* is the degree of its first nonzero term, and its *leading coefficient* is the coefficient of that term. A polynomial is *monic* if its leading coefficient is 1. A polynomial in standard form is *linear* if it is of degree 1, *quadratic* if it is of degree 2, and *cubic* if it is of degree 3. A polynomial is a *binomial* if it has exactly two terms.

Examples 4. $5x^3 - 4x^2 + 7x + 13$ has 4 terms, is cubic, and has leading coefficient 5.
5. $x^4 - 3x^2 + 8x$ has 3 terms, is of degree 4, and is monic.
6. $7x^2 - 14$ is a binomial, is quadratic, and has leading coefficient 7.

■

Polynomials occur in numerous applications throughout this course. To use polynomials effectively, it is necessary to be familiar with their arithmetic. To find the sum or difference of two polynomials, the rules for the arithmetic of real numbers are used. The first step is to remove parentheses. One then has a single polynomial. The next step is to put this polynomial in standard form.

Example 7 Compute $(5x^2 + 4x + 7) + (2x^2 + 5x + 3)$.

Solution
$(5x^2 + 4x + 7) + (2x^2 + 5x + 3)$
$= 5x^2 + 4x + 7 + 2x^2 + 5x + 3$ Remove parentheses.
$= 5x^2 + 2x^2 + 4x + 5x + 7 + 3$ Rearrange terms.
$= 7x^2 + 9x + 10$ Combine like terms.

Example 8 Compute $(8x^3 - 4x^2 + 7x + 1) + (6x^2 - 10x - 9)$.

Solution
$(8x^3 - 4x^2 + 7x + 1) + (6x^2 - 10x - 9)$
$= 8x^3 - 4x^2 + 7x + 1 + 6x^2 - 10x - 9$ Remove parentheses.
$= 8x^3 - 4x^2 + 6x^2 + 7x - 10x + 1 - 9$ Rearrange terms.
$= 8x^3 + 2x^2 - 3x - 8$ Combine like terms.

Example 9 Compute $(4x^3 - 3x^2 + 7) - (5x^4 - 3x^3 - 7x^2 - 2x + 9)$.

Solution
$(4x^3 - 3x^2 + 7) - (5x^4 - 3x^3 - 7x^2 - 2x + 9)$
$= 4x^3 - 3x^2 + 7 - 5x^4 + 3x^3 + 7x^2 + 2x - 9$ Remove parentheses.
$= -5x^4 + 4x^3 + 3x^3 - 3x^2 + 7x^2 + 2x + 7 - 9$ Rearrange terms.
$= -5x^4 + 7x^3 + 4x^2 + 2x - 2$ Combine like terms.

Note how the signs changed in going from the first line of the solution to the second line. The whole expression $5x^4 - 3x^2 - 7x^2 - 2x + 9$ is being subtracted from $4x^3 - 3x^2 + 7$.

■

To multiply a polynomial by a number, simply multiply each coefficient of the polynomial by that number. This is a special case of the distributive law. Thus,

$$5(3x^2 + 4x - 6) = (5 \cdot 3)x^2 + (5 \cdot 4)x - (5 \cdot 6)$$
$$= 15x^2 + 20x - 30$$

To multiply polynomials, use the distributive law and simplify. There are sometimes several different ways to do the problem. This is shown for Example 10.

Example 10 Compute $(x + 3)(x - 2)$

Solution
$(x + 3)(x - 2)$
$= x(x - 2) + 3(x - 2)$ Distributive law.
$= (x^2 - 2x) + (3x - 6)$ Distributive law.
$= x^2 - 2x + 3x - 6$ Remove parentheses.
$= x^2 + x - 6$ Combine like terms.

Example 10 can also be done in another way. Both ways are correct, and neither way is better than the other.

Solution
$(x + 3)(x - 2)$
$= (x + 3)x + (x + 3)(-2)$ Distributive law.
$= (x^2 + 3x) + (-2x - 6)$ Distributive law.
$= x^2 + 3x - 2x - 6$ Remove parentheses.
$= x^2 + x - 6$ Combine like terms.

Example 11 Compute $(2x - 5)(3x + 4)$.

Solution
$(2x - 5)(3x + 4)$
$= 2x(3x + 4) - 5(3x + 4)$ Distributive law.
$= (6x^2 + 8x) - (15x + 20)$ Distributive law.
$= 6x^2 + 8x - 15x - 20$ Remove parentheses.
$= 6x^2 - 7x - 20$ Combine like terms.

Example 12 Compute $(x^2 - 3x + 4)(x + 2)$.

Solution
$(x^2 - 3x + 4)(x + 2)$
$= (x^2 - 3x + 4)x + (x^2 - 3x + 4)2$ Distributive law.
$= (x^3 - 3x^2 + 4x) + (2x^2 - 6x + 8)$ Distributive law.

$$= x^3 - 3x^2 + 4x + 2x^2 - 6x + 8 \qquad \text{Remove parentheses.}$$
$$= x^3 - 3x^2 + 2x^2 + 4x - 6x + 8 \qquad \text{Rearrange terms.}$$
$$= x^3 - x^2 - 2x + 8 \qquad \text{Combine like terms.}$$

Polynomials can also be multiplied by the same procedure used to multiply numbers.

Example 10'
$$\begin{array}{r} x - 2 \\ \times\, x + 3 \\ \hline 3x - 6 \\ x^2 - 2x \\ \hline x^2 + x - 6 \end{array}$$

Example 11'
$$\begin{array}{r} 3x + 4 \\ \times\, 2x - 5 \\ \hline -15x - 20 \\ 6x^2 + 8x \\ \hline 6x^2 - 7x - 20 \end{array}$$

Example 12'
$$\begin{array}{r} x^2 - 3x + 4 \\ \times x + 2 \\ \hline 2x^2 - 6x + 8 \\ x^3 - 3x^2 + 4x \\ \hline x^3 - x^2 - 2x + 8 \end{array}$$

∎

Note that the product of two polynomials can be obtained by multiplying each term of the first polynomial by each term of the second polynomial and adding all the resulting products.

Polynomials are often represented by symbols such as $P(x)$, $Q(r)$, and $S(w)$. If $Q(r) = r^2 - 5r + 4$, the *value* of Q at 2 is obtained by letting r have the value 2.

$$Q(2) = 2^2 - 5 \cdot 2 + 4 = 4 - 10 + 4 = -2$$

Similarly,

$$Q(6) = 6^2 - 5 \cdot 6 + 4 = 36 - 30 + 4 = 10$$

To compute $Q(t)$, substitute t for r in the formula for Q. Then

$$Q(t) = t^2 - 5t + 4$$

To compute $Q(s - 3)$, substitute $s - 3$ for r in the formula for Q. Then simplify the resulting expression.

$$Q(s - 3) = (s - 3)^2 - 5(s - 3) + 4$$
$$= (s^2 - 6s + 9) - (5s - 15) + 4$$
$$= s^2 - 6s + 9 - 5s + 15 + 4$$

$$= s^2 - 6s - 5s + 9 + 15 + 4$$
$$= s^2 - 11s + 28$$

Examples Let $P(x) = 3x + 5$, $Q(x) = x^2 + 4x + 5$, and $R(x) = 4x - 8$

13. $P(7) = 3 \cdot 7 + 5 = 21 + 5 = 26$

14. $P(10) = 3 \cdot 10 + 5 = 35$

15. $Q(2) = 2^2 + 4 \cdot 2 + 5 = 4 + 8 + 5 = 17$

16. $P(2y + 6) = 3(2y + 6) + 5 = 6y + 18 + 5 = 6y + 23$

17. $R(w) = 4w - 8$

Example 18 $P(R(w))$

Solution To compute $P(R(w))$, substitute $R(w)$ for x in the formula for P. Since $R(w) = 4w - 8$,

$$P(R(w)) = P(4w - 8)$$
$$= 3(4w - 8) + 5$$
$$= 12w - 24 + 5$$
$$= 12w - 19$$

Examples 19. $R(Q(1)) = R(1^2 + 4 \cdot 1 + 5) = R(1 + 4 + 5) = R(10) = 4 \cdot 10 - 8 = 40 - 8 = 32$

20. $Q(-x) = (-x)^2 + 4(-x) + 5 = x^2 - 4x + 5$

Example 21 $P(P(x))$

Solution To compute $P(P(x))$ substitute $P(x)$ for x in the formula for P. Since $P(x) = 3x + 5$,

$$P(P(x)) = P(3x + 5)$$
$$= 3(3x + 5) + 5$$

$$= 9x + 15 + 5$$
$$= 9x + 20$$

Polynomials in several variables

So far we have studied polynomials in one variable. Now let's extend this by studying polynomials in several variables.

A *monomial in several variables* is the product of a number and nonnegative integer powers of some of the variables. The following are monomials in b, c, and d: $5b^2c^3$, $3bcd^2$, $-8.3bc^4d^2$, c, bd, $18b^3d^2/5$, and $4c/3$. You should decide why the following are not monomials in b, c, and d: $5b^{1/2}$, $4b^2cx$, $6\sqrt[3]{bcd}$, $3/d$, and $15b^2c^{-5}$. The *coefficient* of a monomial is the numerical factor in it. For example, -12 is the coefficient in $-12b^2cd^5$.

A *polynomial in several variables* is a monomial or the sum of monomials in these variables. Each of the monomials that is added is called a *term* of the polynomial.

Examples of polynomials in x and y are

1. $5x^2y^2 - 7x + 3y + 8$
2. $10x^3y^4 - 7x^2y^5$
3. $12x - 15y + 5$
4. $x^2 - 4x + 3$
5. $x^2 - y^2$

The second and the last of these are binomials.

A polynomial in several variables is in *simplified form* if all parentheses and other grouping symbols have been removed and like terms have been combined. Usually some justification is needed for the steps in which parentheses are removed. The distributive law most often provides this justification. Two terms are *like terms* if each variable has the same exponent in the first term as in the second term.

Examples Simplify each polynomial.

Example 22 $4[3x + 5(y - x)]$

Solution $4[3x + 5(y - x)]$
$= 4[3x + 5y - 5x]$ Distributive law.

$\quad = 4[5y - 2x]$ Combine like terms.
$\quad = 20y - 8x$ Distributive law.

Example 23 $x[4(x - y) + 5(x - 2y)(x + 3)]$

Solution $x[4(x - y) + 5(x - 2y)(x + 3)]$
$= x[4x - 4y + 5x(x + 3) - 10y(x + 3)]$ Distributive law.
$= x[4x - 4y + 5x^2 + 15x - 10xy - 30y]$ Distributive law.
$= x[19x - 34y + 5x^2 - 10xy]$ Combine like terms.
$= 19x^2 - 34xy + 5x^3 - 10x^2y$ Distributive law.

Example 24 $(2x - 3y)[4 + 3(x - y)]$

Solution $(2x - 3y)[4 + 3(x - y)]$
$= (2x - 3y)(4 + 3x - 3y)$ Distributive law.
$= 2x(4 + 3x - 3y) - 3y(4 + 3x - 3y)$ Distributive law.
$= 8x + 6x^2 - 6xy - 12y - 9xy + 9y^2$ Distributive law.
$= 8x + 6x^2 - 15xy - 12y + 9y^2$ Combine like terms.

■

The arithmetic of polynomials in several variables is very similar to the arithmetic of ordinary polynomials. To add or subtract polynomials in several variables, remove parentheses and then combine like terms.

Examples Add or subtract as indicated.

Example 25 $(4x^2 + 3xy - 7z) + (5x^2 + 4y^2 - 10xy + 12)$

Solution $(4x^2 + 3xy - 7z) + (5x^2 + 4y^2 - 10xy + 12)$
$= 4x^2 + 3xy - 7z + 5x^2 + 4y^2 - 10xy + 12$ Remove parentheses.
$= 9x^2 - 7xy - 7z + 4y^2 + 12$ Combine like terms.

Example 26 $(5b^3 - 7c^2 + 8c^3 - 12bc) - (6c^2 + 8c^3 + 7bc - 9c)$

Solution $(5b^3 - 7c^2 + 8c^3 - 12bc) - (6c^2 + 8c^3 + 7bc - 9c)$
$= 5b^3 - 7c^2 + 8c^3 - 12bc - 6c^2 - 8c^3 - 7bc + 9c$ Remove parentheses.
$= 5b^3 - 13c^2 - 19bc + 9c$ Combine like terms.

■

To multiply a polynomial in several variables by a number, simply multiply each coefficient of the polynomial by that number. This is a special case of the distributive law. To multiply polynomials, use the distributive law and simplify. There will be more than one correct way to multiply two polynomials, but all correct ways will give the same answer.

Examples Perform the indicated operations.

Example 27 $10(3x^2 + 7xy - 3z + 8)$

Solution $10(3x^2 + 7xy - 3z + 8)$
$= 30x^2 + 70xy - 30z + 80$ Distributive law.

Example 28 $2(3g + 5h - 7) - 3(4g - 6h + 5)$

Solution $2(3g + 5h - 7) - 3(4g - 6h + 5)$
$= (6g + 10h - 14) - (12g - 18h + 15)$ Distributive law.
$= 6g + 10h - 14 - 12g + 18h - 15$ Remove parentheses.
$= -6g + 28h - 29$ Combine like terms.

Example 29 $(b - 2c)(b + 3c)$

Solution $(b - 2c)(b + 3c)$
$= b(b + 3c) - 2c(b + 3c)$ Distributive law.
$= b^2 + 3bc - 2bc - 6c^2$ Distributive law.
$= b^2 + bc - 6c^2$ Combine like terms.

Exercises 1-4

Write each polynomial in standard form. Write whichever of the following words apply to the polynomial: monic, linear, quadratic, cubic, monomial, binomial.

1. $2 + 3x - 5x^2$
2. $x^2 + 3x - 2 + x^3$
3. $x^2 + 5x + 6 - 3x$
4. $2x^2 - x^2 + 4x - x^2$
5. $x^5 - 3x^2 + 5$
6. $4 - 7x + 3x^2$
7. $x^2 + 2x - 3 - x^2$
8. $x^3 + 2x + 4 - x + 5$
9. $4x^3 + 8x + 7 - 8x$
10. $4x^{10} - x^5$
11. $x^6 - 6x^8 + 1$
12. $x^8 - x^6 - x^8 + x^4$

Do the indicated arithmetic and write the answer in standard form.

13. $(2x^2 + 5x - 7) + (3x^2 - 2x + 8)$
14. $(4x - 7) - (6x - 9)$
15. $(5x - 3) - (3x + 4)$
16. $(x^2 + 3x + 2) - (x^2 - 5x - 6)$
17. $(x + 2)(4x - 5)$
18. $(2x - 1)(3x - 2)$
19. $(x^2 - x + 1)(x + 1)$
20. $(x^3 + 1)(x - 1)$
21. $(x + 5)(x - 5)$
22. $(x^2 + 4)(x^2 - 4)$
23. $8(x^2 + 3x - 2)$
24. $-5(4x^2 - 3x - 6)$
25. $4(x^2 + 3x - 2) - 3(2x^2 + 5x + 4)$
26. $3(x^2 + x - 1) - 2(x^2 - 2x + 4)$
27. $(x + 2)(x - 1) - 3(x - 3)$
28. $(2x - 1)(x - 1) - 4(3x + 2)$

Let $A(w) = 2w + 3$, $B(w) = w^2$, and $C(w) = -w + 5$. Evaluate each of the following.

29. $A(5)$
30. $B(4)$
31. $A(B(x))$
32. $C(B(x))$
33. $B(A(x))$
34. $B(C(x))$
35. $B(q)$
36. $C(v)$
37. $A(3y - 4)$
38. $B(2w + 7)$
39. $C(0)$
40. $C(5)$
41. $C(C(w))$
42. $A(A(w))$
43. $A(-w)$
44. $C(-q)$
45. $B(B(-4))$
46. $B(C(11))$
47. $C(3 - 2x)$
48. $A(5 - 3x)$

Simplify the following.

49. $4(x + 2y)$
50. $x(3x - 5y) + 4$
51. $(3x^2 + 2xy + y) - (x^2 + y^2 - xy)$
52. $(4x^2 + 6xy + 9y^2) - (2x^2 - 5xy - y^2)$
53. $x(x + 3y + 5)$
54. $5(x + 2y) - 2(2x + 5y)$
55. $x[4(x + 1) - y(x + y)]$
56. $y[z(y + x) - x(3 + 2z)]$
57. $(x + y)[2x + 3y + 4(x - y)]$
58. $(y - 2z)[3x + 3(y - x)]$
59. $4x^2 + y^2 + 3xy - 2x^2 - y^2 + xy$
60. $(3 - y)(3 + x) - (3 + y) \cdot (3 - x)$

Perform the indicated arithmetic.

61. $(x + 3y) + (2x - 6y)$
62. $(2b + 5c) - (4b - 2c)$
63. $(b + c)(b - c + 1)$
64. $(y - z)(y + 2z + 3)$
65. $(2x + 5y)(3x - 4y)$
66. $(3b - 2c)(7b + 6c)$
67. $(3x^2 + 5xy + 6y^2) + (x^2 - 6xy - 2y^2)$
68. $(4v^2 + 5vw + w^2) + (-2v^2 + 4vw - w^2)$
69. $(4b^2 - 6bc + 7c^2) - (3b^2 - 2bc - 3c^2)$
70. $(6c^2 - 8cd - 5d^2) - (2c^2 - 12cd - d^2)$
71. $x(2x + 5y) - 2y(x + 3y + 6)$
72. $c(3d - 4g) - d(2c - d + g)$
73. $(-3x^2y) \cdot (4xy^5)$
74. $(2abc^2) \cdot (7a^3b^3)$
75. $-2(3a^2b) \cdot (-4b^2c)$
76. $5(abc) \cdot (-a^3b)$

Section 1-5 Factoring

We now come to the subject of factoring. Factoring is used later in this chapter to help simplify certain complicated expressions and is used in Chapter 2 as one method of solving quadratic equations. So far, we have used the distributive law in the form

$$a(b + c) = ab + ac$$

to multiply out.

Factoring is the opposite of multiplying out. To factor an expression means to write it as a product. Factoring uses the distributive law in the form

$$ab + ac = a(b + c)$$

The first step in factoring is to factor out the largest integer that divides each coefficient and the largest power of each variable that divides every term. This is known as finding a *common factor*.

Examples

1. $2x + 4y = 2(x + 2y)$

2. $6x^2 + 8xy = 2x(3x + 4y)$ Note that you cannot factor out x^2 because the term $8xy$ contains only the first power of x.

3. $7x^3 + 5x^2 = x^2(7x + 5)$

4. $b^2c + bc^2 = bc(b + c)$

5. $24b^2x + 16bx^2 + 32x = 8x(3b^2 + 2bx + 4)$

6. $14abc + 35bxy + 49ab^2 = 7b(2ac + 5xy + 7ab)$

Several forms occur often enough that they should be memorized. These formulas can also be used to make multiplying out easier.

$$a^2 + 2ab + b^2 = (a + b)^2 \quad \text{Square of a sum.}$$
$$a^2 - 2ab + b^2 = (a - b)^2 \quad \text{Square of a difference.}$$
$$a^2 - b^2 = (a - b)(a + b) \quad \text{Difference of two squares.}$$

Examples

7. $x^2 + 6x + 9 = (x + 3)^2$ — Square of a sum with $a = x$ and $b = 3$.

8. $4x^2 + 12x + 9 = (2x + 3)^2$ — Square of a sum with $a = 2x$ and $b = 3$.

9. $9c^2 + 24cd + 16d^2 = (3c + 4d)^2$ — Square of a sum with $a = 3c$ and $b = 4d$.

10. $e^2 - 10e + 25 = (e - 5)^2$ — Square of a difference with $a = e$ and $b = 5$.

11. $25c^2 - 10c + 1 = (5c - 1)^2$ — Square of a difference with $a = 5c$ and $b = 1$.

12. $16b^2 - 25y^2 = (4b - 5y)(4b + 5y)$ — Difference of two squares with $a = 4b$ and $b = 5y$.

13. $e^2 - (f - g)^2 = [e - (f - g)][e + (f - g)]$
    $= (e - f + g)(e + f - g)$ — Difference of two squares with $a = e$ and $b = f - g$.

General quadratic polynomials

Many quadratic expressions can be factored but are not in one of the three forms just discussed. For example, $12x^2 + 7x - 10$ is not the square of a sum or difference, nor is it the difference of two squares. However, $12x^2 + 7x - 10$ can be factored as $(3x - 2)(4x + 5)$.

Now suppose a, b, and c are integers and that all the numbers appearing in factors are required to be integers. If $ax^2 + bx + c = (dx + e)(fx + g)$, $a = df$ and $c = eg$, so that d and f are divisors of a, and e and g are divisors of c. To factor $x^2 + 8x + 7$, note that the only way of factoring 7 is as $7 \cdot 1$. The only possible factorings are $(x + 7)(x + 1)$ and $(x - 7)(x - 1)$. The correct factoring is $x^2 + 8x + 7 = (x + 7)(x + 1)$. The other factoring is not correct because $(x - 7)(x - 1) = x^2 - 8x + 7 \neq x^2 + 8x + 7$.

Note that $x^2 + 9x + 7$ cannot be factored using integer coefficients. The only possible factorings are $(x + 7)(x + 1)$ and $(x - 7)(x - 1)$, and neither possibility works.

To factor $x^2 - 5x + 6$, note that $6 = 6 \cdot 1 = 2 \cdot 3$. The four possible factorings are

$$(x + 6)(x + 1) \qquad (x + 2)(x + 3)$$
$$(x - 6)(x - 1) \qquad (x - 2)(x - 3)$$

The correct factoring, $x^2 - 5x + 6 = (x - 2)(x - 3)$, is obtained by multiplying out each attempted factoring until the correct one is obtained.

To factor $4x^2 - 4x - 15$, note that $4 = 4 \cdot 1 = 2 \cdot 2$ and $15 = 15 \cdot 1 = 5 \cdot 3$. The possible factorings are

$$(4x - 15)(x + 1) \qquad (4x - 1)(x + 15)$$
$$(4x + 15)(x - 1) \qquad (4x + 1)(x - 15)$$
$$(4x - 5)(x + 3) \qquad (2x - 15)(2x + 1)$$
$$(4x + 5)(x - 3) \qquad (2x + 15)(2x - 1)$$
$$(4x - 3)(x + 5) \qquad (2x - 5)(2x + 3)$$
$$(4x + 3)(x - 5) \qquad (2x + 5)(2x - 3)$$

The correct factoring is $4x^2 - 4x - 15 = (2x - 5)(2x + 3)$. After a little practice, it will usually be possible to eliminate many possible factorings without writing them down.

If $a > 0$ and $c < 0$, any factoring of $ax^2 + bx + c$ must be of the form $(\_\_x + \_\_)(\_\_x - \_\_)$, where the blank spaces are filled by counting numbers. If $c > 0$, any factoring must be of the form $(\_\_x + \_\_)(\_\_x + \_\_)$ if $b > 0$ and $(\_\_x - \_\_)(\_\_x - \_\_)$ if $b < 0$.

Examples 14. $x^2 + 8x + 12 = (x + 2)(x + 6)$

15. $x^2 - 7x - 8 = (x - 8)(x + 1)$

16. $x^2 - 9x + 20 = (x-4)(x-5)$

17. $x^2 - 8x + 25$ cannot be factored. The only possible factorings are $(x-1)(x-25)$, $(x+1)(x+25)$, $(x-5)^2$, and $(x+5)^2$. None of these works.

Example 18 Factor $3x^2 + 10x - 8$.

Solution The only factoring of 3 is $3 \cdot 1$. The factorings of 8 are $1 \cdot 8$ and $2 \cdot 4$. The possible factorings are

$$(3x - 1)(x + 8) \qquad (3x - 2)(x + 4)$$
$$(3x + 1)(x - 8) \qquad (3x + 2)(x - 4)$$
$$(3x - 8)(x + 1) \qquad (3x - 4)(x + 2)$$
$$(3x + 8)(x - 1) \qquad (3x + 4)(x - 2)$$

The correct factoring is $(3x - 2)(x + 4)$.

Examples 19. $10x^2 + 11x + 3 = (5x + 3)(2x + 1)$

20. $b^2 + 4bc + 3c^2 = (b + 3c)(b + c)$

21. $b^2 - 6bd - 7d^2 = (b - 7d)(b + d)$

22. $b^2 + 6bd - 7d^2 = (b + 7d)(b - d)$

23. $2c^2 - 7cd - 4d^2 = (c - 4d)(2c + d)$

Examples The following examples involve more than one step.

24. $x^3 + 10x^2 + 21x = x(x^2 + 10x + 21) = x(x + 3)(x + 7)$

25. $16a^2b + 20ab + 4b = 4b(4a^2 + 5a + 1) = 4b(4a + 1)(a + 1)$

26. $x^3 - 4xy^2 = x(x^2 - 4y^2) = x(x - 2y)(x + 2y)$

27. $b^4 - c^4 = (b^2 - c^2)(b^2 + c^2) = (b - c)(b + c)(b^2 + c^2)$

28. $10x^2 + 25x - 15 = 5(2x^2 + 5x - 3) = 5(x + 3)(2x - 1)$

29. $7b^2 + 42bc + 63c^2 = 7(b^2 + 6bc + 9c^2) = 7(b + 3c)^2$

Exercises 1-5

Factor each of the following.

1. $15bc + 12y$
2. $8bc + 21b$
3. $4bc - 8cd$
4. $12xy + 21x^2y^2$
5. $b^2 - 10b + 25$
6. $b^2 - 49$
7. $y^2 - 16k^2$
8. $y^2 + 5y + 4$
9. $x^3 - 4x^2$
10. $b^2c^2 - 1$
11. $x^2 - 9x + 14$
12. $x^2 - 6x - 16$
13. $6x^2 + x - 1$
14. $9x^2 - 30x + 25$
15. $x^2 - x - 2$
16. $y^2 + 3y + 2$
17. $c^2 - 7c + 10$
18. $b^2 + 11bc + 30c^2$
19. $49c^2 - 64d^2$
20. $35x^2y - 21xy + 14y$
21. $b^2c + 12bc + 36c$
22. $x^3 + 10x^2 + 16x$
23. $y^3 - 9y^2 - 36y$
24. $3c^2 - 75b^2$
25. $5x^2 - 30x + 45$
26. $x^3 - 36x$
27. $bx^2 - 7bx - 18b$
28. $bc^2 - 14bc - 32b$
29. $6x^2 - x - 12$
30. $a^2x^2 - 4ax - 21$
31. $12x^2 - 8$
32. $36x^2 - 60x + 25$
33. $15x^2 + 7x - 2$
34. $4y^3 + 16y^2 - 48y$
35. $8b^2 - 2y^2$
36. $5x^2 - 40xy + 80y^2$
*37. $x^3 + 1$
*38. $x^3 - 8$
*39. $x^4 - 16$
*40. $x^4 - 81y^4$
*41. $27x^3 + 8y^3$
*42. $8x^3 - 27y^3$
*43. $b^4 - 2b^2c^2 + c^4$
*44. $r^3 + 3r^2v + 3rv^2 + v^3$

Section 1-6 Rational Expressions

So far, polynomials have been added, subtracted, and multiplied. Each of these operations gives a polynomial as the result. Now it is time to consider quotients of polynomials. Usually the quotient of two polynomials is not a polynomial.

A *rational expression* is the quotient of two polynomials (or of two polynomials in several variables). To simplify a rational expression, factor the numerator and the denominator and divide out—that is, divide both the numerator and the denominator by—their common factors. The remaining factors are usually left as is rather than multiplied out.

Examples 1. $\dfrac{3x-6}{5x-10} = \dfrac{3(x-2)}{5(x-2)} = \dfrac{3}{5} \cdot \dfrac{x-2}{x-2} = \dfrac{3}{5} \cdot 1 = \dfrac{3}{5}$ Divide out $x-2$.

2. $\dfrac{x^2-4}{x^2-5x+6} = \dfrac{(x-2)(x+2)}{(x-2)(x-3)} = \dfrac{x+2}{x-3}$ Divide out $x-2$.

3. $\dfrac{6a^4bx^2}{21ab^2x} = \dfrac{2a^3x}{7b}$ Divide out $3abx$.

Example 4 Simplify $\dfrac{x^3+6x^2+9x}{x^3+7x^2+6x}$

Solution $\dfrac{x^3+6x^2+9x}{x^3+7x^2+6x}$

$= \dfrac{x(x^2+6x+9)}{x(x^2+7x+6)}$ Factor.

$= \dfrac{x(x+3)^2}{x(x+6)(x+1)}$ Factor.

$= \dfrac{(x+3)^2}{(x+6)(x+1)}$ Divide out x.

Example 5 Simplify $\dfrac{4xy^2-16xy-20x}{9by^2-72by+135b}$

Solution $\dfrac{4xy^2-16xy-20x}{9by^2-72by+135b}$

$= \dfrac{4x(y^2-4y-5)}{9b(y^2-8y+15)}$ Factor.

$$= \frac{4x(y-5)(y+1)}{9b(y-5)(y-3)} \qquad \text{Factor.}$$

$$= \frac{4x(y+1)}{9b(y-3)} \qquad \text{Divide out } y-5.$$

∎

To multiply two rational expressions, use the formula $(a/b)\cdot(c/d) = (ac)/(bd)$ for multiplication of fractions. To divide one rational expression by another, use the formula $(a/b) \div (c/d) = (a/b)\cdot(d/c) = (ad)/(bc)$ for division of fractions.

Examples Compute and simplify the product or quotient.

Example 6 $\dfrac{4(x-5)(x+6)}{15(x-3)(x+2)} \cdot \dfrac{5(x+2)(x-1)}{8(x-4)(x+6)}$

Solution $\dfrac{4(x-5)(x+6)}{15(x-3)(x+2)} \cdot \dfrac{5(x+2)(x-1)}{8(x-4)(x+6)}$

$$= \frac{4(x-5)(x+6)\cdot 5(x+2)(x-1)}{15(x-3)(x+2)\cdot 8(x-4)(x+6)} \qquad \begin{array}{l}\text{Multiply the numerators.} \\ \text{Multiply the denominators.}\end{array}$$

$$= \frac{(x-5)(x-1)}{6(x-3)(x-4)} \qquad \text{Divide out } 20(x+6)(x+2).$$

Example 7 $\dfrac{4x-12}{x^2+8x+7} \cdot \dfrac{x^2+6x-7}{9x-27}$

Solution $\dfrac{4x-12}{x^2+8x+7} \cdot \dfrac{x^2+6x-7}{9x-27}$

$$= \frac{4(x-3)}{(x+7)(x+1)} \cdot \frac{(x+7)(x-1)}{9(x-3)} \qquad \text{Factor.}$$

$$= \frac{4(x-1)}{9(x+1)} \qquad \text{Divide out } (x-3)(x+7).$$

To save one step in this example, we did not multiply the numerators and denominators before dividing.

Example 8 $\dfrac{10x+20}{8x+24} \div \dfrac{7x+14}{4x+12}$

Solution $\dfrac{10x+20}{8x+24} \div \dfrac{7x+14}{4x+12}$

$= \dfrac{10x+20}{8x+24} \cdot \dfrac{4x+12}{7x+14}$ To divide fractions, invert the divisor and multiply.

$= \dfrac{10(x+2)}{8(x+3)} \cdot \dfrac{4(x+3)}{7(x+2)}$ Factor.

$= \dfrac{5}{7}$ Divide out $8(x+2)(x+3)$.

Example 9 $\dfrac{8abc^2}{27a^2b^3c} \div \dfrac{4a^2b}{9c^3}$

Solution $\dfrac{8abc^2}{27a^2b^3c} \div \dfrac{4a^2b}{9c^3}$

$= \dfrac{8abc^2}{27a^2b^3c} \cdot \dfrac{9c^3}{4a^2b}$ Invert the divisor and multiply.

$= \dfrac{8 \cdot 9 abc^5}{27 \cdot 4 a^4 b^4 c}$ Multiply.

$= \dfrac{2c^4}{3a^3b^3}$ Divide out $4 \cdot 9abc$.

Example 10 $\dfrac{4bx^2+5bx+b}{5x+15} \div \dfrac{3x^2+4x+1}{7x+21}$

Solution $\dfrac{4bx^2+5bx+b}{5x+15} \div \dfrac{3x^2+4x+1}{7x+21}$

$= \dfrac{4bx^2+5bx+b}{5x+15} \cdot \dfrac{7x+21}{3x^2+4x+1}$ Invert the divisor and multiply.

$= \dfrac{b(4x^2+5x+1)}{5(x+3)} \cdot \dfrac{7(x+3)}{(3x+1)(x+1)}$ Factor.

$$= \frac{7b(4x+1)(x+1)}{5(3x+1)(x+1)} \qquad \text{Factor and also divide out } x+3.$$

$$= \frac{7b(4x+1)}{5(3x+1)} \qquad \text{Divide out } x+1.$$

■

To add or subtract rational expressions with the same denominator, use the formula $(a/c) + (b/c) = (a+b)/c$ for addition of fractions with the same denominator or the formula $(a/c) - (b/c) = (a-b)/c$ for subtraction.

Examples 11. $\dfrac{5}{x} + \dfrac{7}{x} = \dfrac{5+7}{x} = \dfrac{12}{x}$

12. $\dfrac{ab}{c-3} + \dfrac{a^2-5}{c-3} = \dfrac{ab+a^2-5}{c-3}$

13. $\dfrac{2x-3}{x^2-16} + \dfrac{x-9}{x^2-16} = \dfrac{3x-12}{x^2-16} = \dfrac{3(x-4)}{(x-4)(x+4)} = \dfrac{3}{x+4}$

14. $\dfrac{5x}{7} - \dfrac{2x}{7} = \dfrac{5x-2x}{7} = \dfrac{3x}{7}$

15. $\dfrac{4bc-5}{b^2-2} - \dfrac{2bc-3b-7}{b^2-2} = \dfrac{4bc-5-(2bc-3b-7)}{b^2-2}$

$$= \dfrac{4bc-5-2bc+3b+7}{b^2-2} = \dfrac{2bc+3b+2}{b^2-2}$$

Least common denominator

To add or subtract two fractions with different denominators, first change the fractions so that they have the same denominator. One way to do this is to determine the least common denominator. Begin by factoring both denominators—this includes factoring the coefficients. Then form the product of all numbers and polynomials that occur as factors of either denominator. Use as the exponent of a number or polynomial in the least common denominator the larger of its exponents in the two original denominators. Now let's compute

$$\frac{5}{24x^2y^2} + \frac{7}{10xy^5z^8}$$

First factor the denominators completely. Only the coefficients need to be factored in this problem. Since

$$24 = 8 \cdot 3 = 2^3 \cdot 3 \quad \text{and} \quad 10 = 2 \cdot 5$$

the denominators are

$$2^3 \cdot 3 x^2 y^2 \quad \text{and} \quad 2 \cdot 5 x y^5 z^8$$

The least common denominator is of the form:

$$2^?3^?5^?x^?y^?z^?$$

The exponent of 2 must be 3 since 2^3 is in one denominator and $2 = 2^1$ is in the other. The exponent of 3 must be 1 since $3 = 3^1$ is a factor of the first denominator. Similarly the exponent of 5 is 1. Since x^2 is in one denominator and $x = x^1$ is in the other, the exponent of x must be 2. Similarly the exponent of y is 5, since 5 is the larger of 2 and 5. Since z is in the second denominator only and has exponent 8 there, z has exponent 8 in the least common denominator. The least common denominator is

$$2^3 \cdot 3 \cdot 5 x^2 y^5 z^8$$

Now express each fraction as a fraction with the least common denominator as its denominator. Since $2^3 \cdot 3 \cdot 5 x^2 y^5 z^8 = (24x^2y^2) \cdot (5y^3z^8)$, we must multiply the first fraction by $(5y^3z^8)/(5y^3z^8)$ to obtain an equal fraction whose denominator is the least common denominator. Similarly, since $2^3 \cdot 3 \cdot 5 x^2 y^5 z^8 = (10xy^5z^8) \cdot (2^2 \cdot 3x)$, we must multiply the second fraction by $(2^2 \cdot 3x)/(2^2 \cdot 3x) = (12x)/(12x)$. Now both fractions have the same denominator and can be added (or subtracted) as before. Thus

$$\frac{5}{24x^2y^2} + \frac{7}{10xy^5z^8} = \frac{5}{24x^2y^2} \cdot \frac{5y^3z^8}{5y^3z^8} + \frac{7}{10xy^5z^8} \cdot \frac{12x}{12x}$$

$$= \frac{25y^3z^8}{120x^2y^5z^8} + \frac{84x}{120x^2y^5z^8}$$

$$= \frac{25y^3z^8 + 84x}{120x^2y^5z^8}$$

Example 16 Compute $\dfrac{x}{x^2 - 9} + \dfrac{3x - 4}{x^2 - 5x + 6}$.

Solution $\dfrac{x}{x^2 - 9} + \dfrac{3x - 4}{x^2 - 5x + 6}$

$= \dfrac{x}{(x - 3)(x + 3)} + \dfrac{(3x - 4)}{(x - 2)(x - 3)}$ Factor the denominators.

$= \dfrac{x}{(x - 3)(x + 3)} \cdot \dfrac{x - 2}{x - 2} + \dfrac{(3x - 4)}{(x - 2)(x - 3)} \cdot \dfrac{x + 3}{x + 3}$ The least common denominator is $(x - 3)(x + 3) \cdot (x - 2)$.

$= \dfrac{x(x - 2) + (3x - 4)(x + 3)}{(x - 3)(x + 3)(x - 2)}$ Add the numerators.

$= \dfrac{(x^2 - 2x) + (3x^2 + 5x - 12)}{(x - 3)(x + 3)(x - 2)}$

$= \dfrac{4x^2 + 3x - 12}{(x - 3)(x + 3)(x - 2)}$

Example 17 Compute $\dfrac{2x - 3}{12y} - \dfrac{4x + 5}{10y}$.

Solution $\dfrac{2x - 3}{12y} - \dfrac{4x + 5}{10y}$

$= \dfrac{2x - 3}{2^2 \cdot 3y} - \dfrac{4x + 5}{2 \cdot 5y}$ Factor the denominators.

$= \dfrac{2x - 3}{2^2 \cdot 3y} \cdot \dfrac{5}{5} - \dfrac{4x + 5}{2 \cdot 5y} \cdot \dfrac{2 \cdot 3}{2 \cdot 3}$ The least common denominator is $2^2 \cdot 3 \cdot 5y$.

$= \dfrac{10x - 15}{60y} - \dfrac{24x + 30}{60y}$ Multiply out.

$= \dfrac{(10x - 15) - (24x + 30)}{60y}$ Subtract the second numerator from the first.

$= \dfrac{10x - 15 - 24x - 30}{60y}$

$$= \frac{-14x - 45}{60y}$$

Example 18 Compute $\dfrac{6}{x^2 + 8x + 16} - \dfrac{2}{x^2 - 16}$.

Solution $\dfrac{6}{x^2 + 8x + 16} - \dfrac{2}{x^2 - 16}$

$= \dfrac{6}{(x+4)^2} - \dfrac{2}{(x+4)(x-4)}$ Factor the denominators.

$= \dfrac{6}{(x+4)^2} \cdot \dfrac{x-4}{x-4} - \dfrac{2}{(x+4)(x-4)} \cdot \dfrac{x+4}{x+4}$ The least common denominator is $(x+4)^2(x-4)$.

$= \dfrac{6x - 24}{(x+4)^2(x-4)} - \dfrac{2x + 8}{(x+4)^2(x-4)}$ Multiply out.

$= \dfrac{(6x - 24) - (2x + 8)}{(x+4)^2(x-4)}$ Subtract the second numerator from the first.

$= \dfrac{6x - 24 - 2x - 8}{(x+4)^2(x-4)}$

$= \dfrac{4x - 32}{(x+4)^2(x-4)}$

Exercises 1-6

Simplify each of the following.

1. $\dfrac{x^3}{x^2 - x}$

2. $\dfrac{10x - 18}{25x - 45}$

3. $\dfrac{12 - 3x}{36 - 9x}$

4. $\dfrac{x^2 + 3x}{x^4}$

5. $\dfrac{x^2 + 3x + 2}{x^2 - 4}$

6. $\dfrac{c^2 - 9}{c^2 - 6c + 9}$

7. $\dfrac{24a^2bc^3}{48ab^4c}$

8. $\dfrac{x^2 + 7x + 6}{x^2 + 3x - 18}$

9. $\dfrac{x^2 - 3x - 4}{6x + 6}$ 10. $\dfrac{x^3 - x}{x^2 - x}$

11. $\dfrac{9b^2 + 6b + 1}{15b + 5}$ 12. $\dfrac{27v^2w^3}{18vwy}$

Perform the indicated arithmetic and simplify.

13. $\dfrac{4x - 1}{5x + 3} \cdot \dfrac{2x + 1}{8x - 2}$ 14. $\dfrac{x^2 - 9}{x^2 - 16} \cdot \dfrac{x^2 - 8x + 16}{x^2 + 6x + 9}$

15. $\dfrac{4x}{y} + \dfrac{3y}{z}$ 16. $\dfrac{4y}{z} \div \dfrac{3z}{y}$

17. $\dfrac{b^2 - 3b}{b + 4} \div \dfrac{b}{b + 1}$ 18. $\dfrac{x - 5}{x + 3} + \dfrac{2x + 5}{x + 3}$

19. $\dfrac{6x - 2}{8x + 9} \cdot \dfrac{2x + 3}{3x - 1}$ 20. $\dfrac{4a^2b^3c}{9x^2y} \cdot \dfrac{12xy^3}{5abc^2}$

21. $\dfrac{2y - 7}{y - 1} - \dfrac{3y - 12}{y - 1}$ 22. $\dfrac{4z - 3}{10z} - \dfrac{2z - 7}{4z}$

23. $\dfrac{y^2 + 7y + 12}{y^2 - 13y + 30} \cdot \dfrac{y^2 - 11y + 10}{y^2 + 5y + 6}$ 24. $\dfrac{x^2 - 3x - 4}{x^2 - 5x + 6} \div \dfrac{x^2 - 1}{x^2 - 7x + 10}$

25. $\dfrac{x^2 - x}{x + 1} - \dfrac{3x}{x + 2}$ 26. $\dfrac{2x + 1}{x - 1} + \dfrac{5x}{x - 5}$

27. $\dfrac{x^2 + 8x + 12}{x^2 + 6x + 8} + \dfrac{x^2 - 25}{x^2 - 10x + 25}$ 28. $\dfrac{27b^3c^2}{8bd} \div \dfrac{3bc^2d}{2a}$

29. $\dfrac{x^2 - 4}{x^2 + 4x + 4} - \dfrac{x^2 + 3x}{x^2 - 2x}$ 30. $\dfrac{x^3 - x}{x^2 + 3x + 2} \cdot \dfrac{x^2 - 4}{x^2 - 1}$

31. $\dfrac{16x^2y}{4y^3z} \div \dfrac{25xy^3}{14z^2}$ 32. $\dfrac{x^2 - 7x + 6}{x^2 - 9x - 10} \cdot \dfrac{x^2 - 100}{x^2 - 2x - 24}$

33. $\dfrac{4x^2 - x + 2}{x - 3} - \dfrac{2x^2 + x - 2}{x - 3}$ 34. $\dfrac{x^2 + 9x + 20}{x^2 + 2x - 15} + \dfrac{4x - 20}{5x - 25}$

35. $\dfrac{24v^2w^2x}{11x^2y} \cdot \dfrac{33x^4y}{12v^3w^5}$

36. $\dfrac{x^2 - 3x - 10}{x^2 + 4x + 3} \div \dfrac{x^2 + 8x + 7}{x^2 + 3x - 10}$

Section 1-7 Algebraic Expressions

Algebraic expressions are polynomials or rational expressions in which the exponents are allowed to be any rational numbers instead of just counting numbers. The manipulation of algebraic expressions is very similar to the manipulation of polynomials and of rational expressions. To simplify an algebraic expression, remove all grouping symbols and combine like terms. If the expression has both a numerator and a denominator, divide out any common factors. The expression may be left in factored form, but need not be.

Example 1 Simplify $\dfrac{x^{3/2} - x^{1/2}}{x^2 - 1}$.

Solution $\dfrac{x^{3/2} - x^{1/2}}{x^2 - 1}$

$= \dfrac{x^{1/2}(x - 1)}{(x - 1)(x + 1)}$ Factor.

$= \dfrac{x^{1/2}}{x + 1}$ Divide out $x + 1$.

Example 2 Simplify $(4x^2 - x^{1/2})(3x^{3/2} - 3x^{-1/2})$.

Solution $(4x^2 - x^{1/2})(3x^{3/2} - 3x^{-1/2})$

$= 4x^2(3x^{3/2} - 3x^{-1/2}) - x^{1/2}(3x^{3/2} - 3x^{-1/2})$ Distributive law.

$= 12x^{7/2} - 12x^{3/2} - 3x^2 + 3$ Distributive law.

Example 3 Simplify $x^{1/3}[x^{5/3} + (8x)^{2/3} + 27x^{2/3}]$.

Solution $x^{1/3}[x^{5/3} + (8x)^{2/3} + 27x^{2/3}]$

$= x^{1/3}[x^{5/3} + 8^{2/3}x^{2/3} + 27x^{2/3}]$ Remove parentheses. $(ab)^r = a^rb^r$.

$$= x^{1/3}[x^{5/3} + 4x^{2/3} + 27x^{2/3}]$$

$$= x^{1/3}[x^{5/3} + 31x^{2/3}] \qquad \text{Combine like terms.}$$

$$= x^2 + 31x \qquad \text{Multiply out.}$$

■

The following examples illustrate the arithmetic of algebraic expressions.

Example 4 $(3x^{5/2} - 4b\sqrt{c}) + (5x^{5/2} + 2b\sqrt{c})$

$$= 3x^{5/2} - 4b\sqrt{c} + 5x^{5/2} + 2b\sqrt{c}$$

$$= 8x^{5/2} - 2b\sqrt{c}$$

Example 5 $4(\sqrt{bc} - \sqrt{b}) - 2(3\sqrt{bc} - 2\sqrt{b})$

$$= (4\sqrt{bc} - 4\sqrt{b}) - (6\sqrt{bc} - 4\sqrt{b})$$

$$= 4\sqrt{bc} - 4\sqrt{b} - 6\sqrt{bc} + 4\sqrt{b}$$

$$= -2\sqrt{bc}$$

Example 6 $\dfrac{4x^{5/2}y^{7/3}}{9a^2b^{4/3}} \cdot \dfrac{27a^{2/3}b^{2/3}}{2x^2y^{2/3}}$

$$= \dfrac{4 \cdot 27}{9 \cdot 2} \cdot \dfrac{a^{2/3}}{a^2} \cdot \dfrac{b^{2/3}}{b^{4/3}} \cdot \dfrac{x^{5/2}}{x^2} \cdot \dfrac{y^{7/3}}{y^{2/3}}$$

$$= \dfrac{6x^{1/2}y^{5/3}}{a^{4/3}b^{2/3}}$$

Example 7 $\dfrac{18x^{1/2}y^{3/2}}{4b^2c} \div \dfrac{9x^{3/2}y^2}{b^{3/2}c}$

$$= \dfrac{18x^{1/2}y^{3/2}}{4b^2c} \cdot \dfrac{b^{3/2}c}{9x^{3/2}y^2}$$

$$= \frac{18}{4 \cdot 9} \cdot \frac{x^{1/2}}{x^{3/2}} \cdot \frac{y^{3/2}}{y^2} \cdot \frac{b^{3/2}}{b^2} \cdot \frac{c}{c}$$

$$= \frac{1}{2xy^{1/2}b^{1/2}}$$

Example 8 $\quad \dfrac{4}{b^{1/2}} - \dfrac{3}{c^{2/3}}$

$$= \frac{4}{b^{1/2}} \cdot \frac{c^{2/3}}{c^{2/3}} - \frac{3}{c^{2/3}} \cdot \frac{b^{1/2}}{b^{1/2}}$$

$$= \frac{4c^{2/3} - 3b^{1/2}}{b^{1/2} c^{2/3}}$$

Rationalizing denominators or numerators

Sometimes a fraction contains square roots, and it is often desirable to eliminate square roots from either the denominator or the numerator. These procedures (called *rationalizing the denominator* and *rationalizing the numerator*) usually involve using the formula for the difference of two squares in one of two forms:

$$(a + \sqrt{b})(a - \sqrt{b}) = a^2 - b \qquad (\sqrt{a} + \sqrt{b})(\sqrt{a} - \sqrt{b}) = a - b$$

Example 9 Rationalize the denominator in $\dfrac{4}{3 - \sqrt{2}}$.

Solution $\quad \dfrac{4}{3 - \sqrt{2}}$

$$= \frac{4}{3 - \sqrt{2}} \cdot \frac{3 + \sqrt{2}}{3 + \sqrt{2}} \qquad \text{Multiply by } (3 + \sqrt{2})/(3 + \sqrt{2})$$
$$\text{since } (3 - \sqrt{2})(3 + \sqrt{2}) \text{ is rational.}$$

$$= \frac{12 + 4\sqrt{2}}{9 - 2} = \frac{12 + 4\sqrt{2}}{7}$$

Example 10 Rationalize the denominator in $\dfrac{x}{2x - \sqrt{3y}}$.

48 CHAPTER 1 SETS, NUMBERS, AND ALGEBRAIC MANIPULATION

Solution $\dfrac{x}{2x - \sqrt{3y}}$

$= \dfrac{x}{2x - \sqrt{3y}} \cdot \dfrac{2x + \sqrt{3y}}{2x + \sqrt{3y}}$

$= \dfrac{2x^2 + x\sqrt{3y}}{4x^2 - 3y}$

Example 11 Rationalize the denominator in $\dfrac{4b^{1/2}}{\sqrt{2b} + \sqrt{3c}}$.

Solution $\dfrac{4b^{1/2}}{\sqrt{2b} + \sqrt{3c}}$

$= \dfrac{4\sqrt{b}}{\sqrt{2b} + \sqrt{3c}} \cdot \dfrac{\sqrt{2b} - \sqrt{3c}}{\sqrt{2b} - \sqrt{3c}}$

$= \dfrac{4b\sqrt{2} - 4\sqrt{3bc}}{2b - 3c}$

Example 12 Rationalize the numerator in $\dfrac{13 + \sqrt{11}}{5}$.

Solution $\dfrac{13 + \sqrt{11}}{5}$

$= \dfrac{13 + \sqrt{11}}{5} \cdot \dfrac{13 - \sqrt{11}}{13 - \sqrt{11}}$

$= \dfrac{169 - 11}{65 - 5\sqrt{11}}$

$= \dfrac{158}{65 - 5\sqrt{11}}$

Example 13 Rationalize the numerator in $\dfrac{\sqrt{x+h} - \sqrt{x}}{h}$.

Solution $\dfrac{\sqrt{x+h} - \sqrt{x}}{h}$

$= \dfrac{\sqrt{x+h} - \sqrt{x}}{h} \cdot \dfrac{\sqrt{x+h} + \sqrt{x}}{\sqrt{x+h} + \sqrt{x}}$

$$= \frac{(x+h) - x}{h(\sqrt{x+h} + \sqrt{x})}$$

$$= \frac{h}{h(\sqrt{x+h} + \sqrt{x})}$$

$$= \frac{1}{\sqrt{x+h} + \sqrt{x}}$$

Exercises 1-7

Simplify the following algebraic expressions

1. $4x^{3/2} - (9x)^{1/2} + (4x)^{3/2}$
2. $x^{2/3}(6x^{1/3} - 4x^{4/3})$
3. $(2x - 3\sqrt{y})(5x + 4\sqrt{y})$
4. $(2x^{4/3} - x) - x(2x^{1/3} + 1)$
5. $\dfrac{x^{5/3} - 4x^2 + 3x}{x}$
6. $\dfrac{x^{3/2} - 4x + 6x^{1/2}}{5x^{1/2}}$
7. $\dfrac{x^{2/3}}{x^{-1/3}}$
8. $x^{1/2}[x^{3/2} - x(x^{1/2} + 1)]$

Perform the indicated arithmetic and simplify.

9. $(5x^{3/2} + x^{1/3}) \cdot (5x^{3/2} - x^{1/3})$
10. $(x^{1/2} - 3y^{3/2})^2$
11. $(4c^{1/2} + 5c^{3/2}) - 2(3c^{1/2} + 2c^{3/2})$
12. $(10b^2c^{1/3}) \div (5b^{3/2}c)$
13. $(2 + 3\sqrt{x} - 4x) + (5 - \sqrt{x} - 2x)$
14. $(4y^{2/3} + z^{1/3})(2y^{1/3} - z^{2/3})$
15. $(24b^{3/2}c^{1/2}) \div (48b^2c^{3/2})$
16. $[(8x)^{1/3} + 5x^{2/3}] + 2x^{1/3}$

Rationalize the numerator and simplify.

17. $\dfrac{5 - \sqrt{2}}{4}$
18. $\dfrac{10 + 2\sqrt{3}}{7}$
19. $\dfrac{8 - \sqrt{3}}{4 + \sqrt{3}}$
20. $\dfrac{10 - \sqrt{8}}{4 + 3\sqrt{8}}$
21. $\dfrac{x - 2\sqrt{y}}{x + \sqrt{y}}$
22. $\dfrac{\sqrt{c} + \sqrt{2d}}{4}$
23. $\dfrac{\sqrt{5} + \sqrt{x}}{\sqrt{5} - 2\sqrt{x}}$
24. $\dfrac{\sqrt{3x}}{6 + \sqrt{x}}$

Rationalize the denominator and simplify.

25. $\dfrac{4}{\sqrt{5} - \sqrt{3}}$

26. $\dfrac{8 + \sqrt{2}}{6 - 2\sqrt{2}}$

27. $\dfrac{\sqrt{c} + \sqrt{5d}}{\sqrt{c} - \sqrt{5d}}$

28. $\dfrac{3y}{5 - 2\sqrt{y}}$

29. $\dfrac{4}{\sqrt{15}}$

30. $\dfrac{4 - \sqrt{15}}{4 + \sqrt{15}}$

31. $\dfrac{5c}{4 - \sqrt{5c}}$

32. $\dfrac{6}{\sqrt{y}}$

Chapter 1 Review

Are the following statements true or false?
1. $2/3 \in \{\text{rational numbers}\}$
2. $\{4, 7, 9, 4\} = \{9, 7, 4\}$
3. $\{\text{integers}\} \subseteq \{\text{counting numbers}\}$
4. $\{\text{rational numbers}\}$ and $\{\text{irrational numbers}\}$ are disjoint.

Let $U = \{4, 5, 6, 7, 8, 9\}$, $A = \{5, 7, 8\}$, and $B = \{6, 8, 9\}$. Determine each of the following.

5. $A \cup B$
6. $A \cap B$
7. A'
8. $(A' \cap B) \cup (A \cap B')$

Evaluate each expression.

9. $\dfrac{1}{5} + \dfrac{3}{7}$

10. $\left(\dfrac{2}{9}\right) \cdot \left(\dfrac{3}{8}\right)$

11. $\left(\dfrac{4}{3}\right) \div \left(\dfrac{5}{6}\right)$

12. $2|-3| + 3|-2|$

13. $\left(\dfrac{3}{2}\right)^4$

14. $\left(\dfrac{2}{5}\right)^{-3}$

15. $\left(\dfrac{9}{4}\right)^{3/2}$

16. $\left(\dfrac{-8}{27}\right)^{-2/3}$

17. Express 4.75×10^{-2} in standard notation.

18. Express 189,000 in scientific notation.

Simplify each expression.

19. $(3x^2)^4$
20. $\left(\dfrac{x}{y^2}\right)^{-3}$
21. $\left(\dfrac{x^3}{y^6}\right)^{-2/3}$
22. $\dfrac{(x+2y)^{4/3}}{(x+2y)^{-2/3}}$
23. $(x^2 - 4x + 7) + 3(x^2 + 3x - 8)$
24. $(2x - 1)(3x + 4)$
25. $x(3x - 2y) + y(2x + 5y)$
26. $(4b^2 + 7bc - 6c^2) - (3b^2 - 2bc + c^2)$

Let $P(y) = y^2 - y + 4$ and $Q(y) = 3y + 2$. Evaluate each of the following.

27. $P(5)$
28. $P(Q(5))$
29. $Q(x + 6)$
30. $Q(Q(7))$

Factor each of the following.

31. $3x^2 - 18x$
32. $2b^2 - 18c^2$
33. $v^2 + 6vw + 9w^2$
34. $r^2 - 5r - 14$
35. $y^3 - 7y^2 - 18y$
36. $b^2c^2 - 4bc + 4$

Perform the indicated arithmetic and simplify.

37. $\dfrac{2x}{y} + \dfrac{y}{x}$
38. $\dfrac{x^2 - 16}{x^2 + 6x + 9} \cdot \dfrac{x^2 + 5x + 6}{x^2 - 6x + 8}$
39. $\dfrac{w^2 - 6w}{w + 5} \div \dfrac{w^2 - 36}{3w + 15}$
40. $\dfrac{3x - 2}{x^2 - 3x - 10} - \dfrac{2x - 7}{x^2 + 8x + 12}$
41. $(x^{5/2} - 2x^{3/2}) \cdot (x^{3/2} + 4x^{1/2})$
42. $(12b^{3/2}c) \div (18b^{1/2}c^{-2})$
43. $(5 - 2\sqrt{w} + 3w) + 2(3 + 4\sqrt{w} - w)$
44. $(27x)^{1/3} + 4x^{1/3}$

45. Rationalize the denominator and simplify $\dfrac{5 - \sqrt{7}}{2 + \sqrt{7}}$.

46. Rationalize the denominator and simplify $\dfrac{\sqrt{k} + 2\sqrt{3j}}{\sqrt{k} - \sqrt{3j}}$.

47. Rationalize the numerator and simplify $\dfrac{3 - \sqrt{7}}{8}$.

48. Rationalize the numerator and simplify $\dfrac{\sqrt{13} - \sqrt{11}}{4}$.

2

Equations and Inequalities

Section 2-1 Linear Equations in One Variable

Many applications of mathematics involve solving equations. In this chapter, we will apply the algebraic techniques of Chapter 1 to the solution of equations. Consider this first example.

Example 1 Gilbert and Cecilia Rice received a $640 refund on their income tax and decided to invest it in stock in the Wabasco Railroad at $15 a share. For handling the transaction, they must pay their stockbroker $25. How many shares of stock can they buy if they use their whole refund?

Solution Let x be the number of shares of stock bought. Since x shares of stock cost $15x$ dollars, the Rices pay their stockbroker

$$\underbrace{15x}_{\text{(cost of stock)}} + \underbrace{25}_{\text{(stockbroker's commission)}}$$

If the Rices use the whole refund, then the amount they pay the stockbroker is $640. Thus the quantities $15x + 25$ and 640 are equal. This gives the equation

$$15x + 25 = 640$$

To solve this equation, subtract 25 from both sides to obtain

$$15x = 640 - 25$$
$$= 615$$

Now divide both sides by 15 to obtain

$$x = \frac{615}{15} = 41$$

Gilbert and Cecilia Rice can buy 41 shares of stock. ∎

The example is an illustration of the usefulness of equations. Equations are worth further study. Let's begin with some notation, and then consider the simplest and most important kind of equation—the linear equation.

An *equation* is a statement that says two quantities are equal to each other. An equation consists of two expressions with an equal sign ($=$) between them. An equation that does not contain any variables is either a true statement, such as $2 + 3 = 5$, or a false statement, such as $3 + 5 = 12$. If an equation contains a variable, the *solution set* of the equation is the set of those values for the variable that give a true statement when substituted into the equation. For example, the solution set of $y^2 = 4$ is $\{-2, 2\}$, because $(-2)^2 = 4$, $2^2 = 4$, but $y^2 \neq 4$ if y is any number other than -2 or 2.

An equation is *linear* in x if the equation can be manipulated into the form $ax = b$ with $a \neq 0$. You should verify that the equations

$$2x + 5 = 9$$
$$2(x + 3) - 4(x - 5) = 7$$
$$5bx - 4b^2 + cx = 12$$

are linear in x. The equations

$$x^2 = 3x + 2$$
$$x^{1/2} = 4$$
$$\frac{5b}{x} - 4x + 5 = 3b + 8$$

are not linear in x.

To solve a linear equation in x, on the left side of the equation collect all the terms that involve x and on the right side of the equation collect all the terms that do not involve x. For example, to solve

$$6ax - 9b = 5x + 17$$

add $9b - 5x$ to both sides of the equation to obtain

$$6ax - 5x = 9b + 17$$

Next factor out an x on the left side of the equation. In this case, we obtain

$$(6a - 5)x = 9b + 17$$

Then divide both sides of the equation by the coefficient of x. In this case, divide by $6a - 5$ to get

$$x = \frac{9b + 17}{6a - 5}$$

Examples Solve each of the following equations for x.

Example 2 $4x + 5 = 12$

Solution $4x = 7$ Subtract 5 from both sides.

$\quad\quad\quad\quad x = \dfrac{7}{4}$ Divide by 4.

The solution is $7/4$.

Example 3 $2x - 3 = 5 - 6x$

Solution $8x - 3 = 5$ Add $6x$ to both sides.

$\quad\quad\quad\quad 8x = 8$ Add 3 to both sides.

$\quad\quad\quad\quad x = 1$ Divide by 8.

The solution is 1.

Example 4 $ax - b = cx + d$

Solution $ax - cx - b = d$ Subtract cx from both sides.

$\quad\quad\quad\quad ax - cx = b + d$ Add b to both sides.

$\quad\quad\quad\quad (a - c)x = b + d$ Factor out an x.

$$x = \frac{b+d}{a-c} \qquad \text{Divide by } a - c.$$

The solution is $\dfrac{b+d}{a-c}$.

Example 5 $4[2x + 3 - 5(5x - 2)] + 3x - 4 = 2 + 3x$

Solution

$4(2x + 3 - 25x + 10) + 3x - 4 = 2 + 3x$	Distributive law.
$4(-23x + 13) + 3x - 4 = 2 + 3x$	Combine like terms.
$-92x + 52 + 3x - 4 = 2 + 3x$	Distributive law.
$-89x + 48 = 2 + 3x$	Combine like terms.
$-92x = -46$	Subtract $48 + 3x$ from both sides.
$x = \dfrac{-46}{-92} = \dfrac{1}{2}$	Divide by -46.

The solution is $1/2$.

Example 6 Call and Save provides long-distance phone calls for businesses and is in competition with the telephone company. During business hours, Call and Save charges $1.00 plus $0.20 per minute for a call from Boston to Seattle. Mr. Layman just received a bill of $6.80 for a call he made between these cities last week. How long did he talk?

Solution Let m be the number of minutes he spoke. The cost of the call is $1 + 0.20m$. Since the call cost $6.80, we have

$1 + 0.20m = 6.80$	
$0.20m = 5.80$	Subtract 1 from both sides.
$m = \dfrac{5.80}{0.20} = 29$	Divide by 0.20.

Mr. Layman spoke for 29 minutes. ∎

Each of the previous equations had exactly one solution. *A linear equation has exactly one solution.* You should always check for errors by substituting the solution (or solutions) of an equation into the equation and verifying that a true statement is obtained. For instance, in Example 5, both sides of the equation equal $3\frac{1}{2}$ when $\frac{1}{2}$ is substituted for x.

Some equations appear to be linear but are not because the variable drops out when you attempt to solve the equation. For example, attempting to solve $2(3x + 3) = 3(2x + 1)$ gives $6x + 6 = 6x + 3$ and then $0 = -3$. This is a false statement. The equation $2(3x + 3) = 3(2x + 1)$ is not linear and has no solutions. An equation is *inconsistent* if it has no solutions. Attempting to solve $5(2x + 4) = 2(5x + 10)$ gives $10x + 20 = 10x + 20$, and then $0 = 0$. This is a true statement. The equation $5(2x + 4) = 2(5x + 10)$ is not linear and has all real numbers as solutions. An *identity* is an equation that has as solutions all real numbers for which all the terms in the equation are defined.

You should verify that the equations

$$x + 1 = x + 2$$
$$x^2 + 4 = 0$$

are inconsistent. You should also verify that the equations

$$x^2 - 3x + 2 = (x - 1)(x - 2)$$
$$x = x$$
$$\frac{x^2 - 4}{x - 2} = x + 2$$

are identities.

Exercises 2-1

Write the solution(s) of each equation. Indicate when an equation is inconsistent or is an identity.

1. $4x = 10$
2. $6x = 30$
3. $2x + 5 = 4x - 9$
4. $3x + 7 = 5 - 2x$
5. $5x + 1 = 6x + 2$
6. $8x - 3 = 4x - 9$
7. $7x + 6 = 7 + 7x$
8. $4x - 2 = 2 - 4x$
9. $3x + 5 = 10x - 7$
10. $9x - 1 = 9x - 1$
11. $2(x - 3) = 5(4 - x)$
12. $3(x - 2) = 5(2x + 1)$
13. $4(x - 1) = 2(2x - 2)$
14. $6(2x + 1) = 4(5x + 2)$
15. $3(4 + 2x) - 4(2 + 3x) = 0$
16. $5(2x + 3) - 3(5x + 2) = 0$
17. $6(3x + 1) = 3(6x + 4)$
18. $3x - 4 = 5(2x + 8)$
19. $8x + 16 = 6x + 8$
20. $5x + 4 = 9(2 - 5x)$
21. $3x + 1 - 2(x - 1) = 0$
22. $-5(3 - 2x) + 4x + 1 = 0$

23. $4(3x + 7) - 6(2x + 5) = 0$

24. $7[2x + 3(x - 1)] = 25x + 104$

25. $2[3(4 - x) - 2(3x + 1)] = x - 18$

26. $8(3x - 2) + 4(3 - 6x) = 0$

27. $3[(x + 2) - 4(x - 1)] = 9(2 - x)$

28. $4[2(x + 3) - (3x + 5)] = 5x - 5$

29. $(x + 2)(x + 3) = (x + 4)(x + 5)$

30. $x^2 - 4 = (x - 3)^2$

31. The Hampton Electric Company charges $7 each month for service plus 5¢ for each kilowatt-hour. George Bayer's electric bill for January was $27.45. How many kilowatt-hours of electricity did he use?

32. Ralph Auld is a salesman. He receives a weekly salary of $90 plus 4% commission. How much did he sell last week, if he earned $246?

33. Carol Duncan rents a car in Dayton for $20 a day plus 30¢ a mile. Seven days later, she returns the car in Toledo and is charged $387. This includes a charge of $25 for returning the car in a different city than she rented it in. How many miles did she drive?

34. A barbell set has 380 lb of weights. There are ten 20-lb weights and twelve 10-lb weights. The remaining weights are 5 lb each. How many 5-lb weights are in the barbell set?

35. Inez Brown received a bill from C and G Arco for $35.97. Of this amount, $16.43 was for an oil change, oil filter, and lube. The remainder was for a fill-up with unleaded gas at $1.83 per gallon. How many gallons of gas were put into her car?

36. Eric Hetman is using his $1200 in savings to start a furniture business in his basement. He will make and sell tables. He needs to buy $850 worth of tools before he can begin. After he has made some tables, he must spend $70 on advertising. How many tables can he make, if the materials for a table cost him $40?

37. The formula $(P + a)V = (P_0 - P)b$ gives an approximate relationship between the load P being raised by a muscle and the speed V with which the muscle shortens. Here a, b, and P_0 are constants depending on the particular muscle. Solve this equation for V. Also solve for P.

38. The amount of material needed to make a can is equal to the can's

surface area S. S is given by the formula $S = 2\pi rh + 2\pi r^2$, where h is the height of the can and r is the radius of the top (or bottom) of the can. The material used to make a can costs 2¢ per square foot. A large can has radius 1.5 feet and requires 62.8¢ of material. How tall is the can?

39. Frank Scavone has grades of 87%, 68%, and 75% on the three tests in his accounting course. To earn a "B" in the course, he needs an average of 80%. What is the lowest score on the final exam that will earn him a "B" in the course? The final counts twice as much as each of the other tests.

40. Calco expects calculator sales in the United States next year to reach 18 million, up 20% from this year. How many calculators are being sold this year?

41. Lucy Robin is running for city council. She has a campaign budget of $40,000 and has decided to spend $25,000 on newspaper ads and $7000 on handbills and posters. The remainder is to be spent on 60-sec radio ads that cost $40 each. How many radio ads can she buy?

42. Lucy's opponent, Sally Weiser, has $50,000 to spend and has decided to spend $20,000 on newspaper ads and $12,000 on handbills and posters. The remainder is to be spent on 30-sec radio ads that cost $25 each. How many radio ads can she buy?

The fine for speeding in North Luxor is $10 plus $3 for each mile over the speed limit of 25 mph.

43. A taxi driver had to pay a $61 fine for speeding. How fast was the driver going?

44. A delivery truck was stopped for speeding and the driver later had to pay a $46 fine. How fast was the truck going?

Solve each of the following equations.

45. $2.13x = 9.72$

46. $5.71x + 8.39 = 6.45$

47. $4.312x - 9.861 = 5.741x + 2.379$

48. $2.32(x - 1.41) = 4.17(x - 2.94)$

Section 2-2 Quadratic Equations

So far we have studied linear equations, that is, equations that involve only the first power of the variable. Now it is time to study equations that also involve the square of the variable.

A *quadratic equation* is an equation that can be manipulated into the form $ax^2 + bx + c = 0$ with $a \neq 0$.

The simplest way to solve quadratic equations is by *extraction of roots*. If a quadratic equation is of the form $q^2 = r$ and $r > 0$, then either $q = \sqrt{r}$ or $q = -\sqrt{r}$.

Example 1 $x^2 = 9$

Solution $x = 3$ or $x = -3$

The solutions are -3 and 3.

Example 2 $4x^2 - 25 = 0$

Solution $4x^2 = 25$

$$x^2 = \frac{25}{4}$$

$$x = \frac{5}{2} \quad \text{or} \quad x = -\frac{5}{2}$$

The solutions are $-5/2$ and $5/2$.

Example 3 $(x + 3)^2 = 36$

Solution $x + 3 = 6$ or $x + 3 = -6$
$x = 3$ or $x = -9$

The solutions are -9 and 3.
■

The method of extracting roots usually does not work since most quadratic equations do not have the needed simple form. In cases where the coefficients are rational numbers and the solutions are rational, we can use the method of *factoring*.

Example 4 $x^2 - 5x - 6 = 0$

Solution $(x - 6)(x + 1) = 0$ Factor.

$x - 6 = 0$ or $x + 1 = 0$ Equate each factor to 0.

$x = 6$ or $x = -1$

The solutions are -1 and 6.
∎

The principle behind the factoring method is the property: *If $pq = 0$, then either $p = 0$ or $q = 0$.* That is, if the product of two numbers is 0, then at least one of the numbers must be 0.

Example 5 $4x^2 - 16x + 15 = 0$

Solution $(2x - 3)(2x - 5) = 0$ Factor.

$2x - 3 = 0$ or $2x - 5 = 0$ Equate each factor to 0.

$2x = 3$ or $2x = 5$

$x = \dfrac{3}{2}$ or $x = \dfrac{5}{2}$

The solutions are $3/2$ and $5/2$.

Example 6 $x(2x - 3) = -1$

Solution $2x^2 - 3x = -1$

$2x^2 - 3x + 1 = 0$

$(2x - 1)(x - 1) = 0$ Factor.

$2x - 1 = 0$ or $x - 1 = 0$ Equate each factor to 0.

$2x = 1$ or $x = 1$

$x = \dfrac{1}{2}$ or $x = 1$

The solutions are $1/2$ and 1.
∎

In Example 6, the first step was to multiply out the left side of the equation and add 1 to each side in order to put the equation into the form

2-2 QUADRATIC EQUATIONS

$ax^2 + bx + c = 0$. We could then factor and set each factor equal to 0. The original equation in Example 6 has a number other than 0 on the right side. When this happens, one cannot merely set each factor equal to zero, or to any other number. To see this, note that if $pq = -1$, it is possible to have $p = 5$ and $q = -1/5$, or $p = 1/2$ and $q = -2$. In fact, if p is any number besides zero and $q = -1/p$, then $pq = -1$.

Now let's try to solve the equation

$$x^2 + 6x + 4 = 0$$

The left-hand side cannot be factored, so the factoring method will not work. Next try to use the method of extraction of roots. The equation cannot be put in the form $x^2 = r$, so let's try to put the equation into the form $(x + k)^2 = r$ and then take the square root. Then

$$x^2 + 2kx + k^2 = r$$
$$x^2 + 2kx + (k^2 - r) = 0$$

If this is to be the same equation as $x^2 + 6x + 4 = 0$, then $2k = 6$ and $k^2 - r = 4$. Since $2k = 6$, $k = 3$. Then $3^2 - r = 4$, so $r = 5$.

Now rewrite the original equation as $(x + 3)^2 = 5$. Take the square root of both sides to obtain:

$$x + 3 = -\sqrt{5} \quad \text{or} \quad x + 3 = \sqrt{5}$$
$$x = -3 - \sqrt{5} \quad \text{or} \quad x = -3 + \sqrt{5}$$

The solutions of the equation are $-3 - \sqrt{5}$ and $-3 + \sqrt{5}$.

Now we will go through this same procedure in general in order to obtain a formula, called the *quadratic formula*, which gives the solutions of the general quadratic equation $ax^2 + bx + c = 0$, $a \neq 0$, in terms of a, b, and c. Thus the quadratic formula can be used to solve *any* quadratic equation. Begin by transforming the general quadratic equation $ax^2 + bx + c = 0$ into the form $(x + k)^2 = r^2$.

$$ax^2 + bx + c = 0$$
$$x^2 + \frac{b}{a}x + \frac{c}{a} = 0 \qquad \text{Divide by } a.$$
$$x^2 + \frac{b}{a}x = \frac{-c}{a} \qquad \text{Subtract } c/a \text{ from both sides.}$$

Now add a number to the left-hand side of the equation in order to have a perfect square there. This process is called *completing the square*. Since $(x + p)^2 = x^2 + 2px + p^2$, we must have

$$2p = \frac{b}{a}, \quad p = \frac{b}{2a}, \quad \text{and} \quad p^2 = \frac{b^2}{4a^2}$$

Then

$$x^2 + \frac{b}{a}x + \frac{b^2}{4a^2} = \frac{b^2}{4a^2} - \frac{c}{a} \qquad \text{Add } \frac{b^2}{4a^2} \text{ to both sides.}$$

$$\left(x + \frac{b}{2a}\right)^2 = \frac{b^2}{4a^2} - \frac{c}{a}$$

$$= \frac{b^2 - 4ac}{4a^2}$$

The number $b^2 - 4ac$ is called the *discriminant* of the equation. Three cases must now be considered.

Case I $b^2 - 4ac < 0$

Since $\left(x + \frac{b}{2a}\right)^2 \geq 0$, the equation has no real numbers as solutions and is inconsistent. (In this book, we are using the real numbers as the universal set when solving equations. There is a larger number system, called the complex numbers, that contains the real numbers, and in which this equation has solutions.)

Case II $b^2 - 4ac = 0$

Then $\left(x + \frac{b}{2a}\right)^2 = 0$, so that $x + \frac{b}{2a} = 0$ and $x = \frac{-b}{2a}$.

The equation has one solution. In this case, the solution is called a *double solution* of the equation.

Case III $b^2 - 4ac > 0$

$\left(x + \frac{b}{2a}\right)^2 = \frac{b^2 - 4ac}{4a^2}$. By extraction of roots,

$$x + \frac{b}{2a} = \frac{\sqrt{b^2 - 4ac}}{2a} \quad \text{or} \quad x + \frac{b}{2a} = \frac{-\sqrt{b^2 - 4ac}}{2a}$$

$$x = \frac{-b + \sqrt{b^2 - 4ac}}{2a} \quad \text{or} \quad x = \frac{-b - \sqrt{b^2 - 4ac}}{2a}$$

This last line is usually shortened by writing

$$\boxed{x = \frac{-b \pm \sqrt{b^2 - 4ac}}{2a}} \quad \text{Quadratic formula}$$

The quadratic formula gives the solutions of the equation in case II and case III. In case I, the right-hand side of the quadratic formula is undefined and the equation has no solutions.

Example 7 $x^2 + 5x - 6 = 0$

Solution Use the quadratic formula with $a = 1$, $b = 5$, and $c = -6$.

$$x = \frac{-5 \pm \sqrt{5^2 - 4(1)(-6)}}{2(1)} = \frac{-5 \pm \sqrt{49}}{2} = \frac{-5 \pm 7}{2}$$

$$x = \frac{-5 + 7}{2} = 1 \quad \text{or} \quad x = \frac{-5 - 7}{2} = -6$$

The solutions are -6 and 1.

Example 8 $3x^2 + 5x - 4 = 0$

Solution Use the quadratic formula with $a = 3$, $b = 5$, and $c = -4$.

$$x = \frac{-5 \pm \sqrt{5^2 - 4(3)(-4)}}{2(3)} = \frac{-5 \pm \sqrt{73}}{6}$$

$$x = \frac{-5 - \sqrt{73}}{6} \quad \text{or} \quad x = \frac{-5 + \sqrt{73}}{6}$$

The solutions are $\frac{-5 - \sqrt{73}}{6}$ and $\frac{-5 + \sqrt{73}}{6}$.

Example 9 $4x^2 + 3x + 2 = 0$

Solution $$x = \frac{-3 \pm \sqrt{3^2 - 4(4)(2)}}{2(4)}$$

$$= \frac{-3 \pm \sqrt{-23}}{8}$$

There are no solutions since the discriminant is negative. The equation is inconsistent.

Example 10 $9x^2 + 30x + 25 = 0$

Solution $$x = \frac{-30 \pm \sqrt{30^2 - 4(9)(25)}}{2(9)}$$

$$= \frac{-30 \pm \sqrt{0}}{18} = \frac{-30}{18} = \frac{-5}{3}$$

The solution is $-5/3$. There is only one solution since the discriminant is 0.

Example 11 $x^2 + 10x + 7 = 0$

Solution $$x = \frac{-10 \pm \sqrt{10^2 - 4(1)(7)}}{2(1)}$$

$$= \frac{-10 \pm \sqrt{72}}{2} = \frac{-10 \pm \sqrt{36 \cdot 2}}{2}$$

$$= \frac{-10 \pm 6\sqrt{2}}{2} = -5 \pm 3\sqrt{2}$$

The solutions are $-5 - 3\sqrt{2}$ and $-5 + 3\sqrt{2}$.

Exercises 2-2

Solve the following equations by extraction of roots.

1. $x^2 = 16$
2. $(x - 2)^2 = 49$
3. $(2x + 1)^2 = 25$
4. $x^2 - 64 = 0$
5. $x^2 + 3 = 84$
6. $5x^2 = 20$

Solve the following equations by factoring.

7. $x^2 - 9x - 10 = 0$
8. $x^2 - 6x - 27 = 0$

9. $x^2 - 5x + 4 = 0$
10. $x^2 + 12x + 20 = 0$
11. $x^2 + 14x + 49 = 0$
12. $x^2 - 8x + 15 = 0$
13. $6x^2 + 5x + 1 = 0$
14. $x^2 - 12x + 36 = 0$
15. $x^2 + 15x + 50 = 0$
16. $6x^2 - x - 2 = 0$
17. $4x^2 - 4x + 1 = 0$
18. $2x^2 + 5x - 12 = 0$

Solve the following equations by any method.

19. $x^2 + 5x + 3 = 0$
20. $x^2 - 4x + 2 = 0$
21. $x^2 - 6x - 10 = 0$
22. $x^2 + 3x + 1 = 0$
23. $2x^2 - 5x + 7 = 0$
24. $3x^2 + 10x + 2 = 0$
25. $3x^2 + 30x + 75 = 0$
26. $x^2 - 3x - 5 = 0$
27. $5x^2 - 45 = 0$
28. $x^2 + 4x + 8 = 0$
29. $2x^2 + 3x - 1 = 0$
30. $4x^2 + 2x - 5 = 0$
31. $x^2 + 8x + 5 = 0$
32. $x^2 - 6x + 9 = 0$
33. $x^2 + 12x - 7 = 0$
34. $x^2 - 8x - 15 = 0$
35. $(x + 3)^2 = (x + 4)(x - 2)$
36. $(x - 4)^2 = (x + 3)(x + 6)$
37. $x^2 + 5x + 4 = 2x + 1$
38. $2x^2 + 4x + 8 = x^2 + x + 6$
39. $(x - 2)^2 = (x + 1)^2 - 6x + 3$
40. $5 - 12x + 3x^2 = 0$
41. $4x^2 - 7x - 12 = 3x^2 - 5x + 3$
42. $x^2 - 3x + 2 = x + 4$
43. $(x - 1)(x - 2) = 1$
44. $(4 - x)^2 = 15$
45. $10 + 8x - x^2 = 0$
46. $3x^2 + 5x - 2 = x - 8$
47. $(2x + 3)^2 = 49$
48. $(x - 2)(x + 3) = 14$

If an object near the earth's surface is thrown upward with an initial speed of v_0 from a height s_0, its height s is given by $s = -16t^2 + v_0 t + s_0$. This formula is valid until the object lands. Here t is measured in seconds and s is measured in feet.

49. An object is released ($v_0 = 0$) at a height of 64 ft. When does the object land?

50. An object is thrown up at a speed of 48 ft/sec from an initial height of 64 ft. When does the object land?

51. A ball on the ground is thrown up with an initial speed of 32 ft/sec. At the same time, a second ball is released 32 ft directly above the first ball. How high are the balls when they collide?

52. An object is thrown downwards with an initial speed of 20 ft/sec from a height of 80 ft. When does the object land?

53. To cover an area on the ground with concrete costs $4.75 a square meter. Kenneth Geiger wants to cover a circular region in his backyard with concrete and erect a birdbath on it. How large will the radius of

this region be if he spends $8.00 on concrete?

54. Instead, suppose Mr. Geiger wanted a square region. What would be the length of its side?

A congressional candidate's campaign managers believe that the candidate's percentage P of the vote is related to campaign expenditures C (measured in millions of dollars) by the equation $P = 30 + 6C + 3C^2/8$.

55. How much must be spent if the candidate is to win comfortably with 60% of the vote?

56. How much must be spent if the candidate is to squeak by with 51% of the vote?

Solve the following equations.

*57. $x^4 - 9 = 0$

*58. $x^4 - 256 = 0$

*59. $x^4 - 5x^2 + 4 = 0$

*60. $x^4 - 7x^2 + 10 = 0$

*61. $x^4 - 23x^2 - 50 = 0$

*62. $x^4 + 7x^2 + 12 = 0$

Solve each of the following equations.

C63. $5.73x^2 + 8.94x + 2.31 = 0$

C64. $2.34x^2 - 5.71x - 6.12 = 0$

C65. $0.21x^2 + 3.25x + 2.17 = 0$

C66. $4.83x^2 + 1.85x - 2.18 = 0$

Section 2-3 More General Equations

In this section, equations that are more complicated than quadratic equations are studied. These equations may involve the third and higher powers of the variable, rational expressions, or square roots. Several different solution techniques will be studied. First, we will extend the factoring method. To solve an equation $p(x) = 0$, where p is a polynomial, factor p into the product of linear and quadratic polynomials. Then set each factor equal to 0.

Example 1 $x^3 - 8x^2 + 12x = 0$

Solution $x(x^2 - 8x + 12) = 0$ Factor.

$x(x - 2)(x - 6) = 0$ Factor.

$x = 0$ or $x - 2 = 0$ or $x - 6 = 0$ Set each factor equal to 0.

$x = 0$ or $x = 2$ or $x = 6$

The solutions are 0, 2, and 6.

Example 2 $x^4 - x^3 = 0$

Solution $x^3(x-1) = 0$ Factor.

$x^3 = 0$ or $x - 1 = 0$ Set each factor equal to 0.

$x = 0$ or $x = 1$

The solutions are 0 and 1.

Example 3 $x^3 + 7x^2 - 3x = 0$

Solution $x(x^2 + 7x - 3) = 0$ Factor.
$x = 0$ or $x^2 + 7x - 3 = 0$ Set each factor equal to 0.
$x = 0$ or $x = \dfrac{-7 - \sqrt{61}}{2}$ or $x = \dfrac{-7 + \sqrt{61}}{2}$ Use the quadratic formula.

The solutions are 0, $(-7 - \sqrt{61})/2$ and $(-7 + \sqrt{61})/2$.

Example 4 $(x + 2)(x - 3)(x - 5)^2 = 0$

Solution $x + 2 = 0$ or $x - 3 = 0$ or $(x - 5)^2 = 0$ Set each factor equal to 0.

$x = -2$ or $x = 3$ or $x = 5$

The solutions are -2, 3, and 5.
■

We will now study equations involving rational expressions. In an equation involving fractions with variables in the denominators, multiply both sides of the equation by the least common denominator in order to eliminate the fractions.

Example 5 $\dfrac{4}{x} + \dfrac{5}{x^2} = 3$

Solution $\qquad 4x + 5 = 3x^2$ Multiply by x^2.

$-3x^2 + 4x + 5 = 0$

$x = \dfrac{-4 \pm \sqrt{76}}{-6} = \dfrac{-4 \pm 2\sqrt{19}}{-6} = \dfrac{2 \pm \sqrt{19}}{3}$

The solutions are $(2 - \sqrt{19})/3$ and $(2 + \sqrt{19})/3$.

Example 6 $\dfrac{2}{x-2} + \dfrac{3}{x+3} = 1$

Solution $2(x+3) + 3(x-2) = (x+3)(x-2)$ Multiply by $(x+3)(x-2)$.
$$2x + 6 + 3x - 6 = x^2 + x - 6$$
$$5x = x^2 + x - 6$$
$$0 = x^2 - 4x - 6$$
$$x = \dfrac{4 \pm \sqrt{40}}{2}$$
$$= \dfrac{4 \pm 2\sqrt{10}}{2} = 2 \pm \sqrt{10}$$

The solutions are $2 - \sqrt{10}$ and $2 + \sqrt{10}$.

Example 7 $\dfrac{5}{x+4} + 3 = \dfrac{8}{x+4}$

Solution $5 + 3(x+4) = 8$ Multiply by $x+4$.
$$5 + 3x + 12 = 8$$
$$3x + 17 = 8$$
$$3x = -9$$
$$x = -3$$

The solution is -3.

Example 8 $\dfrac{2x}{x-5} + 3 = \dfrac{10}{x-5}$

Solution $2x + 3(x-5) = 10$ Multiply by $x-5$.
$$2x + 3x - 15 = 10$$
$$5x - 15 = 10$$
$$5x = 25$$
$$x = 5$$

However, $x = 5$ is not a solution of this equation because 5 cannot be substituted for x in the expression $2x/(x-5)$. The equation has no solutions and is therefore inconsistent. An *extraneous root* is a number that is not a solution of the original equation but is a solution of an equation obtained in the process of solving the original equation. In this example, 5 is an extraneous root. In the process of solution, we multiplied both sides of the equation by $x - 5$, which is 0 when $x = 5$. When both sides of an equation are multiplied by an expression involving a variable, there may be extraneous roots.

■

Now we will solve certain equations involving roots. To solve an equation of the form $\sqrt{p(x)} = q(x)$, square both sides. There may be extraneous roots, so it is necessary to check each possible answer by substituting it into the original equation and determining if a true statement is obtained.

Example 9 $\sqrt{x^2 - 5x + 4} = 2$

Solution

$x^2 - 5x + 4 = 4$ Square both sides.

$x^2 - 5x = 0$

$x(x - 5) = 0$

$x = 0$ or $x - 5 = 0$ Check the answers:

$$\sqrt{0^2 - 5 \cdot 0 + 4} = \sqrt{4} = 2$$

$$\sqrt{5^2 - 5 \cdot 5 + 4} = \sqrt{4} = 2$$

$x = 0$ or $x = 5$

The solutions are 0 and 5.

Example 10 $\sqrt{x^2 - 5x + 4} = -2$

Solution The same process as in the previous example gives 0 and 5. These are both extraneous roots. This equation is inconsistent because it has no solutions.

Example 11 $\sqrt{2x^2 - x - 2} = x$

Solution $2x^2 - x - 2 = x^2$ Square both sides.

$x^2 - x - 2 = 0$

$(x - 2)(x + 1) = 0$

$x - 2 = 0$ or $x + 1 = 0$

$x = 2$ or $x = -1$

The solution is 2. Substituting -1 for x in the original equation gives the false statement $1 = -1$, so that -1 is an extraneous root. ∎

Sometimes, a relation involving several variables is given and you need to solve for one variable in terms of the other variables. The same techniques are used as in problems involving only one variable.

Example 12 $a(x + 5y) = b$. Solve for y.

Solution $\quad x + 5y = \dfrac{b}{a}$ $\qquad$ Divide by a.

$\qquad 5y = \dfrac{b}{a} - x = \dfrac{b - ax}{a}$ $\qquad$ Subtract x from both sides.

$\qquad y = \dfrac{b - ax}{5a}$ $\qquad$ Divide by 5.

Example 13 $xy + 3x = 12$. Solve for x.

Solution $\quad x(y + 3) = 12$ $\qquad$ Factor.

$\qquad x = \dfrac{12}{y + 3}$ $\qquad$ Divide by $y + 3$.

Example 14 $S = P(1 + r)^n$. Solve for r. ($S > 0, P > 0, r > 0$)

Solution $\quad (1 + r)^n = \dfrac{S}{P}$ $\qquad$ Divide by P.

$\qquad 1 + r = \left(\dfrac{S}{P}\right)^{1/n}$ $\qquad$ Take the nth root of both sides.

$\qquad r = \left(\dfrac{S}{P}\right)^{1/n} - 1$

Example 15 $x^2 + 4x + 4 = y$. Solve for x.

Solution $\quad x^2 + 4x + (4 - y) = 0$

$$x = \frac{-4 \pm \sqrt{4^2 - 4 \cdot 1 \cdot (4-y)}}{2 \cdot 1}$$

$$= \frac{-4 \pm \sqrt{16 - (16 - 4y)}}{2}$$

$$= \frac{-4 \pm \sqrt{4y}}{2} = \frac{-4 \pm 2\sqrt{y}}{2} = -2 \pm \sqrt{y}$$

$$x = -2 - \sqrt{y} \quad \text{or} \quad x = -2 + \sqrt{y}$$

Use the quadratic formula with $a = 1$, $b = 4$, and $c = 4 - y$.

Exercises 2-3

Solve each of the following equations.

1. $x^4 - 9x^2 = 0$
2. $x^3 + 8x^2 + 7x = 0$
3. $x(x-1)(x+2)(x+5) = 0$
4. $x(2x-3)^2 = 0$
5. $\dfrac{4}{x} = 8$
6. $\dfrac{2}{x} + 5 = 12$
7. $x + \dfrac{1}{x} = 2$
8. $\sqrt{2x+3} = 4$
9. $\dfrac{3x}{x+2} + 4 = \dfrac{-6}{x+2}$
10. $\dfrac{6}{x-3} - \dfrac{12}{x+1} = 1$
11. $1 = \dfrac{6}{x} - \dfrac{5}{x^2}$
12. $x - \dfrac{3}{x} = 2$
13. $\sqrt{5 - 2x} = -3$
14. $(x^2 + 4)(x^2 - 3x - 2) = 0$
15. $x^4 - 5x^3 - 14x^2 = 0$
16. $\sqrt{x^2 + 5x - 2} = 3 - x$
17. $\sqrt{x^2 - 4x + 5} = x - 3$
18. $x^3 + 7x^2 + 15x = 0$
19. $(x - 4)^2(x^2 + 8x + 11) = 0$
20. $1 = \dfrac{5}{x} - \dfrac{8}{x^2}$
21. $\sqrt{x^2 - 4x + 8} = x$
22. $\dfrac{2}{x^2 - 1} = \dfrac{1}{x - 1} - \dfrac{1}{x + 1}$
23. $\dfrac{4}{x-1} + \dfrac{5}{x+2} = 3$
24. $(x^2 + 9)(x^2 - 2x + 6) = 0$

Solve each equation for every variable in the equation. Each letter represents a variable.

25. $S = \dfrac{a}{1-r}$

26. $3v - \pi r^2 h = 0,\ r > 0$

27. $\dfrac{1}{r} = \dfrac{1}{a} + \dfrac{1}{b}$

28. $\dfrac{y-3}{x+2} = 5$

29. $p = a + (n-1)d$

30. $Fs = \dfrac{mv^2}{2},\ v > 0$

31. $x^2 + 12x + 8 = 2 - y$

32. $x^2 - 6x - 9 = 3 - 2y$

33. According to special relativity, if a particle has rest mass (mass when not moving) m_0 and is moving at a speed v, then its mass m is given by the formula

$$m = \dfrac{m_0}{\sqrt{1 - v^2/c^2}}$$

where c is the speed of light. In metric units, c is about 3.00×10^8 m/sec. Solve this equation for v. If the mass of a particle is ten times its rest mass, how fast is the particle moving?

34. In a gas, molecules are constantly colliding with each other. The mean free path L is the average distance a molecule travels between collisions. The equation for L is

$$L = \dfrac{1}{\pi\sqrt{2}\,nd^2}$$

Here n, the number of molecules per unit volume, is about 3×10^{19} molecules/cm³, and d is the diameter of a molecule. Determine d if $L = 1.2 \times 10^{-5}$ cm.

35. The van der Waals equation for a gas at high density is

$$\left(p + \dfrac{a}{v^2}\right)(v - b) = RT$$

Here a, b, and R are constants, p is the pressure, T is the temperature, and v is the volume per mole. A mole is a very convenient unit in chemistry. One mole is about 6×10^{23} molecules. Solve this equation for p.

Suppose we have a long cylindrical tube of length L and radius a. Assume there is a steady flow of liquid through the tube, and that the pressure is p_1

at one end of the tube and p_2 at the other end, with $p_1 > p_2$. Let v be the viscosity of the liquid. (Viscosity is a measure of how "thick" the liquid is and of how much the liquid resists flowing.) The speed of flow s of the liquid depends on the distance r from the center of the tube and is given by the formula

$$s = \frac{p_1 - p_2}{4vL} \cdot (a^2 - r^2)$$

36. Solve this equation for r.
37. Solve this equation for p_2.

The amount Q of liquid per unit time leaving the tube is given by the formula

$$Q = \frac{\pi(p_1 - p_2)a^4}{8vL}$$

This formula is called Poiseuille's formula after the French physician who discovered it. Poiseuille's formula is used to study blood flow and to find the viscosity of blood under various conditions.

38. Solve Poiseuille's formula for v.
39. Solve Poiseuille's formula for a.

Solve each of the following equations.

*40. $x^3 - 4x^2 + x + 6 = 0$ *41. $x^3 - x^2 - 10x - 8 = 0$
*42. $x^3 - 4x^2 - 3x = -18$ *43. $x^3 + x^2 - 1 = x$
*44. $(x + 3)^{1/3} = x - 3$ *45. $2(3x + 6)^{1/3} = x - 1$

Section 2-4 Inequalities

Until now, you have only been concerned with equations. Inequalities are also important. The entire subject of linear programming (Chapter 4) is concerned with inequalities. As an example consider the following situation.

Example 1 Rafael Martinez is opening an ice cream stand. His supplier gives him a choice of two purchasing agreements. He can pay $1.70 for each gallon of ice cream. The other option is to pay $250 each week for 100 gal of ice cream and then purchase additional ice cream for $1.50 a gallon. For what amounts of weekly ice cream sales is the second option better?

Solution Let x be the number of gallons of ice cream sold each week. With the first option, the weekly cost of ice cream is $1.70x$. If we assume $x \geq 100$, the weekly cost with the second option is $250 + 1.50(x - 100)$. The second option will be better if it results in a lower cost than the first option, that is, if

$$250 + 1.50(x - 100) < 1.70x$$

Now solve this inequality for x.

$250 + 1.50x - 150 < 1.70x$	Distributive law.
$1.50x + 100 < 1.70x$	Combine like terms.
$100 < 0.20x$	Subtract $1.50x$ from both sides.
$500 < x$	Multiply by 5.

Thus, the second option is better if weekly ice cream sales exceed 500 gal.

Linear inequalities

Let's consider inequalities further. A *linear inequality* in one variable involves only the first power of the variable. Linear inequalities can be solved algebraically. The steps used are very similar to the steps used in solving linear equations.

Example 2 $4x + 3 > 2x - 5$

Solution	
$2x > -8$	Subtract $2x + 3$ from both sides.
$x > -4$	Divide by 2.

Example 3 $12x - 3 \leq 15x - 7$

Solution	
$4 \leq 3x$	Add $7 - 12x$ to both sides.
$\dfrac{4}{3} \leq x$	Divide by 3.

Example 4 $1 + 3(x - 2) < 5(x + 1)$

Solution	
$1 + 3x - 6 < 5x + 5$	Distributive law.
$3x - 5 < 5x + 5$	Combine like terms.

$$-2x < 10 \qquad \text{Add } 5 - 5x \text{ to both sides.}$$
$$x > -5 \qquad \text{Divide by } -2.$$

■

Note that the direction of the inequality sign changed in Example 4 when we divided by -2. Recall that if $ac < bc$, then

$$a < b \quad \text{if} \quad c > 0 \quad \text{but} \quad a > b \quad \text{if} \quad c < 0$$

If an inequality is multiplied or divided by a negative number, the direction of the inequality sign must be changed.

Example 5 $2 + 5(x + 1) \geq 8(x + 3)$

Solution
$$2 + 5x + 5 \geq 8x + 24 \qquad \text{Distributive law.}$$
$$5x + 7 \geq 8x + 24 \qquad \text{Combine like terms.}$$
$$-3x \geq 17 \qquad \text{Subtract } 8x + 7 \text{ from both sides.}$$
$$x \leq -\frac{17}{3} \qquad \text{Divide by } -3. \text{ Notice that the inequality sign changes direction.}$$

Example 6 Mr. and Mrs. Moore are going to buy a refrigerator. They are considering a Sears model which costs $662 and is expected to use 90 kWh of electricity each month, and a Whirlpool which costs $500 and is expected to use 100 kWh each month. For which electric rates would the total cost (purchase price plus cost of electricity) over an expected 15-year lifetime be less for the Sears refrigerator than the Whirlpool?

Solution Let r be the cost (in cents) of a kilowatt hour of electricity. Since 15 years is $15 \cdot 12 = 180$ months, the Sears refrigerator would use $90 \cdot 180 = 16,200$ kWh of electricity over its lifetime at a cost of $16,200r$¢, or $162r$. The total cost of this refrigerator (in dollars) is then $662 + 162r$. Similarly, the total cost of the Whirlpool is $500 + 180r$.

The total cost over the expected lifetime is less for the Sears model if

$$662 + 162r < 500 + 180r$$

Now solve this inequality for r.

$$162 < 18r \qquad \text{Subtract } 500 + 162r \text{ from both sides.}$$

$$\frac{162}{18} < r \qquad \text{Divide by 18.}$$

$$9 < r$$

The Sears model will cost less over a lifetime if electricity costs more than 9¢ per kWh. ∎

The statement $2 < 3x + 7 < 15$ is a pair of inequalities: $2 < 3x + 7$ and $3x + 7 < 15$. The two inequalities can be solved at the same time.

$$2 < 3x + 7 < 15$$
$$2 - 7 < (3x + 7) - 7 < 15 - 7 \qquad \text{Subtract 7.}$$
$$-5 < 3x < 8 \qquad \text{Simplify.}$$
$$-\frac{5}{3} < x < \frac{8}{3} \qquad \text{Divide by 3.}$$

Sign graphs

Inequalities involving higher powers of x or rational expressions are often difficult to solve algebraically. A simpler way to solve such inequalities is to draw a *sign graph*.

To solve $x^2 + 4x + 3 > 0$, start by drawing a number line. Factor $x^2 + 4x + 3 = 0$ as $(x + 1)(x + 3)$. On the number line, indicate the numbers (-3 and -1) where $x^2 + 4x + 3 = 0$. These two numbers break the line up into three parts. Within a part, each factor of $x^2 + 4x + 3$ always keeps the same sign.

For $x < -3$, the factor $x + 1$ is negative, so a minus sign $(-)$ is recorded. For $-3 < x < -1$, the factor $x + 1$ is negative, so a minus sign is recorded. One way to determine this is to evaluate $x + 1$ at a number that satisfies the inequality $-3 < x < -1$. If -2 is chosen, then $x + 1 = -2 + 1 = -1$, which is negative. For $x > -1$, the factor $x + 1$ is positive, so a plus $(+)$ sign is recorded. Now determine and record the signs for the factor $x + 3$.

Use the signs of the factors to determine the sign of $x^2 + 4x + 3$ inside each part. Then $x^2 + 4x + 3 > 0$ when $x < -3$ or $x > -1$. Finally, test the numbers -3 and -1 to see if they satisfy the inequality. Some of the examples and exercises have numbers in which the expression in the inequality is undefined. These numbers are also indicated on the number line and are used to break the line up into parts.

$$x^2 + 4x + 3 > 0$$

Solution $x < -3$ or $x > -1$

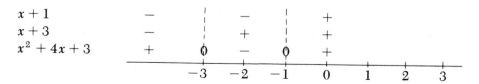

Examples Solve each inequality with a sign graph.

Example 7 $x^3 - 3x^2 - 10x \geq 0$
$x(x^2 - 3x - 10) \geq 0$
$x(x - 5)(x + 2) \geq 0$

Solution $-2 \leq x \leq 0$ or $5 \leq x$

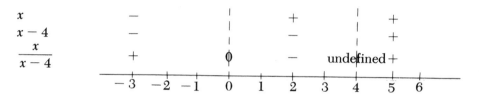

Example 8 $\dfrac{x}{x - 4} \leq 0$

Solution $0 \leq x < 4$

Example 9 $(x + 3)(x + 4)^2 < 0$

Solution $x < -4$ or $-4 < x < -3$

```
x + 3              −       | − |   +
(x + 4)²           +       | + |   +
(x + 3)(x + 4)²    −       ⊙ − ⊙   +
         ─┼──┼──┼──┼──┼──┼──┼──┼──┼─
         −6 −5 −4 −3 −2 −1  0  1  2  3
```

■

> In order to use a sign graph, 0 must be on one side of the inequality sign.

To solve $x^2 + 3 > -4x$, add $4x$ to both sides to obtain $x^2 + 4x + 3 > 0$. The inequality $x^2 + 4x + 3 > 0$ was solved earlier in this section and the solution is $x < -3$ or $x > -1$. To solve $2 \geq (3x - 8)/(x - 4)$, subtract 2 from both sides to obtain

$$0 \geq \frac{3x - 8}{x - 4} - 2 = \frac{3x - 8}{x - 4} - \frac{2(x - 4)}{x - 4} = \frac{(3x - 8) - (2x - 8)}{x - 4} = \frac{x}{x - 4}$$

The inequality $0 \geq x/(x - 4)$ is the same as $x/(x - 4) \leq 0$ and was solved in Example 8.

Exercises 2-4

Solve the following linear inequalities algebraically.

1. $3x + 2 > 8$
2. $2x - 5 \leq 7$
3. $4 - 2x \geq 5 - 3x$
4. $8 + 3x < 5 - x$
5. $10 - x > 5 + 4x$
6. $4(1 + 3x) + 2 > 6 - 6x$
7. $4 - 2x \leq 1 + 3(2 - 5x)$
8. $6 - 5x \geq 9 - 2x$
9. $2(x - 8) < 3(2 - 3x)$
10. $3x + 2 > -4(2 + x)$
11. $4 - 3x < -2(3 - x)$
12. $5(4 - 3x) > 3(2 - 4x) - 1$

Solve each inequality by using a sign graph.

13. $x^2 - 5x + 6 < 0$
14. $x^2 + 3x + 2 \geq 0$
15. $(x - 1)(x - 2)(x - 3) < 0$
16. $x(x - 1)^2(x + 3) \leq 0$
17. $\dfrac{x(x + 3)}{x - 1} \geq 0$
18. $x(x - 2)(x + 4) > 0$
19. $(x - 4)^2(x + 2)(x - 5) > 0$
20. $x^2 + 4x - 21 < 0$
21. $x^2 - 5x - 14 \geq 0$
22. $\dfrac{(x - 1)(x + 2)}{x} < 0$

23. $\dfrac{x-2}{(x-3)(x+1)} < 0$

24. $\dfrac{x^2}{x+3} > 0$

25. $x^3 - 5x^2 - 14x > 0$

26. $x^2 - 6x < 0$

27. $x^4 - 25x^2 < 0$

28. $\dfrac{(x-4)}{x+2} > 0$

29. $x^2 - 4 \geq 0$

30. $x^3 + 8x^2 + 12x \geq 0$

31. Lois Scharf is opening a Great Burger restaurant. This is a franchise chain, and she must make an irrevocable choice of one of two franchise fee arrangements. Choice A is to pay $100 a week plus 7% of gross income. Choice B is to pay $200 a week plus 5% of gross income. For which values of the weekly gross income is choice B better?

32. Ralph Davis has just been employed as a book salesman by the Sherlock Publishing Company. He must make a choice between two salary arrangements. He can be paid a straight 8% commission, or he can have a monthly salary of $1200 plus a 2% commission. For which values of monthly sales will he earn more on straight commission?

33. E-Z Car Rental has two plans. The Low-Mileage Plan is $18 each day plus 25¢ a mile. The High-Mileage Plan is $30 each day plus 15¢ a mile. How high must the daily mileage be in order for the high mileage plan to be cheaper?

34. Sidney West is tired of shoveling snow and has decided to buy a Toro snowblower. He can buy a small one for $200 or a large one for $500. He expects either one to last for ten years. Each year he expects to use his snowblower 15 times. With the small snowblower, he expects to take 45 min to clear his driveway, and with the large snowblower, 20 min. To make his decision, he must decide how much his time is worth. For what values of his time is he better off with the smaller snowblower?

35. The weekly profit at the E, E, and G Company is $500q - 40{,}000 - q^2$, where q is the number of oscilloscopes sold during the week. For what values of q does E, E, and G make a profit?

36. The monthly profit at Rapko Furniture is $130q - 4000 - q^2$, where q

is the number of bedroom sets made during the month. For what values of q does Rapko make a profit?

37. A toy rocket on the ground is thrown upward with an initial speed of 112 ft/sec. When is the rocket more than 160 ft high? The height s (in feet) of the rocket is given by $s = -16t^2 + 112t$, where t is the time (in seconds).

38. If an object weighs w pounds at the earth's surface, its weight W at an altitude of x miles is approximately $w(4000)^2/(4000 + x)^2$. At what altitudes is $W < w/4$?

Solve each inequality by using a sign graph.
*39. $x^3 + 3x^2 - 10x - 24 \geq 0$ *40. $x^3 + 6x^2 + 3x - 10 \geq 0$
*41. $x^4 + 7x^2 - 144 \leq 0$ *42. $x^4 - 3x^2 - 4 > 0$

Solve each inequality algebraically.
*43. $|2x + 1| > 4$ *44. $|3x + 2| < 8$
*45. $|4 - 3x| < 13$ *46. $|5 - 2x| > 15$

Section 2-5   Word Problems

So far in this chapter, we have learned how to solve certain equations and inequalities and have studied some applications of these topics. Now we will consider in detail the use of equations in solving several specific types of word problems.

> *Procedure for Solving Word Problems*
> 1. Read the problem very carefully.
> 2. Determine what the variables are—these are the unknown quantities.
> 3. Look for equations relating the variables to each other and to the other data in the problem.
> 4. Solve these equations.
> 5. Decide which of the solutions apply to the original problem.
> 6. Be sure to answer the question that was asked!

Investment problems

    Example 1   Ed Wu has an annual income of $21,800 from two investments. He has

$5000 more invested at 16% interest than at 12% interest. How much does Ed have invested?

Solution There are two unknown quantities: the amount invested at 12% and the amount invested at 16%. However, there is a relationship between these quantities: the amount invested at 16% is $5000 more than the amount invested at 12%. Either of these quantities could be the variable x. Since a choice is necessary, let x be the amount invested at 12%. Then the amount invested at 16% is $5000 more than x, or $x + 5000$.

The income from the amount x deposited at 12% is $0.12x$. The income from the amount $x + 5000$ deposited at 16% is $0.16(x + 5000)$. The total investment income is the sum of the incomes from the two investments, or $0.12x + 0.16(x + 5000)$. The total investment income is given to be $21,800. This gives the equation

$$0.12x + 0.16(x + 5000) = 21{,}800$$
$$0.12x + 0.16x + 800 = 21{,}800$$
$$0.28x = 21{,}000$$
$$x = \frac{21{,}000}{0.28} = 75{,}000$$

Ed has $75,000 invested at 12% and $75,000 + $5000 = $80,000 invested at 16%. His total investment is $75,000 + $80,000 = $155,000.

Example 7 Mary O'Brian has $25,000 invested, some at 10% and the remainder at 15%. Her annual income from both investments is $3350. How much does she have invested at 10%?

Solution Let x be the amount invested at 10%. Then $25{,}000 - x$ is the amount invested at 15%. Her total investment income is then

$$0.10x + 0.15(25{,}000 - x)$$

Since her total investment income is given as $3350,

$$0.10x + 0.15(25{,}000 - x) = 3350$$
$$0.10x + 3750 - 0.15x = 3350$$
$$-0.05x = 3350 - 3750 = -400$$
$$x = \frac{-400}{-0.05} = 8000$$

82   CHAPTER 2   EQUATIONS AND INEQUALITIES

Mary has $8000 invested at 10%.
∎

Distance problems

Example 3 In a motorcycle race, the winner averaged 100 miles per hour and finished 1/4 hour ahead of the loser. The loser averaged 95 miles per hour. How many miles long was the race?

Solution The unknown quantity, the variable, is the length x of the race. We must set up an equation involving x. Besides the average speeds of the contestants, the only data given is the difference in finishing times. This suggests finding the time each contestant required to complete the race. Since distance = speed × time, time = distance/speed.

The winner required $x/100$ hours to complete the race. The loser required $x/95$ hours to complete the race. Since the winner won by 1/4 hour, the loser's time exceeds the winner's time by 1/4 hour. This gives the equation:

$$\frac{x}{95} - \frac{x}{100} = \frac{1}{4}$$

Now solve this equation. Note that 1900 is the least common denominator.

$$\frac{x}{95} - \frac{x}{100} = \frac{1}{4}$$
$$20x - 19x = 475 \qquad \text{Multiply by 1900.}$$
$$x = 475$$

Since $x = 475$, the race was 475 miles long. Notice that the winner took $475/100 = 4.75$ hours and the loser took $475/95 = 5$ hours.

Example 4 At 3 P.M., a bus averaging 45 miles per hour leaves San Francisco. At 5 P.M., a car averaging 55 miles per hour leaves San Francisco and takes the same route. How far will the vehicles travel before the car passes the bus?

Solution Let t be the time in hours since 5 P.M. At time t, the bus has been traveling for $t + 2$ hours and has gone $45(t + 2)$ miles. The car has gone $55t$ miles. The car passes the bus when

$$55t = 45(t + 2)$$
$$55t = 45t + 90$$
$$10t = 90$$
$$t = 9$$

The car passes the bus when $t = 9$. In nine hours, the car will travel $9 \cdot 55 = 495$ miles.

∎

Mixture problems

Example 5 Hilltop Dairy has milk containing 3% butterfat and has 1000 gal of milk containing 6% butterfat. How many gallons of the 3% butterfat milk must be mixed with the 6% butterfat milk in order to have 4% butterfat milk?

Solution The variable x is the number of gallons of 3% milk to be used. To obtain an equation, equate the percentage of butterfat in the mixture with 4%. This is the same as saying

$$\frac{\text{number of gallons of butterfat in mixture}}{\text{number of gallons of mixture}} = 0.04$$

Now x gallons of 3% milk contain $0.03x$ gal of butterfat, and 1000 gal of 6% milk contain $0.06(1000) = 60$ gal of butterfat. Thus, the mixture contains $0.03x + 60$ gal of butterfat. The number of gallons in the mixture is x (the number of gal of 3% milk) + 1000 (the number of gal of 6% milk). This gives the equation

$$\frac{0.03x + 60}{x + 1000} = 0.04$$

Now solve the equation.

$$\frac{0.03x + 60}{x + 1000} = 0.04$$
$$0.03x + 60 = 0.04(x + 1000) \quad \text{Multiply by } x + 1000.$$
$$0.03x + 60 = 0.04x + 40$$
$$20 = 0.01x$$
$$2000 = x$$

Hilltop Dairy must mix 2000 gal of 3% milk with the 1000 gal of 6% milk and will obtain 3000 gal of 4% milk.

Example 6 Ramon has 10 lb of premium coffee which costs $7.50 a pound. He wants to mix this with ordinary coffee which costs $4.00 a pound in order to make a mixture which costs $5.50 a pound. How much of the mixture will he make?

Solution Let x be the number of pounds of ordinary coffee he uses. Then the total cost of the coffee is

$$\underset{\text{(premium)}}{7.50 \cdot 10} + \underset{\text{(ordinary)}}{4 \cdot x} = 75 + 4x$$

The weight of the mixture is $10 + x$ pounds, and so the cost of the mixture is

$$5.50 \cdot (10 + x) = 55 + 5.5x$$

Thus,

$$55 + 5.5x = 75 + 4x$$
$$1.5x = 20$$
$$x = \frac{20}{1.5} = 20 \div \frac{3}{2} = 20 \cdot \frac{2}{3} = \frac{40}{3} = 13\tfrac{1}{3}$$

Ramon will use $13\tfrac{1}{3}$ lb of ordinary coffee. Since he also uses 10 lb of premium coffee, the weight of the mixture is $13\tfrac{1}{3} + 10 = 23\tfrac{1}{3}$ lb. ∎

Work problem

Example 7 A man and his son working together can paint their house in six days. The man working alone can paint the house in nine days. How long would it take the son to paint the house alone?

Solution The only variable is the number of days it would take the son to paint the house alone. Let this quantity be x.

Since the son could paint the house in x days, in one day he does $1/x$ of the total job. Since the father could paint the house in nine days, in one day he does $1/9$ of the job.

In one day, the father and son together do $1/x + 1/9$ of the job. Since they can do the whole job together in six days, in one day they do $1/6$ of the job. This allows us to write the equation

$$\frac{1}{x} + \frac{1}{9} = \frac{1}{6}$$

Now solve the equation.

$$18 + 2x = 3x \qquad \text{Multiply by } 18x.$$
$$18 = x$$

The son can do the job alone in 18 days. In six days, the son does $6/18 = 1/3$ of the job, and the father does $6/9 = 2/3$ of the job. ∎

Area problem

Example 8 A theater that is rectangular in shape holds 1500 people. There would be five fewer rows if each row held ten more seats. How many rows are there?

Solution First, let $x =$ the number of rows. Since the theater has 1500 seats, there are $1500/x$ seats in each row. If there were five fewer rows, the number of rows would be $x - 5$. If each row held ten more seats, there would be $10 + 1500/x$ seats in each row. Since the theater would still hold 1500 people, the new number of rows times the new number of seats per row must equal 1500. This gives the equation:

$$(x - 5)(10 + 1500/x) = 1500$$

Now solve this equation for x.

$$10x + 1450 - 7500/x = 1500 \qquad \text{Multiply out.}$$
$$10x^2 + 1450x - 7500 = 1500x \qquad \text{Multiply by } x.$$
$$10x^2 - 50x - 7500 = 0$$
$$x^2 - 5x - 750 = 0$$
$$(x - 30)(x + 25) = 0$$
$$x - 30 = 0 \quad \text{or} \quad x + 25 = 0$$
$$x = 30 \quad \text{or} \quad x = -25$$

There are two solutions of the equation. The solution $x = -25$ must be rejected because the number of rows in a theater cannot be negative. Therefore, there are 30 rows. We will now check this answer. There are $1500/30 = 50$ seats in each row. If there were five fewer rows, there would be $30 - 5 = 25$ rows and if each row had ten more seats, there would be $50 + 10 = 60$ seats in each row. Then the theater would hold $25 \cdot 60 = 1500$ people. Thus, our answer is correct.

∎

Exercises 2-5

1. A pension fund has some money invested at 8% and has twice as much invested at 11%. Annual income from both investments is $3,000,000. How much does the pension fund have invested at 11%?

2. Mr. Ito has an annual income of $1180 from two investments. He has $6000 more invested at 7% than at 12%. How much does he have invested at 12%?

3. A sum of $17,000 is invested, part at 10% interest and part at 8% interest. The total annual income from the $17,000 is $1560. How much is invested at 8%?

4. A sum of $20,000 is invested, part at 12% and part at 10%. The total annual income from both investments is $2240. How much is invested at 10%?

5. In an auto race, one car averaged 120 mph and another car averaged 100 mph. The faster car finished 30 min ahead of the slower car. How many miles long was the race?

6. At noon, a train traveling 30 mph leaves New York City for Cleveland. At 2:00 P.M. a train traveling 50 mph leaves Cleveland for New York City. The two cities are 500 mi apart. When do the trains meet?

7. A bus traveling 40 mph leaves on a trip three hours before a car traveling 55 mph. After how many miles will the car overtake the bus?

8. Joe has $1.35 in pennies and nickels. He has 15 more pennies than nickels. How many pennies does he have?

9. How much of a solution that is 80% antifreeze must be added to 5 qt of a solution that is 30% antifreeze in order for the mixture to be 50% antifreeze?

10. How much water must be added to 6 qt of a solution that is 20% dye to obtain a solution that is 8% dye?

11. At Green Stadium, admission to a soccer game is $3.00 for adults and $1.50 for children. One day, 145 people paid $330 to see a game. How many children watched the game?

12. A ticket to a movie costs $2.00 for an adult and $1.25 for a child. One afternoon, 100 tickets were sold for a total amount of $155. How many children attended the movie?

13. Cheryl Ross wishes to receive an average return of 8% on her total investments. She has $7000 invested at 5%. How much additional money would she have to invest at 10%?

14. Bill can clean the house in 16 hr and Sally can clean the house in 12 hr. How long will it take them to clean the house if they both work?

15. An electrician and his apprentice can rewire a house in 36 hr. The apprentice working alone would take four times as long to rewire the house as the electrician working alone. How long would the electrician need to rewire the house alone?

16. A rectangle is three times as long as it is wide. Its area is 48 sq in. What are its dimensions?

17. A box with a square bottom is half as high as it is long. Its volume is 500 cu in. What is its height?

18. The product of two numbers is 35. The larger number is three less than twice the smaller number. What are the numbers?

The Fahrenheit and Celsius temperature scales are related by the equation $C = 5(F - 32)/9$, where C is the temperature in degrees Celsius and F is the temperature in degrees Fahrenheit. (The Celsius temperature scale is sometimes called the centigrade temperature scale.)

19. Normal body temperature is 98.6° F. What is normal body temperature in degrees Celsius?

20. Is $-10°$ F colder than $-10°$ C?

21. If the temperature is 20° C, what is the temperature in degrees Fahrenheit?

22. At what temperatures (measured in degrees Celsius) does water freeze and boil?

Section 2-6 Coordinates and Graphs

Until now we have mainly considered equations with only a few solutions. In this section, equations that involve two variables and have infinitely many solutions are discussed. Since infinitely many solutions cannot be listed, we will draw a diagram, called a graph, instead. First, however, coordinates must be defined.

Choose a horizontal line, make it into a number line, and call it the *x-axis*. Draw a vertical line through the origin of the *x*-axis. Make this vertical line into a number line with up as the positive direction and the same origin as the *x*-axis. This second line is called the *y-axis*. The *x*-axis and the *y*-axis are called the *coordinate axes*. See Diagram 1.

Now associate an ordered pair of real numbers with each point in the plane. Pick any point P, and draw a line through P perpendicular to the *x*-axis. This line meets the *x*-axis in a point Q. Let b be the number which corresponds to Q. Draw a line through P perpendicular to the *y*-axis. This line meets the *y*-axis in a point R. Let c be the number that corresponds to R. The point P corresponds to the ordered pair (b, c). In Diagram 2, P corresponds to $(2, 5)$.

> **Definition.** If the point P corresponds to the ordered pair (b, c), then b and c are the coordinates of P. We say that b is the *first coordinate* or *x-coordinate* of P and that c is the *second coordinate* or *y-coordinate* of P.

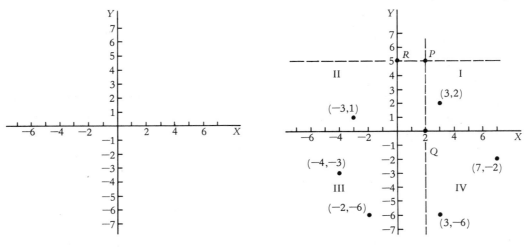

Diagram 1.   **Diagram 2.**

In Diagram 2, several points and their coordinates are shown. The coordinate axes divide the plane into four regions called *quadrants*. Quadrant I, for example, is $\{(x, y) | x > 0 \text{ and } y > 0\}$.

Now let's define the graph of an equation.

> **Definition.** Assume an equation has x and y as variables. The *graph* of the equation is the set of points whose coordinates satisfy the equation.

To draw the graph of the equation $2x + 3y = 18$, first find some points whose coordinates satisfy this equation. One point is $(3, 4)$, since if 3 is substituted for x and 4 for y, the true statement $2(3) + 3(4) = 18$ is obtained. Several other points are shown in Diagram 3.

This equation has infinitely many points on its graph. To verify this, solve the equation for y to obtain $y = 6 - 2x/3$. Thus, if x is any real number, and $y = 6 - 2x/3$, then the point (x, y) is on the graph of the equation.

After plotting several points, the graph of the equation is obtained by drawing a curve connecting the points. In this case, the graph appears to be a straight line. In Chapter 9, we will study lines in detail, and we will demonstrate that

> The graph of the equation $ax + by = c$ is a line, provided a and b are not both 0.

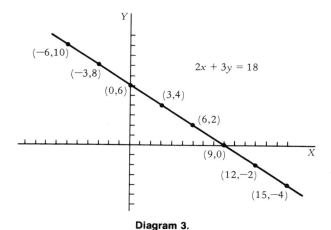

Diagram 3.

Knowing that the graph of an equation, such as $4x - y = 8$, is a line makes graphing that equation very easy. Since two points determine a line, we only need to plot two points. The best points to plot are the points, called *intercepts*, where the line meets the coordinate axes.

Now let's find the intersection of the graph of $4x - y = 8$ with the x-axis. At every point on the x-axis, $y = 0$. Thus, at the intersection point, the equation $4x - 0 = 8$ is true, so $x = 2$. The intersection point must be $(2, 0)$.

To find the intersection of the line $4x - y = 8$ with the y-axis, recall that at every point on the y-axis, $x = 0$. Thus, at the intersection point the equation $4(0) - y = 8$ holds, so that $-y = 8$ and $y = -8$. The intersection point is $(0, -8)$. The graph of $4x - y = 8$ is shown in Diagram 4.

Some lines, such as $2x - 5y = 0$, go through the origin and thus do not have two intercepts. To graph this line, pick a value, say 2, for y. Then determine the value for x (5, in this case) that gives a solution of the equation. Now plot the points $(0, 0)$ and $(5, 2)$ and draw the line through them. (See Diagram 5.)

To graph the line $x = 4$, note that if $y = 0$, then $x = 4$. If $x = 0$, no value for y gives a solution of the equation. However, if $x = 4$, any value for y gives a solution of the equation. Pick a value, say 5, for y and plot the point $(4, 5)$. The graph is shown in Diagram 6. Note that the line is vertical. If k is a constant, the graph of $x = k$ is a vertical line.

Similarly, the graph of $y = k$ is a horizontal line. The graph of $y = 3$ is shown in Diagram 7.

In Chapter 4, a very important technique called linear programming is discussed. One way of solving linear programming problems is called the

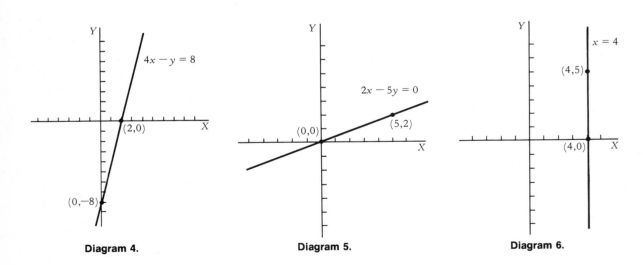

Diagram 4. **Diagram 5.** **Diagram 6.**

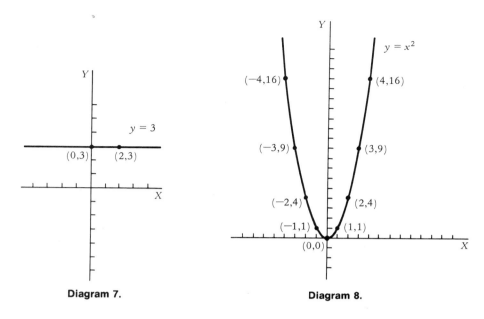

Diagram 7.

Diagram 8.

graphical method. To use this method it is necessary to draw the graphs of lines.

Now let's do a few examples of graphs that are not lines. In each case, first plot a few points and then connect the points. Later on, in Chapter 9, you will learn how to draw the graphs of several types of equations.

Example 1 Draw a graph of $y = x^2$.

Solution In Diagram 8, several points are plotted. These points are obtained by letting x take on small integer values, both positive and negative. These points are then connected.

Example 2 Draw the graph of $y = \sqrt{x}$.

Solution In Diagram 9, several points are plotted and then connected. These points are obtained by letting $x = 0$, and by letting x equal small counting numbers with integer square roots. Note that x cannot take negative values because the square root of a negative number is not defined.

Example 3 Draw the graph of $y = |x|$.

Solution In Diagram 10, several points are plotted and then connected. These points

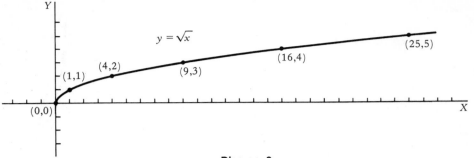

Diagram 9.

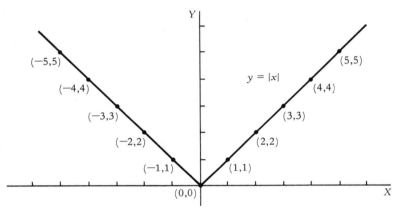

Diagram 10.

are obtained by letting *x* take small integer values. The graph consists of two half-lines joined at the origin.

Exercises 2-6

Draw the graph of each line.

1. $3x + 5y = 15$
2. $4x + 7y = 14$
3. $2x - y = 4$
4. $3x - 2y = 6$
5. $-x + 3y = 6$
6. $-2x + y = 4$
7. $2x - 3y = 0$
8. $2x + 5y = 0$
9. $-3x + y = 0$
10. $4x - 3y = 0$
11. $2x = 14$
12. $y = 5$
13. $y = -2$
14. $x = 9$
15. $4x - 3y = 24$
16. $4x + 5y = 40$

Draw the graph of each equation.

17. $y = \dfrac{x^2}{4}$
18. $y = 1 + x^2$
19. $y = (x + 3)^2$
20. $y = (x - 1)^2$
21. $y = 4\sqrt{x}$
22. $y = \sqrt{x - 2}$
23. $y = \sqrt{x + 1}$
24. $y = 2|x|$
25. $y = 1 + |x|$
26. $y = |1 + x|$
*27. $x^2 + y^2 = 25$
*28. $xy = 1$
*29. $|x| + |y| = 6$
*30. $y = x^2 - 4x + 3$
*31. $y = x^2 + 6x + 8$
*32. $x - y^2 = 0$

*33. Discuss the graph of $ax + by = c$ when
    (a) $a = b = 0$ and $c \neq 0$ (b) $a = b = c = 0$

Chapter 2 Review

Solve the following equations and inequalities.

1. $4y + 3 = 7y - 12$
2. $2 - 3c = c + 10$
3. $5(2x - 1) = 6(12 - x) + 83$
4. $6(7 + 4x) = 8(3x + 4) + 10$
5. $2[6x - 3(x - 4)] = 9(x + 3)$
6. $3x^2 = 192$
7. $(3x + 2)^2 = 25$
8. $x^2 + 4x - 21 = 0$
9. $x^2 + 22x + 121 = 0$
10. $x^2 - 6x - 16 = 0$
11. $x^2 + 7x + 11 = 0$
12. $x^2 - 4x - 6 = 0$
13. $x^2 - 5x + 10 = 0$
14. $(x - 2)(x - 4) = 8$
15. $(x + 2)(x + 3) + 1 = 0$
16. $(x + 4)(x - 14)^3(x^2 + 9) = 0$
17. $x^3 + 10x^2 - 75x = 0$
18. $4 + \left(\dfrac{5}{x}\right) = 19$
19. $\dfrac{x}{x + 3} - 1 = \dfrac{2 - x}{x + 3}$
20. $\dfrac{10}{x + 3} + \dfrac{6}{x} = 5$
21. $\sqrt{4 + 5x} = \sqrt{6x - 5}$
22. $\sqrt{x^2 + 3x - 3} = 2 - x$
23. $12 + 7x > 9 - 2x$
24. $4(5 - 2x) < -3(2x - 1)$
25. $x^2 - 4x - 32 \geq 0$
26. $(x + 4)(x - 2)(x - 5) < 0$
27. $\dfrac{x + 1}{x - 8} \geq 0$
28. $\dfrac{(x + 2)(x + 4)}{x - 1} \leq 0$

29. The Sandusky Water Company sends bills every three months. The charge is $10.00 for service plus $0.80 for every thousand cubic feet of

water used. How much water did Jeffrey Taylor and his family use, if their bill was for $38.00?

30. Freda Scarne sells pharmaceutical supplies. She receives a monthly salary of $800 plus a 2% commission. Last year, she earned $17,400. What was the total amount of her sales for the year?

31. A low-option medical insurance plan costs $30 a month and pays half of all expenses. A high-option plan costs $54 a month and pays 80% of all expenses. For what values of monthly expenses is the high-option plan better?

32. In one production run at Blair Aluminum, up to 1000 storm windows can be made. The cost of making x storm windows is $25x + 5000 - x^2/100$. How many storm windows could be made, if at most $15,000 is available to pay for the manufacturing cost?

Draw the graph of each equation.

33. $3x + 7y = 14$

34. $5x - 4y = 0$

35. $y = \left| \dfrac{x}{2} \right|$

36. $y = 2\sqrt{x + 4}$

3

Linear Algebra

Many situations in business, government, and the social sciences involve large quantities of data. When the data form a rectangular array—for example, the number of hours worked each day by each employee, or an income tax table—the information can be conveniently stored in a matrix. Matrices are also used in controlling inventories and in national economic planning. In this chapter, we study matrices and systems of linear equations, and then proceed to use matrices to solve systems of linear equations efficiently.

Section 3-1 Matrices

A *matrix* is a rectangular array of numbers. Matrices provide a convenient and concise way to store certain data. For this reason, matrices occur frequently in computer applications.

An *inventory matrix* is a matrix used to store inventory data. For example, suppose Field's has two warehouses, A and B. Current inventory is 10 boxes of shirts, 20 boxes of ties, and 5 boxes of sweaters at warehouse A, and 30 boxes of shirts, no ties, and 10 boxes of sweaters at warehouse B. This information could be expressed as a chart

	Warehouse A	Warehouse B
Shirts	10	30
Ties	20	0
Sweaters	5	10

or in a matrix

$$\begin{bmatrix} 10 & 30 \\ 20 & 0 \\ 5 & 10 \end{bmatrix}$$

In this book, small matrices are used for the sake of convenience. In real applications, matrices can be enormous. For example, Field's has three different items stored and has two warehouses. Sears sells thousands of different items and has stores in every city, as well as numerous warehouses. The inventory matrix for Sears would fill a book by itself.

Further examples of matrices are:

$$\begin{bmatrix} 4 & 1 & 3 \\ 2 & 7 & 1.3 \\ -6 & 8.2 & 5 \end{bmatrix} \quad \begin{bmatrix} 1 & 7.2 & 6 \\ 0 & -3 & 9 \end{bmatrix} \quad \begin{bmatrix} 7 \\ 3 \\ 9 \end{bmatrix} \quad \begin{bmatrix} -2 & 1 & 0 & 8 & 5 \end{bmatrix}$$

The matrix

$$A = \begin{bmatrix} 4.1 & 7 \\ -2 & 3 \\ 6 & 0 \end{bmatrix}$$

has three *rows* and two *columns*. The numbers in the first row are 4.1 and 7, the numbers in the second row are -2 and 3, and the numbers in the third row are 6 and 0. A *row* of a matrix is always written horizontally. The matrix A has two columns. The numbers in the first column are 4.1, -2, and 6. The numbers in the second column are 7, 3, and 0. A *column* of a matrix is always written vertically. A is a 3×2 (read 3 by 2) matrix, or has *size* 3×2. The *size* of a matrix is the number of rows by the number of columns. The numbers in a matrix are called the *entries*. The entry in the ith row and jth column of a matrix A is denoted $a_{i,j}$ and is said to be in position i, j in A. For example $a_{1,1} = 4.1$, $a_{3,1} = 6$, and $a_{3,2} = 0$.

Two matrices are *equal* if they have the same size and if the entries in corresponding positions are equal. For example,

$$\begin{bmatrix} 2 & 1 \\ 4 & 7 \end{bmatrix} = \begin{bmatrix} 1+1 & 9-8 \\ 2+2 & 14 \div 2 \end{bmatrix}$$

but

$$\begin{bmatrix} 1 & 3 \\ 5 & 4 \end{bmatrix} \neq \begin{bmatrix} 1 & 5 \\ 3 & 4 \end{bmatrix} \text{ and } \begin{bmatrix} 2 & -9 & 3 \\ 4 & 6 & 1 \end{bmatrix} \neq \begin{bmatrix} 2 & -9 \\ 4 & 6 \end{bmatrix}$$

Now let's begin our study of matrix algebra.

Two matrices can be added if they have the same size. To *add* two matrices, add corresponding entries. Thus

$$\begin{bmatrix} 1 & 4 & -3 \\ 2 & 6 & 1 \\ -8 & 4 & 5 \end{bmatrix} + \begin{bmatrix} -6 & 9 & 2 \\ 8 & 3 & 5 \\ -2 & 1 & 7 \end{bmatrix} = \begin{bmatrix} 1-6 & 4+9 & -3+2 \\ 2+8 & 6+3 & 1+5 \\ -8-2 & 4+1 & 5+7 \end{bmatrix}$$

$$= \begin{bmatrix} -5 & 13 & -1 \\ 10 & 9 & 6 \\ -10 & 5 & 12 \end{bmatrix}$$

However, if two matrices do not have the same size, they cannot be added. For example,

$$\begin{bmatrix} 4 & -3 & 9 \\ 6 & -1 & 2 \\ 4 & 8 & 1 \end{bmatrix} + \begin{bmatrix} 4 & 10 \\ 8 & -6 \\ -9 & 1 \end{bmatrix}$$

is undefined.

One matrix can be subtracted from another matrix if they have the same size. To *subtract* a matrix B from a matrix A, subtract each entry of B from the corresponding entry of A. Thus

$$\begin{bmatrix} 4 & -2 \\ 6 & 1 \end{bmatrix} - \begin{bmatrix} 3 & 5 \\ -2 & 4 \end{bmatrix} = \begin{bmatrix} 4-3 & -2-5 \\ 6-(-2) & 1-4 \end{bmatrix} = \begin{bmatrix} 1 & -7 \\ 8 & -3 \end{bmatrix}$$

If A and B do not have the same size, $A - B$ is undefined.

Let us return to Field's. In the inventory matrix, the first column corresponds to warehouse A and the second column corresponds to warehouse B. The first row corresponds to shirts, the second row to ties, and the third row to sweaters. The matrix notation is convenient because it involves very little writing.

Field's now makes a shipment to its customers from the warehouses. The shipment has matrix

$$\begin{bmatrix} 8 & 15 \\ 12 & 0 \\ 3 & 10 \end{bmatrix}$$

The 12 in position 2, 1 in the matrix means that 12 boxes of ties are shipped from warehouse A. After the shipment, the inventory is given by the matrix

$$\begin{bmatrix} 10 & 30 \\ 20 & 0 \\ 5 & 10 \end{bmatrix} - \begin{bmatrix} 8 & 15 \\ 12 & 0 \\ 3 & 10 \end{bmatrix} = \begin{bmatrix} 2 & 15 \\ 8 & 0 \\ 2 & 0 \end{bmatrix}$$

The warehouses are resupplied by a shipment with matrix

$$\begin{bmatrix} 25 & 10 \\ 10 & 20 \\ 15 & 35 \end{bmatrix}$$

The inventory is now given by the matrix

$$\begin{bmatrix} 2 & 15 \\ 8 & 0 \\ 2 & 0 \end{bmatrix} + \begin{bmatrix} 25 & 10 \\ 10 & 20 \\ 15 & 35 \end{bmatrix} = \begin{bmatrix} 27 & 25 \\ 18 & 20 \\ 17 & 35 \end{bmatrix}$$

Any matrix can be multiplied by any real number. To multiply a matrix by a number, multiply each entry in the matrix by the number. For example,

$$5 \begin{bmatrix} 1 & 4 & -7 \\ 2 & 6 & 9 \end{bmatrix} = \begin{bmatrix} 5 \times 1 & 5 \times 4 & 5 \times (-7) \\ 5 \times 2 & 5 \times 6 & 5 \times 9 \end{bmatrix} = \begin{bmatrix} 5 & 20 & -35 \\ 10 & 30 & 45 \end{bmatrix}$$

Suppose Field's decides to double its current inventory. Its inventory matrix after doubling is

$$2 \begin{bmatrix} 27 & 25 \\ 18 & 20 \\ 17 & 35 \end{bmatrix} = \begin{bmatrix} 54 & 50 \\ 36 & 40 \\ 34 & 70 \end{bmatrix}$$

The *negative of A*, denoted $-A$, is $(-1)A$. Thus $-\begin{bmatrix} 2 & -3 \\ 4 & 6 \end{bmatrix} = \begin{bmatrix} -2 & 3 \\ -4 & -6 \end{bmatrix}$. Then $A - B = A + (-B)$. For example,

$$\begin{bmatrix} 2 & 3 & -1 \\ 4 & -6 & 2 \end{bmatrix} - \begin{bmatrix} 3 & -5 & 10 \\ -2 & 4 & 1 \end{bmatrix}$$

$$= \begin{bmatrix} 2 & 3 & -1 \\ 4 & -6 & 2 \end{bmatrix} + \begin{bmatrix} -3 & 5 & -10 \\ 2 & -4 & -1 \end{bmatrix}$$

$$= \begin{bmatrix} -1 & 8 & -11 \\ 6 & -10 & 1 \end{bmatrix}$$

Example 1 $2\begin{bmatrix} 4 & 1 & 7 \\ -3 & 2 & 5.4 \\ 6 & 1 & 0 \end{bmatrix} + 5\begin{bmatrix} 1 & 3 & 0 \\ 0 & 4 & 2.1 \\ -9 & 1.3 & 7 \end{bmatrix}$

$$= \begin{bmatrix} 8 & 2 & 14 \\ -6 & 4 & 10.8 \\ 12 & 2 & 0 \end{bmatrix} + \begin{bmatrix} 5 & 15 & 0 \\ 0 & 20 & 10.5 \\ -45 & 6.5 & 35 \end{bmatrix}$$

$$= \begin{bmatrix} 13 & 17 & 14 \\ -6 & 24 & 21.3 \\ -33 & 8.5 & 35 \end{bmatrix}$$

Example 2 $10\begin{bmatrix} -3 & 2 \\ 1.6 & 4 \end{bmatrix} - 3\begin{bmatrix} 1 & 5 \\ 7 & 2 \end{bmatrix} = \begin{bmatrix} -30 & 20 \\ 16 & 40 \end{bmatrix} - \begin{bmatrix} 3 & 15 \\ 21 & 6 \end{bmatrix}$

$$= \begin{bmatrix} -33 & 5 \\ -5 & 34 \end{bmatrix}$$

Example 2 can also be done in another way.

$10\begin{bmatrix} -3 & 2 \\ 1.6 & 4 \end{bmatrix} - 3\begin{bmatrix} 1 & 5 \\ 7 & 2 \end{bmatrix}$

$= 10\begin{bmatrix} -3 & 2 \\ 1.6 & 4 \end{bmatrix} + (-3)\begin{bmatrix} 1 & 5 \\ 7 & 2 \end{bmatrix}$ This step would usually not be shown.

$$= \begin{bmatrix} -30 & 20 \\ 16 & 40 \end{bmatrix} + \begin{bmatrix} -3 & -15 \\ -21 & -6 \end{bmatrix}$$

$$= \begin{bmatrix} -33 & 5 \\ -5 & 34 \end{bmatrix}$$

∎

A *zero matrix* is any matrix in which every entry is 0. The $m \times n$ zero matrix is called $0_{m \times n}$. Thus

$$0_{3 \times 2} = \begin{bmatrix} 0 & 0 \\ 0 & 0 \\ 0 & 0 \end{bmatrix} \text{ and } 0_{2 \times 2} = \begin{bmatrix} 0 & 0 \\ 0 & 0 \end{bmatrix}$$

Properties of Matrix Algebra
A, B, and C denote $m \times n$ matrices and b and c denote numbers.

1. $(A + B) + C = A + (B + C) \quad (bc)(A) = b(cA)$ Associative laws

2. $(b + c)A = bA + cA \quad b(A + B) = bA + bB$ Distributive laws

3. $A + B = B + A$ Commutative law

4. $1A = A \quad 0_{j \times m} A = 0_{j \times n}$ Properties of 1 and 0

5. $A + 0_{m \times n} = 0_{m \times n} + A = A$ Identity law

6. $A + (-A) = (-A) + A = 0_{m \times n}$ Inverse law

So far, matrix algebra is just like ordinary algebra, except that two matrices can be added (or one subtracted from the other) only if they have the same size. In the next section, we will study matrix multiplication and see that matrix multiplication is not commutative, that is, there are matrices A and B with $AB \neq BA$.

Exercises 3-1

Determine the size of each matrix.

1. $\begin{bmatrix} 4 & 3 & -6 \\ 2 & -7 & -9 \end{bmatrix}$

2. $\begin{bmatrix} -4 & 7 & 1 \\ 9 & -2 & 4 \\ -8 & 6 & -3 \end{bmatrix}$

3. $\begin{bmatrix} 4 & 8 & -5 \\ -2 & 6 & 3 \\ -1 & 2 & -9 \\ 0 & 0 & 0 \end{bmatrix}$
4. $\begin{bmatrix} -3 & 1 & 2 & 5 & 6 \\ 4 & 9 & -2 & 7 & 1 \end{bmatrix}$

5. $\begin{bmatrix} 4 & 8 \\ 10 & 2 \end{bmatrix}$
6. $\begin{bmatrix} 3 & 1 & 7 & 2 \end{bmatrix}$

Let $B = \begin{bmatrix} -4 & 1 & 3 & 2.7 \\ 9 & 2.5 & -1 & 6 \\ 3 & 2 & 1 & 8 \end{bmatrix}$

7. $b_{1,1}$
8. $b_{3,2}$
9. $b_{1,3}$
10. $b_{3,1}$
11. $b_{3,3}$
12. $b_{2,3}$
13. $b_{2,4}$
14. $b_{1,4}$

Perform the indicated arithmetic, if possible.

15. $\begin{bmatrix} 1 & -7 & 3 \\ 2 & 6 & -4 \end{bmatrix} + \begin{bmatrix} 2 & 3 & 7 \\ 9 & -4 & 2 \end{bmatrix}$

16. $\begin{bmatrix} 2 & -3 \\ -6 & -1 \\ 4 & 2 \end{bmatrix} - \begin{bmatrix} 3 & 2 \\ -5 & -1 \\ -4 & 7 \end{bmatrix}$

17. $\begin{bmatrix} 2 & 1 \\ 4 & -3 \end{bmatrix} - \begin{bmatrix} -6 & 1 \\ 3 & -9 \end{bmatrix}$

18. $6\begin{bmatrix} 1 & 3 & -2 \\ 0 & 2 & 5 \end{bmatrix}$

19. $-3\begin{bmatrix} 7 & 4 \\ -1 & -6 \\ 2 & 3 \end{bmatrix}$

20. $\begin{bmatrix} 3 & -6 & 2 \\ 1 & 11 & -9 \\ 2 & 0 & 5 \\ 4 & 1 & 3 \end{bmatrix} + \begin{bmatrix} -3 & 6 & 2 \\ 11 & 2 & 3 \\ 8 & 7 & -6 \end{bmatrix}$

21. $5\begin{bmatrix} 1 & -3 \\ 2 & 6 \end{bmatrix} + 3\begin{bmatrix} 8 & 7 \\ 4 & 1 \end{bmatrix}$

22. $2\begin{bmatrix} 4 & -3 & 2 \\ 5 & -8 & -8 \end{bmatrix} + 4\begin{bmatrix} 9 & 7 & -1 \\ 2 & 4 & 3 \end{bmatrix}$

23. $7\begin{bmatrix} 1 & 2 & 0 \\ 0 & 4 & 6 \\ -3 & 2 & 5 \end{bmatrix} + 3\begin{bmatrix} 4 & 1 & 2 \\ 2 & 7 & 1 \\ 8 & 3 & 5 \\ 5 & 2 & 6 \end{bmatrix}$

24. $2\begin{bmatrix} 7 & 6 \\ -3 & 1 \\ 2 & 5 \end{bmatrix} - 8\begin{bmatrix} 1 & 3 \\ 0 & 2 \\ 5 & 4 \end{bmatrix}$

25. $2\begin{bmatrix} 3 & -1 & 2 \\ 4 & 7 & 0 \end{bmatrix} - 4\begin{bmatrix} 2 & 1 & 3 \\ -6 & 8 & -5 \end{bmatrix}$

26. $\begin{bmatrix} 2 & 1 \\ 4 & 7 \end{bmatrix} - 3\begin{bmatrix} -1 & 2 \\ 5 & 4 \end{bmatrix}$

27. $5\begin{bmatrix} -8 & 1 & 3 \\ -2 & -9 & 4 \\ -6 & 0 & 7 \end{bmatrix} - 2\begin{bmatrix} 3 & -1 & 5 \\ 4 & -2 & -9 \\ 7 & 6 & 8 \end{bmatrix}$

28. $3\begin{bmatrix} 2 & -1 & 7 \\ -9 & -2 & 4 \\ -6 & 1 & 3 \end{bmatrix} - 2\begin{bmatrix} -1 & 3 & -7 \\ -4 & 2 & 1 \\ 5 & 0 & -6 \end{bmatrix}$

At Stop and Shop, Mark bought four loaves of bread, two quarts of milk, three pounds of plums, and one chicken. Helga bought two loaves of bread, three quarts of milk, one pound of plums, and two chickens.

29. Write this information in a 2 × 4 matrix.

30. If Mark and Helga each bought three times as much of each item, what would the matrix be?

31. Suppose Mark goes back for three more quarts of milk and Helga goes back for another loaf of bread. Write this information in a 2 × 4 matrix and use matrix addition to write a matrix that represents the purchases for the day.

Stratford Records makes records and cassette tapes at an east coast plant E, a Chicago plant C, and a west coast plant W. The following matrices give its production costs in dollars for each item at each plant.

$$\begin{array}{c} \text{Plant E} \\ \begin{array}{cc} \text{Record} & \text{Tape} \end{array} \\ \begin{array}{c} \text{Materials} \\ \text{Labor} \end{array} \begin{bmatrix} 1.31 & 1.73 \\ 2.45 & 2.35 \end{bmatrix} \end{array} \quad \begin{array}{c} \text{Plant C} \\ \begin{array}{cc} \text{Record} & \text{Tape} \end{array} \\ \begin{bmatrix} 1.36 & 1.70 \\ 2.36 & 2.25 \end{bmatrix} \end{array} \quad \begin{array}{c} \text{Plant W} \\ \begin{array}{cc} \text{Record} & \text{Tape} \end{array} \\ \begin{bmatrix} 1.32 & 1.70 \\ 2.21 & 2.15 \end{bmatrix} \end{array}$$

32. Write a matrix that represents the cost of producing 100 of an item at the Chicago plant.

33. Suppose any plant makes as many of each item as any other plant. Write a matrix that represents the average production costs of each item.

34. Suppose labor costs increase by $0.10 for each item at plant E and material costs increase by $0.07 for a record and $0.15 for a tape. Write this information as a matrix, and use matrix addition to obtain the new matrix for plant E.

35. Economists expect inflation to be 12% annually. Suppose all costs for Stratford are expected to increase at the rate of inflation. Write the expected production cost matrices for Stratford, as of one year later. Round off all entries to the nearest cent.

36. A new chemical, drothygon, is being tested for use as a tranquilizer. An experiment on rats gives the following results.

$$\begin{array}{c} \text{Tranquilized} \\ \begin{array}{cc} \text{Yes} & \text{No} \end{array} \\ \begin{array}{c} \text{Drothygon} \\ \text{Placebo} \end{array} \begin{bmatrix} 17 & 8 \\ 3 & 22 \end{bmatrix} \end{array}$$

Since only 50 rats can be tested at once and 500 rats must be tested, the experiment is repeated and the additional nine matrices are obtained as follows.

$$\begin{bmatrix} 16 & 9 \\ 5 & 20 \end{bmatrix} \text{ is obtained four times.}$$

$$\begin{bmatrix} 16 & 9 \\ 4 & 21 \end{bmatrix} \text{ is obtained twice.}$$

$$\begin{bmatrix} 15 & 10 \\ 3 & 22 \end{bmatrix} \text{ is obtained three times.}$$

Write one matrix that contains the results on all 500 rats.

Section 3-2 Matrix Multiplication

In this section, we continue our study of matrix algebra by considering the multiplication of matrices. Matrix multiplication is a much more complicated operation than matrix addition and subtraction and does not satisfy the same algebraic laws as the multiplication of numbers.

To define matrix multiplication, we will first define row and column matrices and then define their multiplication.

> A *row matrix* is a matrix with only one row.
> A *column matrix* is a matrix with only one column.

Thus $[3 \quad 9 \quad 2]$ and $[-1 \quad 4 \quad 6 \quad 5]$ are row matrices, and $\begin{bmatrix} 4 \\ 7 \\ 1 \end{bmatrix}$ and $\begin{bmatrix} 3 \\ -2 \\ 6 \\ 9 \end{bmatrix}$ are column matrices.

If A is a row matrix and B is a column matrix, the *product AB* is defined precisely when A and B have the same number of entries. The product AB is a number and is obtained by multiplying each entry of A by a corresponding entry of B and adding the products.

Example 1 Compute the product $[4 \quad 2 \quad 5] \begin{bmatrix} 7 \\ 3 \\ 1 \end{bmatrix}$.

Solution Multiply the first entry of the row matrix and the first entry of the column matrix to form the product 4×7. Then multiply the second entries to form the product 2×3. Next multiply the third entries to form the product 5×1. Finally, add these three products to obtain the product of the row matrix with the column matrix. Thus,

$$[4 \quad 2 \quad 5] \begin{bmatrix} 7 \\ 3 \\ 1 \end{bmatrix} = (4 \times 7) + (2 \times 3) + (5 \times 1) = 28 + 6 + 5 = 39$$

Examples 2. $[-2 \quad 7]\begin{bmatrix} 2 \\ 6 \end{bmatrix} = (-2 \times 2) + (7 \times 6) = -4 + 42 = 38$

3. $[3 \quad 1 \quad 2 \quad 5]\begin{bmatrix} 6 \\ 1 \\ 6 \\ 8 \end{bmatrix} = (3 \times 6) + (1 \times 1) + (2 \times 6) + (5 \times 8)$

$$= 18 + 1 + 12 + 40 = 71$$

4. $[5 \quad 2 \quad 6]\begin{bmatrix} 4 \\ 1 \\ 3 \\ 9 \end{bmatrix}$

is undefined because the row matrix does not have the same number of entries as the column matrix.

Example 5 At Pick and Pay, I buy four loaves of bread, two dozen eggs, and one quart of milk. The prices are $0.45 for a loaf of bread, $0.79 for a dozen eggs, and $0.53 for a quart of milk. Express the total cost as a product of a row matrix with a column matrix.

Solution Use the number of each item purchased to form the following row matrix A and use the unit cost of each item to form the column matrix B.

$$A = \begin{matrix} \text{Bread} & \text{Eggs} & \text{Milk} \\ [4 & 2 & 1] \end{matrix} \qquad B = \begin{bmatrix} 0.45 \\ 0.79 \\ 0.53 \end{bmatrix} \begin{matrix} \text{Bread} \\ \text{Eggs} \\ \text{Milk} \end{matrix}$$

The total cost is the product $AB = \$3.91$. Note how the product AB is formed and determine the significance of each of the numbers 4×0.45, 2×0.79, and 1×0.53.

∎

Let's move now to the problem of defining multiplication of matrices that are not necessarily row or column matrices.

If A and B are matrices, A and B are *conformable* if the number of columns of A equals the number of rows of B. If A has size $i \times j$ and B has size $m \times n$, A and B are conformable precisely when $j = m$. If A and B are conformable, the product AB will be an $i \times n$ matrix; that is, AB has as many rows as A and as many columns as B. If A and B are not conformable, AB is undefined.

Example 6 What size is the product $\begin{bmatrix} 7 & 1 \\ -2 & 3 \\ -5 & 9 \end{bmatrix} \begin{bmatrix} 1 & 7 & 6 & 2 \\ 10 & 9 & 2 & 0 \end{bmatrix}$?

Solution The first matrix has size 3×2 and the second matrix has size 2×4. The matrices are conformable because of the twos, and the product has size 3×4. Note that the product $[7 \quad 1] \begin{bmatrix} 1 \\ 10 \end{bmatrix}$ of the row $[7 \quad 1]$ of the first matrix and the column $\begin{bmatrix} 1 \\ 10 \end{bmatrix}$ of the second matrix is defined.

Example 7 What size is the product $\begin{bmatrix} 1 & 7 & 6 & 2 \\ 10 & 9 & 2 & 0 \end{bmatrix} \begin{bmatrix} 7 & 1 \\ -2 & 3 \\ 5 & 9 \end{bmatrix}$?

Solution The first matrix has size 2×4 and the second matrix has size 3×2. The matrices are not conformable because $4 \neq 3$. Thus the product of the two matrices is not defined. Note that the product of the row $[1 \quad 7 \quad 6 \quad 2]$ of the first matrix and the column $\begin{bmatrix} 7 \\ -2 \\ 5 \end{bmatrix}$ of the second matrix is not defined, because the row and the column do not have the same number of entries. ■

Examples 6 and 7 show that the product of AB of two matrices may be defined while the product BA is undefined. Now we will define the product of two matrices.

> If A and B are conformable matrices, the *product AB* has as its entry in position i, j the product of the ith row of A with the jth column of B.

3-2 MATRIX MULTIPLICATION

The condition that A and B are conformable is the same as the condition that the product of a row of A with a column of B be defined.

Example 8 Compute $\begin{bmatrix} 1 & 2 \\ 3 & 4 \end{bmatrix} \begin{bmatrix} 5 & 6 \\ 7 & 8 \end{bmatrix}$.

Solution Since the matrices are both 2×2, the product is defined and is a 2×2 matrix. To compute the entry in position i, j of the product, multiply the ith row of the first matrix and the jth column of the second matrix. For example, the entry in position 2, 1 of the product is $\begin{bmatrix} 3 & 4 \end{bmatrix} \begin{bmatrix} 5 \\ 7 \end{bmatrix} = (3 \times 5) + (4 \times 7) = 43$. Thus,

$$\begin{bmatrix} 1 & 2 \\ 3 & 4 \end{bmatrix} \begin{bmatrix} 5 & 6 \\ 7 & 8 \end{bmatrix} = \begin{bmatrix} (1 \times 5) + (2 \times 7) & (1 \times 6) + (2 \times 8) \\ (3 \times 5) + (4 \times 7) & (3 \times 6) + (4 \times 8) \end{bmatrix} = \begin{bmatrix} 19 & 22 \\ 43 & 50 \end{bmatrix}$$

Example 9 $\begin{bmatrix} 5 & 6 \\ 7 & 8 \end{bmatrix} \begin{bmatrix} 1 & 2 \\ 3 & 4 \end{bmatrix} = \begin{bmatrix} (5 \times 1) + (6 \times 3) & (5 \times 2) + (6 \times 4) \\ (7 \times 1) + (8 \times 3) & (7 \times 2) + (8 \times 4) \end{bmatrix} = \begin{bmatrix} 23 & 34 \\ 31 & 46 \end{bmatrix}$.

■

Examples 8 and 9 show that matrix multiplication is not commutative.

Example 10 $\begin{bmatrix} 2 & -4 \\ -6 & 12 \end{bmatrix} \begin{bmatrix} 6 & 2 \\ 3 & 1 \end{bmatrix} = \begin{bmatrix} (2 \times 6) - (4 \times 3) & (2 \times 2) - (4 \times 1) \\ (-6 \times 6) + (12 \times 3) & (-6 \times 2) + (12 \times 1) \end{bmatrix}$

$$= \begin{bmatrix} 0 & 0 \\ 0 & 0 \end{bmatrix} = 0_{2 \times 2}$$

■

Example 10 shows that the product of two matrices can be a zero matrix, even though neither of the matrices is a zero matrix.

Example 11 $\begin{bmatrix} 6 & -3 \\ -4 & 2 \end{bmatrix} \begin{bmatrix} 3 & 2 \\ 9 & 6 \end{bmatrix} = \begin{bmatrix} (6 \times 3) - (3 \times 9) & (6 \times 2) - (3 \times 6) \\ (-4 \times 3) + (2 \times 9) & (-4 \times 2) + (2 \times 6) \end{bmatrix}$

$$= \begin{bmatrix} -9 & -6 \\ 6 & 4 \end{bmatrix}$$

Example 12 $\begin{bmatrix} 15 & -6 \\ -1 & 1 \end{bmatrix} \begin{bmatrix} 3 & 2 \\ 9 & 6 \end{bmatrix} = \begin{bmatrix} (15 \times 3) - (6 \times 9) & (15 \times 2) - (6 \times 6) \\ (-1 \times 3) + (1 \times 9) & (-1 \times 2) + (1 \times 6) \end{bmatrix}$

$$= \begin{bmatrix} -9 & -6 \\ 6 & 4 \end{bmatrix}.$$

■

Examples 11 and 12 show that it is possible to have $AC = BC$ even though $A \neq B$ and $C \neq 0$.

Example 13 $\begin{bmatrix} 1 & 3 & 5 \\ 10 & 0 & 2 \end{bmatrix} \begin{bmatrix} 4 & 1 \\ 6 & 8 \\ 7 & 5 \end{bmatrix}$

$$= \begin{bmatrix} (1 \times 4) + (3 \times 6) + (5 \times 7) & (1 \times 1) + (3 \times 8) + (5 \times 5) \\ (10 \times 4) + (0 \times 6) + (2 \times 7) & (10 \times 1) + (0 \times 8) + (2 \times 5) \end{bmatrix}$$

$$= \begin{bmatrix} 57 & 50 \\ 54 & 20 \end{bmatrix}$$

Example 14 $\begin{bmatrix} 4 & 1 \\ 6 & 8 \\ 7 & 5 \end{bmatrix} \begin{bmatrix} 1 & 3 & 5 \\ 10 & 0 & 2 \end{bmatrix}$

$$= \begin{bmatrix} (4 \times 1) + (1 \times 10) & (4 \times 3) + (1 \times 0) & (4 \times 5) + (1 \times 2) \\ (6 \times 1) + (8 \times 10) & (6 \times 3) + (8 \times 0) & (6 \times 5) + (8 \times 2) \\ (7 \times 1) + (5 \times 10) & (7 \times 3) + (5 \times 0) & (7 \times 5) + (5 \times 2) \end{bmatrix}$$

$$= \begin{bmatrix} 14 & 12 & 22 \\ 86 & 18 & 46 \\ 57 & 21 & 45 \end{bmatrix}$$

■

Examples 13 and 14 show that when the same two matrices are multiplied in different orders, the resulting products can have different sizes.

Example 15
$$\begin{bmatrix} 3 & 7 & 1 \\ 2 & 4 & 9 \\ 6 & 8 & 5 \end{bmatrix} \begin{bmatrix} 5 & 8 & 4 \\ 9 & 2 & 6 \\ 1 & 3 & 7 \end{bmatrix}$$

$$= \begin{bmatrix} (3 \times 5) + (7 \times 9) + (1 \times 1) & (3 \times 8) + (7 \times 2) + (1 \times 3) & (3 \times 4) + (7 \times 6) + (1 \times 7) \\ (2 \times 5) + (4 \times 9) + (9 \times 1) & (2 \times 8) + (4 \times 2) + (9 \times 3) & (2 \times 4) + (4 \times 6) + (9 \times 7) \\ (6 \times 5) + (8 \times 9) + (5 \times 1) & (6 \times 8) + (8 \times 2) + (5 \times 3) & (6 \times 4) + (8 \times 6) + (5 \times 7) \end{bmatrix}$$

$$= \begin{bmatrix} 79 & 41 & 61 \\ 55 & 51 & 95 \\ 107 & 79 & 107 \end{bmatrix}$$

■

Matrix multiplication does not satisfy the same algebraic rules as the multiplication of numbers. The examples show that it is possible to have:

1. $AB \neq BA$ even though both products are defined and have the same size.
2. AB and BA both defined but with different sizes.
3. One of the products AB and BA defined and the other not defined.
4. $AB = 0$ but $A \neq 0$ and $B \neq 0$.
5. $AC = BC$ even though $A \neq B$ and $C \neq 0$.

Matrix multiplication does satisfy certain algebraic rules, however. In the following list, A, B and C are matrices, 0 is a zero matrix, and p is a real number. Assume the sizes of the matrices are chosen so that all expressions in this list are defined.

Properties of Matrix Multiplication

1. $A(BC) = (AB)C$ Associative laws
   $p(AB) = (pA)B = A(pB)$

2. $A(B + C) = AB + AC$ Distributive laws
   $(A + B)C = AC + BC$

3. $A0 = 0$ $0A = 0$ Properties of zero matrix

One difficulty with matrix arithmetic is that sometimes an operation cannot be performed because the result would be undefined. For example, a 2×2 matrix cannot be added to a 3×4 matrix. For square matrices of the same size, this problem cannot happen.

> A matrix is *square* if it has the same number of rows as of columns.

If n is a fixed integer, any expression involving sums and products of $n \times n$ matrices is defined and is an $n \times n$ matrix. As long as only $n \times n$ matrices are considered, *sums and products are always defined*.

We now define certain square matrices for which multiplication is very easy.

> The *diagonal* of a square matrix goes from the upper left entry to the lower right entry. A square matrix is a *diagonal matrix* if every entry not on the diagonal is 0.

Example 16 The following matrices are diagonal:

$$\begin{bmatrix} 3 & 0 \\ 0 & 7 \end{bmatrix} \quad \begin{bmatrix} 4 & 0 & 0 \\ 0 & 10 & 0 \\ 0 & 0 & -3 \end{bmatrix} \quad \begin{bmatrix} -3 & 0 \\ 0 & 0 \end{bmatrix} \quad \begin{bmatrix} 5 & 0 & 0 & 0 \\ 0 & 5 & 0 & 0 \\ 0 & 0 & 5 & 0 \\ 0 & 0 & 0 & 5 \end{bmatrix}$$

$$\begin{bmatrix} 0 & 0 & 0 \\ 0 & 0 & 0 \\ 0 & 0 & 0 \end{bmatrix}$$

Example 17 The following matrices are not diagonal:

$$\begin{bmatrix} 2 & ① \\ 0 & 10 \end{bmatrix} \quad \underbrace{\begin{bmatrix} 3 & 0 & 0 \\ 0 & 4 & 0 \end{bmatrix}}_{\text{not square}} \quad \begin{bmatrix} 5 & 0 & 0 & 0 \\ 0 & 5 & 0 & 0 \\ 0 & 0 & 5 & ① \\ 0 & 0 & 0 & 5 \end{bmatrix}$$

The circled numbers would have to be 0 for the matrix to be diagonal. ∎

If A and B are diagonal $n \times n$ matrices, AB is a diagonal $n \times n$ matrix, $AB = BA$, and the entry in i,i position in AB is $a_{i,i}b_{i,i}$. To multiply two diagonal matrices, merely multiply the entries in corresponding positions on the diagonals.

Example 18
$$\begin{bmatrix} 2 & 0 & 0 \\ 0 & 7 & 0 \\ 0 & 0 & 3 \end{bmatrix} \begin{bmatrix} 4 & 0 & 0 \\ 0 & 10 & 0 \\ 0 & 0 & 12 \end{bmatrix} = \begin{bmatrix} 2 \times 4 & 0 & 0 \\ 0 & 7 \times 10 & 0 \\ 0 & 0 & 3 \times 12 \end{bmatrix} = \begin{bmatrix} 8 & 0 & 0 \\ 0 & 70 & 0 \\ 0 & 0 & 36 \end{bmatrix}$$
∎

The $n \times n$ *identity matrix*, denoted I_n, is the $n \times n$ diagonal matrix with a 1 in every position on the diagonal. Thus

$$I_2 = \begin{bmatrix} 1 & 0 \\ 0 & 1 \end{bmatrix}, \quad I_3 = \begin{bmatrix} 1 & 0 & 0 \\ 0 & 1 & 0 \\ 0 & 0 & 1 \end{bmatrix}, \quad I_4 = \begin{bmatrix} 1 & 0 & 0 & 0 \\ 0 & 1 & 0 & 0 \\ 0 & 0 & 1 & 0 \\ 0 & 0 & 0 & 1 \end{bmatrix}$$

Identity matrices are to matrix multiplication as the number 1 is to the multiplication of real numbers. For example,

$$\begin{bmatrix} 3 & 2 \\ -5 & 7 \end{bmatrix} \begin{bmatrix} 1 & 0 \\ 0 & 1 \end{bmatrix} = \begin{bmatrix} 3 & 2 \\ -5 & 7 \end{bmatrix}$$

and

$$\begin{bmatrix} 1 & 0 & 0 \\ 0 & 1 & 0 \\ 0 & 0 & 1 \end{bmatrix} \begin{bmatrix} 2 & 4 & 1 & 3 \\ 5 & 10 & -8 & 7 \\ 3 & 9 & 2 & -6 \end{bmatrix} = \begin{bmatrix} 2 & 4 & 1 & 3 \\ 5 & 10 & -8 & 7 \\ 3 & 9 & 2 & -6 \end{bmatrix}$$

Thus the product of a matrix A with an identity matrix is A, provided the product is defined.

> If A is an $m \times n$ matrix, then
> $AI_n = A \quad I_m A = A \quad$ Identity laws

Here are some applications of matrix multiplication.

Example 19 In the previous section, the matrix $\begin{bmatrix} 10 & 30 \\ 20 & 0 \\ 5 & 10 \end{bmatrix}$ described the original inventories for Field's. Suppose each box of shirts costs Field's $1000, each box of ties costs $1200, and each box of sweaters costs $900. What is the total cost of the inventory?

Solution Form the product

$$\begin{array}{c} \text{Shirts} \quad \text{Ties} \quad \text{Sweaters} \\ [1000 \quad 1200 \quad 900] \end{array} \begin{array}{c} \text{A} \quad \text{B} \\ \begin{bmatrix} 10 & 30 \\ 20 & 0 \\ 5 & 10 \end{bmatrix} \begin{array}{l} \text{Shirts} \\ \text{Ties} \\ \text{Sweaters} \end{array} \end{array} = \begin{array}{c} \text{A} \quad \text{B} \\ (38500 \quad 39000) \end{array}$$

The cost of the items stored in warehouse A is $38,500 and in warehouse B $39,000. Since there is only one warehouse A and one warehouse B, the matrix $\begin{bmatrix} 1 \\ 1 \end{bmatrix}$ describes the number of warehouses. The product

$$[38500 \quad 39000]\begin{bmatrix} 1 \\ 1 \end{bmatrix} = \$77{,}500$$ is the total cost of the inventory.

Example 20 Farm Dairy Company makes vanilla and chocolate ice cream at plants I, II, III, and IV. The matrix D

$$\begin{array}{c} \quad \text{I} \quad \;\; \text{II} \quad \;\;\text{III} \quad \text{IV} \\ \begin{array}{c} \text{Vanilla} \\ \text{Chocolate} \end{array} \begin{bmatrix} 150 & 350 & 80 & 250 \\ 100 & 500 & 140 & 200 \end{bmatrix} = D \end{array}$$

describes how many gallons of each flavor are made daily at the various plants. For example, 140 gal of chocolate ice cream are made daily at plant III. All plants use the same recipes. The amount of each ingredient needed is given in the matrix R.

$$\begin{array}{c} \quad \text{Vanilla} \quad \text{Chocolate} \\ \begin{array}{c} \text{Milk} \\ \text{Cream} \\ \text{Sugar} \\ \text{Vanilla extract} \\ \text{Baking chocolate} \end{array} \begin{bmatrix} 4 & 4 \\ 1 & 1 \\ 1.5 & 2 \\ 5 & 1.6 \\ 0 & 4 \end{bmatrix} = R \end{array}$$

3-2 MATRIX MULTIPLICATION

In the recipes, milk and cream are measured in pints, sugar is measured in cups, vanilla extract is measured in fluid ounces, and baking chocolate is measured in ounces.

PROBLEM A How much of each ingredient is needed every day at each plant?

Solution

$$RD = \begin{bmatrix} 4 & 4 \\ 1 & 1 \\ 1.5 & 2 \\ 5 & 1.6 \\ 0 & 4 \end{bmatrix} \begin{bmatrix} 150 & 350 & 80 & 250 \\ 100 & 500 & 140 & 200 \end{bmatrix}$$

$$= \begin{matrix} \text{Milk} \\ \text{Cream} \\ \text{Sugar} \\ \text{Vanilla extract} \\ \text{Baking chocolate} \end{matrix} \begin{bmatrix} \text{I} & \text{II} & \text{III} & \text{IV} \\ 1000 & 3400 & 880 & 1800 \\ 250 & 850 & 220 & 450 \\ 425 & 1525 & 400 & 775 \\ 910 & 2550 & 624 & 1570 \\ 400 & 2000 & 560 & 800 \end{bmatrix}$$

PROBLEM B Suppose the cost matrix is given by

$$\begin{matrix} \text{Milk} & \text{Cream} & \text{Sugar} & \text{Vanilla extract} & \text{Baking chocolate} \\ [0.27 & 0.48 & 0.07 & 0.04 & 0.16] \end{matrix} = C$$

What is the cost of the ingredients used at each plant in a day?

Solution

$$C(RD) = [0.27 \quad 0.48 \quad 0.07 \quad 0.04 \quad 0.16] \begin{bmatrix} 1000 & 3400 & 880 & 1800 \\ 250 & 850 & 220 & 450 \\ 425 & 1525 & 400 & 775 \\ 910 & 2550 & 624 & 1570 \\ 400 & 2000 & 560 & 800 \end{bmatrix}$$

$$= \begin{matrix} \text{I} & \text{II} & \text{III} & \text{IV} \\ [520.15 & 1854.75 & 485.76 & 947.05] \end{matrix}$$

PROBLEM C What is the total daily cost of ingredients for the Farm Dairy Company?

Solution This is the sum of the costs for the individual plants. This could be expressed as a matrix product

$$\begin{matrix} \text{I} & \text{II} & \text{III} & \text{IV} \\ [520.15 & 1854.75 & 485.76 & 947.05] \end{matrix} \begin{bmatrix} 1 \\ 1 \\ 1 \\ 1 \end{bmatrix} \begin{matrix} \text{I} \\ \text{II} \\ \text{III} \\ \text{IV} \end{matrix} = 3807.71$$

The matrix $\begin{bmatrix} 1 \\ 1 \\ 1 \\ 1 \end{bmatrix}$ is used because there is one of each plant. The total daily cost of ingredients for the Farm Dairy Company is $3807.71.

PROBLEM D What is the cost of the ingredients for one gallon of ice cream?

Solution This is the product

$$CR = [0.27 \quad 0.48 \quad 0.07 \quad 0.04 \quad 0.16] \begin{bmatrix} 4 & 4 \\ 1 & 1 \\ 1.5 & 2 \\ 5 & 1.6 \\ 0 & 4 \end{bmatrix}$$

$$= \begin{matrix} \text{Vanilla} & \text{Chocolate} \\ [1.865 & 2.404] \end{matrix}$$

If we round off to the nearest cent, the ingredients for a gallon of vanilla ice cream cost $1.87 and the ingredients for a gallon of chocolate ice cream cost $2.40.

Exercises 3-2

Multiply the matrices, if possible.

1. $[3 \quad -9 \quad 2] \begin{bmatrix} 4 \\ 10 \\ 3 \end{bmatrix}$

2. $[5 \quad 6] \begin{bmatrix} 8 \\ 4 \end{bmatrix}$

3. $[2 \ -1 \ 3 \ 5] \begin{bmatrix} 4 \\ 3 \\ 1 \end{bmatrix}$
4. $[3 \ 1 \ 5 \ 2] \begin{bmatrix} -6 \\ 0 \\ 7 \\ 3 \end{bmatrix}$

Suppose A is a 3×5 matrix, B is a 4×3 matrix, C is a 3×3 matrix, D is a 5×3 matrix, and E is a 3×2 matrix. Determine whether or not each product is defined. If a product is defined, what is its size?

5. AB
6. BA
7. BC
8. DE
9. AD
10. DA
11. ADE
12. BCA

Compute each expression, if it is defined.

13. $\begin{bmatrix} 2 & 3 \\ -5 & 7 \end{bmatrix} \begin{bmatrix} 4 & -2 \\ 6 & 3 \end{bmatrix}$
14. $\begin{bmatrix} 3 & -2 \\ 1 & 5 \end{bmatrix} \begin{bmatrix} -7 & 2 \\ 6 & 10 \end{bmatrix}$

15. $\begin{bmatrix} 3 & -9 \\ 7 & 1 \\ 2 & 4 \end{bmatrix} \begin{bmatrix} 7 & 1 \\ -3 & 5 \\ -6 & 2 \end{bmatrix}$
16. $\begin{bmatrix} 2 & -5 & 7 \\ -9 & 3 & 1 \end{bmatrix} \begin{bmatrix} -4 & 6 \\ 3 & -1 \\ 2 & 4 \end{bmatrix}$

17. $\begin{bmatrix} 3 & -1 & 2 \\ 4 & 1 & 5 \\ 6 & 3 & -7 \end{bmatrix} \begin{bmatrix} 1 & -5 & 2 \\ 9 & 6 & 4 \\ 3 & 8 & -1 \end{bmatrix}$
18. $\begin{bmatrix} -4 & 6 & 1 \\ -2 & 5 & 9 \\ 3 & 7 & 3 \end{bmatrix} \begin{bmatrix} 1 & 5 & 3 \\ 7 & 9 & -2 \\ 8 & 4 & -1 \end{bmatrix}$

19. $\begin{bmatrix} 4 & 3 & 5 \\ 6 & 2 & 0 \\ 0 & 0 & 6 \end{bmatrix} \begin{bmatrix} 1 & 2 & 4 \\ 5 & 9 & 1 \\ 3 & 8 & 2 \end{bmatrix}$
20. $\begin{bmatrix} 3 & 7 & 1 \\ 4 & 2 & 6 \\ 5 & 8 & 2 \end{bmatrix} \begin{bmatrix} 1 & 4 & 3 \\ 3 & 2 & 6 \\ 2 & 1 & 7 \end{bmatrix}$

21. $[8 \ 7 \ 2] \begin{bmatrix} 10 & 4 & 3 \\ 6 & 12 & 4 \\ 1 & 3 & 7 \end{bmatrix}$
22. $\begin{bmatrix} 8 & 9 & 4 \\ 3 & 7 & 9 \\ 2 & 6 & 1 \end{bmatrix} \begin{bmatrix} 2 \\ 1 \\ 5 \end{bmatrix}$

23. $\begin{bmatrix} 4 & 0 & 0 & 0 \\ 0 & 3 & 0 & 0 \\ 0 & 0 & 6 & 0 \\ 0 & 0 & 0 & 8 \end{bmatrix} \begin{bmatrix} 5 & 0 & 0 & 0 \\ 0 & -6 & 0 & 0 \\ 0 & 0 & -9 & 0 \\ 0 & 0 & 0 & 12 \end{bmatrix}$

24. $\begin{bmatrix} 1 & 4 \\ 2 & 3 \\ 6 & 5 \\ 7 & 9 \end{bmatrix} \begin{bmatrix} 4 & 2 \\ 5 & 1 \end{bmatrix}$

25. $\begin{bmatrix} 2 & 3 \\ 5 & 7 \end{bmatrix} \begin{bmatrix} 4 & 1 \\ 8 & 2 \end{bmatrix} - 5 \begin{bmatrix} 3 & 7 \\ 2 & 9 \end{bmatrix}$

26. $4 \begin{bmatrix} 5 & 2 \\ 6 & 1 \end{bmatrix} + \begin{bmatrix} 1 & 3 \\ 5 & 2 \end{bmatrix} \begin{bmatrix} 2 & 6 \\ 8 & 0 \end{bmatrix}$

27. $\begin{bmatrix} 3 & 1 \\ 2 & 4 \end{bmatrix} \begin{bmatrix} 1 & 2 \\ 3 & 6 \end{bmatrix} - \begin{bmatrix} 1 & 2 \\ 3 & 6 \end{bmatrix} \begin{bmatrix} 3 & 1 \\ 2 & 4 \end{bmatrix}$

28. $\begin{bmatrix} 4 & 1 & 2 \\ 3 & 5 & 7 \end{bmatrix} \begin{bmatrix} 6 & 1 \\ 2 & 5 \\ 3 & 9 \end{bmatrix} - \begin{bmatrix} 6 & 1 \\ 2 & 5 \\ 3 & 9 \end{bmatrix} \begin{bmatrix} 4 & 1 & 2 \\ 3 & 5 & 7 \end{bmatrix}$

29. $\begin{bmatrix} 1 & 2 & 5 \\ 5 & 3 & 2 \\ 6 & 7 & 1 \end{bmatrix} \begin{bmatrix} 4 & 3 & 2 \\ 1 & 5 & 0 \\ 0 & 0 & 3 \end{bmatrix} - 6 \begin{bmatrix} 2 & 1 & 9 \\ 4 & 3 & 7 \\ 2 & 0 & 8 \end{bmatrix}$

30. $\begin{bmatrix} 7 & 0 & 3 \\ 4 & 1 & 2 \\ 0 & 2 & 5 \end{bmatrix} \begin{bmatrix} 0 & 0 & 3 \\ 4 & 1 & 6 \\ 0 & 2 & 0 \end{bmatrix} + 4 \begin{bmatrix} 1 & 3 & 9 \\ 2 & 6 & 1 \\ 3 & 8 & 2 \end{bmatrix}$

Kingburger has three stores in Hanover. Each store sells burgers, beverages, and fries. The matrix S shows how many of each item is sold at each store on a typical day.

$$\begin{array}{c} \\ \text{Burgers} \\ \text{Beverages} \\ \text{Fries} \end{array} \begin{array}{ccc} \text{I} & \text{II} & \text{III} \\ \begin{bmatrix} 800 & 1200 & 1000 \\ 600 & 900 & 850 \\ 500 & 750 & 600 \end{bmatrix} \end{array} = S$$

The entries in matrix P are the prices at which the items are sold.

$$\begin{array}{cccc} & \text{Burgers} & \text{Beverages} & \text{Fries} \\ P = & [1.19 & 0.35 & 0.58] \end{array}$$

31. Use matrix multiplication to find a matrix in which each entry is the

typical daily revenue at a store.

32. Use matrix multiplication to determine the typical daily revenue for Kingburger in Hanover.

Belko Manufacturing makes three products: A, B, and C. Each of these products requires subassemblies I and II. The subassembly matrix S describes how many of each subassembly are required to produce one of each product.

$$\text{Product} \begin{array}{c} \\ A \\ B \\ C \end{array} \begin{array}{c} \text{Subassembly} \\ \text{I} \quad \text{II} \\ \begin{bmatrix} 5 & 7 \\ 3 & 1 \\ 2 & 4 \end{bmatrix} = S \end{array}$$

For example, 3 units of subassembly I are needed to make one of product B. Each subassembly is made from parts a, b, c, and d. The parts matrix P tells how many of each part are needed to make one of a subassembly.

$$\text{Subassembly} \begin{array}{c} \\ \text{I} \\ \text{II} \end{array} \begin{array}{c} \text{Part} \\ a \quad b \quad c \quad d \\ \begin{bmatrix} 1 & 3 & 2 & 1 \\ 4 & 2 & 0 & 3 \end{bmatrix} = P \end{array}$$

*33. Find a matrix that describes how many of each part are needed for each product.

*34. A customer places an order with matrix

$$\begin{array}{c} \text{Product} \\ \text{A} \quad \text{B} \quad \text{C} \\ \begin{bmatrix} 10 & 20 & 6 \end{bmatrix} = D \end{array}$$

Write a matrix that describes how many of each part are needed to fill the order.

The cost of each part is given by the matrix

$$\text{Part} \begin{array}{c} a \\ b \\ c \\ d \end{array} \begin{bmatrix} 2.00 \\ 3.40 \\ 5.00 \\ 1.20 \end{bmatrix} = C$$

*35. What is the total cost of the parts needed to fill the customer's order?

*36. Write a matrix whose entries describe the cost of the parts needed for each subassembly.

*37. Write a matrix whose entries describe the cost of the parts needed for each product.

*38. Use the answer to exercise 37 to compute the cost of the parts needed to fill the customer's order. This exercise must have the same answer as exercise 35. Which law of matrix arithmetic guarantees this?

One tool used by social scientists in studying the structure of organizations is called the chain of command matrix. If M is a chain of command matrix, $m_{i,j} = 1$ if person j receives orders directly from person i, and $m_{i,j} = 0$ otherwise. Use the matrix

$$M = \begin{bmatrix} 0 & 0 & 0 & 1 & 1 \\ 0 & 0 & 1 & 1 & 0 \\ 0 & 0 & 0 & 1 & 0 \\ 0 & 0 & 0 & 0 & 0 \\ 0 & 1 & 1 & 0 & 0 \end{bmatrix}$$

in exercises 39–46.

39. Who is in command? (*Hint:* No one gives orders to the boss.)

40. Who cannot give orders?

Compute the matrix M^2. The entry in position i, j in M^2 is the number of ways that person j can receive orders from person i indirectly in exactly two steps.

41. Who cannot give orders in exactly two steps?

42. In how many ways can person 5 give orders to person 4 in exactly two steps?

43. To whom can the commander give orders in exactly two steps?

Compute the matrix $M + M^2$. The entry in position i, j in this matrix is the number of ways that person i can give orders to person j in two or fewer steps.

44. In how many ways can person 3 receive orders in one or two steps?

45. In how many ways can the commander give orders in one or two steps?

46. In how many ways can person 5 give orders in one or two steps?

Compute each expression.

47. $\begin{bmatrix} 2.13 & 4.75 \\ 6.19 & 2.79 \end{bmatrix} \begin{bmatrix} 12.72 & -9.81 \\ 5.48 & 6.93 \end{bmatrix}$

48. $\begin{bmatrix} 4.12 & -3.74 \\ 6.68 & 5.73 \end{bmatrix} \begin{bmatrix} 8.21 & 10.93 \\ -7.27 & -1.28 \end{bmatrix} + 2.73 \begin{bmatrix} 9.10 & 8.36 \\ 5.74 & -2.48 \end{bmatrix}$

49. $\begin{bmatrix} 7.35 & -2.61 & 4.93 \\ 3.61 & -8.42 & 3.11 \\ 5.79 & 4.32 & -1.38 \end{bmatrix} \begin{bmatrix} 3.14 & 1.41 & -1.73 \\ -2.36 & 6.42 & 8.01 \\ 8.27 & 3.62 & 7.91 \end{bmatrix}$

Section 3-3 Two Linear Equations in Two Unknowns

Let's now leave matrices for a while in order to consider another part of linear algebra—systems of linear equations. In this section, systems of two linear equations in two unknowns are discussed. In the next section, we will study systems of three linear equations in three unknowns and learn how to use matrices to solve such systems.

A linear equation $ax + by = c$ in two unknowns x and y has as its graph a line in the plane. If two linear equations in two unknowns are given, three cases can occur.

1. *The two lines intersect in one point.* (See Diagram 1.)

In this case, the system of equations has a unique solution, that is, there is exactly one point (x, y) whose coordinates are solutions of both equations. For example, to solve the system

$$x + 2y = 11$$
$$2x + 6y = 30$$

subtract twice the first equation from the second equation.

$$2x + 6y = 30$$
$$-\underline{2x + 4y = 22} \quad \text{Subtract.}$$
$$2y = 8$$

Then divide by 2 to obtain $y = 4$. Substitute 4 for y in either of the original equations and obtain $x = 3$. The solution is $x = 3$, $y = 4$.

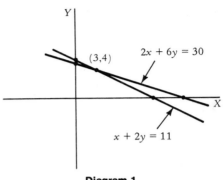

Diagram 1

2. *The two lines are parallel but distinct.* (See Diagram 2.)

In this case, the system of equations has no solution. For example, to solve the system

$$2x + 3y = 15$$
$$8x + 12 = 40$$

subtract the second equation from four times the first equation.

$$8x + 12y = 60$$
$$-\quad \underline{8x + 12y = 40} \quad \text{Subtract.}$$
$$0 = 20$$

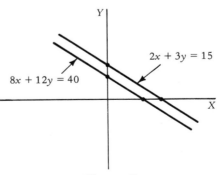

Diagram 2

This gives $0 = 20$, which is a false statement. This means that the system is inconsistent and has no solution.

3. *The two lines coincide.* (See Diagram 3.)

In this case, the system of equations has infinitely many solutions. The coordinates of each point on the line are a solution. For example, to solve the system

$$x - y = 5$$
$$2x - 2y = 10$$

subtract the second equation from twice the first equation.

$$2x - 2y = 10$$
$$-\underline{2x - 2y = 10}\qquad \text{Subtract.}$$
$$0 = 0$$

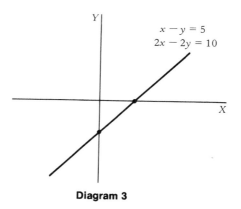

Diagram 3

The statement $0 = 0$ is an identity, and this means that the two equations represent the same line. Thus, any pair of numbers that satisfies one of the equations must also satisfy the other equation.

Elimination method

We will study two procedures, the *elimination method* and the *substitution method*, for solving two linear equations in two unknowns. Either procedure will determine which of the three cases holds, and will result in finding the unique solution in case 1. First, let's study the elimination method. This is the method we have used until now. Here we multiply the equations by suitable numbers, so that in the new equations, one variable has (ignoring sign) the same coefficient both times. Then add or subtract the new equations in order to obtain an equation in which this variable does not occur, that is, has been eliminated. For example, to solve

$$4x - 3y = 5$$
$$3x + 2y = 8$$

multiply the first equation by 2 and the second equation by 3 to obtain

$$8x - 6y = 10$$
$$9x + 6y = 24$$

Add these equations to obtain $17x = 34$, so that $x = 2$. Substitute 2 for x in $4x - 3y = 5$ to obtain

$$8 - 3y = 5$$
$$-3y = -3$$
$$y = 1$$

The answer should be checked by substituting $x = 2$, $y = 1$ in both of the original equations. This system could also be solved by multiplying the first equation by 3, the second equation by 4, and subtracting.

Example 1 Solve $2x - y = 4$
$x + 2y = 7$

Solution

$$\begin{aligned} 4x - 2y &= 8 \\ + \quad x + 2y &= 7 \\ \hline 5x &= 15 \end{aligned}$$

Multiply by 2.
Add.
So $x = 3$.

$$\begin{aligned} 2 \times 3 - y &= 4 \\ 6 - y &= 4 \\ y &= 2 \end{aligned}$$

Substitute $x = 3$ in $2x - y = 4$

Example 2 Solve $5x + 3y = 14$
$2x + 4y = 14$

3-3 TWO LINEAR EQUATIONS IN TWO UNKNOWNS

Solution

$$10x + 6y = 28 \quad \text{Multiply by 2.}$$
$$-\underline{10x + 20y = 70} \quad \text{Multiply by 5 and subtract.}$$
$$-14y = -42 \quad \text{So } y = 3.$$

$$5x + (3 \times 3) = 14 \quad \text{Substitute } y = 3 \text{ in } 5x + 3y = 14.$$
$$5x = 5$$
$$x = 1$$

Example 3 Solve $2x + 5y = 8$
$6x + 15y = 18$

Solution

$$6x + 15y = 24 \quad \text{Multiply by 3.}$$
$$\underline{-6x + 15y = 18} \quad \text{Subtract.}$$
$$0 = 6 \quad \text{No solution.}$$

Example 4 Andrew's stock in Exxon and General Mills is worth $16,000. Exxon currently sells for $70 a share and General Mills for $30 a share. If Exxon stock triples in value and General Mills stock goes up 50%, Andrew's stock will be worth $34,500. How many shares of each stock does he have?

Solution Let e denote the number of shares of Exxon that Andrew owns and g the number of shares of General Mills. The current value of his stock is $70e + 30g$, so that

$$70e + 30g = 16{,}000$$

If Exxon triples in value, each share would be worth $210. If General Mills goes up 50%, each share of its stock would be worth $45. Andrew's stock would then be worth $210e + 45g$, so that

$$210e + 45g = 34{,}500$$

Now solve these equations.

$$210e + 90g = 48{,}000$$
$$-\quad 210e + 45g = 34{,}500$$
$$\overline{45g = 13{,}500}$$
$$g = \frac{13{,}500}{45} = 300$$

Multiply the first equation by 3.
Subtract.

$$210e + 45 \cdot 300 = 34{,}500$$
$$210e + 13{,}500 = 34{,}500$$
$$210e = 21{,}000$$
$$e = \frac{21{,}000}{210} = 100$$

Substitute $g = 300$ in $210e + 45g = 34{,}500$.

Andrew has 100 shares of Exxon and 300 shares of General Mills.

Substitution method

We now come to the next method of solving two linear equations in two unknowns—the *substitution method*. Here you solve an equation for one of the variables. For example, you might solve the first equation for y in terms of x. Then you substitute this expression involving x for y in the other equation to obtain an equation that has only one variable. Thus to solve

$$5x + y = 11$$
$$3x + 2y = 8$$

solve the first equation for y to obtain $y = 11 - 5x$. Substitute $11 - 5x$ for y in the second equation to obtain $3x + 2(11 - 5x) = 8$, so we have eliminated y. Now

$$3x + 22 - 10x = 8$$
$$-7x = 8 - 22$$
$$-7x = -14$$
$$x = 2$$
$$y = 11 - 5x = 11 - (5 \times 2) = 1$$

To solve

$$y = 4x - 1$$
$$y = 2x + 5$$

equate the two values of y to obtain

$$4x - 1 = 2x + 5 \quad \text{Now solve for } x \text{ and then find } y.$$
$$2x = 6$$
$$x = 3$$
$$y = 4x - 1 = (4 \times 3) - 1 = 11$$

Example 5 $2x + 3y = 5$
$3x - y = 13$

Solution

$$y = 3x - 13 \quad \text{Solve the second equation for } y;$$
$$2x + 3(3x - 13) = 5 \quad \text{substitute in the first equation.}$$
$$2x + 9x - 39 = 5$$
$$11x = 5 + 39$$
$$11x = 44$$
$$x = 4$$
$$y = (3 \times 4) - 13 = -1$$

Now check the answer by substituting $x = 4$ and $y = -1$ in the original equations. If true statements are obtained, the answer is correct. If even one false statement is obtained, the answer is wrong and there is a mistake in the solution.

$$2(4) + 3(-1) = 8 - 3 = 5 \checkmark \quad \text{The answer is correct.}$$
$$3(4) - (-1) = 12 + 1 = 13 \checkmark$$

Example 6 Solve $y = 5x - 7$
$y = 3x - 1$

Solution $5x - 7 = 3x - 1 \quad$ Equate the two values of y.
$$2x = 6$$
$$x = 3$$
$$y = (5 \times 3) - 7 = 8$$

Example 7 A county contains a city and its suburbs. The population of the city is 3,500,000 and is decreasing by 200,000 each year. The population of the suburbs is 1,800,000 and is increasing by 100,000 each year. When will the city and the suburbs have equal populations?

Solution Let p stand for population and t for time (measured in years). At time t, the populations of the city and suburbs are given by

$$p = 3{,}500{,}000 - 200{,}000t \quad \text{City}$$
$$p = 1{,}800{,}000 + 100{,}000t \quad \text{Suburbs}$$

The city and the suburbs will have equal populations when the two values of p are equal.

$$3{,}500{,}000 - 200{,}000t = 1{,}800{,}000 + 100{,}000t$$

$1{,}700{,}000 = 300{,}000t$ — Add $200{,}000t - 1{,}800{,}000$ to both sides.

$\dfrac{1{,}700{,}000}{300{,}000} = 5\tfrac{2}{3} = t$ — Divide by 300,000.

The city and the suburbs will have equal populations in $5\tfrac{2}{3}$ years, that is, in 5 years and 8 months.

Exercises 3-3

Determine whether each system of equations has no solution, exactly one solution, or infinitely many solutions. Whenever a system has one solution, find the solution and check your answer.

1. $2x - 3y = 6$
   $x + y = 8$

2. $4x + 2y = 18$
   $3x - y = 1$

3. $4x + y = 5$
   $8x + 2y = 12$

4. $4a - b = 10$
   $3a + b = 11$

5. $3a + 5b = 9$
   $5a - 3b = -2$

6. $2x + 6y = 9$
   $5x + 15y = 22.5$

7. $2x - 3y = 7$
   $7x + 10y = 4$

8. $5x + 8y = 1$
   $2x + 3y = 8$

9. $4x + 5y = 3$
   $3x - 2y = -15$

10. $6x + y = 7$
    $8x + 3y = 1$

11. $2x - 3y = 6$
    $12x - 18y = 36$

12. $x + y = 1$
    $5x + 2y = 3$

13. $6x + 5y = 8$
    $10x - 7y = -2$

14. $3x + 5y = 15$
    $12x + 20y = 64$

15. $6x - 3y = 12$
    $10x - 5y = 18$

16. $3x + y = 9$
    $x + 3y = 7$

17. $p + q = 0$
    $p - q = 5$

18. $y = 7x - 4$
    $y = 3x + 4$

19. $y = 3x + 5$
    $y = x + 3$

20. $y = 2x + 6$
    $y = 4x - 1$

21. Henry has $2 in nickels and dimes. If each of his nickels turned into a quarter, he would have $4.80. How many dimes does he have?

22. An alloy containing 40% silver and an alloy containing 70% silver are to be mixed to produce 50 lb of an alloy containing 60% silver. How much of each alloy is needed?

23. Kate has some money invested at 6% interest and some at 10% interest. Her annual income from both investments is $6980. Twice the amount she has invested at 10% exceeds three times the amount she has invested at 6% by $1000. What is the total amount she has invested?

24. At Buy-High Foods, 20 lb of flour and 10 pounds of rice cost $9. Also, 30 lb of flour and 12 lb of rice cost $12. What would 10 lb of flour and 40 lb of rice cost?

25. Wilco Company makes picture frames. A metal frame requires $3.00 of materials and three-quarters of an hour of labor. A wooden frame requires $4.00 of materials and half of an hour of labor. One day, $122 of materials and 25 hours of labor were used. How many of each kind of frame were made?

26. Goodlawn Gardening takes care of yards. An ordinary treatment requires one hour and costs $15. A deluxe treatment requires 2.5 hr and costs $40. Max worked 48 hr last week and collected $750 for Goodlawn for his work. How many ordinary treatments and how many deluxe treatments had he done?

27. At Kane Field, reserved seats for baseball games cost $4.00 and general admission is $2.50. For a recent important game, all the tickets were sold and receipts totaled $210,000. For an unimportant game, 25% of the reserved seats and 40% of the general admission seats were filled and receipts totaled $60,000. How many seats are there in the stadium?

28. Animals in an experiment must be kept on a rigid diet. An animal is to receive 124 g of carbohydrate and 50 g of fat daily, along with other nutrients. Two animal feed mixes are available. Each ounce of mix 1 contains 8 g of carbohydrate and 5 g of fat. Each ounce of mix 2 contains 12 g of carbohydrate and 2 g of fat. How many ounces of each mix should the animal be fed daily?

29. Bill's Men's Shop sells woolen suits and cotton suits. If Bill's sold its entire inventory of suits, receipts would be $43,800. However the manager expects to sell only half of the woolen suits and two-thirds of the cotton suits, for a total of $26,700. How many woolen suits and how many cotton suits are in the store? A woolen suit sells for $150 and a cotton suit sells for $120.

30. Rafael Santos has some money invested in stock and some in bonds. Last year, the stock paid a 2% dividend, the bond paid 8% interest, and he received a total of $780. This year, the stock pays a 1% dividend, the bond pays 10% interest, and he receives $810. How much does he have invested in stocks? In bonds?

31. Mr. Altgeld is told by his doctor to take 2000 mg of vitamin C and 57 mg of iron daily in vitamin pills. He decides to do this by taking Vitaful and Orangerust pills. How many pills of each type should he take a day? Each Vitaful pill contains 200 mg of vitamin C and 4 mg of iron, while each Orangerust pill contains 80 mg of vitamin C and 5 mg of iron.

32. After six months, Mr. Altgeld's doctor changes the prescription to 1000 mg of vitamin C and 54 mg of iron. How many pills of each type should Mr. Altgeld take each day now?

33. Duncan McDuff is running for the school board and wants to spend $5750 on newspaper ads at $100 each and radio spots at $50 each. He wants to spend $1250 more on newspaper ads than on radio spots. How many newspaper ads does he buy?

34. Duncan's opponent, Laura Green, has $5750 to spend but wants to spend $750 more on radio spots than on newspaper ads. How many radio spots does she buy?

Solve each system of equations. Note that in each system, at least one of the equations is not linear.

*35. $x^2 + y^2 = 16$
$x^2 + 2y^2 = 25$

*36. $x^2 + y^2 = 7$
$x^2 - y^2 = 1$

*37. $y = x + 1$
$25 = x^2 + y^2$

*38. $x^2 + y^2 = 16$
$y = x + 6$

*39. $x^2 - 2y = 0$
$x^2 - y^2 - 1 = 0$

*40. $x^2 - y^2 - 9 = 0$
$2x^2 + y^2 - 18 = 0$

Solve each system of equations.

41. $1.47x + 2.93y = 12.41$
$5.41x - 1.37y = 7.12$

42. $2.39x + 2.13y = 11.85$
$1.93x + 3.48y = 11.43$

Section 3-4 Systems of Three Linear Equations in Three Unknowns

A linear equation $ax + by + cz = d$ in three unknowns x, y, and z is the equation of a plane in space. Solving a system of three linear equations in three unknowns is equivalent to finding the intersection of three planes. Diagram 1 shows the possibilities.

The intersection of three planes can be a point (Diagram 1a), a line (Diagrams 1b and c), a plane (Diagram 1d), or the empty set (Diagrams 1e

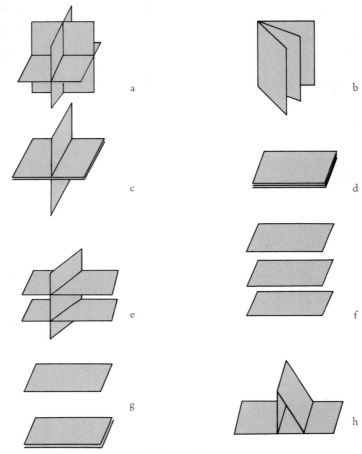

Diagram 1

through 1h). Thus a system of three linear equations in three unknowns can have no solution, exactly one solution, or infinitely many solutions. Similarly a system of n linear equations in n unknowns can have no solution, exactly one solution, or infinitely many solutions.

> A system of n linear equations in n unknowns is *independent* if it has exactly one solution. Otherwise the system is *dependent*.

A system of three linear equations in three unknowns can be solved by an elimination method like the one used in the previous section to solve two linear equations in two unknowns. However we will instead use a matrix method, called *Gauss–Jordan elimination*, that is more efficient. This procedure will tell us whether or not the system is independent and will give the solution if the system is independent. A slight variation of this procedure, designed for better numerical accuracy, is frequently used on computers to solve independent systems of linear equations. An extension of Gauss–Jordan elimination, which is not included in this book, allows us to find all solutions of any system of linear equations in any number of unknowns.

Now let's associate two matrices with each system of linear equations.

Definition. The *matrix of coefficients* of the system of linear equations

$$a_1 x + b_1 y + c_1 z = d_1$$
$$a_2 x + b_2 y + c_2 z = d_2$$
$$a_3 x + b_3 y + c_3 z = d_3$$

is

$$\begin{bmatrix} a_1 & b_1 & c_1 \\ a_2 & b_2 & c_2 \\ a_3 & b_3 & c_3 \end{bmatrix}$$

The *augmented matrix* is

$$\left[\begin{array}{ccc|c} a_1 & b_1 & c_1 & d_1 \\ a_2 & b_2 & c_2 & d_2 \\ a_3 & b_3 & c_3 & d_3 \end{array}\right]$$

The augmented matrix has one more column than the coefficient matrix. For convenience, this last column is usually separated from the rest of the matrix by a vertical line.

Example 1 The coefficient matrix of the system

$$x + y + z = 1$$
$$x + y - 2z = 3$$
$$2x + y + z = 2$$

is

$$\begin{bmatrix} 1 & 1 & 1 \\ 1 & 1 & -2 \\ 2 & 1 & 1 \end{bmatrix}$$

and the augmented matrix is

$$\left[\begin{array}{ccc|c} 1 & 1 & 1 & 1 \\ 1 & 1 & -2 & 3 \\ 2 & 1 & 1 & 2 \end{array}\right]$$

Example 2 The coefficient matrix of the system

$$2x + 3z = 0$$
$$y - 5z = 0$$
$$3y + 6x + 4z = 12$$

is $\begin{bmatrix} 2 & 0 & 3 \\ 0 & 1 & -5 \\ 6 & 3 & 4 \end{bmatrix}$

and the augmented matrix is

$$\begin{bmatrix} 2 & 0 & 3 & | & 0 \\ 0 & 1 & -5 & | & 0 \\ 6 & 3 & 4 & | & 12 \end{bmatrix}$$

Note that a variable is considered to have a coefficient of 0 if it does not occur in an equation. The variables must be written in the same order in each equation, so that $3y + 6x + 4z = 12$ is not confused with $3x + 6y + 4z = 12$.

Example 3 Given that the augmented matrix of a system of linear equations is

$$\begin{bmatrix} 3 & 1 & 5 & | & 6 \\ 4 & 2 & 9 & | & 7 \\ 0 & 3 & 8 & | & 2 \end{bmatrix}$$

the equations are

$$3x + y + 5z = 6$$
$$4x + 2y + 9z = 7$$
$$3y + 8z = 2$$

∎

Row operations

We now consider the problem of solving a system of linear equations by working with its augmented matrix. This is accomplished by performing certain operations on the rows of the augmented matrix. These operations correspond to procedures that may be used in solving a system of linear equations. The procedures for linear equations are:

1. Multiply an equation by a nonzero real number.

2. Interchange two equations.
3. Add a multiple of one equation to another equation.

The corresponding *row operations* on the augmented matrix are the following.

Row Operations
1. Multiply a row of the matrix by a nonzero real number.
2. Interchange two rows.
3. Add a multiple of one row to another row.

Example 4 Multiply by 3 the first row of

$$\begin{bmatrix} 1 & 2 & 30 \\ 2 & 1 & 3 \\ 4 & 7 & 1 \end{bmatrix} \text{ to obtain } \begin{bmatrix} 3 & 6 & 90 \\ 2 & 1 & 3 \\ 4 & 7 & 1 \end{bmatrix}$$

Example 5 Interchange the second and third rows of

$$\begin{bmatrix} 3 & 6 \\ 1 & 9 \\ -4 & 2 \end{bmatrix} \text{ to obtain } \begin{bmatrix} 3 & 6 \\ -4 & 2 \\ 1 & 9 \end{bmatrix}$$

Example 6 To the second row, add -2 times the first row of

$$\begin{bmatrix} 1 & 3 & 5 & 6 \\ 2 & 4 & 1 & 13 \\ 7 & 9 & 4 & 1 \end{bmatrix} \text{ to obtain } \begin{bmatrix} 1 & 3 & 5 & 6 \\ 0 & -2 & -9 & 1 \\ 7 & 9 & 4 & 1 \end{bmatrix}$$

To perform this row operation, to each number in the second row add -2 times the number just above it in the first row.
■

Suppose the matrix A is the augmented matrix of a system of linear equations, and that B is obtained from A by one row operation. Then it can be shown that the new system of equations, which corresponds to B, has the same solution as the old system of equations, which corresponds to A. That is, any new system obtained from the old system either by multiplying an equation by a nonzero number, or by interchanging two equations, or by adding a multiple of one equation to another equation will have the same solution as the old system. In fact, if P is obtained from A by a sequence of row operations, the corresponding systems of equations will have the same solution.

We will show in the next section that the augmented matrix from Example 1 of this section can be reduced by row operations to the matrix

$$\begin{bmatrix} 1 & 0 & 0 & | & 1 \\ 0 & 1 & 0 & | & \frac{2}{3} \\ 0 & 0 & 1 & | & -\frac{2}{3} \end{bmatrix}$$

The system of equations that corresponds to this new augmented matrix is

$$1 \cdot x + 0 \cdot y + 0 \cdot z = 1 \qquad x = 1$$
$$0 \cdot x + 1 \cdot y + 0 \cdot z = \frac{2}{3} \quad \text{or} \quad y = \frac{2}{3}$$
$$0 \cdot x + 0 \cdot y + 1 \cdot z = -\frac{2}{3} \qquad z = -\frac{2}{3}$$

As noted, this system of equations has the same solution as the original system of equations in Example 1. Hence this system and the original system both have the solution $x = 1$, $y = 2/3$, and $z = -2/3$.

The procedure for solving a system of n linear equations in n unknowns is first to write the augmented matrix, and then to apply row operations so as to reduce the coefficient part of the augmented matrix to the $n \times n$ identity matrix. The numbers in the last column of this reduced matrix are the solution of the original system of equations. In the next section, we learn how to perform the matrix reduction.

Exercises 3-4

Write the augmented matrix of each system of linear equations.

1. $2x - y = 5$
   $3x + 2y = 6$

2. $3x - 5y + 4z = 2$
   $2x + y - 3z = 8$

3. $2x + 3y + 6z = 9$
   $4x + 3y + 6z = 11$
   $4x + 3y + 6z = 11$

4. $3a + 2b + 5c = 10$
   $2a - b - 4c = -6$
   $-5a + 2b + 7c = 9$

5. $\begin{aligned} x - y - 3z &= 7 \\ -2x + 5z &= 12 \\ 3y + 2z &= 9 \\ 4x + 2y &= 15 \end{aligned}$

6. $\begin{aligned} 5x - 3y + 2z &= 0 \\ -2x + 4y + 8z &= 17 \\ -3x - 2y - z &= 10 \end{aligned}$

7. $\begin{aligned} -2a - 4b + 9c &= 5 \\ 3a + 7b + 2c &= 11 \\ 4a - 5b &= 15 \end{aligned}$

8. $\begin{aligned} 12x + 3y - 6z &= 5 \\ x - 2y - 5z &= 7 \\ -9x - 2y + 11z &= 13 \end{aligned}$

The augmented matrix of a system of linear equations is given. The unknowns are x, y, and z. Write the equations.

9. $\begin{bmatrix} 4 & 1 & 2 & -3 \\ 3 & -5 & 7 & 5 \\ 0 & 2 & 6 & 9 \end{bmatrix}$

10. $\begin{bmatrix} 8 & 7 & -6 & 11 \\ -2 & 1 & 3 & 14 \\ 5 & 10 & 0 & -6 \end{bmatrix}$

11. $\begin{bmatrix} 1 & 3 & 7 & 15 \\ -9 & -4 & 2 & 11 \\ -6 & 3 & 8 & 2 \end{bmatrix}$

12. $\begin{bmatrix} 3 & -1 & 5 & 0 \\ 6 & -2 & 4 & 17 \\ -5 & 1 & 3 & 12 \end{bmatrix}$

Let $A = \begin{bmatrix} 1 & 3 & 6 \\ 4 & 8 & 12 \\ 2 & 1 & 3 \end{bmatrix}$. *Write the matrix that is obtained by performing the indicated row operation on a.*

13. Multiply the third row by 3.
14. Multiply the second row by 1/4.
15. Interchange the first and second rows.
16. Interchange the first and third rows.
17. Add 3 times the first row to the third row.
18. Add 4 times the third row to the second row.
19. Add -2 times the first row to the third row.
20. Add -4 times the first row to the second row.

Section 3-5 Gauss–Jordan Elimination

We now proceed to the matrix reduction procedure, called *Gauss–Jordan elimination*, that is used in the solution of systems of linear equations. To reduce the coefficient part of the augmented matrix to the identity matrix, use row operations to make the *first* column of the matrix the same as the *first* column of the identity matrix. Then go on to make the *second* column of the matrix the same as the *second* column of the identity matrix, and so on. The procedure can be stated in the following manner.

Gauss–Jordan Elimination

1. If the entry in position 1,1 is 0, interchange the first row with the highest row below it that does not have a 0 in the first column.

2. Now divide the new first row by the entry in position 1,1 (or multiply by the reciprocal of this number). The matrix now has a 1 in position 1,1.

3. Add suitable multiples of the first row to each other row so that the first column of the matrix is the same as the first column of the identity matrix.

4. Now repeat the process using the second row and position 2,2. Make the second column of the matrix the same as the second column of the identity matrix. Continue until the coefficient matrix has been reduced to the identity matrix.

Note that once the ith column of the augmented matrix is made into the ith column of the identity matrix, it does not change again.

We will now solve

$$x + y + z = 1$$
$$x + y - 2z = 3$$
$$2x + y + z = 2$$

by Gauss–Jordan elimination. This system of equations was discussed in the previous section. The augmented matrix is

$$\begin{bmatrix} 1 & 1 & 1 & | & 1 \\ 1 & 1 & -2 & | & 3 \\ 2 & 1 & 1 & | & 2 \end{bmatrix}$$

Since there is a 1 in position 1,1, go directly to step 3. Add -1 times the first row to the second row and -2 times the first row to the third row in order to make the first column $\begin{bmatrix} 1 \\ 0 \\ 0 \end{bmatrix}$. This gives the matrix

$$\begin{bmatrix} 1 & 1 & 1 & | & 1 \\ 0 & 0 & -3 & | & 2 \\ 0 & -1 & -1 & | & 0 \end{bmatrix}$$

Since the entry in position 2,2 is 0, interchange the second and third rows to obtain

$$\begin{bmatrix} 1 & 1 & 1 & | & 1 \\ 0 & -1 & -1 & | & 0 \\ 0 & 0 & -3 & | & 2 \end{bmatrix}$$

Note that only rows 2 and 3 may be interchanged here. Rows 2 and 1 may not be interchanged because row 1 is not below row 2. Now multiply the second row by -1 so that the entry in position 2,2 becomes 1. This gives

$$\begin{bmatrix} 1 & 1 & 1 & | & 1 \\ 0 & 1 & 1 & | & 0 \\ 0 & 0 & -3 & | & 2 \end{bmatrix}$$

Now add -1 times the second row to the first row to obtain

$$\begin{bmatrix} 1 & 0 & 0 & | & 1 \\ 0 & 1 & 1 & | & 0 \\ 0 & 0 & -3 & | & 2 \end{bmatrix}$$

The second column is now $\begin{bmatrix} 0 \\ 1 \\ 0 \end{bmatrix}$. To obtain a 1 in position 3,3, divide the third row by -3. This gives

$$\begin{bmatrix} 1 & 0 & 0 & | & 1 \\ 0 & 1 & 1 & | & 0 \\ 0 & 0 & 1 & | & -\dfrac{2}{3} \end{bmatrix}$$

Now subtract the third row from the second row to obtain

$$\begin{bmatrix} 1 & 0 & 0 & | & 1 \\ 0 & 1 & 0 & | & \frac{2}{3} \\ 0 & 0 & 1 & | & -\frac{2}{3} \end{bmatrix}$$

The coefficient part of the matrix has now been reduced to the identity matrix, so that the matrix reduction is complete. The solution of the system of equations is $x = 1$, $y = 2/3$, $z = -2/3$. The answer should be checked by substituting the solution in all of the original equations and verifying that true statements are obtained:

$$1 + \frac{2}{3} + \left(-\frac{2}{3}\right) \stackrel{?}{=} 1 \checkmark$$

$$1 + \frac{2}{3} - 2\left(-\frac{2}{3}\right) \stackrel{?}{=} 3 \checkmark$$

$$2(1) + \frac{2}{3} + \left(-\frac{2}{3}\right) \stackrel{?}{=} 2 \checkmark$$

If the coefficient part of the matrix reduces to the identity matrix, the system of equations is independent. Otherwise the system is dependent. If at any time in the elimination process it is impossible to perform a step because a nonzero number is not available, the system is dependent. If at any time all entries in any row in the coefficient part of the matrix are 0, the system is dependent.

Example 1 Solve

$$\begin{aligned} 3y + z &= 4 \\ x + y + 2z &= -1 \\ 3x + y + z &= 3 \end{aligned}$$

Solution The augmented matrix is

$$\begin{bmatrix} 0 & 3 & 1 & | & 4 \\ 1 & 1 & 2 & | & -1 \\ 3 & 1 & 1 & | & 3 \end{bmatrix}$$

Note the 0 in position 1, 1.

3-5 GAUSS–JORDAN ELIMINATION

$$\begin{bmatrix} 1 & 1 & 2 & | & -1 \\ 0 & 3 & 1 & | & 4 \\ 3 & 1 & 1 & | & 3 \end{bmatrix}$$

Interchange row 1 with row 2. Now a 1 is in position 1,1.

$$\begin{bmatrix} 1 & 1 & 2 & | & -1 \\ 0 & 3 & 1 & | & 4 \\ 0 & -2 & -5 & | & 6 \end{bmatrix} \quad \text{row } 3 - 3 \times \text{row } 1$$

The first column is now the same as the first column of the identity matrix.

$$\begin{bmatrix} 1 & 1 & 2 & | & -1 \\ 0 & 1 & \frac{1}{3} & | & \frac{4}{3} \\ 0 & -2 & -5 & | & 6 \end{bmatrix} \quad \left(\frac{1}{3}\right) \text{row } 2$$

Now a 1 is in position 2,2.

$$\begin{bmatrix} 1 & 0 & \frac{5}{3} & | & -\frac{7}{3} \\ 0 & 1 & \frac{1}{3} & | & \frac{4}{3} \\ 0 & 0 & -\frac{13}{3} & | & \frac{26}{3} \end{bmatrix} \quad \begin{array}{l} \text{row } 1 - \text{row } 2 \\ \\ \text{row } 3 + 2 \times \text{row } 2 \end{array}$$

The second column is now the same as the second column of the identity matrix.

$$\begin{bmatrix} 1 & 0 & \frac{5}{3} & | & -\frac{7}{3} \\ 0 & 1 & \frac{1}{3} & | & \frac{4}{3} \\ 0 & 0 & 1 & | & -2 \end{bmatrix} \quad \left(-\frac{3}{13}\right) \text{row } 3$$

Now a 1 is in position 3,3.

$$\begin{bmatrix} 1 & 0 & 0 & | & 1 \\ 0 & 1 & 0 & | & 2 \\ 0 & 0 & 1 & | & -2 \end{bmatrix} \quad \begin{array}{l} \text{row } 1 - \left(\frac{5}{3}\right) \text{row } 3 \\ \\ \text{row } 2 - \left(\frac{1}{3}\right) \text{row } 3 \end{array}$$

The coefficient part of the matrix has been reduced to the identity matrix.

$$x = 1, \; y = 2, \; z = -2$$

Example 2 Solve

$$2x + 4y + 8z = 12$$
$$3x + 6y + 9z = 16$$
$$6x + 12y + 27z = 38$$

Solution
$$\begin{bmatrix} 2 & 4 & 8 & | & 12 \\ 3 & 6 & 9 & | & 16 \\ 6 & 12 & 27 & | & 38 \end{bmatrix}$$ This is the augmented matrix.

$$\begin{bmatrix} 1 & 2 & 4 & | & 6 \\ 3 & 6 & 9 & | & 16 \\ 6 & 12 & 27 & | & 38 \end{bmatrix}$$ $\left(\frac{1}{2}\right)$ row 1

$$\begin{bmatrix} 1 & 2 & 4 & | & 6 \\ 0 & 0 & -3 & | & -2 \\ 0 & 0 & 3 & | & 2 \end{bmatrix}$$ row 2 − 3 × row 1
row 3 − 6 × row 1

It is impossible to get any number but 0 in position 2,2 by the procedures allowed—multiplying the second row by a nonzero number or interchanging the second row with the third row. This means that the system of equations is dependent.

Example 3 Ed Wong, an old friend of mine, has a total of $70,000 invested in common stock, corporate bonds, and municipal bonds. To limit his risk, Ed has twice as much invested in corporate bonds as in stock. Last year the common stock paid a 2% dividend, the corporate bonds paid 10% interest, and the municipal bonds paid 6% interest. Ed's total income last year from his investments was $4800. How much does he have invested in bonds?

Solution Let

$S =$ amount invested in common stock.
$C =$ amount invested in corporate bonds.
$M =$ amount invested in municipal bonds.

We then have the equations

$$\begin{aligned} S + C + M &= 70{,}000 & \text{Total amount invested.} \\ C &= 2S & \text{Limit risk.} \\ 0.02S + 0.10C + 0.06M &= 4{,}800 & \text{Last year's income.} \end{aligned}$$

Rewrite the equation $C = 2S$ as $-2S + C = 0$. This is necessary because all the variables must be on the left side of the equation in order to form the augmented matrix. The system of equations now is

$$\begin{aligned} S + C + M &= 70{,}000 \\ -2S + C &= 0 \\ 0.02S + 0.10C + 0.06M &= 4{,}800 \end{aligned}$$

3-5 GAUSS–JORDAN ELIMINATION

The augmented matrix is

$$\begin{bmatrix} 1 & 1 & 1 & | & 70{,}000 \\ -2 & 1 & 0 & | & 0 \\ 0.02 & 0.10 & 0.06 & | & 4{,}800 \end{bmatrix}$$

Now use Gauss–Jordan elimination. Note that there is already a 1 in position 1, 1.

$$\begin{bmatrix} 1 & 1 & 1 & | & 70{,}000 \\ 0 & 3 & 2 & | & 140{,}000 \\ 0 & 0.08 & 0.04 & | & 3{,}400 \end{bmatrix} \quad \begin{array}{l} \text{row } 2 + (2 \times \text{row } 1) \\ \text{row } 3 - (0.02 \times \text{row } 1) \end{array}$$

$$\begin{bmatrix} 1 & 1 & 1 & | & 70{,}000 \\ 0 & 1 & \dfrac{2}{3} & | & \dfrac{140{,}000}{3} \\ 0 & 0.08 & 0.04 & | & 3{,}400 \end{bmatrix} \quad \left(\dfrac{1}{3}\right) \text{row } 2$$

$$\begin{bmatrix} 1 & 0 & \dfrac{1}{3} & | & \dfrac{70{,}000}{3} \\ 0 & 1 & \dfrac{2}{3} & | & \dfrac{140{,}000}{3} \\ 0 & 0 & -\dfrac{1}{75} & | & -\dfrac{1{,}000}{3} \end{bmatrix} \quad \begin{array}{l} \text{row } 1 - \text{row } 2 \\ \\ \text{row } 3 - (0.08 \times \text{row } 2) \end{array}$$

$$\begin{bmatrix} 1 & 0 & \dfrac{1}{3} & | & \dfrac{70{,}000}{3} \\ 0 & 1 & \dfrac{2}{3} & | & \dfrac{140{,}000}{3} \\ 0 & 0 & 1 & | & 25{,}000 \end{bmatrix} \quad -75 \text{ row } 3$$

$$\begin{bmatrix} 1 & 0 & 0 & | & 15{,}000 \\ 0 & 1 & 0 & | & 30{,}000 \\ 0 & 0 & 1 & | & 25{,}000 \end{bmatrix} \quad \begin{array}{l} \text{row } 1 - \left(\dfrac{1}{3} \times \text{row } 3\right) \\ \text{row } 2 - \left(\dfrac{2}{3} \times \text{row } 3\right) \end{array}$$

$S = 15{,}000$, $C = 30{,}000$, $M = 25{,}000$.

Ed thus has \$30,000 + \$25,000 = \$55,000 invested in bonds.

∎

The Gauss–Jordan elimination procedure we are using can also be used to solve any system of n linear equations in n unknowns. The same steps are used in the same order, but the process takes longer because the matrix is larger.

Note that Gauss–Jordan elimination is a very rigid procedure. At each stage of the process, there is only one possibility for the next step. Thus if several people each solve the same system of linear equations by Gauss–Jordan elimination, each person would perform precisely the same computations and obtain precisely the same matrices throughout the solution process.

Exercises 3-5

Determine whether or not each system of equations is independent. Use Gauss–Jordan elimination to solve each system that is independent.

1. $x + 3y - 2z = 11$
   $2x + 8y + 4z = 8$
   $4x + 5y - 3z = 27$

2. $x - y + 3z = -2$
   $4x + y + 22z = 17$
   $2x - 3y + 5z = -8$

3. $2x + 6y + 4z = 38$
   $3x + 9y + 7z = 59$
   $5x - 2y + 8z = 6$

4. $2y + 3z = 3$
   $x - y + 2z = 10$
   $2x + y + z = 2$

5. $x + 2y - 4z = 6$
   $3x + 6y - 8z = 15$
   $5x + 10y + 3z = 9$

6. $x + 3y + z = 9$
   $x - 2y - 3z = -12$
   $2x + 5y + z = 13$

7. $x - 2y + 2z = 2$
   $x - 2z = -16$
   $5x + 3y + z = -5$

8. $2x - 8y + 10z = 14$
   $x + 5y - 3z = 9$
   $4x + 2y + 4z = 32$

9. $2y - 4z = 12$
   $3x - 6y + 6z = -6$
   $4x + 5y + 8z = -8$

10. $x - 5y + 4z = 8$
    $2x - 9y + 6z = 12$
    $5x - 23y + 16z = 28$

11. $x - 6y + z = -1$
    $2x + 8y - z = 2$
    $-6x - 5y + 3z = 32$

12. $x + y + 3z = 16$
    $-3x + 5y - 25z = 32$
    $5x + 7y + 12z = 100$

13. $x + 3y - 6z = 12$
    $2x + 8y - 14z = 16$
    $4x + 12y - 24z = 48$

14. $x + 3y - 2z = 0$
    $-2x - 5y + 6z = 25$
    $4x + 2y - 3z = 0$

15. $\begin{aligned} 2x + 4y - 2z &= -4 \\ x + 3y + 2z &= 20 \\ -4x + 6y + 3z &= 15 \end{aligned}$

16. $\begin{aligned} x - 3y - z &= 4 \\ 4x - 11y &= 25 \\ 5x - 20z &= 5 \end{aligned}$

17. The Wigbee Company makes furniture. Each week, there are 3200 work-hours available in the assembly department, 3300 in the finishing department, and 3400 in the packaging department. Wigbee makes bookcases, tables, and armchairs. Each bookcase requires 2 hr of assembling, 1 hr of finishing, and 4 hr of packaging. Each table requires 4 hr of assembling, 3 hr of finishing, and 6 hr of packaging. Each armchair requires 6 hr of assembling, 7 hr of finishing, and 5 hr of packaging. How many of each item are made each week, if all available time is used in all departments?

18. Suppose the labor costs at Wigbee are $45 for a bookcase, $81 for a table, and $106 for an armchair. What is the labor cost for one work-hour in each department?

19. Juanita Westcott had a total income of $2640 last year from investments in bonds of grades AA, A, and B. The grade AA bonds paid 8% interest, the grade A bonds paid 10% interest, and the grade B bonds paid 12% interest. She had twice as much invested in grade AA and A bonds combined as in grade B bonds. She had a total of $27,000 invested. How much did she have invested in each grade of bond?

20. Tickets to a concert on campus cost $1.00 each for students, $1.50 for faculty, and $2.00 for others. Total attendance was 460 with total receipts of $570. Three times as many students as faculty attended. How many students attended?

Solve each system of equations by Gauss–Jordan elimination.

*21. $\begin{aligned} x_1 + x_2 + x_3 + x_4 &= 12 \\ x_1 + 2x_2 + 4x_3 - x_4 &= -1 \\ 2x_1 + 6x_2 + 13x_3 + 3x_4 &= 36 \\ -x_1 + 2x_2 + 6x_3 + x_4 &= 7 \end{aligned}$

*22. $\begin{aligned} 2x_1 + 4x_2 + 2x_3 + 2x_4 &= 0 \\ 2x_1 + 5x_2 + 3x_3 - x_4 &= -17 \\ 5x_1 + 11x_2 + 7x_3 + 2x_4 &= -15 \\ -2x_1 + 3x_2 - 5x_3 + 6x_4 &= 6 \end{aligned}$

Section 3-6 Inverse of a Matrix

If $bx = by$ and $b \neq 0$, both sides of the equation can be divided by b to obtain $x = y$. Another way to do this is to multiply both sides of $bx = by$ by $1/b$ or b^{-1}. If $BX = BY$ is a matrix equation, we can never divide both sides of the equation by B because division by a matrix is not defined. Under certain conditions, however, we can multiply both sides of the equation by a matrix called B^{-1} to obtain $X = Y$.

> **Definition.** An $n \times n$ matrix B is *invertible* if there is a matrix C such that $BC = CB = I_n$. The matrix C is called the *inverse* of B and is denoted B^{-1}.

If C is the inverse of B, then C is invertible and $C^{-1} = B$. Thus, B and C are inverses of each other. In particular, if B is invertible, then B^{-1} is invertible and

$$(B^{-1})^{-1} = B$$

According to the definition, to verify that two $n \times n$ matrices B and C are inverses of each other, it is necessary to verify that both $BC = I_n$ and $CB = I_n$. However, it can be shown that if one of these two equations is valid, the other equation must also be valid and B and C are inverses of each other. Thus,

$$\begin{bmatrix} 2 & 3 & 1 \\ 1 & 4 & 2 \\ 5 & 6 & 4 \end{bmatrix} \text{ and } \begin{bmatrix} \frac{4}{12} & -\frac{6}{12} & \frac{2}{12} \\ \frac{6}{12} & \frac{3}{12} & -\frac{3}{12} \\ -\frac{14}{12} & \frac{3}{12} & \frac{5}{12} \end{bmatrix}$$

are inverses of each other because

$$\begin{bmatrix} 2 & 3 & 1 \\ 1 & 4 & 2 \\ 5 & 6 & 4 \end{bmatrix} \begin{bmatrix} \frac{4}{12} & -\frac{6}{12} & \frac{2}{12} \\ \frac{6}{12} & \frac{3}{12} & -\frac{3}{12} \\ -\frac{14}{12} & \frac{3}{12} & \frac{5}{12} \end{bmatrix} = \begin{bmatrix} 1 & 0 & 0 \\ 0 & 1 & 0 \\ 0 & 0 & 1 \end{bmatrix}$$

Many matrices do not have inverses. For example, $\begin{bmatrix} 1 & 3 \\ 2 & 6 \end{bmatrix}$ has no inverse. To see this, suppose

$$\begin{bmatrix} 1 & 3 \\ 2 & 6 \end{bmatrix}^{-1} = \begin{bmatrix} w & x \\ y & z \end{bmatrix}$$

Then

$$\begin{bmatrix} 1 & 3 \\ 2 & 6 \end{bmatrix}\begin{bmatrix} w & x \\ y & z \end{bmatrix} = \begin{bmatrix} 1 & 0 \\ 0 & 1 \end{bmatrix}$$

so that

$$\begin{bmatrix} w + 3y & x + 3z \\ 2w + 6y & 2x + 6z \end{bmatrix} = \begin{bmatrix} 1 & 0 \\ 0 & 1 \end{bmatrix}$$

Equating entries gives four equations, two of which are:

$$w + 3y = 1$$
$$2w + 6y = 0$$

This pair of equations has no solution.

Going through this procedure for a general 2×2 matrix gives the following result:

The matrix $\begin{bmatrix} a & b \\ c & d \end{bmatrix}$ has no inverse if $ad - bc = 0$. If $ad - bc \neq 0$, then $\begin{bmatrix} a & b \\ c & d \end{bmatrix}^{-1} = \frac{1}{ad - bc}\begin{bmatrix} d & -b \\ -c & a \end{bmatrix}$.

Examples 1. $\begin{bmatrix} 4 & 5 \\ 2 & 3 \end{bmatrix}^{-1} = \frac{1}{(4 \times 3) - (5 \times 2)}\begin{bmatrix} 3 & -5 \\ -2 & 4 \end{bmatrix} = \frac{1}{2}\begin{bmatrix} 3 & -5 \\ -2 & 4 \end{bmatrix}$.

2. $\begin{bmatrix} 6 & 2 \\ 9 & 3 \end{bmatrix}^{-1}$ is not defined because $(6 \times 3) - (2 \times 9) = 0$.

3. $\begin{bmatrix} 3 & 8 \\ 4 & 15 \end{bmatrix}^{-1} = \frac{1}{(3 \times 15) - (8 \times 4)}\begin{bmatrix} 15 & -8 \\ -4 & 3 \end{bmatrix} = \frac{1}{13}\begin{bmatrix} 15 & -8 \\ -4 & 3 \end{bmatrix}$.

Now let's look at the relationship between matrix inversion and other matrix algebra.

If A and B have inverses, $A + B$ and $A - B$ need not have inverses, but AB must be invertible. Since $(AB)(B^{-1}A^{-1}) = A(BB^{-1})A^{-1} = AIA^{-1} = AA^{-1} = I$, it follows that $(AB)^{-1} = B^{-1}A^{-1}$.

In general, the product of invertible matrices is invertible. The inverse of the product is the product, in reverse order, of the inverses. For example, if A, B, and C are invertible $n \times n$ matrices, $(ABC)^{-1} = C^{-1}B^{-1}A^{-1}$.

If k is a nonzero real number and A is an invertible matrix, kA is invertible and $(kA)^{-1} = (1/k)A^{-1}$. For example, $(5A)^{-1} = (1/5)A^{-1}$.

Solving equations using the inverse matrix

We will now study the use of the inverse of a matrix in solving equations involving matrices, and in solving systems of linear equations.

Suppose $AX = AY$ and A has an inverse. Multiply both sides of this equation on the left by A^{-1}. Remember that multiplication of matrices is not commutative.

$$A^{-1}(AX) = A^{-1}(AY)$$
$$(A^{-1}A)X = (A^{-1}A)Y \quad \text{Associative law}$$
$$IX = IY$$
$$X = Y$$

Suppose A is invertible.

1. If $AX = AY$, then $X = Y$.
2. If $XA = YA$, then $X = Y$.
3. If $AX = YA$, no conclusion can be drawn.

Consider the system of linear equations

$$2x + 2y + 3z = 3$$
$$y + z = 2$$
$$x + y + z = 4$$

This system of equations may be written in the form

$$\begin{bmatrix} 2 & 2 & 3 \\ 0 & 1 & 1 \\ 1 & 1 & 1 \end{bmatrix} \begin{bmatrix} x \\ y \\ z \end{bmatrix} = \begin{bmatrix} 3 \\ 2 \\ 4 \end{bmatrix}$$

This is a matrix equation of the form $AX = B$, where A is a square matrix and X and B are column matrices. Multiply on the left by A^{-1} to solve for X.

$$AX = B$$
$$A^{-1}(AX) = A^{-1}B$$
$$(A^{-1}A)X = A^{-1}B$$
$$IX = A^{-1}B$$
$$X = A^{-1}B$$

Since

$$\begin{bmatrix} 2 & 2 & 3 \\ 0 & 1 & 1 \\ 1 & 1 & 1 \end{bmatrix}^{-1} = \begin{bmatrix} 0 & -1 & 1 \\ -1 & 1 & 2 \\ 1 & 0 & -2 \end{bmatrix},$$

(You should multiply the two matrices to verify this.)

the solution of the equation is

$$\begin{bmatrix} x \\ y \\ z \end{bmatrix} = \begin{bmatrix} 0 & -1 & 1 \\ -1 & 1 & 2 \\ 1 & 0 & -2 \end{bmatrix} \begin{bmatrix} 3 \\ 2 \\ 4 \end{bmatrix} = \begin{bmatrix} 2 \\ 7 \\ -5 \end{bmatrix}$$

that is, $x = 2$, $y = 7$, $z = -5$.

Any system of n linear equations in n unknowns can be written $AX = B$. Here A is the matrix of coefficients, X is the column matrix of unknowns, and B is the column matrix whose entries are the numbers on the right sides of the equations. If A is invertible, the system of equations is independent and has the solution $X = A^{-1}B$. If A is not invertible, the system of equations is dependent.

Therefore, to solve

$$2x + 3y = 12$$
$$5x - 2y = 11$$

by matrix inversion, write the system of equations as the single matrix equation

$$\begin{bmatrix} 2 & 3 \\ 5 & -2 \end{bmatrix} \begin{bmatrix} x \\ y \end{bmatrix} = \begin{bmatrix} 12 \\ 11 \end{bmatrix}$$

The inverse of $\begin{bmatrix} 2 & 3 \\ 5 & -2 \end{bmatrix}$ is

$$\frac{1}{2 \times (-2) - (3 \times 5)} \begin{bmatrix} -2 & -3 \\ -5 & 2 \end{bmatrix} = \frac{1}{-19} \begin{bmatrix} -2 & -3 \\ -5 & 2 \end{bmatrix}$$

$$= \begin{bmatrix} \frac{2}{19} & \frac{3}{19} \\ \frac{5}{19} & -\frac{2}{19} \end{bmatrix}$$

The solution of the equations is

$$\begin{bmatrix} x \\ y \end{bmatrix} = \begin{bmatrix} \frac{2}{19} & \frac{3}{19} \\ \frac{5}{19} & -\frac{2}{19} \end{bmatrix} \begin{bmatrix} 12 \\ 11 \end{bmatrix} = \begin{bmatrix} 3 \\ 2 \end{bmatrix}$$

that is, $x = 3$, $y = 2$. You should verify this by checking the answer in the original equations.

It is usually easier to solve a system of equations by the Gauss–Jordan method than by using the inverse, so that in practice the preceding method is not used to solve single systems of linear equations. However, in applications there is frequently a fixed matrix A and it is necessary to solve $AX = B$ for many different values of B. In this case, it is easier to solve the problem by inverting A. The reason is that A needs to be inverted only once, and then for each value of B it is only necessary to compute $A^{-1}B$. If the Gauss–Jordan method were used, it would be necessary to start over for each different value of B.

Finding the inverse of an invertible matrix

Now let's derive a procedure for finding the inverse of any invertible matrix. This procedure is very similar to Gauss–Jordan elimination. Suppose

$\begin{bmatrix} w & x \\ y & z \end{bmatrix}$ is the inverse of $\begin{bmatrix} a & b \\ c & d \end{bmatrix}$. Then

$$\begin{bmatrix} a & b \\ c & d \end{bmatrix}\begin{bmatrix} w & x \\ y & z \end{bmatrix} = \begin{bmatrix} 1 & 0 \\ 0 & 1 \end{bmatrix}$$

This last equation can be written as two equations:

$$\begin{bmatrix} a & b \\ c & d \end{bmatrix}\begin{bmatrix} w \\ y \end{bmatrix} = \begin{bmatrix} 1 \\ 0 \end{bmatrix} \quad \text{and} \quad \begin{bmatrix} a & b \\ c & d \end{bmatrix}\begin{bmatrix} x \\ z \end{bmatrix} = \begin{bmatrix} 0 \\ 1 \end{bmatrix}$$

This is true because when two matrices are multiplied, a column of the second matrix is used to obtain a column of the product. Now the matrix equation

$$\begin{bmatrix} a & b \\ c & d \end{bmatrix}\begin{bmatrix} w \\ y \end{bmatrix} = \begin{bmatrix} 1 \\ 0 \end{bmatrix}$$

is the same as the system of equations

$$aw + by = 1$$
$$cw + dy = 0$$

and has the augmented matrix $\begin{bmatrix} a & b & | & 1 \\ c & d & | & 0 \end{bmatrix}$.

The matrix equation $\begin{bmatrix} a & b \\ c & d \end{bmatrix}\begin{bmatrix} x \\ z \end{bmatrix} = \begin{bmatrix} 0 \\ 1 \end{bmatrix}$ has augmented matrix $\begin{bmatrix} a & b & | & 0 \\ c & d & | & 1 \end{bmatrix}$. To solve each of these two systems of equations, the matrix $\begin{bmatrix} a & b \\ c & d \end{bmatrix}$ must be reduced to $\begin{bmatrix} 1 & 0 \\ 0 & 1 \end{bmatrix}$. To save work and writing, augment the matrix $\begin{bmatrix} a & b \\ c & d \end{bmatrix}$ with two columns to obtain $\begin{bmatrix} a & b & | & 1 & 0 \\ c & d & | & 0 & 1 \end{bmatrix}$. When $\begin{bmatrix} a & b \\ c & d \end{bmatrix}$ is reduced to $\begin{bmatrix} 1 & 0 \\ 0 & 1 \end{bmatrix}$, this new augmented matrix is reduced to $\begin{bmatrix} 1 & 0 & | & w & x \\ 0 & 1 & | & y & z \end{bmatrix}$.

> To find the inverse of an $n \times n$ matrix A, form the augmented matrix $[A \mid I_n]$. If A reduces to I_n, then $[A \mid I_n]$ reduces to $[I_n \mid A^{-1}]$. If A does not reduce to I_n, then A is not invertible.

Example 4 Find $\begin{bmatrix} 4 & 5 \\ 2 & 3 \end{bmatrix}^{-1}$.

Solution
$\begin{bmatrix} 4 & 5 & | & 1 & 0 \\ 2 & 3 & | & 0 & 1 \end{bmatrix}$ Set up the proper augmented matrix.

$\begin{bmatrix} 1 & \frac{5}{4} & | & \frac{1}{4} & 0 \\ 2 & 3 & | & 0 & 1 \end{bmatrix}$ $\left(\frac{1}{4}\right)$ row 1

$\begin{bmatrix} 1 & \frac{5}{4} & | & \frac{1}{4} & 0 \\ 0 & \frac{2}{4} & | & -\frac{2}{4} & 1 \end{bmatrix}$ row 2 $- 2 \times$ row 1

$\begin{bmatrix} 1 & \frac{5}{4} & | & \frac{1}{4} & 0 \\ 0 & 1 & | & -1 & 2 \end{bmatrix}$ $2 \times$ row 2

$\begin{bmatrix} 1 & 0 & | & \frac{3}{2} & -\frac{5}{2} \\ 0 & 1 & | & -1 & 2 \end{bmatrix}$ row 1 $- \left(\frac{5}{4}\right)$ row 2

$\begin{bmatrix} 4 & 5 \\ 2 & 3 \end{bmatrix}^{-1} = \begin{bmatrix} \frac{3}{2} & -\frac{5}{2} \\ -1 & 2 \end{bmatrix}$

Example 5 Find $\begin{bmatrix} 1 & 2 & 1 \\ 1 & 3 & 2 \\ 1 & 0 & 1 \end{bmatrix}^{-1}$.

Solution
$\begin{bmatrix} 1 & 2 & 1 & | & 1 & 0 & 0 \\ 1 & 3 & 2 & | & 0 & 1 & 0 \\ 1 & 0 & 1 & | & 0 & 0 & 1 \end{bmatrix}$ Set up the proper augmented matrix.

$\begin{bmatrix} 1 & 2 & 1 & | & 1 & 0 & 0 \\ 0 & 1 & 1 & | & -1 & 1 & 0 \\ 0 & -2 & 0 & | & -1 & 0 & 1 \end{bmatrix}$ row 2 $-$ row 1
row 3 $-$ row 1

$$\begin{bmatrix} 1 & 0 & -1 & | & 3 & -2 & 0 \\ 0 & 1 & 1 & | & -1 & 1 & 0 \\ 0 & 0 & 2 & | & -3 & 2 & 1 \end{bmatrix} \quad \begin{array}{l} \text{row } 1 - 2 \times \text{row } 2 \\ \\ \text{row } 3 + 2 \times \text{row } 2 \end{array}$$

$$\begin{bmatrix} 1 & 0 & -1 & | & 3 & -2 & 0 \\ 0 & 1 & 1 & | & -1 & 1 & 0 \\ 0 & 0 & 1 & | & -\frac{3}{2} & 1 & \frac{1}{2} \end{bmatrix} \quad \left(\frac{1}{2}\right) \text{row } 3$$

$$\begin{bmatrix} 1 & 0 & 0 & | & \frac{3}{2} & -1 & \frac{1}{2} \\ 0 & 1 & 0 & | & \frac{1}{2} & 0 & -\frac{1}{2} \\ 0 & 0 & 1 & | & -\frac{3}{2} & 1 & \frac{1}{2} \end{bmatrix} \quad \begin{array}{l} \text{row } 1 + \text{row } 3 \\ \\ \text{row } 2 - \text{row } 3 \end{array}$$

$$\begin{bmatrix} 1 & 2 & 1 \\ 1 & 3 & 2 \\ 1 & 0 & 1 \end{bmatrix}^{-1} = \begin{bmatrix} \frac{3}{2} & -1 & \frac{1}{2} \\ \frac{1}{2} & 0 & -\frac{1}{2} \\ -\frac{3}{2} & 1 & \frac{1}{2} \end{bmatrix} \quad \begin{array}{l} \text{You should check this answer} \\ \text{by verifying that } AA^{-1} = I_3. \end{array}$$

Example 6 Find $\begin{bmatrix} 4 & 12 & 16 \\ 2 & 7 & 9 \\ 8 & 26 & 34 \end{bmatrix}^{-1}$

Solution

$$\begin{bmatrix} 4 & 12 & 16 & | & 1 & 0 & 0 \\ 2 & 7 & 9 & | & 0 & 1 & 0 \\ 8 & 26 & 34 & | & 0 & 0 & 1 \end{bmatrix} \quad \text{Set up the proper augmented matrix.}$$

$$\begin{bmatrix} 1 & 3 & 4 & | & \frac{1}{4} & 0 & 0 \\ 2 & 7 & 9 & | & 0 & 1 & 0 \\ 8 & 26 & 34 & | & 0 & 0 & 1 \end{bmatrix} \quad \left(\frac{1}{4}\right) \text{row } 1$$

$$\begin{bmatrix} 1 & 3 & 4 & | & \frac{1}{4} & 0 & 0 \\ 0 & 1 & 1 & | & -\frac{1}{2} & 1 & 0 \\ 0 & 2 & 2 & | & -2 & 0 & 1 \end{bmatrix} \quad \begin{array}{l} \text{row } 2 - 2 \times \text{row } 1 \\ \\ \text{row } 3 - 8 \times \text{row } 1 \end{array}$$

$$\begin{bmatrix} 1 & 0 & 1 \\ 0 & 1 & 1 \\ 0 & 0 & 0 \end{bmatrix} \begin{array}{|ccc} \frac{7}{4} & -3 & 0 \\ -\frac{1}{2} & 1 & 0 \\ -1 & -2 & 1 \end{array} \quad \begin{array}{l} \text{row } 1 - 3 \times \text{row } 2 \\ \\ \text{row } 3 - 2 \times \text{row } 2 \end{array}$$

The third row has all zeros to the left of the vertical line. Therefore the original matrix does not reduce to the identity matrix and is not invertible. ∎

Exercises 3-6

Find the inverse of each matrix, if possible

1. $\begin{bmatrix} 7 & 3 \\ -2 & 4 \end{bmatrix}$
2. $\begin{bmatrix} 6 & -2 \\ 5 & -8 \end{bmatrix}$

3. $\begin{bmatrix} -4 & 8 \\ 5 & 10 \end{bmatrix}$
4. $\begin{bmatrix} 3 & 2 \\ -1 & -6 \end{bmatrix}$

5. $\begin{bmatrix} -4 & 12 \\ -10 & 30 \end{bmatrix}$
6. $\begin{bmatrix} 3 & -7 \\ 2 & 8 \end{bmatrix}$

7. $\begin{bmatrix} 5 & 8 \\ 3 & 6 \end{bmatrix}$
8. $\begin{bmatrix} 14 & -2 \\ 7 & -1 \end{bmatrix}$

9. $\begin{bmatrix} 1 & 4 & 6 \\ 2 & 9 & 12 \\ 3 & 15 & 19 \end{bmatrix}$
10. $\begin{bmatrix} 1 & -2 & 3 \\ -1 & 3 & 5 \\ 2 & 6 & 10 \end{bmatrix}$

11. $\begin{bmatrix} -1 & 2 & 5 \\ 2 & -3 & 8 \\ 5 & -8 & 12 \end{bmatrix}$
12. $\begin{bmatrix} -1 & 4 & 7 \\ 1 & -5 & -6 \\ 3 & -12 & -20 \end{bmatrix}$

13. $\begin{bmatrix} 1 & 2 & 4 \\ -2 & 5 & 9 \\ -3 & 3 & 5 \end{bmatrix}$
14. $\begin{bmatrix} 0 & 1 & 3 \\ 1 & 0 & 5 \\ 3 & 4 & 2 \end{bmatrix}$

15. $\begin{bmatrix} 0 & 2 & 4 \\ 1 & 1 & 3 \\ 2 & 4 & 0 \end{bmatrix}$
16. $\begin{bmatrix} 1 & -2 & 1 \\ -1 & 4 & 3 \\ -2 & 12 & 14 \end{bmatrix}$

Suppose A, B, C, and D are invertible matrices. Write a formula for each of the following.

17. $(ABCD)^{-1}$
18. $(5A)^{-1}$
19. $(4BD)^{-1}$
20. $(B^{-1}C)^{-1}$

Solve each system of equations by finding the inverse of the matrix of coefficients.

21. $4x + 3y = 27$
    $3x - 2y = -1$

22. $2x - 3y = 2$
    $-x + 5y = 20$

23. $x - y + 2z = 12$
    $3x - 2y + 5z = 30$
    $2x + 3y + 7z = 26$

24. $x + 2y - 3z = 1$
    $2x + 3y + 5z = 23$
    $x - y - 4z = -4$

25. Solve exercise 17 of the previous section using matrix inversion.

The following exercises refer to exercise 17 of the previous section. The management of the Wigbee Company is considering slight changes in the sizes of the departments. One of the factors involved in the decision is the effect this would have on output. In exercises 26–30, determine the number of bookcases, tables, and armchairs produced each week. The three numbers given are the numbers of work-hours available each week in the assembly, finishing, and packaging departments, respectively.

26. 3400, 3500, 3600
27. 3000, 3100, 3200
28. 3200, 3300, 3380
29. 3260, 3400, 3400
30. 3240, 3350, 3400

*31. A square matrix A satisfies the equation $3A^2 + 5A + 2I = 0$. Show that A is invertible and find A^{-1}.

*32. A square matrix B satisfies $B^3 = 0$. Show that B is not invertible.

Find the inverse of each matrix.

33. $\begin{bmatrix} 2.75 & 4.93 \\ 1.86 & 10.17 \end{bmatrix}$

34. $\begin{bmatrix} 43.1 & -26.2 \\ 18.5 & 19.9 \end{bmatrix}$

◆ 35. $\begin{bmatrix} -0.124 & 0.376 \\ 0.485 & 0.216 \end{bmatrix}$ ◆ 36. $\begin{bmatrix} 3.92 & 85.48 \\ 1.36 & 37.25 \end{bmatrix}$

Section 3-7 Input-Output Analysis

Input-Output Analysis is concerned with the following problem. Any economic system is required to produce certain quantities of consumer goods. What must each industry do to accomplish this? Input-output analysis was developed by the economist Wassily Leontief. He received the Nobel Prize for Economics in 1973 for this work. (See *Time*, October 29, 1973, pages 113–114.)

To simplify the problem, assume that the economy has only two industries, called P and Q. The information necessary to do input-output analysis is contained in the *technological matrix* A:

$$\begin{array}{c} \\ \text{Producer P} \\ \text{Producer Q} \end{array} \begin{array}{cc} \text{User P} & \text{User Q} \\ \begin{bmatrix} 0.30 & 0.40 \\ 0.50 & 0.20 \end{bmatrix} \end{array} = A$$

Each industry is a producer as well as a consumer, since industries need raw materials, tools, energy, and so on. Each row of the technological matrix corresponds to an industry acting as a producer. Each column of the technological matrix corresponds to an industry acting as a consumer or user. The technological matrix is interpreted this way: to produce each dollar's worth of output, industry P needs to use $0.30 of its own output and $0.50 of the output of industry Q. To produce each dollar's worth of output, industry Q needs to use $0.40 of the output of industry P and $0.20 of its own output. Suppose that nonindustrial demand is for $180 of the output of industry P and for $360 of the output of industry Q. Let D be the *demand matrix*. D is the column matrix $\begin{bmatrix} 180 \\ 360 \end{bmatrix}$. Suppose industry P must produce output x_1 and industry Q output x_2 to meet this demand. Let

$$X = \begin{bmatrix} x_1 \\ x_2 \end{bmatrix}$$

Now total demand consists of two parts: industrial demand and nonindustrial demand. Now industrial demand for the output of industry P is $0.30x_1 + 0.40x_2$ and for industry Q is $0.50x_1 + 0.20x_2$. Thus total demand is $0.30x_1 + 0.40x_2 + 180$ for the output of industry P and $0.50x_1 + 0.20x_2 + 360$ for the output of industry Q. This gives the system of equations

3-7 INPUT-OUTPUT ANALYSIS

$$x_1 = 0.30x_1 + 0.40x_2 + 180$$
$$x_2 = 0.50x_1 + 0.20x_2 + 360$$

if the amount produced is to equal the amount demanded for each industry. In general, the matrix equation

$$X = AX + D$$

equates the amount produced with the amount demanded. In this equation, AX is the industrial demand, D is the nonindustrial demand, and X is the output matrix. Now solve this equation for X. The identity matrix of the same size as A is denoted by I.

$$X = AX + D$$
$$X - AX = D$$
$$IX - AX = D$$
$$(I - A)X = D$$
$$X = (I - A)^{-1}D$$

For our problem,

$$I - A = \begin{bmatrix} 1 & 0 \\ 0 & 1 \end{bmatrix} - \begin{bmatrix} 0.30 & 0.40 \\ 0.50 & 0.20 \end{bmatrix} = \begin{bmatrix} 0.7 & -0.4 \\ -0.5 & 0.8 \end{bmatrix}$$

$$(I - A)^{-1} = \begin{bmatrix} 0.7 & -0.4 \\ -0.5 & 0.8 \end{bmatrix}^{-1}$$

$$= \frac{1}{(0.7)(0.8) - (-0.4)(-0.5)} \begin{bmatrix} 0.8 & 0.4 \\ 0.5 & 0.7 \end{bmatrix}$$

$$= \frac{1}{0.36} \begin{bmatrix} 0.8 & 0.4 \\ 0.5 & 0.7 \end{bmatrix} = \frac{1}{36} \begin{bmatrix} 80 & 40 \\ 50 & 70 \end{bmatrix}$$

and then

$$X = \frac{1}{36} \begin{bmatrix} 80 & 40 \\ 50 & 70 \end{bmatrix} \begin{bmatrix} 180 \\ 360 \end{bmatrix} = \begin{bmatrix} 800 \\ 950 \end{bmatrix}$$

Industry P must produce $800 of output and industry Q must produce $950 of output.

Suppose the matrix D is changed to $\begin{bmatrix} 270 \\ 180 \end{bmatrix}$. Now

$$X = (I-A)^{-1}D = \frac{1}{36}\begin{bmatrix} 80 & 40 \\ 50 & 70 \end{bmatrix}\begin{bmatrix} 270 \\ 180 \end{bmatrix} = \begin{bmatrix} 800 \\ 725 \end{bmatrix}$$

Once the matrix $[I-A]^{-1}$ has been calculated, it is easy to determine X for many different values of D.

Example 1 In Northland, the economy has two industries: agriculture and manufacturing. Each dollar of agricultural output requires 25¢ of agricultural output and 60¢ of manufacturing output. Each dollar of manufacturing output requires 50¢ of agricultural output and 40¢ of manufacturing output. What should be produced, if nonindustrial demand is $1200 of agricultural output and $1500 of manufacturing output?

Solution The technological matrix A is $\begin{bmatrix} 0.25 & 0.50 \\ 0.60 & 0.40 \end{bmatrix}$.

$$I - A = \begin{bmatrix} 1 & 0 \\ 0 & 1 \end{bmatrix} - \begin{bmatrix} 0.25 & 0.50 \\ 0.60 & 0.40 \end{bmatrix} = \begin{bmatrix} 0.75 & -0.50 \\ -0.60 & 0.60 \end{bmatrix}$$

and

$$(I-A)^{-1} = \begin{bmatrix} 4 & \frac{10}{3} \\ 4 & 5 \end{bmatrix}$$

$$X = (I-A)^{-1}D = \begin{bmatrix} 4 & \frac{10}{3} \\ 4 & 5 \end{bmatrix}\begin{bmatrix} 1200 \\ 1500 \end{bmatrix} = \begin{bmatrix} 9800 \\ 12300 \end{bmatrix}$$

Production should be $9800 of agricultural output and $12,300 of manufacturing output. ∎

Exercises 3-7

Exercises 1 and 2 refer to Example 1.

1. What should be produced in Northland if nonindustrial demand is $1000 of agricultural output and no manufacturing output?

2. What should be produced in Northland if nonindustrial demand is $500 of agricultural output and $600 of manufacturing output?

3. What must industries A and B produce if the technological matrix is

$\begin{bmatrix} 0.1 & 0.7 \\ 0.4 & 0.2 \end{bmatrix}$ and nonindustrial demand is $600 of the output of industry A and $200 of the output of industry B?

4. Do exercise 3 if nonindustrial demand changes to $1000 of the output of industry A and $2000 of the output of industry B.

5. To produce $1 of industry C output requires $0.30 of its own output and $0.20 of industry D output. To produce $1 of industry D output requires $0.40 of industry C output and $0.10 of its own output. How much must each industry produce to meet a nonindustrial demand of $500 for the output of each industry?

6. To produce $1 of industry R output requires $0.30 of industry S output. To produce $1 of industry S output requires $0.20 of its own output and $0.10 of industry R output. How much must each industry produce, if nonindustrial demand is $800 for industry R output and $500 for industry S output?

Chapter 3 Review

Let $M = \begin{bmatrix} 6 & 3.9 & -3 & 5 \\ \pi & -2 & 4 & 4.7 \end{bmatrix}$

1. What size is M?
2. What is $m_{2,3}$?

Let $A = \begin{bmatrix} 2 & 4 \\ 3 & 8 \end{bmatrix}$, $B = \begin{bmatrix} 4 & 3 & 4 \\ 1 & 6 & 2 \\ 2 & 0 & 5 \end{bmatrix}$, $C = \begin{bmatrix} 1 & 0 & 2 \\ 2 & 1 & 3 \\ 5 & 2 & 9 \end{bmatrix}$, and

$D = \begin{bmatrix} 4 & 3 & 7 \\ 1 & 0 & 2 \end{bmatrix}$. Compute whichever of the following are defined.

3. $2B - C$
4. $A + 2D$
5. AD
6. BC
7. DA
8. A^{-1}
9. C^{-1}
10. D^{-1}

Solve each system of equations by Gauss–Jordan elimination.

11. $x + 3y = 9$
    $2x + 10y = 22$

12. $2x + 4y = 22$
    $5x + 3y = 20$

13. $x + y + z = 15$
    $x - y + 2z = 7$
    $2x + 5y - 6z = 3$

14. $4x + 8y = -12$
    $2x + 3y + z = 0$
    $3x + 5y + 4z = 3$

15. Use your answer to exercise 9 to solve the system of equations

$$x + 2z = 6$$
$$2x + y + 3z = 14$$
$$5x + 2y + 9z = 1$$

16. An economy with two industries, P and Q, has technological matrix $\begin{bmatrix} 6/14 & 9/14 \\ 2/14 & 3/14 \end{bmatrix}$. Nonindustrial demand is $10,000 of industry P output and $12,000 of industry Q output. How much does each industry need to produce?

The inventory matrix for John's Pet Stores is

	Downtown	East	West
Puppies	20	30	25
Kittens	15	25	35
Birds	10	20	15

The downtown store sells 6 puppies and 2 birds. The east side store sells 8 puppies, 5 kittens, and 6 birds. The west side store sells 12 puppies and 15 kittens.

17. Use matrix algebra to compute the inventory matrix after these sales.

18. Each puppy costs $100 wholesale, each kitten costs $50, and each bird costs $30. Use matrix algebra to write a matrix whose entries represent the cost of the original inventory at the different stores.

Linear Programming

Section 4-1 Systems of Linear Inequalities

Linear programming is a method of solving certain maximum and minimum problems. Linear programming may be used, for example, to determine how many of each possible item a factory should make in order to use its facilities and the available labor most profitably, or how much of each possible ingredient in chickenfeed should be used in order to feed the chickens properly and as cheaply as possible. We will study two methods of solving linear programming problems. The first method is the *graphical method*. The second method is a matrix procedure called the *simplex method*.

The material in this section, systems of linear inequalities, is needed for the graphical method and also has applications of its own.

Example 1 The Texxon Oil plant at Forestview makes regular and premium gasoline. To make 1 unit (100,000 gal) of regular requires 2 hr of cracking and 3 hr of refining. To make 1 unit of premium requires 3 hr of cracking and 3.5 hr of refining. Each day the cracking plant can be operated for up to 12 hr and the refinery for up to 16 hr. What are the possible outputs at Forestview?

Solution Let R = number of units of regular produced each day, and let P = number of units of premium produced each day. To produce R units of regular and P units of premium requires $2R + 3P$ hr of cracking and $3R + 3.5P$ hr of refining. Since only 12 hr of cracking and 16 hr of refining are available each day, R and P must satisfy the two inequalities:

$$2R + 3P \leq 12 \quad \text{Cracking}$$

$$3R + 3.5P \leq 16 \quad \text{Refining}$$

Since the plant does not manufacture negative amounts of gasoline, R and P must both be nonnegative:

$$R \geq 0$$
$$P \geq 0$$

The possible outputs at Forestview must thus satisfy four inequalities. Since this system of inequalities has infinitely many solutions, we will draw a graph of the solutions rather than attempt to list them. This is done at the end of the section.
∎

Before drawing the graph of a system of linear inequalities, let's determine the solution of a single linear inequality in two variables.

For example, to solve $3x + 4y \leq 24$, first draw the graph of $3x + 4y = 24$ by finding the intercepts. Now solve the inequality for y.

$$3x + 4y \leq 24$$

$$4y \leq 24 - 3x$$

$$y \leq 6 - \frac{3}{4}x$$

If a is any number, the point $(a, 6 - (3/4)a)$ lies on the line $3x + 4y = 24$. The point (a, y) satisfies the inequality exactly when $y \leq 6 - (3/4)a$. Thus a point on the vertical line $x = a$ satisfies the less than inequality when the point is below the point $(a, 6 - (3/4)a)$. This is the same as saying that when the point is below the line $3x + 4y = 24$. The solution of the inequality $3x + 4y \leq 24$ is thus the line $3x + 4y = 24$ together with all points below this line. The solution is shaded in Diagram 1.

> **Definition.** A *closed half-plane* is a line together with all points in the plane on one side of the line.

4-1 SYSTEMS OF LINEAR INEQUALITIES

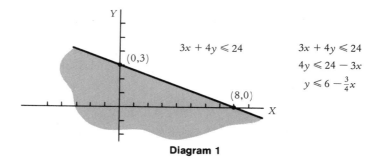

Diagram 1

$3x + 4y \leq 24$
$4y \leq 24 - 3x$
$y \leq 6 - \frac{3}{4}x$

The solution of a linear inequality $ax + by \leq c$ or $ax + by \geq c$ in two variables is always a closed half-plane. To solve the inequality $3x - 4y \geq 12$, draw the line $3x - 4y = 12$ by finding the intercepts.

The solution must be the line $3x - 4y = 12$ and all points on one side of the line. To decide which side of the line, test the origin. The origin does not satisfy the inequality because $3(0) - 4(0) \geq 12$ is false. The solution contains the side of the line not containing the origin and is shaded in Diagram 2. Any point not on the line $3x - 4y = 12$ could be tested instead of the origin. However, the arithmetic is always easiest when the origin is used. Also, it is easy to tell from the graph which side of the line contains the origin. From the graph can you easily decide which side of the line the point $(2, -2)$ is on?

To solve the linear inequality $2x + y \geq 0$, draw the line $2x + y = 0$ by plotting the origin and one other point. Since the origin is on this line, some

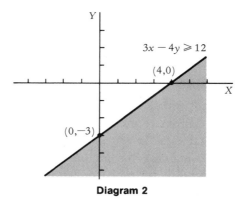

Diagram 2

other point must be tested to decide which side of the line satisfies the inequality. It is convenient to pick a point on the x-axis, for example, $(5,0)$. Since $2(5) + 0 > 0$, the solution contains the side of the line $2x + y = 0$ containing $(5,0)$. The solution is shaded in Diagram 3.

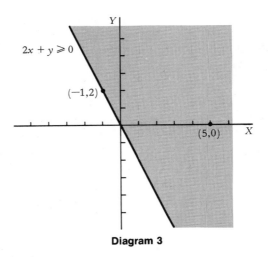

Diagram 3

Now we will study systems of linear inequalities. Our systems will always include the two inequalities $x \geq 0$ and $y \geq 0$, because in most applications of linear programming the variables cannot take negative values. These two inequalities are called *nonnegativity constraints*. Since the points satisfying $x \geq 0$ and $y \geq 0$ are exactly the points in the first quadrant, along with its boundary, we only need to look in the first quadrant.

Consider the system of linear inequalities.

$$3x + y \leq 6$$
$$2x + y \leq 5$$
$$x \geq 0, y \geq 0$$

First, draw the graphs of $3x + y = 6$ and $2x + y = 5$ on the same coordinate axes. See Diagram 4. Shade the part of the first quadrant that satisfies $3x + y \leq 6$ with vertical lines. This is the side of $3x + y = 6$ containing the origin. Shade the part of the first quadrant that satisfies $2x + y \leq 5$ with horizontal lines. This is the side of $2x + y = 5$ containing the origin. The solution of the system of linear inequalities is the intersection of the two shaded regions—the region shaded twice in Diagram 4. The *corners* or *vertices* of this region are $(0,0)$, $(2,0)$, $(1,3)$, and $(0,5)$. The vertex $(1,3)$ is found by solving the system of equations

$$3x + y = 6$$
$$2x + y = 5$$

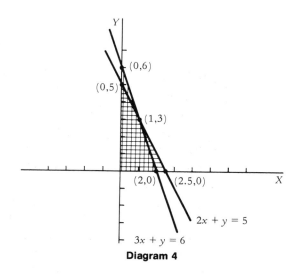

Diagram 4

Example 2 Solve the system of inequalities

$$2x + 5y \geq 10$$
$$3x + 2y \geq 7$$
$$x \geq 0, y \geq 0$$

Solution In Diagram 5, the lines $2x + 5y = 10$ and $3x + 2y = 7$ are drawn. The side of each of these lines opposite the origin is shaded. The solution is the doubly shaded region in Diagram 5. Note that the solution is unbounded. Two corners of the solution region are $(5,0)$ and $(0, 7/2)$. To obtain the other corner, solve

$$
\begin{aligned}
2x + 5y &= 10 \\
3x + 2y &= 7 \\
6x + 15y &= 30 \quad &\text{Multiply by 3.} \\
6x + 4y &= 14 \quad &\text{Multiply by 2 and subtract.} \\
\hline
11y &= 16
\end{aligned}
$$

164 CHAPTER 4 LINEAR PROGRAMMING

$$y = \frac{16}{11}$$

$$2x + 5\left(\frac{16}{11}\right) = 10 \qquad \text{Substitute } y = \frac{16}{11} \text{ in } 2x + 5y = 10.$$

$$2x + \frac{80}{11} = 10$$

$$2x = 10 - \frac{80}{11}$$

$$2x = \frac{110}{11} - \frac{80}{11} = \frac{30}{11}$$

$$x = \frac{15}{11}$$

The third corner is $(15/11, 16/11)$.

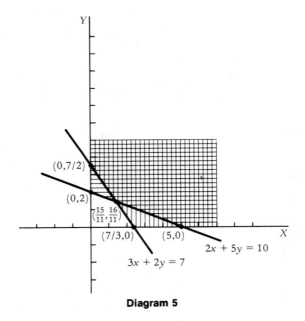

Diagram 5

Example 3 Solve the system of inequalities

$$x + 2y \leq 20$$
$$x - y \leq 8$$
$$y \leq 8$$
$$x \geq 0, y \geq 0$$

Solution The graphs of the lines $x + 2y = 20$, $x - y = 8$, and $y = 8$ are shown in Diagram 6. The origin satisfies the inequalities $x + 2y \leq 20$, $x - y \leq 8$, and $y \leq 8$. The side of each line containing the origin is shaded. The solution is the triply shaded region. The corners of that region are $(0, 8)$, $(0, 0)$, $(8, 0)$, A, and B. Point A is the solution of the system

$$y = 8$$
$$x + 2y = 20$$

and is $(4, 8)$. Point B is the solution of the system

$$x - y = 8$$
$$x + 2y = 20$$

and is $(12, 4)$.

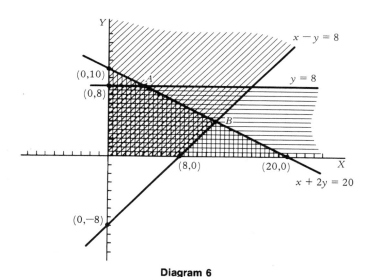

Diagram 6

∎

The solution of Example 1 is shown in Diagram 7. Every point in the shaded region represents possible outputs at Forestview. The points outside the shaded region are not possible outputs at Forestview. You should check that the lines, the region, and the corners are correct as shown. It would be a good idea to do this before beginning the exercises.

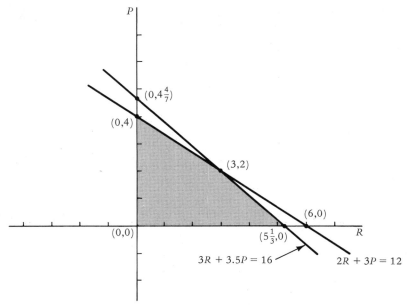

Diagram 7

Exercises 4-1

Draw the graph of each inequality.

1. $3x - 2y \leq 18$
2. $2x + 5y \leq 10$
3. $2x + 4y \geq 8$
4. $3x + y \geq 6$
5. $2x - y \leq 4$
6. $4x - 3y \leq 24$
7. $5x - 2y \leq 0$
8. $-2x + 3y \geq 0$

Solve each system of inequalities. On your graph, label each corner of the solution region with its coordinates. You should save your solutions for the next section.

9. $2x + y \leq 5$
   $x + 2y \leq 5$
   $x \geq 0, y \geq 0$

10. $4x + y \leq 9$
    $2x + 3y \leq 7$
    $x \geq 0, y \geq 0$

11. $3x + y \geq 6$
    $x + 2y \geq 7$
    $x \geq 0, y \geq 0$

12. $2x + 3y \geq 12$
    $x + y \geq 5$
    $x \geq 0, y \geq 0$

13. $2x + y \geq 8$
    $4x - y \leq 4$
    $x \geq 0, y \geq 0$

14. $x + y \geq 6$
    $x - 2y \leq 3$
    $x \geq 0, y \geq 0$

15. $3x + 2y \leq 18$
    $2x + 4y \leq 20$
    $x \geq 0, y \geq 0$

16. $x + 2y \geq 5$
    $3x + 4y \geq 11$
    $x \geq 0, y \geq 0$

17. $y \geq 2x$
    $2x + y \leq 8$
    $x \geq 0, y \geq 0$

18. $x + 4y \leq 9$
    $2x + 3y \leq 8$
    $x \geq 0, y \geq 0$

19. $3x + 2y \geq 12$
    $3x + 2y \leq 18$
    $y \leq 3x$
    $x \geq 0, y \geq 0$

20. $x + y \leq 6$
    $3x + 2y \leq 12$
    $x + 2y \leq 10$
    $x \geq 0, y \geq 0$

21. $x + 4y \leq 8$
    $4x + 5y \leq 20$
    $6x + y \leq 6$
    $x \geq 0, y \geq 0$

22. $x + 2y \leq 15$
    $x + 2y \geq 10$
    $y \geq 2x$
    $x \geq 0, y \geq 0$

23. $4x + y \geq 8$
    $3x + 2y \geq 12$
    $x + y \geq 5$
    $x \geq 0, y \geq 0$

24. $2x + y \leq 12$
    $x + 3y \leq 16$
    $2x + 3y \leq 24$
    $x \geq 0, y \geq 0$

In exercises 25–28, determine what are the variables. Write a system of inequalities, and draw the graph. You should save your solutions for the next section.

25. Pascal Toys manufactures a standard toy truck and a deluxe model. A standard truck takes 15 min to assemble and a deluxe truck takes 20 min to assemble. Painting takes 10 min for a standard truck and 30 min for a deluxe truck. What are the possible daily outputs, if 8 hr of assembly time and 9.5 hr of painting time are available daily?

26. Fred Cross owns a very large farm. He wants to plant beets and tomatoes. Seed, fertilizer, and other costs are $20 for an acre of beets and $10 for an acre of tomatoes. Each acre of beets requires one day of labor, but each acre of tomatoes requires two days of labor. Mr. Cross has $1400 and 160 days of labor available. How many acres of each plant could he grow?

27. Ms. Armbrewster is told by her doctor to supplement her diet with at least 20 mg of thiamine and 60 mg of iron daily. Rather than buy thiamine and iron separately, she instead decides to buy the vitamin

pills Ferritol and Vitamint. Each Ferritol pill contains 3 mg of thiamine and 8 mg of iron. Each Vitamint pill contains 1 mg of thiamine and 6 mg of iron. How many of each pill could she take daily?

28. John Toliver has 1000 cu ft of space available in his warehouse and $12,000 to invest. He expects the prices of motor oil and coffee to increase dramatically in the next year. Thus, he is planning to buy these commodities, store them in his warehouse for a year, and then sell them at a large profit. A case of motor oil costs $10 wholesale and requires 4 cu ft of space. A container of coffee costs $100 wholesale and occupies 2 cu ft of space. How much oil and coffee could John invest in?

Solve each system of inequalities. On your graph, label each corner of the solution region with its coordinates.

*29. $2x + 3y \leq 6$
$x + 5y \leq 5$
$4y \leq x$
$x \geq 0, y \geq 0$

*30. $4x + 2y \geq 8$
$x + y \geq 3$
$4x + 6y \geq 17$
$x \geq 0, y \geq 0$

Section 4-2 The Graphical Method

In Section 4-1, we learned how to draw the graph of the solution of a system of linear inequalities. In this section, the graphs are used to solve linear programming problems. The procedure will work because only a certain type of region can be the solution of a system of linear inequalities. A region S in the plane is *convex* if S contains the line segment that joins any two points of S. The solution of any system of linear inequalities is convex. The reasons are these. First, a half-plane is convex. Second, the solution of a system of linear inequalities is the intersection of half-planes. Third, the intersection of convex sets is convex.

The shaded regions in Diagram 1 could each be the solution of a system of linear inequalities. The shaded regions in Diagram 2 could not be because they are not convex.

Definition. A *linear programming* problem asks for the maximum or minimum value of an *objective function* $F = ax + by$ on the solution region of a system of linear inequalities. Each of the linear inequalities is called a *constraint*.

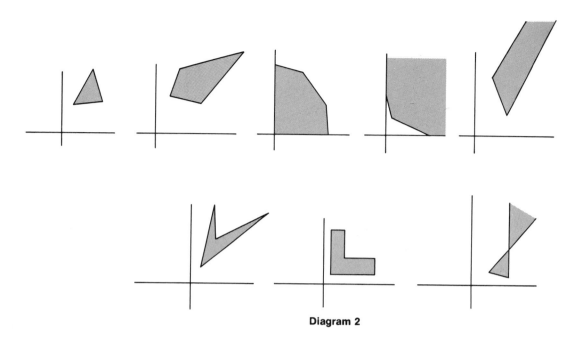

Diagram 2

In applications, the objective function often represents profit, revenue, or costs.

To solve such a problem by the *graphical method*, first determine the solution region S of the system of inequalities. Consider the problem

$$\text{Maximize } F = 4x + 3y$$
$$\text{subject to } 3x + y \leq 6$$
$$2x + y \leq 5$$
$$x \geq 0, y \geq 0$$

In the previous section, we determined the region S. In Diagram 3, S is shown, along with the lines $4x + 3y = k$ for several values of k. As k gets larger, the line $4x + 3y = k$ gets higher. The objective function F takes the value k on S precisely when the line $4x + 3y = k$ intersects S.

We can restate our problem as: What is the largest value of k for which the line $4x + 3y = k$ intersects S? Since S is convex, the line $4x + 3y = k$ intersects S in a line segment or in a point. In this specific problem, if a line $4x + 3y = k_1$ intersects S in a line segment, the line $4x + 3y = k_2$ intersects S if k_2 is a little larger than k_1, so that k_1 cannot be the maximum. The only possibilities for the maximum are the values k for which the line $4x + 3y = k$ intersects S in only one point. If a line intersects S in only one point, that

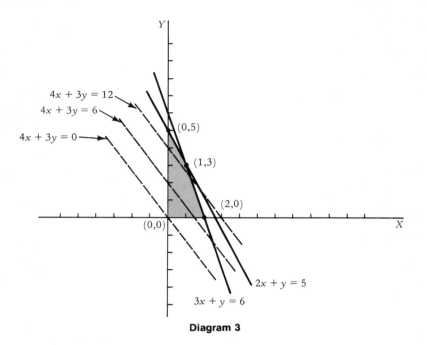

Diagram 3

point must be a corner of S. Thus the maximum value of the objective function is attained at a corner of S. Now look at Table 1. The corner points of S are listed, along with the values of F at these points. The maximum value of F is 15 and is attained at $(0,5)$. For further evidence that the maximum value of F is attained at a corner of S, you should evaluate F at some other points of S, such as $(1,2)$, $(0.5, 4)$, $(1.5, 1)$, $(1.5, 1.5)$, and $(0.2, 4.6)$.

TABLE 1

Corner Point	$F = 4x + 3y$
$(0,0)$	0
$(2,0)$	8
$(1,3)$	13
$(0,5)$	15

If the objective function were $G = 6x + 3y$, the highest line $6x + 3y = k$ to intersect S would be the line $6x + 3y = 15$. This line would intersect S in the line segment joining $(0,5)$ to $(1,3)$. The maximum value of 15 would now be attained at the corners $(0,5)$ and $(1,3)$ as well as at every point of the line segment joining them. See Diagram 4.

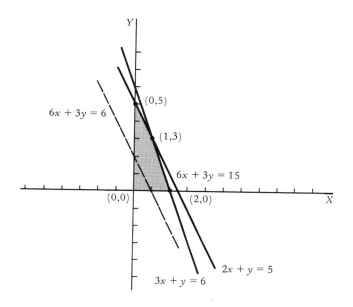

The general procedure for solving a linear programming problem by the graphical method is:

1. Draw the graph of the system of linear inequalities. The solution region is called the *feasible region*.
2. List the corners of the feasible region.
3. Evaluate the objective function at each corner.

The largest value of the objective function at a corner is the maximum. If this value is obtained at two corners, then it is attained at every point on the line segment joining them. Similarly, the smallest value of the objective function at a corner is the minimum.

Example 1 Maximize $3x + 7y$
subject to $6x + 5y \leq 100$
$8x + 5y \leq 120$
$x \geq 0, y \geq 0$

Solution Diagram 5 shows the feasible region.

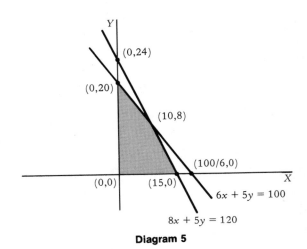

Diagram 5

The maximum is 140 and is attained at $(0, 20)$.

Example 2 Minimize $F = 3x + 2y$
subject to $11x + 5y \geq 75$
$\phantom{\text{subject to }}4x + 15y \geq 80$
$\phantom{\text{subject to }}x \geq 0,\ y \geq 0$

Solution

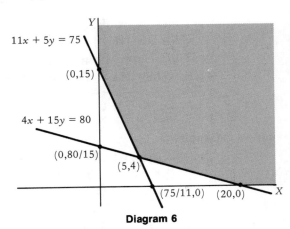

Diagram 6

Corner	$F = 3x + 2y$
$(0, 15)$	30
$(5, 4)$	23
$(20, 0)$	60

The minimum is 23 and is attained at $(5, 4)$.

Example 3 Maximize $F = 3x + 3y$
subject to $x + y \leq 12$
$\qquad\qquad 3x - y \leq 8$
$\qquad\qquad x \geq 0,\ y \geq 0$

Solution

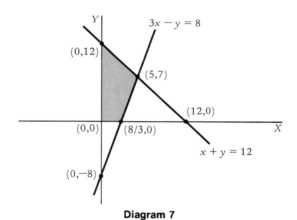

Diagram 7

Corner	$F = 3x + 3y$
$(0, 0)$	0
$\left(\dfrac{8}{3}, 0\right)$	8
$(5, 7)$	36
$(0, 12)$	36

The maximum is 36 and is attained at every point of the line segment joining $(5, 7)$ and $(0, 12)$.

Example 4 In Example 1 of Section 4-1, the output of the Texxon Oil plant at Forestview was discussed. The output was subject to constraints. These constraints formed a system of linear inequalities, which was graphed in

Diagram 7 of Section 4-1. Now suppose Texxon makes a profit of $20,000 on each unit of regular gasoline and $24,000 on each unit of premium. How many units of each kind of gasoline should be made each day in order to maximize profit?

Solution The profit F is given by

$$F = 20{,}000R + 24{,}000P$$

All that is necessary is to evaluate F at the corners of the feasible region. These corners were found in Section 4-1.

Corner	$F = 20{,}000R + 24{,}000P$
$(0,0)$	0
$(5\frac{1}{3},0)$	106,666.67
$(3,2)$	108,000
$(0,4)$	96,000

Texxon should make 3 units of regular and 2 units of premium at Forestview each day in order to make the maximum possible profit of $108,000.

■

Exercises 4-2

Could the region shown be the feasible region for a linear programming problem?

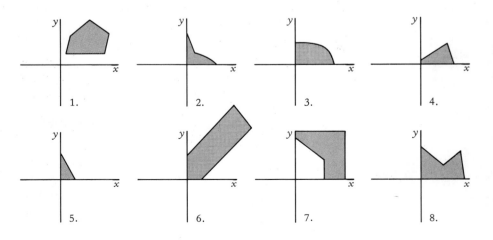

Find the maximum or minimum value of each objective function. In each exercise, the constraints are the inequalities in the problem with the same number in Exercises 4-1.

9. Maximize $6x + 24y$
10. Maximize $4x + 3y$
11. Minimize $10x + y$
12. Minimize $8x + 8y$
13. Minimize $5x + 2y$
14. Minimize $2x + y$
15. Maximize $2x + 4y$
16. Minimize $5x + 4y$
17. Maximize $5x + 2y$
18. Maximize $x + 2y$
19. Minimize $5x - y$
20. Maximize $2x + 4y$
21. Maximize $23x + 46y$
22. Maximize $3x + 2y$
23. Minimize $10x + 5y$
24. Maximize $x + y$
25. Maximize daily profit if profit is the profit is $5 on a standard toy truck and $7 on a deluxe toy truck.
26. Maximize profit if profit is $80 on an acre of beets and $70 on an acre of tomatoes.
27. Minimize daily cost if each Ferritol pill costs 8¢ and each Vitamint pill costs 5¢.
28. Maximize profit if motor oil sells for $15.00 a case one year later and coffee sells for $135.00 a container then.
*29. Maximize $6x + y$
*30. Minimize $3x + 2y$

Section 4-3 Applications

When an applied problem in linear programming is given, the first step is to identify the variables. These are the quantities whose values are to be selected by a decision maker. Then the constraints must be listed. There are two types of constraint. One type of constraint says that only a certain amount of something is available, for example, that 160 man-hours of labor are available each day. This type of constraint may instead say that a certain amount of something is needed, for example, that 50 chairs must be manufactured each day to meet a prior commitment. The other type of constraint is a nonnegativity constraint. This says that a variable cannot be negative. For example, a factory cannot manufacture a negative number of tables. The next step is to write the objective function and to determine whether the objective function is to be maximized or minimized.

The problem has now been set up and must be solved. In this section, we will use the graphical method. The graphical method works only when there are two variables. If there are more than two variables, the graph of the system of inequalities cannot be drawn in the plane—in this case, the simplex method must be used.

Steps in Solving an Applied Linear Programming Problem

1. Identify the variables.
2. List the constraints, including nonnegativity constraints.
3. Write the objective function. Determine whether the objective function is to be maximized or minimized.
4. Solve the problem by the graphical method or the simplex method.

Example 1 Ski Unlimited makes two kinds of skis: racing and cross-country. A pair of racing skis requires 6 hours of fabrication and 1 hour of finishing. A pair of cross-country skis requires 4 hours of fabrication and 1 hour of finishing. Profit is $40 on a pair of racing skis and $30 on a pair of cross-country skis. Because of other products being made, only 108 hours of fabrication time and 24 hours of finishing time are available each week. How many pairs of each kind of skis should be made each week in order to maximize profit?

Solution 1. The variables are

r = number of pairs of racing skis made each week.
c = number of pairs of cross-country skis made each week.

2. The amount of fabrication time needed to make these skis is $6r + 4c$.

One constraint, then, is $6r + 4c \leq 108$. This says that no more fabrication time is used than is available.
The amount of finishing time needed to make these skis is $r + c$. Another constraint is $r + c \leq 24$. This says that no more finishing time is used than is available.
The nonnegativity constraints are $r \geq 0$ and $c \geq 0$. These say that the plant cannot make a negative number of either type of skis.

3. The objective function is profit. Call this objective function P. Then $P = 40r + 30c$. The problem is

Maximize $P = 40r + 30c$
subject to $6r + 4c \leq 108$ Fabrication
$r + c \leq 24$ Finishing
$r \geq 0, c \geq 0$ Nonnegativity

4. The feasible region is shown in Diagram 1.

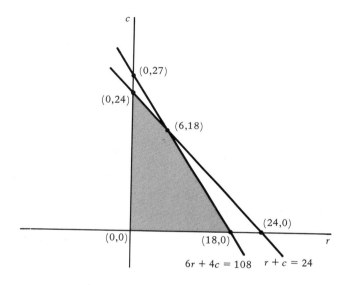

The maximum possible profit is $780. This is achieved by making 6 pairs of racing skis and 18 pairs of cross-country skis each week.

If profit is maximized ($r = 6$, $c = 18$), all available fabrication time and finishing time is used and there is no slack.

Example 2 Suppose now that demand changes so that prices change. The profit is now $50 on a pair of racing skis but remains at $30 on a pair of cross-country skis. How many pairs of each kind of skis should now be made each week in order to maximize profit?

Solution The variables, constraints, and feasible region remain the same. The only change is in the objective function. The objective function P is now $P = 50r + 30c$. A new diagram is not needed.

Corner	$P = 50r + 30c$
(0, 0)	0
(18, 0)	900
(6, 18)	840
(0, 24)	720

The maximum profit possible now is $900. This is achieved by making 18 pairs of racing skis each week and no cross-country skis. If this is done, $6(18) + 4(0) = 108$ hr of production time are used each week. This is all of the available production time. $18 + 0 = 18$ hr of finishing time are used each week. Since 24 hr of finishing time are available, 6 hr of finishing time are available but are not used. We say that these 6 hr of finishing time are *slack*.

Example 3 Suppose the profit remains at $30 on a pair of cross-country skis. What is the minimum profit on a pair of racing skis, at which Ski Unlimited should make racing skis only?

Solution Suppose p is the profit on a pair of racing skis. If p is $40, we found that Ski Unlimited should make both kinds of skis. If p is $50, Ski Unlimited should make racing skis only.

The profit is $18p$ if only racing skis are made and $6p + 540$ if both kinds of skis are made. Racing skis only should be made if

$$18p > 6p + 540$$
$$12p > 540$$
$$p > 45$$

If $p > 45$, Ski Unlimited should make racing skis only. If $p = 45$, Ski Unlimited could make 18 pairs of racing skis, or 6 pairs of racing skis and 18 pairs of cross-country skis, or Ski Unlimited could also use any point on the line segment joining (18, 0) with (6, 18) to determine what to make.
∎

The solution of the last problem shows a certain instability that occurs in economics. Assume the profit on a pair of cross-country skis remains at $30. Suppose the profit on a pair of racing skis fluctuates, so that it is rarely exactly $45 but is usually close to $45. Ski Unlimited will then keep changing from making racing skis only (the point $(18, 0)$ on the graph) to making both types of skis (the point $(6, 18)$ on the graph). This is a large change. Ski Unlimited will rarely, if ever, adopt a compromise.

Example 4 An animal food is to be composed of two ingredients: grain and silage. Each pound of grain contains 2400 calories, 20 g of protein, and costs 12¢. Each pound of silage contains 600 calories, 2 g of protein, and costs 2¢. An animal requires at least 17,400 calories and 130 g of protein each day. How cheaply can the animal be fed?

Solution 1. The variables are

$$g = \text{number of pounds of grain each day.}$$
$$s = \text{number of pounds of silage each day.}$$

2. The constraints are

$$2400g + 600s \geq 17400 \quad \text{Calories}$$
$$20g + 2s \geq 130 \quad \text{Protein}$$
$$g \geq 0, s \geq 0 \quad \text{Nonnegativity}$$

3. The objective function to be minimized is cost: $C = 12g + 2s$. See

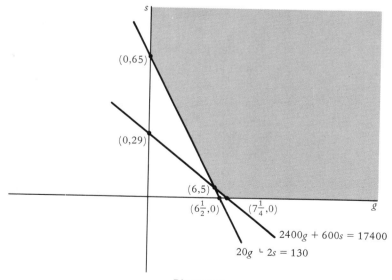

Diagram 2

Diagram 2 for the graph.

4.
Corner	$C = 12g + 2s$
$(0, 65)$	130
$(6, 5)$	82
$(7\frac{1}{4}, 0)$	87

The animal should be fed 6 lb of grain and 5 lb of silage each day at a cost of 82¢.

∎

Exercises 4-3

1. Dennis makes furniture in his spare time. He needs two hours to make a magazine rack and three hours to make a small bookcase. He needs one plywood board to make a magazine rack and two plywood boards to make a bookcase. He has seven plywood boards available and has 12 hours to work. His profit is $5 on a magazine rack and $8 on a bookcase. He wishes to maximize profit.
    a. How many magazine racks and how many bookcases should he make?
    b. Does he have any plywood or any time left over?
    c. How much profit does he make?
    d. If profit increases to $6 on a magazine rack but remains at $8 on a bookcase, how many magazine racks and how many bookcases should Dennis make? Does he have any plywood or any time left over? How much profit does he make?

2. Farmer Jones has 100 acres of land in which to plant corn and soybeans. Each acre of corn requires 2 bushels of storage space and yields a profit of $40. Each acre of soybeans requires 3 bushels of storage space and yields a profit of $80. Farmer Jones has 260 bushels of storage space available. He wishes to maximize profit.
    a. How many acres of each crop should he plant?
    b. How much profit does he make?
    c. Does he have any unused land or storage space?
    d. Suppose his profit now changes to $50 per acre for both crops. How much profit does he make now? How many acres of each crop should he plant?

3. A fertilizer is to contain two ingredients, Grow and Bloom. Each pound

of Grow contains 2 oz of nitrogen and 2 oz of phosphates. Each pound of Bloom contains 2 oz of nitrogen and 5 oz of phosphates. The final product must contain at least 70 oz of nitrogen and 100 oz of phosphates. Each pound of Grow costs $1 and each pound of Bloom costs $1.50. It is desired to minimize the cost of the fertilizer.
a. What is the minimum possible cost?
b. How many pounds of each ingredient are used?
c. Suppose the cost of Grow remains at $1 per pound. How high can the cost of Bloom go before you are forced to use Grow only?

4. Repeat exercise 3 if the formula for Grow is changed so that one pound of Grow now contains 4 oz of nitrogen and 2 oz of phosphates.

5. The Euclid Nut Company has 60 lb of cashews and 100 lb of peanuts to use to make 5-lb boxes of mixed nuts. A box of Deluxe Mix requires 4 lb of peanuts and 1 lb of cashews and sells for $8.00. A box of Elegant Mix requires 2 lb of peanuts and 3 lb of cashews and sells for $12.50. How many boxes of each mix must be made in order to maximize revenue? Are any nuts left over?

6. A garbage truck carries radioactive wastes in sealed containers. Each container from Northeast Nuclear weighs 20 lb and is 5 cu ft in volume. Each container from First Nuclear weighs 20 lb and is 3 cu ft in volume. The truck can only carry 800 lb of radioactive waste containers, and only 150 cu ft of space is available. The charge for carrying a container from Northeast Nuclear is $250 and the charge for carrying a container from First Nuclear is $220. What is the maximum possible revenue for the garbage truck? Is there any empty space remaining in the truck?

7. Lutz Watercraft makes two types of inflatable boats, a two-person boat and a four-person boat. Each two-person boat requires 9 work-hours in the cutting department and 8 work-hours in the assembly department. Each four-person boat requires 18 work-hours in the cutting department and 12 work-hours in the assembly department. Each month, there are 3600 work-hours available in the cutting department and 2800 work-hours available in the assembly department. Profit is $500 on a two-person boat and $700 on a four-person boat. The manager wishes to maximize profit.
a. What is the maximum possible profit?

b. To maximize profit, what should be made?

c. If profit is maximized, are any available resources not used?

8. Repeat exercise 7 if profit decreases to $400 on a two-person boat.

9. Greensod lawn seed mixture contains bluegrass seeds and rye seeds. Bluegrass seeds cost 30¢ an ounce and rye seeds cost 15¢ an ounce. Greensod has already agreed to deliver 600 lb of the mixture to a customer. The mixture must contain at least 60% bluegrass. What is the minimum possible cost of the mixture?

10. Downgoods has 90 lb of down and 160 hours of labor available each day. To make a sleeping bag requires 2 lb of down and 3 hr of labor. To make a parka requires 1 lb of down and 2 hr of labor. How many sleeping bags and parkas should Downgoods manufacture each day in order to maximize profit? Profit is $25 on a sleeping bag and $20 on a parka. Is any down left over? Is all the labor used?

11. Farmer Smith has 1000 acres in which to plant alfalfa and wheat. Government regulations prohibit him for planting more than 400 acres of wheat. To help preserve his soil, he wants to plant at least as many acres in alfalfa as in wheat. His profit is $10 on each acre of alfalfa and $24 on each acre of wheat. He desires to maximize his profit.

a. How many acres of wheat should he plant?

b. What is his profit?

c. Without government regulations, how many acres of wheat would he plant and how much profit would he make?

12. Union Typewriter makes manual and electric typewriters. The company can make up to 250 manual typewriters and up to 200 electric typewriters daily, but cannot make more than 300 typewriters in a day. Each day, 700 man-hours of labor are available. To make an electric typewriter requires 3 man-hours of labor but a manual typewriter requires only 2. Profit is $30 on a manual typewriter and $50 on an electric typewriter. The company wishes to maximize profit.

a. How many manual typewriters and how many electric typewriters should be made each day?

b. How many available man-hours of labor are not used each day?

c. What is the profit?

Section 4-4 The Simplex Method

The *simplex method* is a procedure used to solve linear programming problems. Unlike the graphical method, this method can be used in problems involving any number of variables. The simplex method is a matrix method and is easily used on a computer. In this section, maximization problems in which each constraint is a nonnegativity constraint or is of the form (expression involving variables) $\leq$ (nonnegative constant) are considered.

Because the simplex method is long, we will study it one step at a time.

Step 1 The first step in the simplex method is to convert inequalities to equalities by introducing *slack variables*.

Example 1 Maximize $F = x + 5y + 7z$
subject to $2x + y + 3z \leq 40$
$x + 2y + z \leq 45$
$x \geq 0, y \geq 0, z \geq 0$

Solution Two slack variables u and v are needed. One slack variable is needed for each constraint other than a nonnegativity constraint. The first two inequalities can now be written

$$2x + y + 3z + u = 40$$
$$x + 2y + z + v = 45$$

The variables u and v are called slack variables because they "take up the slack." An inequality like $x + 2y + z \leq 45$ in a linear programming problem might state that 45 units of some commodity are available. The variable v represents the difference between 45 and the number of units actually used. That is, v is the number of available units that are not used. The variables u and v must be nonnegative in order that x, y, and z satisfy the original inequalities.

The equation $F = x + 5y + 7z$ can be rewritten for later convenience as $F - x - 5y - 7z = 0$. The system for this example is now written

$$2x + y + 3z + u = 40$$
$$x + 2y + z + v = 45$$
$$F - x - 5y - 7z = 0$$
$$x \geq 0, y \geq 0, z \geq 0, u \geq 0, v \geq 0$$

∎

Step 2 The second step in the simplex method is to write the *initial simplex matrix*. To obtain the initial simplex matrix, ignore the nonnegativity constraints and the symbol F and make the remaining numbers into a matrix. This matrix is essentially the augmented matrix of a system of linear equations. If a variable does not appear in an equation, it has a coefficient of 0 in that equation. At the top of the matrix, we have written for convenience the variable involved in each column for Example 1.

$$\begin{array}{ccccc} x & y & z & u & v \\ \end{array}$$
$$\begin{bmatrix} 2 & 1 & 3 & 1 & 0 & | & 40 \\ 1 & 2 & 1 & 0 & 1 & | & 45 \\ -1 & -5 & -7 & 0 & 0 & | & 0 \end{bmatrix}$$

The last row in the matrix is called the *objective row* and means $F - x - 5y - 7z = 0$, or $F = x + 5y + 7z$.

In each of the following linear-programming examples, we introduce slack variables and then write the initial simplex matrix.

Example 2 Maximize $F = 3x + y$
subject to $x + 2y \le 7$
$4x + 2y \le 10$
$2x + 3y \le 12$
$x \ge 0, y \ge 0$

Solution Introduce three slack variables u, v, and w and rewrite the system.

$$x + 2y + u = 7$$
$$4x + 2y + v = 10$$
$$2x + 3y + w = 12$$
$$F - 3x - y = 0$$
$$x \ge 0, y \ge 0, u \ge 0, v \ge 0, w \ge 0$$

The initial simplex matrix is:

$$\begin{bmatrix} x & y & u & v & w & \\ 1 & 2 & 1 & 0 & 0 & 7 \\ 4 & 2 & 0 & 1 & 0 & 10 \\ 2 & 3 & 0 & 0 & 1 & 12 \\ -3 & -1 & 0 & 0 & 0 & 0 \end{bmatrix}$$

Example 3 Maximize $F = x + 10y + 5z$
subject to $2x + y + 4z \leq 18$
$x + 2y + 6z \leq 45$
$5x + 3y + 7z \leq 60$
$x \geq 0,\ y \geq 0,\ z \geq 0$

Solution Introduce three slack variables and rewrite the system.

$$2x + y + 4z + u = 18$$
$$x + 2y + 6z + v = 45$$
$$5x + 3y + 7z + w = 60$$
$$F - x - 10y - 5z = 0$$
$$x \geq 0,\ y \geq 0,\ z \geq 0,\ u \geq 0,\ v \geq 0,\ w \geq 0$$

The initial simplex matrix is

$$\begin{bmatrix} x & y & z & u & v & w & \\ 2 & 1 & 4 & 1 & 0 & 0 & 18 \\ 1 & 2 & 6 & 0 & 1 & 0 & 45 \\ 5 & 3 & 7 & 0 & 0 & 1 & 60 \\ -1 & -10 & -5 & 0 & 0 & 0 & 0 \end{bmatrix}$$

■

Step 3 The third step in the simplex method is to locate an entry called the *pivot* or *pivotal entry*. To do this,

1. Find the most negative number in the objective row. This means the negative number with largest absolute value. The column containing this number is called the *pivotal column*.

   For the initial simplex matrix of Example 2, the first column is the

pivotal column because -3 is the most negative of the numbers $-3, -1,$ 0, 0, 0, and 0.
2. For each positive element in the pivotal column, form the ratio of the entry in the same row in the last column to the element in the pivotal column. The row in which this ratio is smallest is the *pivotal row*.

For Example 2, form the ratios $7/1$, $10/4$, and $12/2$. The smallest of these ratios is $10/4$, so that the second row is the pivotal row.
3. The *pivotal entry* or *pivot* is the entry in the intersection of the pivotal column and the pivotal row.

For the initial simplex matrix of Example 2, the pivot is 4. This is the entry in the second row and the first column, that is, in position 2, 1.

In Example 3, the y-column is the pivotal column because -10 is the most negative of the numbers $-1, -10, -5, 0, 0, 0,$ and 0. The ratios $18/1$, $45/2$, and $60/3$ are formed. The smallest ratio is $18/1$, so that the first row is the pivotal row and the 1 in position 1, 2 is the pivot.

Step 4 The fourth step in the simplex method is to perform a procedure called *pivoting*. This procedure is a combination of several row operations. These are the same row operations that we used in Chapter 3 to solve systems of linear equations by Gauss–Jordan elimination.

1. Divide the pivotal row by the pivotal entry. There is now a 1 in the pivotal position.
2. Add suitable multiples of the (new) pivotal row to each other row so that each entry in the pivotal column, except the pivot, becomes 0.

For the initial simplex matrix of Example 1,

$$\begin{array}{ccccc} x & y & z & u & v \\ \end{array}$$
$$\begin{bmatrix} 2 & 1 & \boxed{3} & 1 & 0 & | & 40 \\ 1 & 2 & 1 & 0 & 1 & | & 45 \\ -1 & -5 & -7 & 0 & 0 & | & 0 \end{bmatrix}$$

the pivotal column is the z-column. Form the ratios $40/3$ and $45/1$. Since $40/3$ is smaller, the pivotal entry is 3. Now perform the pivoting for Example 1.

4-4 THE SIMPLEX METHOD 187

$$\begin{bmatrix} x & y & z & u & v & \\ \frac{2}{3} & \frac{1}{3} & 1 & \frac{1}{3} & 0 & \frac{40}{3} \\ \frac{1}{3} & \boxed{\frac{5}{3}} & 0 & -\frac{1}{3} & 1 & \frac{95}{3} \\ \frac{11}{3} & -\frac{8}{3} & 0 & \frac{7}{3} & 0 & \frac{280}{3} \end{bmatrix} \begin{array}{l} \left(\frac{1}{3}\right) \text{ row 1} \\ \text{row 2} - \text{row 1} \\ \text{row 3} + 7 \text{ row 1} \end{array}$$

Step 5 The fifth step is to repeat steps 3 and 4 until the objective row contains no negative numbers. Since $-8/3$ is the only negative number in the last row of the matrix we formed in step 4, the y-column is the pivotal column. Form the ratios $(40/3)/(1/3) = 40$ and $(95/3)/(5/3) = 19$. Since 19 is smaller, the second row is the pivotal row and $5/3$ is the pivot. Now do the pivoting.

$$\begin{bmatrix} x & y & z & u & v & \\ \frac{3}{5} & 0 & 1 & \frac{2}{5} & -\frac{1}{5} & 7 \\ \frac{1}{5} & 1 & 0 & -\frac{1}{5} & \frac{3}{5} & 19 \\ \frac{21}{5} & 0 & 0 & \frac{9}{5} & \frac{8}{5} & 144 \end{bmatrix} \begin{array}{l} \text{row 1} - \left(\frac{1}{3}\right) \text{ row 2} \\ \left(\frac{3}{5}\right) \text{ row 2} \\ \text{row 3} + \left(\frac{8}{3}\right) \text{ row 2} \end{array}$$

Since the objective row now contains no negative entries, step 5 is complete and the matrix just obtained is the *final simplex matrix*.

Step 6 The sixth and last step is to interpret the final simplex matrix. The last entry in the objective row, 144 in this matrix, is the maximum possible value of the objective function. Find the value of each variable when this maximum is obtained as follows.

1. If a column is all zeros except for a single 1, the last entry in the row containing this 1 is the value of the corresponding variable.
2. Otherwise the variable has the value 0.

Thus $x = 0$, $y = 19$, $z = 7$, $u = 0$, and $v = 0$.

In Example 2, the initial simplex matrix was found to be

$$\begin{bmatrix} x & y & u & v & w & \\ 1 & 2 & 1 & 0 & 0 & 7 \\ \boxed{4} & 2 & 0 & 1 & 0 & 10 \\ 2 & 3 & 0 & 0 & 1 & 12 \\ -3 & -1 & 0 & 0 & 0 & 0 \end{bmatrix}$$

and the number 4 was the pivot. Pivoting gives the matrix

$$\begin{bmatrix} x & y & u & v & w & \\ 0 & \frac{3}{2} & 1 & -\frac{1}{4} & 0 & \frac{9}{2} \\ 1 & \frac{1}{2} & 0 & \frac{1}{4} & 0 & \frac{5}{2} \\ 0 & 2 & 0 & -\frac{1}{2} & 1 & 7 \\ 0 & \frac{1}{2} & 0 & \frac{3}{4} & 0 & \frac{15}{2} \end{bmatrix} \begin{array}{l} \text{row 1} - \text{row 2} \\ \left(\frac{1}{4}\right) \text{row 2} \\ \text{row 3} - 2(\text{row 2}) \\ \text{row 4} + 3(\text{row 2}) \end{array}$$

Since there are no negative numbers in the objective row, this is the final simplex matrix. The maximum value of the objective function is $15/2$. This is attained when $x = 5/2$ and $y = 0$. At this point, $u = 9/2$, $v = 0$, and $w = 7$.

In Example 3, the initial simplex matrix was found to be

$$\begin{bmatrix} x & y & z & u & v & w & \\ 2 & \boxed{1} & 4 & 1 & 0 & 0 & 18 \\ 1 & 2 & 6 & 0 & 1 & 0 & 45 \\ 5 & 3 & 7 & 0 & 0 & 1 & 60 \\ -1 & -10 & -5 & 0 & 0 & 0 & 0 \end{bmatrix}$$

and the pivot was the 1 in position 1, 2. Pivoting gives the matrix

$$\begin{bmatrix} x & y & z & u & v & w & \\ 2 & 1 & 4 & 1 & 0 & 0 & 18 \\ -3 & 0 & -2 & -2 & 1 & 0 & 9 \\ -1 & 0 & -5 & -3 & 0 & 1 & 6 \\ 19 & 0 & 35 & 10 & 0 & 0 & 180 \end{bmatrix} \begin{array}{l} \text{row 1} \\ \text{row 2} - 2(\text{row 1}) \\ \text{row 3} - 3(\text{row 1}) \\ \text{row 4} + 10(\text{row 1}) \end{array}$$

This is the final simplex matrix. The maximum possible value of the objective function is 180. When the objective function is maximized, $x = 0$, $y = 18$, $z = 0$, $u = 0$, $v = 9$, and $w = 6$.

Summary

Let's summarize the steps in the simplex method.

Simplex Method

1. Introduce slack variables and rewrite the system conveniently.
2. Write the initial simplex matrix.
3. Find the pivot.
4. Perform the pivoting.
5. Repeat steps 3 and 4 until the objective row contains no negative numbers.
6. Interpret the final simplex matrix.

Now we will do a problem all at once.

Example 4 Maximize $F = x + 3y + 4z$
subject to
$$3y + z \leq 6$$
$$3x + 12y + 6z \leq 40$$
$$x \geq 0, y \geq 0, z \geq 0$$

Step 1 Introduce slack variables u and v and write the system as

$$3y + z + u = 6$$
$$3x + 12y + 6z + v = 40$$
$$F - x - 3y - 4z = 0$$
$$x \geq 0, y \geq 0, z \geq 0, u \geq 0, v \geq 0$$

Step 2 The initial simplex matrix is

$$\begin{array}{ccccc} x & y & z & u & v \\ \end{array}$$
$$\begin{bmatrix} 0 & 3 & \boxed{1} & 1 & 0 & | & 6 \\ 3 & 12 & 6 & 0 & 1 & | & 40 \\ -1 & -3 & -4 & 0 & 0 & | & 0 \end{bmatrix}$$

Step 3 The pivotal column is column three because -4 is the most negative of the numbers $-1, -3, -4, 0, 0,$ and 0. The first row is the pivotal row because $6/1$ is smaller then $40/6$. The pivot is the 1 in position $1, 3$.

Step 4 Pivoting gives the matrix

$$\begin{array}{c} \begin{array}{ccccc} x & y & z & u & v \end{array} \\ \left[\begin{array}{ccccc|c} 0 & 3 & 1 & 1 & 0 & 6 \\ \boxed{3} & -6 & 0 & -6 & 1 & 4 \\ -1 & 9 & 0 & 4 & 0 & 24 \end{array}\right] \begin{array}{l} \text{row 1} \\ \text{row 2} - 6 \text{ row 1} \\ \text{row 3} + 4 \text{ row 1} \end{array} \end{array}$$

Step 5 Further pivoting is needed because the objective row contains the negative number -1. The first column is the pivotal column. Since the 3 in position $2, 1$ is the only positive entry in the pivotal column, this entry must be the pivot. Pivoting again gives

$$\begin{array}{c} \begin{array}{ccccc} x & y & z & u & v \end{array} \\ \left[\begin{array}{ccccc|c} 0 & 3 & 1 & 1 & 0 & 6 \\ 1 & -2 & 0 & -2 & \frac{1}{3} & \frac{4}{3} \\ 0 & 7 & 0 & 2 & \frac{1}{3} & \frac{76}{3} \end{array}\right] \begin{array}{l} \text{row 1} \\ \left(\frac{1}{3}\right) \text{row 2} \\ \text{row 3} + \text{row 2} \end{array} \end{array}$$

Since there are now no negative entries in the objective row, pivoting is complete.

Step 6 The maximum possible value of F is $76/3$. When $F = 76/3$, $x = 4/3$, $y = 0$, $z = 6$, $u = 0$, and $v = 0$.
■

Here is an outline of what the simplex method does. In Example 4, there was a system of five inequalities in three unknowns. The solution of each inequality is a plane together with all points on one side of the plane. The feasible region is the intersection of these solutions. The maximum must occur at a corner of the feasible region. At a corner, three of the planes meet, so that three of the inequalities become equalities.

Any simplex matrix gives values to all the variables. The initial simplex matrix gives $x = 0$, $y = 0$, and $z = 0$. Initially we are at the origin, which is a corner of the feasible region, and $F = 0$. The rules for pivoting are chosen so that after pivoting, we are at another corner of the feasible region, and the value of F at the new corner is greater than or equal to the value of F at the previous corner. After pivoting, $x = 0$, $y = 0$, and $z = 6$. This corner is the point where the planes $x = 0$, $y = 0$, and $3y + z = 6$ meet. At this point, $F = 24$.

Now pivot again to obtain the point $x = 4/3$, $y = 0$, and $z = 6$. This corner is the point where the planes $y = 0$, $3y + z = 6$, and $3x + 12y + 6z = 40$ meet. At this point $F = 76/3$. This is the maximum possible value of F. To see this, note that the objective row in the simplex matrix would give the equation $F + 7y + 2u + v/3 = 76/3$. Since y, u, and v must be nonnegative, $F \leq 76/3$, and $F = 76/3$ is possible only when $y = u = v = 0$.

Example 5 Mayapple Games Company produces pool tables and construction sets. To produce a pool table requires 1 hour of labor and $30 of materials. To produce a construction set requires 4 hours of labor and $20 of materials. The company has 1600 hours of labor available and can spend up to $18,000 on materials. How many pool tables and how many construction sets should the Mayapple Games Company produce if profit is $10 on a pool table and $8 on a construction set?

Solution Assume that the company desires to maximize profit. Let p = number of pool tables produced, c = number of construction sets produced.
Then the problem is to

$$\begin{aligned}
\text{Maximize } P &= 10p + 8c \\
\text{subject to } \quad p + 4c &\leq 1600 \quad \text{Labor} \\
30p + 20c &\leq 18{,}000 \quad \text{Materials} \\
p \geq 0, c &\geq 0 \quad \text{Nonnegativity}
\end{aligned}$$

This problem could be solved by the graphical method, but let's use the simplex method instead. Introduce slack variables and rewrite the system as

$$\begin{aligned}
p + 4c + u &= 1600 \\
30p + 20c \quad\quad + v &= 18{,}000 \\
P - 10p - 8c \quad\quad\quad &= 0 \\
p \geq 0, c \geq 0, u \geq 0, v &\geq 0
\end{aligned}$$

The initial simplex matrix is

$$\begin{bmatrix} p & c & u & v & \\ 1 & 4 & 1 & 0 & 1600 \\ \boxed{30} & 20 & 0 & 1 & 18{,}000 \\ -10 & -8 & 0 & 0 & 0 \end{bmatrix}$$

Pivoting gives

$$\begin{bmatrix} p & c & u & v & & \\ 0 & \boxed{\dfrac{10}{3}} & 1 & -\dfrac{1}{30} & 1000 & \text{row 1} - \text{row 2} \\ 1 & \dfrac{2}{3} & 0 & \dfrac{1}{30} & 600 & \left(\dfrac{1}{30}\right)\text{row 2} \\ 0 & -\dfrac{4}{3} & 0 & \dfrac{1}{3} & 6000 & \text{row 3} + 10\text{ row 2} \end{bmatrix}$$

Pivoting again gives the matrix

$$\begin{bmatrix} p & c & u & v & & \\ 0 & 1 & \dfrac{3}{10} & -\dfrac{1}{100} & 300 & \left(\dfrac{3}{10}\right)\text{row 1} \\ 1 & 0 & -\dfrac{1}{5} & \dfrac{1}{25} & 400 & \text{row 2} - \left(\dfrac{2}{3}\right)\text{row 1} \\ 0 & 0 & \dfrac{2}{5} & \dfrac{8}{25} & 6400 & \text{row 3} + \left(\dfrac{4}{3}\right)\text{row 1} \end{bmatrix}$$

This is the final simplex matrix since there are no negative numbers in the objective row. The maximum possible profit is $6400. To make a profit of $6400, the company must manufacture 400 pool tables and 300 construction sets. Since $u = v = 0$, all of the available labor is used and all of the money available to buy materials is used.

Example 6 A chemical plant can make three products: glaze, solvent, and clay. A pound of glaze sells for $5, a pound of solvent sells for $3, and a pound of clay sells for $4. During the manufacturing process, 20 units of air pollution are released into the air for each pound of glaze produced, 15 units for each pound of solvent, and 10 units for each pound of clay. During the manufacturing process, each pound of glaze requires 30 min in the evaporation tank, each pound of solvent requires 40 min, and each pound of clay requires 10 min. What is the maximum possible daily revenue, if a maximum of 300 units of air pollution may be produced daily and the evaporation tank is available for only 4 hr each day?

Solution There are three variables.

$$g = \text{number of pounds of glaze produced each day.}$$
$$s = \text{number of pounds of solvent produced each day.}$$
$$c = \text{number of pounds of clay produced each day.}$$

The problem is to maximize revenue. The objective function is

$$R = 5g + 3s + 4c$$

There are two constraints besides the three nonnegativity constraints.

$$20g + 15s + 10c \leq 300 \quad \text{Air pollution}$$
$$30g + 40s + 10c \leq 240 \quad \text{Evaporation tank time}$$
$$g \geq 0, s \geq 0, c \geq 0 \quad \text{Nonnegativity}$$

Note the number 240 in the evaporation tank time constraint (4 hr is 240 min). Since time is measured in minutes when describing the manufacturing process, time must also be measured in minutes when describing the available resources.

Since there are three variables in this problem, the simplex method must be used because the graphical method is not applicable.

Now introduce slack variables and rewrite the system as

$$20g + 15s + 10c + u = 300$$
$$30g + 40s + 10c + v = 240$$
$$R - 5g - 3s - 4c = 0$$
$$g \geq 0, s \geq 0, c \geq 0, u \geq 0, v \geq 0$$

The initial simplex matrix is

$$\begin{bmatrix} g & s & c & u & v & \\ 20 & 15 & 10 & 1 & 0 & 300 \\ \boxed{30} & 40 & 10 & 0 & 1 & 240 \\ -5 & -3 & -4 & 0 & 0 & 0 \end{bmatrix}$$

The number 30 is the pivot. Pivoting gives

$$\begin{bmatrix} g & s & c & u & v & \\ 0 & -\dfrac{35}{3} & \dfrac{10}{3} & 1 & -\dfrac{2}{3} & 140 \\ 1 & \dfrac{4}{3} & \boxed{\dfrac{1}{3}} & 0 & \dfrac{1}{30} & 8 \\ 0 & \dfrac{11}{3} & -\dfrac{7}{3} & 0 & \dfrac{1}{6} & 40 \end{bmatrix} \begin{array}{l} \text{row 1} - 20 \text{ row 2} \\ \left(\dfrac{1}{30}\right) \text{row 2} \\ \text{row 3} + 5(\text{row 2}) \end{array}$$

The number $1/3$ is the pivot. Pivoting gives

$$\begin{bmatrix} g & s & c & u & v & \\ -10 & -25 & 0 & 1 & -1 & 60 \\ 3 & 4 & 1 & 0 & \dfrac{1}{10} & 24 \\ 7 & 13 & 0 & 0 & \dfrac{2}{5} & 96 \end{bmatrix} \begin{array}{l} \text{row 1} - \left(\dfrac{10}{3}\right) \text{row 2} \\ 3(\text{row 2}) \\ \text{row 3} + \left(\dfrac{7}{3}\right) \text{row 2} \end{array}$$

This is the final simplex matrix. The maximum possible daily revenue is $96. To obtain the maximum possible revenue, 24 lb of clay should be made daily. No glaze or solvent should be made. This is because the final values are $g = 0$, $s = 0$, and $c = 24$. Since $u = 60$, the plant gives off 60 fewer units of air pollution than the maximum permissible amount. Since $v = 0$, all available time in the evaporation tank is used. ■

Exercises 4-4

Introduce slack variables, write the initial simplex matrix, and determine the first pivot.

1. Maximize $F = 5x + 2y$
   subject to $2x + y \leq 8$
   $x + 4y \leq 12$
   $x \geq 0$, $y \geq 0$

2. Maximize $F = 3x + 7y$
   subject to $3x + 4y \leq 24$
   $2x + 5y \leq 20$
   $x \geq 0$, $y \geq 0$

3. Maximize $F = 2x + 8y$
   subject to $x + y \leq 12$
   $3x + 2y \leq 18$
   $2x + y \leq 14$
   $x \geq 0, y \geq 0$

4. Maximize $F = 3x + 6y$
   subject to $4x + 3y \leq 60$
   $2x + y \leq 40$
   $3x - 2y \leq 6$
   $x \geq 0, y \geq 0$

5. Maximize $F = x + 4y + 5z$
   subject to $3x + 2y + 4z \leq 34$
   $x + 5y + 2z \leq 42$
   $x \geq 0, y \geq 0, z \geq 0$

6. Maximize $F = 3x + 2y + z$
   subject to $x + 3y + 5z \leq 19$
   $3x + y + 2z \leq 12$
   $x \geq 0, y \geq 0, z \geq 0$

7. Maximize $F = 4x + 2y + 7z$
   subject to $x + y + 2z \leq 12$
   $3x + 5y + \leq 16$
   $2x + 3y + 6z \leq 18$
   $x \geq 0, y \geq 0, z \geq 0$

8. Maximize $F = 6x + 12y + 14z$
   subject to $2x + 3y + 4z \leq 12$
   $x + y + z \leq 5$
   $x + 2y + 10z \leq 20$
   $x \geq 0, y \geq 0, z \geq 0$

9. Solve exercise 9 of Section 4-2 by the simplex method.
10. Solve exercise 10 of Section 4-2 by the simplex method.
11. Solve exercise 18 of Section 4-2 by the simplex method.
12. Solve exercise 21 of Section 4-2 by the simplex method.

Solve each problem by the simplex method:

13. Maximize $F = 6x + 3y + 2z$
    subject to $x + 2y + 4z \leq 12$
    $2x + 3y + 6z \leq 60$
    $2x + 5y + z \leq 30$
    $x \geq 0, y \geq 0, z \geq 0$

14. Maximize $F = 2x + 3y + z$
    subject to $3x + 4y + 8z \leq 40$
    $x + y + z \leq 18$
    $3x + 2y + 5z \leq 24$
    $x \geq 0, y \geq 0, z \geq 0$

15. Maximize $2x + 4y + 5z$
    subject to $3x + y + 2z \leq 20$
    $2x + 4y + z \leq 8$
    $x \geq 0, y \geq 0, z \geq 0$

16. Maximize $2x + 5y - z$
    subject to $2x + 3y + 4z \leq 18$
    $3x + y + z \leq 5$
    $x \geq 0, y \geq 0, z \geq 0$

17. Maximize $x + 2y + z$
    subject to $x + 3y + 2z \leq 10$
    $2x + y + z \leq 16$
    $4x + 2y + z \leq 24$
    $x \geq 0, y \geq 0, z \geq 0$

18. Maximize $2x + 3y + z$
    subject to $5x - 2y + z \leq 8$
    $2x + y + z \leq 16$
    $3x + 2y + 6z \leq 40$
    $x \geq 0, y \geq 0, z \geq 0$

19. Maximize $x + 3y + 2z$
    subject to $2x + 3y + z \leq 12$
    $x + y + 4z \leq 16$
    $x \geq 0, y \geq 0, z \geq 0$

20. Maximize $2x + 4y + z$
    subject to $4x + 2y + 3z \leq 24$
    $x + y + 2z \leq 10$
    $x \geq 0, y \geq 0, z \geq 0$

21. The Knockdown Furniture Company manufactures stools, chairs, and coffee tables. To manufacture a stool requires two boards, two man-hours of labor, and two cans of finish. To manufacture a chair requires two boards, three man-hours of labor, and five cans of finish. To manufacture a coffee table requires one board, three man-hours of labor, and one can of finish. The company makes a profit of $12 on each stool and each coffee table, and a profit of $18 on each chair. Each day, 20 boards, 30 man-hours of labor, and 40 cans of finish are available.
    a. What is the maximum possible profit?
    b. How many stools, chairs, and coffee tables should be manufactured each day in order to maximize profit?
    c. Are any boards, labor, or finish not used?

22. The Cavity Candy Store makes a deluxe bridge mix, an elegant bridge mix, and a luxurious bridge mix. The profit is $3 on a box of deluxe bridge mix, $2 on a box of elegant bridge mix, and $1 on a box of luxurious bridge mix. A box of deluxe bridge mix contains 2 lb of crunchies and 2 lb of chewies. A box of elegant bridge mix contains one pound of crunchies and 2 lb of chewies. A box of luxurious bridge mix contains 2 lb of crunchies and 3 lb of chewies. Each day, 400 lb of crunchies and 600 lb of chewies are available for use.
    a. What is the maximum possible daily profit?
    b. How many boxes of each mix should be made each day?
    c. Are all the available crunchies and chewies used?

23. Squire Baggins wishes to plant corn, lettuce, and tomatoes on his farm. Each acre of corn requires one hour of labor at planting time and three hours of labor at harvest time. Each acre of lettuce requires two hours of labor at planting time and one hour at harvest time. Each acre of tomatoes requires two hours of labor at planting time and three hours at harvest time. Squire Baggins is willing to work up to 100 hr at planting time and up to 100 hr at harvest time.

    After harvesting his crop, he can exchange the corn from one acre for five enormous birthday cakes (Squire Baggins has a weakness for pastries and the like), the lettuce from one acre for four enormous birthday cakes, and the tomatoes from one acre for six enormous birthday cakes. He wants as many enormous birthday cakes as possible. How many acres should he plant of each crop? How many hours will he work at planting time? At harvest time?

24. Nuclear Boat is deciding what ships to offer to build. The profit is $20,000,000 on a cruiser, $30,000,000 on a submarine, and $10,000,000 on a frigate. The company has 25 labor teams available. One labor team is assigned to a cruiser or a frigate, but two labor teams must be assigned to each submarine. Thirty storage sites are available. Three storage sites must be assigned to a submarine or cruiser, but a frigate requires only two storage sites. How many of each kind of ship should Nuclear Boat offer to build? If the offers are all accepted, how much profit would Nuclear Boat make? Are all the labor teams and storage sites used?

25. A commercial fishery owns Moose Pond and stocks it with bass, trout, and whitefish. After a year, the fish are caught and sold. Each bass sells for $7, each trout for $5, and each whitefish for $6. There are two foods, F and G, available in Moose Pond. Each month, 2000 lb of F and 1200 lb of G are available. Each bass needs 10 lb of F and 5 lb of G a month. Each trout needs 6 lb of F and 11 lb of G a month. Each whitefish needs 12 lb of F and 4 lb of G a month. The fishery manager wishes to maximize revenue. How many of each fish should he stock Moose Pond with?

26. Valencia Orchards sells fruit baskets. A deluxe box of fruit contains 3 papayas, 1 orange, and 6 mangos. A grade A box of fruit contains 5 papayas, 3 oranges, and 2 mangos. A tropical box of fruit contains 3 papayas, 2 oranges, and 5 mangos. Valencia Orchards has available 72,000 papayas, 32,400 oranges, and 108,000 mangos. Profit is $6 on a deluxe box, $8 on a grade A box, and $1 on a tropical box. How many boxes of each type should be made, in order to maximize profit? How much fruit is left over?

*27. The Regional Transit Agency (RTA) may purchase up to 36 vehicles for Community Responsive Transit. This service provides door-to-door transportation at very low, subsidized rates for the elderly and the handicapped. The RTA can purchase cars, vans, and minibuses. A car can carry four passengers; a van, six; and a minibus, eight. Since minibuses are the most expensive vehicles to operate, the number of minibuses cannot exceed the total number of other vehicles. Since a vehicle must often be sent for only one passenger, there may be at most twice as many vans as cars. How many vehicles of each type should the RTA buy in order to maximize the number of passengers who can be carried at once?

Section 4-5 The Dual Problem

In Section 4-4, the simplex method was used to solve maximization problems. Minimization is also important, however. Many applications of linear programming involve minimizing cost. In this section, the simplex method is used to solve minimization problems in which each constraint is a nonnegativity constraint or is of the form

(expression involving variables) $\geq$ (nonnegative constant)

A notion from matrix theory is needed first.

> **Definition.** If B is a matrix, the *transpose* of B, written B^t, is the matrix defined by $(B^t)_{i,j} = B_{j,i}$.

If B is an $m \times n$ matrix, B^t is an $n \times m$ matrix. Each column of B is a row of B^t, and each row of B is a column of B^t. To form B^t, make the first row of B into the first column of B^t, the second row of B into the second column of B^t, and so on.

For example let

$$B = \begin{bmatrix} 1 & 3 & 5 & 7 \\ 2 & 4 & 6 & 8 \\ 9 & 12 & 11 & 17 \end{bmatrix}$$

B is a 3×4 matrix, so B^t is a 4×3 matrix. The first row of B is [1 3 5 7], so the first column of B^t is

$$\begin{bmatrix} 1 \\ 3 \\ 5 \\ 7 \end{bmatrix}$$

The second row of B is [2 4 6 8] so that the second column of B^t is

$$\begin{bmatrix} 2 \\ 4 \\ 6 \\ 8 \end{bmatrix}$$

The third row of B is [9 12 11 17] and so the third column of B^t is

$$\begin{bmatrix} 9 \\ 12 \\ 11 \\ 17 \end{bmatrix}$$

Thus,

$$B^t = \begin{bmatrix} 1 & 2 & 9 \\ 3 & 4 & 12 \\ 5 & 6 & 11 \\ 7 & 8 & 17 \end{bmatrix}$$

Examples

1. $\begin{bmatrix} 1 & 3 \\ 10 & 9 \end{bmatrix}^t = \begin{bmatrix} 1 & 10 \\ 3 & 9 \end{bmatrix}$

2. $[4 \quad 2 \quad 7 \quad 1]^t = \begin{bmatrix} 4 \\ 2 \\ 7 \\ 1 \end{bmatrix}$

3. $\begin{bmatrix} 5 & 3 \\ -2 & 6 \\ 9 & -1 \end{bmatrix}^t = \begin{bmatrix} 5 & -2 & 9 \\ 3 & 6 & -1 \end{bmatrix}$

4. $\begin{bmatrix} -2 \\ 6 \\ 1 \end{bmatrix}^t = \begin{bmatrix} -2 & 6 & 1 \end{bmatrix}$

■

Now we return to the simplex method. Consider the problem

$$\text{Minimize } F = 2x + 3y + 5z$$
$$\text{subject to } 3x + 2y + 6z \geq 19$$
$$4x + y + 3z \geq 12$$
$$x \geq 0, y \geq 0, z \geq 0$$

Write this problem in matrix form as:

$$\text{Minimize } F = \begin{bmatrix} 2 & 3 & 5 \end{bmatrix} \begin{bmatrix} x \\ y \\ z \end{bmatrix}$$

$$\text{subject to } \begin{bmatrix} 3 & 2 & 6 \\ 4 & 1 & 3 \end{bmatrix} \begin{bmatrix} x \\ y \\ z \end{bmatrix} \geq \begin{bmatrix} 19 \\ 12 \end{bmatrix} \quad \text{and} \quad \begin{bmatrix} x \\ y \\ z \end{bmatrix} \geq \begin{bmatrix} 0 \\ 0 \\ 0 \end{bmatrix}$$

The $\geq$ relation between two matrices means that each entry in the first matrix $\geq$ the corresponding entry in the second matrix. Call the matrix $\begin{bmatrix} x \\ y \\ z \end{bmatrix}$ of variables X. Let P denote the matrix $\begin{bmatrix} 2 & 3 & 5 \end{bmatrix}$, M denote the matrix $\begin{bmatrix} 3 & 2 & 6 \\ 4 & 1 & 3 \end{bmatrix}$, and C denote the matrix $\begin{bmatrix} 19 \\ 12 \end{bmatrix}$.
The problem is then

$$\text{Minimize } F = PX$$
$$\text{subject to } MX \geq C \quad X \geq 0$$

> Any minimization problem has associated with it a maximization problem, called the *dual problem*, which has the same solution.

We will now write this dual problem. The original problem was:

$$\text{Minimize } F = PX$$
$$\text{subject to } MX \geq C \quad X \geq 0$$

4-5 THE DUAL PROBLEM

> **Definition.** The *dual problem* is:
>
> Maximize $G = C^t Y$
>
> subject to $M^t Y \leq P^t$ and $Y \geq 0$

Here Y is the matrix $\begin{bmatrix} p \\ q \end{bmatrix}$ and p and q are variables. The matrix Y will have as many entries as there are constraints (other than nonnegativity constraints) in the original problem. The zero matrix is chosen to have the same size as the matrix Y. Since the original and the dual problem have the same solution, the minimum possible value of F is the same as the maximum possible value of G.

For our problem, the dual problem in matrix form is

$$\text{Maximize } G = \begin{bmatrix} 19 & 12 \end{bmatrix} \begin{bmatrix} p \\ q \end{bmatrix}$$

$$\text{subject to } \begin{bmatrix} 3 & 4 \\ 2 & 1 \\ 6 & 3 \end{bmatrix} \begin{bmatrix} p \\ q \end{bmatrix} \leq \begin{bmatrix} 2 \\ 3 \\ 5 \end{bmatrix} \text{ and } \begin{bmatrix} p \\ q \end{bmatrix} \geq \begin{bmatrix} 0 \\ 0 \end{bmatrix}$$

The dual problem is

$$\text{Maximize } G = 19p + 12q$$
$$\text{subject to } 3p + 4q \leq 2$$
$$2p + q \leq 3$$
$$6p + 3q \leq 5$$
$$p \geq 0, q \geq 0$$

Examples Write the problem in matrix form, and write the dual problem in matrix form.

Example 5 Minimize $5x + 2y$

subject to $3x + y \geq 8$

$2x + 2y \geq 5$

$x \geq 0, y \geq 0$

Solution In matrix form, this is

$$\text{Minimize } \begin{bmatrix} 5 & 2 \end{bmatrix} \begin{bmatrix} x \\ y \end{bmatrix}$$

$$\text{subject to } \begin{bmatrix} 3 & 1 \\ 2 & 2 \end{bmatrix} \begin{bmatrix} x \\ y \end{bmatrix} \geq \begin{bmatrix} 8 \\ 5 \end{bmatrix} \text{ and } \begin{bmatrix} x \\ y \end{bmatrix} \geq \begin{bmatrix} 0 \\ 0 \end{bmatrix}$$

The dual problem is

$$\text{Maximize } \begin{bmatrix} 8 & 5 \end{bmatrix} \begin{bmatrix} p \\ q \end{bmatrix}$$

$$\text{subject to } \begin{bmatrix} 3 & 2 \\ 1 & 2 \end{bmatrix} \begin{bmatrix} p \\ q \end{bmatrix} \leq \begin{bmatrix} 5 \\ 2 \end{bmatrix} \text{ and } \begin{bmatrix} p \\ q \end{bmatrix} \geq \begin{bmatrix} 0 \\ 0 \end{bmatrix}$$

Example 6 Minimize $4x + 3y + 5z$
subject to $2x + 4y + 7z \geq 25$
$8x + 2y + z \geq 19$
$x \geq 0, y \geq 0, z \geq 0$

Solution In matrix form, the problem is

$$\text{Minimize } \begin{bmatrix} 4 & 3 & 5 \end{bmatrix} \begin{bmatrix} x \\ y \\ z \end{bmatrix}$$

$$\text{subject to } \begin{bmatrix} 2 & 4 & 7 \\ 8 & 2 & 1 \end{bmatrix} \begin{bmatrix} x \\ y \\ z \end{bmatrix} \geq \begin{bmatrix} 25 \\ 19 \end{bmatrix} \text{ and } \begin{bmatrix} x \\ y \\ z \end{bmatrix} \geq \begin{bmatrix} 0 \\ 0 \\ 0 \end{bmatrix}$$

The dual problem is

$$\text{Maximize } \begin{bmatrix} 25 & 19 \end{bmatrix} \begin{bmatrix} p \\ q \end{bmatrix}$$

$$\text{subject to } \begin{bmatrix} 2 & 8 \\ 4 & 2 \\ 7 & 1 \end{bmatrix} \begin{bmatrix} p \\ q \end{bmatrix} \leq \begin{bmatrix} 4 \\ 3 \\ 5 \end{bmatrix} \text{ and } \begin{bmatrix} p \\ q \end{bmatrix} \geq \begin{bmatrix} 0 \\ 0 \end{bmatrix}$$

∎

Let's do a complete minimization problem.

Example 7 Minimize $F = x + y + 3z$
subject to $x + 2y + 4z \geq 12$
$2x + y + z \geq 8$
$x \geq 0, y \geq 0, z \geq 0$

Solution In matrix form, this problem is

$$\text{Minimize } \begin{bmatrix} 1 & 1 & 3 \end{bmatrix} \begin{bmatrix} x \\ y \\ z \end{bmatrix}$$

$$\text{subject to } \begin{bmatrix} 1 & 2 & 4 \\ 2 & 1 & 1 \end{bmatrix} \begin{bmatrix} x \\ y \\ z \end{bmatrix} \geq \begin{bmatrix} 12 \\ 8 \end{bmatrix} \text{ and } \begin{bmatrix} x \\ y \\ z \end{bmatrix} \geq \begin{bmatrix} 0 \\ 0 \\ 0 \end{bmatrix}$$

The dual problem in matrix form is

$$\text{Maximize } G = \begin{bmatrix} 12 & 8 \end{bmatrix} \begin{bmatrix} p \\ q \end{bmatrix}$$

$$\text{subject to } \begin{bmatrix} 1 & 2 \\ 2 & 1 \\ 4 & 1 \end{bmatrix} \begin{bmatrix} p \\ q \end{bmatrix} \leq \begin{bmatrix} 1 \\ 1 \\ 3 \end{bmatrix} \text{ and } \begin{bmatrix} p \\ q \end{bmatrix} \geq \begin{bmatrix} 0 \\ 0 \end{bmatrix}$$

The dual problem is

$$\text{Maximize } G = 12p + 8q$$
$$\text{subject to } p + 2q \leq 1$$
$$2p + q \leq 1$$
$$4p + q \leq 3$$
$$p \geq 0, q \geq 0$$

We now proceed to solve the dual problem by the simplex method. Introduce slack variables u, v, and w and write the system as

$$p + 2q + u = 1$$
$$2p + q + v = 1$$
$$4p + q + w = 3$$
$$G - 12p - 8q = 0$$
$$p \geq 0, q \geq 0, u \geq 0, v \geq 0, w \geq 0$$

The initial simplex matrix is

$$\begin{array}{ccccc} p & q & u & v & w \end{array}$$
$$\left[\begin{array}{ccccc|c} 1 & 2 & 1 & 0 & 0 & 1 \\ \boxed{2} & 1 & 0 & 1 & 0 & 1 \\ 4 & 1 & 0 & 0 & 1 & 3 \\ -12 & -8 & 0 & 0 & 0 & 0 \end{array} \right]$$

Pivoting gives

$$\begin{bmatrix} & p & q & u & v & w & \\ 0 & \boxed{\tfrac{3}{2}} & 1 & -\tfrac{1}{2} & 0 & \tfrac{1}{2} \\ 1 & \tfrac{1}{2} & 0 & \tfrac{1}{2} & 0 & \tfrac{1}{2} \\ 0 & -1 & 0 & -2 & 1 & 1 \\ 0 & -2 & 0 & 6 & 0 & 6 \end{bmatrix} \begin{array}{l} \text{row 1} - \text{row 2} \\ \left(\tfrac{1}{2}\right)\text{row 2} \\ \text{row 3} - 4\text{ row 2} \\ \text{row 4} + 12\text{ row 2} \end{array}$$

Note that in determining that $3/2$ is the pivot the ratio $1/(-1)$ is not considered because -1 is not positive. Pivoting again gives

$$\begin{bmatrix} & p & q & u & v & w & \\ 0 & 1 & \tfrac{2}{3} & -\tfrac{1}{3} & 0 & \tfrac{1}{3} \\ 1 & 0 & -\tfrac{1}{3} & \tfrac{2}{3} & 0 & \tfrac{1}{3} \\ 0 & 0 & \tfrac{2}{3} & -\tfrac{7}{3} & 1 & \tfrac{4}{3} \\ 0 & 0 & \underset{x}{\tfrac{4}{3}} & \underset{y}{\tfrac{16}{3}} & \underset{z}{0} & \tfrac{20}{3} \end{bmatrix} \begin{array}{l} \left(\tfrac{2}{3}\right)\text{row 1} \\ \text{row 2} - \left(\tfrac{1}{2}\right)\text{row 1} \\ \text{row 3} + \text{row 1} \\ \text{row 4} + 2\text{ row 1} \end{array}$$

This is the final simplex matrix. The solution of the dual problem is: the maximum of G is $20/3$. This is attained when $p = 1/3$ and $q = 1/3$.

The solution of the original problem is: the minimum of F is $20/3$. Furthermore, this is attained when $x = 4/3$, $y = 16/3$, and $z = 0$.

The numbers in the objective row of the final simplex matrix of the dual problem, in the columns for slack variables, give the point at which the minimum value of the objective function for the original problem is attained.

For example, suppose that in a minimization problem, there are an objective function F, two variables x and y, and three constraints other than nonnegativity constraints. Suppose the final simplex matrix for the dual problem is

4-5 THE DUAL PROBLEM

$$\begin{array}{cccccc} p & q & r & v & w & \\ \begin{bmatrix} 1 & 4 & 0 & 0 & 10 & | & 5 \\ 0 & -3 & 1 & 1 & 2 & | & 15 \\ 0 & 12 & 10 & 0 & 6 & | & 17 \end{bmatrix} \\ & & \nearrow & \nearrow & & \\ & & x & y & & \end{array}$$

Then the solution of the original problem is that the minimum value of F is 17 and is attained when $x = 0$ and $y = 6$.

Example 8 A power plant is required to generate at least 100,000 kW of power. The plant can burn low sulfur coal, high sulfur coal, and oil. A ton of low sulfur coal costs \$8, generates 20,000 kWh of power, and releases 50 units of sulfur dioxide into the air. A ton of high sulfur coal costs \$6, generates 20,000 kWh of power, and releases 100 units of sulfur dioxide into the air. A barrel of oil costs \$12, generates 30,000 kWh of power, and releases 100 units of sulfur dioxide into the air. The company is required by law to use at least 2 tons of low sulfur coal each hour, and to pay a fine of 2¢ for each unit of sulfur dioxide released into the air. How much of each fuel should the company use each hour to minimize the total cost? Total cost is the cost of buying the fuel plus the fines for pollution.

Solution Let l = number of tons of low sulfur coal burned each hour.
h = number of tons of high sulfur coal burned each hour.
i = number of barrels of oil burned each hour.
The objective function to be minimized is total cost.

$$C = 8l + 6h + 12i + 50(0.02)l + 100(0.02)h + 100(0.02)i$$
$$= 9l + 8h + 14i$$

The constraints are

$$\begin{array}{ll} l \geq 2 & \text{Law} \\ 20{,}000l + 20{,}000h + 30{,}000i \geq 100{,}000 & \text{Amount of power} \\ l \geq 0, h \geq 0, i \geq 0 & \text{Nonnegativity} \end{array}$$

Since this is a minimization problem, we must set up and solve the dual problem. In matrix form, the problem is

$$\text{Minimize } \begin{bmatrix} 9 & 8 & 14 \end{bmatrix} \begin{bmatrix} l \\ h \\ i \end{bmatrix}$$

subject to $\begin{bmatrix} 1 & 0 & 0 \\ 20{,}000 & 20{,}000 & 30{,}000 \end{bmatrix} \begin{bmatrix} l \\ h \\ i \end{bmatrix} \geq \begin{bmatrix} 2 \\ 100{,}000 \end{bmatrix}$

and $\begin{bmatrix} l \\ h \\ i \end{bmatrix} \geq \begin{bmatrix} 0 \\ 0 \\ 0 \end{bmatrix}$

The dual problem in matrix form is

Maximize $\begin{bmatrix} 2 & 100{,}000 \end{bmatrix} \begin{bmatrix} x \\ y \end{bmatrix}$

subject to $\begin{bmatrix} 1 & 20{,}000 \\ 0 & 20{,}000 \\ 0 & 30{,}000 \end{bmatrix} \begin{bmatrix} x \\ y \end{bmatrix} \leq \begin{bmatrix} 9 \\ 8 \\ 14 \end{bmatrix}$ and $\begin{bmatrix} x \\ y \end{bmatrix} \geq \begin{bmatrix} 0 \\ 0 \end{bmatrix}$

This can be written

Maximize $G = 2x + 100{,}000y$
subject to $\quad x + 20{,}000y \leq 9$
$\qquad\qquad\quad 20{,}000y \leq 8$
$\qquad\qquad\quad 30{,}000y \leq 14$

$x \geq 0,\ y \geq 0$

Introduce slack variables and write the system as

$$x + 20{,}000y + u \qquad\qquad = 9$$
$$20{,}000y \qquad + v \qquad = 8$$
$$30{,}000y \qquad\qquad + w = 14$$
$$G - 2x - 100{,}000y \qquad\qquad\qquad = 0$$

$x \geq 0,\ y \geq 0,\ u \geq 0,\ v \geq 0,\ w \geq 0$

The initial simplex matrix is

$$\begin{array}{ccccc} x & y & u & v & w \end{array}$$
$$\begin{bmatrix} 1 & 20{,}000 & 1 & 0 & 0 & | & 9 \\ 0 & \boxed{20{,}000} & 0 & 1 & 0 & | & 8 \\ 0 & 30{,}000 & 0 & 0 & 1 & | & 14 \\ -2 & -100{,}000 & 0 & 0 & 0 & | & 0 \end{bmatrix}$$

Pivoting gives

$$\begin{bmatrix} x & y & u & v & w & & \\ 1 & 0 & 1 & -1 & 0 & | & 1 \\ 0 & 1 & 0 & \frac{1}{20{,}000} & 0 & | & \frac{1}{2500} \\ 0 & 0 & 0 & -\frac{3}{2} & 1 & | & 2 \\ -2 & 0 & 0 & 5 & 0 & | & 40 \end{bmatrix} \begin{array}{l} \text{row 1} - 20{,}000 \text{ row 2} \\ \left(\frac{1}{20{,}000}\right) \text{row 2} \\ \text{row 3} - 30{,}000 \text{ row 2} \\ \text{row 4} + 100{,}000 \text{ row 2} \end{array}$$

Pivoting again gives

$$\begin{bmatrix} x & y & u & v & w & & \\ 1 & 0 & 1 & -1 & 0 & | & 1 \\ 0 & 1 & 0 & \frac{1}{20{,}000} & 0 & | & \frac{1}{2500} \\ 0 & 0 & 0 & -\frac{3}{2} & 1 & | & 2 \\ 0 & 0 & 2 & 3 & 0 & | & 42 \end{bmatrix} \begin{array}{l} \text{row 1} \\ \text{row 2} \\ \text{row 3} \\ \text{row 4} + 2 \text{ row 1} \end{array}$$

Now interpret this final simplex matrix for the original problem. The minimum possible cost is $42. This minimum cost occurs when the plant burns 2 tons of low sulfur coal, 3 tons of high sulfur coal, and no oil each hour. ∎

Exercises 4-5

Write the dual problem for each of the following. Do not solve the problems.

1. Minimize $2x + 5y$
   subject to $3x + y \geq 8$
   $x + 2y \geq 4$
   $x \geq 0, y \geq 0$

2. Minimize $3x + y$
   subject to $x + 5y \geq 12$
   $2x + 3y \geq 10$
   $x \geq 0, y \geq 0$

3. Minimize $x + 3y + 2z$
   subject to $2x + y + 4z \geq 12$
   $x + 3y + 2z \geq 16$
   $x \geq 0, y \geq 0, z \geq 0$

4. Minimize $2x + 8y + 3z$
   subject to $x + 3y + 6z \geq 19$
   $2x + y + z \geq 17$
   $x \geq 0, y \geq 0, z \geq 0$

5. Minimize $3x + 2y$
   subject to $2x + y \geq 12$
   $x + 3y \geq 14$
   $x + y \geq 8$
   $x \geq 0, y \geq 0$

6. Minimize $2x + 5y$
   subject to $3x + 2y \geq 9$
   $x + y \geq 4$
   $2x + 3y \geq 6$
   $x \geq 0, y \geq 0$

7. Minimize $x + 2y + 5x$
   subject to $x + 5y \geq 11$
   $2x + 3y + z \geq 9$
   $x + y + z \geq 6$
   $x \geq 0, y \geq 0, z \geq 0$

8. Minimize $4x + 3y + 8z$
   subject to $3x + 2y + z \geq 16$
   $x + 8y + 5z \geq 42$
   $2x + y + 6z \geq 32$
   $x \geq 0, y \geq 0, z \geq 0$

9. Solve exercise 11 of Section 4-2 by the simplex method.
10. Solve exercise 16 of Section 4-2 by the simplex method.
11. Solve exercise 23 of Section 4-2 by the simplex method.

12. Minimize $x + 2y$
    subject to $2x + y \geq 6$
    $x + 3y \geq 13$
    $x \geq 0, y \geq 0$

13. Minimize $3x + 2y$
    subject to $x + y \geq 9$
    $4x + 3y \geq 30$
    $x \geq 0, y \geq 0$

14. Minimize $2x + y + 3z$
    subject to $x + 3y + z \geq 12$
    $3x + y + 4z \geq 8$
    $x \geq 0, y \geq 0, z \geq 0$

15. Minimize $3x + 4y + 2z$
    subject to $2x + y + z \geq 6$
    $x + 3y + 2z \geq 8$
    $x \geq 0, y \geq 0, z \geq 0$

16. Minimize $x + 5y + 3z$
    subject to $x + y + z \geq 8$
    $3x + 2y + z \geq 14$
    $x \geq 0, y \geq 0, z \geq 0$

17. Minimize $3x + y + 2z$
    subject to $4x + 2y + z \geq 12$
    $x + 2y + 5z \geq 10$
    $z \geq 0, y \geq 0, z \geq 0$

18. Minimize $2x + y + 4z$
    subject to $x + y + z \geq 4$
    $x + 3y + 2z \geq 6$
    $3x + y + 3z \geq 8$
    $x \geq 0, y \geq 0, z \geq 0$

19. Minimize $x + 2y + 5z$
    subject to $x + y + 6z \geq 10$
    $3x + 2y + z \geq 5$
    $2x + y + 4z \geq 7$
    $x \geq 0, y \geq 0, z \geq 0$

20. Minimize $4x + 3y + z$
    subject to $2x + 2y + z \geq 12$
    $x + y + 2z \geq 8$
    $x + 2y + z \geq 6$
    $x \geq 0, y \geq 0, z \geq 0$

21. Minimize $3x + y + 6z$
    subject to $x + y + z \geq 4$
    $2x + y + 3z \geq 8$
    $3x + 4y + z \geq 10$
    $x \geq 0, y \geq 0, z \geq 0$

4-5 THE DUAL PROBLEM   209

22. The Wolf Coal Company owns the Easton Mine, the Union Mine, and the Prairie Mine. During a day of operation, the Easton Mine produces 10 tons of high sulfur coal and 5 tons of low sulfur coal. During a day of operation, the Union Mine produces 10 tons of high sulfur coal and 20 tons of low sulfur coal. During a day of operation, the Prairie Mine produces 10 tons of high sulfur coal and 10 tons of low sulfur coal. The Wolf Coal Company has signed contracts which require it to produce 120 tons of high sulfur coal and 150 tons of low sulfur coal. It costs $200 to keep the Easton Mine open for a day, $400 to keep the Union Mine open for a day, and $300 to keep the Prairie Mine open for a day.
   a. How many days should each mine be open, in order to fulfill the contracts at the minimum possible cost?
   b. How cheaply can the contracts be fulfilled?
   c. If the cost of fulfilling the contracts is minimized, is more coal produced than is needed?

23. Birdfood is to be prepared by mixing seeds, grain, and nuts. The birdfood is required to contain at least 20 units of vitamins, 9 oz of protein, and 15,000 calories. Each pound of seeds contains 3 units of vitamins, 1 oz of protein, and 3000 calories. Each pound of nuts contains 5 units of vitamins, 3 oz of protein, and 1000 calories. Each pound of grain contains 3 units of vitamins, 2 oz of protein, and 2500 calories. A pound of seeds costs 60¢, a pound of nuts costs 80¢, and a pound of grain costs 10¢.
   a. How much of each ingredient should be used, if the birdfood is to cost as little as possible?
   b. What is the minimum possible cost of the birdfood?
   c. Will the birds receive any more vitamins, protein, or calories than they need?

24. Roberta Greenleaf of Lincoln Electronics is placing an order for radios with her wholesaler. The wholesaler will not accept orders for less than 90 radios in total. The wholesaler sells AM radios for $10, FM radios for $15, and clock radios for $20. The retailer wants to buy at least as many clock radios as other radios. How many radios of each kind should she buy, if she wants to minimize the cost of the order?

25. Sebastian Holmes owns a farm. This year, he will plant corn, soybeans, and tomatoes. To qualify for government payments, he must plant crops

on at least 1000 acres of land and must plant at least twice as many acres of soybeans as of corn. He must plant at least 340 acres of tomatoes in order to fulfill a contract. To grow and harvest one acre of corn costs $10, of soybeans $15, and of tomatoes $25. How many acres of each crop should he plant in order to minimize costs?

Chapter 4 Review

Solve each problem by the graphical method.

1. Maximize $F = 7x + 3y$
   subject to $5x + 2y \leq 30$
   $2x + y \leq 14$
   $x \geq 0, y \geq 0$

2. Minimize $9x + 12y$
   subject to $3x + 4y \geq 24$
   $2x + 3y \geq 17$
   $x \geq 0, y \geq 0$

3. Wolf Shoes specializes in made to order dress shoes and hiking boots. To make a pair of dress shoes requires 2 hr of skilled labor and 1 hr of unskilled labor. To make a pair of hiking boots requires 3 hr of skilled labor and 2 hr of unskilled labor. Each week, 190 hr of skilled labor and 120 hr of unskilled labor are available. Each pair of dress shoes sells for $62 and each pair of hiking boots sells for $95. How many pairs of shoes and boots should be made each week in order to maximize revenue? Is all the available labor used?

4. Do exercise 3 if the price of dress shoes goes to $70.

Solve each problem by the simplex method.

5. Maximize $7x + 6y$
   subject to $x + 2y \leq 5$
   $3x + 8y \leq 18$
   $x \geq 0, y \geq 0$

6. Maximize $3x + y + 4z$
   subject to $2x + 5y + z \leq 12$
   $x + 3y + 2z \leq 8$
   $x \geq 0, y \geq 0, z \geq 0$

7. Minimize $5x + 2y + 4z$
   subject to $2x + y + z \geq 8$
   $x + 4y + 2z \geq 16$
   $x \geq 0, y \geq 0, z \geq 0$

8. Minimize $65x + 120y + 70z$
   subject to $x + y + 2z \geq 5$
   $x + 2y + 2z \geq 6$
   $2x + 4y + z \geq 8$
   $x \geq 0, y \geq 0, z \geq 0$

9. A metal refinery can process three different kinds of ore. To process one ton of ore A requires 1 hr of heating, 3 hr of separating, and 1 hr of purifying. To process one ton of ore B requires 2 hr of heating, 3 hr of separating, and 5 hr of purifying. To process one ton of ore C requires 2

hr of heating, 2 hr of separating, and 2 hr of purifying. Each week, furnaces are available for 120 hr of heating, centrifuges are available for 180 hr of separating, and the chemical section has available 240 hr of purifying. The charge for refining one ton of ore A or ore C is $60 and for ore B $90. How many tons of each ore should the refinery process each week in order to maximize revenue? Are all facilities used to capacity?

5

Mathematics of Finance

Very few people are in the position to buy a new car with cash. Most will need to finance a new car with a small down payment and several years of monthly payments. The total amount you pay will exceed the price of the car by an amount called *interest*. By making equal monthly payments, you are *amortizing* the debt on the car. In this chapter, you will study interest and amortization, as well as various types of annuities and other items from what is called the mathematics of finance.

Section 5-1 Interest

Few things in life are free, especially the use of someone else's money. If you want to borrow money, you must pay for its use.

Interest is the fee paid for the use of money. The amount of interest paid depends on the *interest rate r*. Interest rate is a percentage, and we recall that *percent* means parts per one hundred. Thus,

$$5\% = \frac{5}{100} = 0.05, \quad 6\% = \frac{6}{100} = 0.06, \quad \text{and} \quad 12\% = \frac{12}{100} = 0.12$$

Now the amount borrowed on a loan is called the *principal* and is denoted P. The time that the principal is kept is denoted t and is measured in years. Interest is denoted I. For short loans, it is customary to use *simple interest*, which is given by the formula

$$I = Prt \quad \text{Simple interest formula}$$

When the loan expires, the *amount A* to be repaid is the principal plus the interest. Thus $A = P + I = P + Prt = P(1 + rt)$, so we have

$$A = P(1 + rt) \quad \text{Amount at simple interest}$$

Example 1 If $1000 is borrowed from a bank for one year at 10% interest, the interest is $(1000)(0.10)(1) = \$100$. If the loan is for six months, the interest is $(1000)(0.10)(0.5) = \$50$.

Example 2 If $500 is borrowed at 9% interest for one month, the interest is $(500)(0.09)(1/12) = \$3.75$ and the amount to be repaid is $503.75.

Example 3 If $10,000 is borrowed at 8% interest for four months, the interest is $(10,000)(0.08)(4/12) = \$266.67$ and the amount to be repaid is $10,266.67.

Example 4 If $400 is borrowed at 20% interest for one and a half years, the interest is $(400)(0.20)(1.5) = \$120$ and the amount to be repaid is $520.

Example 5 A new bridge is needed in the town of Hamilton. In order to pay for it, $10,000,000 of 20-year bonds are issued, paying 8% interest. Each year, every bondholder is mailed a check for the interest owed. How much interest does Hamilton pay each year?

Solution The amount of interest paid each year is

$$(10{,}000{,}000)(0.08)(1) = \$800{,}000$$

At the end of 20 years, the bonds are paid off. In the next section, we will learn how Hamilton could collect the money to pay off the bonds.

Example 6 National Nuclear needs to borrow money on a short-term basis. Its financial vice-president places an ad in business publications offering to pay back $102,500 three months after it receives your $100,000. What interest rate is National Nuclear offering?

Solution Use the formula $A = P(1 + rt)$ with

$$A = 102{,}500$$
$$P = 100{,}000$$
$$t = \frac{1}{4} \quad 3 \text{ months} = \frac{1}{4} \text{ year}$$
$$102{,}500 = 100{,}000\left(1 + \frac{r}{4}\right)$$
$$1.025 = 1 + \frac{r}{4}$$
$$0.025 = \frac{r}{4}$$
$$0.1 = r \quad \text{so} \quad r = 10\%$$

National Nuclear is offering to pay 10% interest.

■

Compound interest

Simple interest is usually used for short-term loans, or for loans in which the interest is paid regularly and the principal is paid off all at once. In most other situations, *compound* interest is used.

> *Compounding of interest* is the process of converting the interest on an account into principal at regular time intervals.

Once interest becomes principal, it starts earning interest. For example, suppose $1000 is borrowed at 8% interest compounded quarterly. At the end of one quarter, the interest is $(1000)(0.08)(0.25) = \$20.00$ and the amount owed is now $1020.00. The $20.00 in interest is now principal. Interest for the second quarter is $(1020)(0.08)(1/4) = \$20.40$, so that the amount owed after two quarters is $1020.00 + \$20.40 = \$1040.40$. The interest for the third quarter is $(1040.40)(0.08)(1/4) = \$20.81$, so that the amount owed after three quarters is $1040.40 + \$20.81 = \$1061.21$. The interest for the fourth quarter is $(1061.21)(0.08)(1/4) = \$21.22$, so that the amount owed after one year is $1061.21 + \$21.22 = \$1082.43$. Note that at the end of each quarter, the interest is added to the principal.

The amount owed on this loan can be computed more directly. To do so, note that the rate of interest i for each quarterly compounding period is given by

$$r = \frac{0.08}{4} = 0.02 = 2\%$$

At the end of one quarter, the amount owed is 1000(1.02). During the second quarter, this entire amount earns interest, so that the amount owed at the end of two quarters is $1000(1.02) \cdot (1.02) = 1000(1.02)^2$. Similarly, the amount owed at the end of three quarters is $1000(1.02)^3$ and, after one year, $1000(1.02)^4$ is owed. This last quantity can be computed using a calculator. You could instead use Appendix Table 1 with $i = 2\%$ and $n = 4$ to obtain

$$(1.02)^4 = 1.082432$$

The amount owed is then $1000(1.082432) = 1082.432$. Since a penny is the smallest unit of currency, this number is rounded off to $1082.43.

Suppose interest is compounded n times per year, at an annual interest rate r. Then r is called the *nominal rate of interest*, $i = r/n$ is the interest rate for each compounding period, and in t years interest is compounded nt times.

Formula for Compound Interest

$$A = P\left(1 + \frac{r}{n}\right)^{nt} = P(1 + i)^{nt} \quad \text{Amount at compound interest}$$

Example 7 A six-year certificate of deposit pays 8% interest compounded semiannually (twice a year). The value of a $2000 certificate of deposit at maturity is $(2000)(1 + 0.08/2)^{2 \times 6} = 2000(1 + 0.04)^{12} = \$3202.06$. The interest paid is $\$3202.06 - \$2000 = \$1202.06$.

Example 8 A $100 bond pays 8% interest for five years. The value of the bond after five years is:

$(100)\left[1 + (0.08)(5)\right] = \$140.00$ if this is simple interest.

$(100)(1 + 0.08)^5 = \$146.93$ if interest is compounded annually.

$(100)\left(1 + \dfrac{0.08}{2}\right)^{2 \times 5} = \$148.02$ if interest is compounded semiannually.

$(100)\left(1 + \dfrac{0.08}{4}\right)^{4 \times 5} = \$148.59$ if interest is compounded quarterly.

$$(100)\left(1 + \frac{0.08}{12}\right)^{12\times 5} = \$148.98 \quad \text{if interest is compounded monthly.}$$

$$(100)\left(1 + \frac{0.08}{365}\right)^{365\times 5} = \$149.18 \quad \text{if interest is compounded daily, and each year is assumed to have 365 days.}$$

■

Observe that as interest is compounded more frequently, more interest is paid. This happens because as soon as interest is compounded, interest already earned becomes principal and starts earning interest. Note that much less interest is paid if simple interest is used than if interest is compounded annually. However, there is little difference between monthly and daily compounding of interest.

The way to obtain the most interest at a given interest rate is to have interest compounded not every day, not every hour, not every minute or second, but at every instant. This is called *continuously compounded interest*. The computation of continuously compounded interest is postponed to Chapter 9, since it involves the exponential function, which is studied there.

Effective interest rate

Since it is difficult to compare different ways of computing interest, the Truth in Lending Act requires that contracts and advertisements involving loans state the *effective interest rate*.

> The *effective interest rate* r_E on a loan is the rate of interest, compounded annually, that results in the same amount of interest as the given compounding procedure.

We will now derive a formula for r_E in terms of r and the number n of times interest is compounded each year. The amount that a \$1 deposit becomes after one year is $(1 + r/n)^n$ and is also $1 + r_E$. Equate these two expressions and solve for r_E to obtain the formula for effective interest rate.

$$\boxed{r_E = \left(1 + \frac{r}{n}\right)^n - 1} \quad \text{Effective interest rate}$$

Example 9 The effective interest rate for 6% compounded

$$\text{semiannually is} \quad \left(1 + \frac{0.06}{2}\right)^2 - 1 = 6.09\%$$

$$\text{quarterly is} \quad \left(1 + \frac{0.06}{4}\right)^4 - 1 = 6.136\%$$

$$\text{monthly is} \quad \left(1 + \frac{0.06}{12}\right)^{12} - 1 = 6.168\%$$

These last two are correct to the nearest one-thousandth of a percent.

Example 10 Bill McFee, owner of the Bath Boutique, is contracting for $100,000 in improvements. He must pay for these improvements on January 15. Since he is short of cash, he wants to borrow the money and repay it, along with all the interest, on the next January 15, using revenue from next year's holiday season.

He visits the banks he does business with. Eighth Federal would charge 10% simple interest. Citizens Savings and Loan charges 9.8% compounded quarterly. National County charges 9.7% compounded monthly. At which bank would he pay the least interest?

Solution Compute the effective interest rates.

$$\text{Eighth Federal} \quad r_E = 10\%$$

$$\text{Citizens Savings and Loan} \quad r_E = \left(1 + \frac{0.098}{4}\right)^4 - 1 = 10.166\%$$

$$\text{National County} \quad r_E = \left(1 + \frac{0.097}{12}\right)^{12} - 1 = 10.143\%$$

He should use Eighth Federal to obtain the lowest effective interest rate and thus pay the least interest. Note that Eighth Federal had the highest nominal interest rate.
■

By now you have noticed that calculating interest involves a great deal of numerical computation. For convenience, tables are provided in the Appendix of the book. Most of the exercises can be done with the aid of these tables. If you wish to, however, you may use your calculator for all of the exercises.

The computations in this chapter, and of the answers for this chapter, were done using the tables, whenever possible. Some calculators may give slightly different answers for the same computations. Thus, if you do a problem and your answer is very close to the answer in the book, you probably did the problem correctly.

There are some problems for which tables are very convenient *under any circumstances*. These problems involve determining when the total in an account has reached a certain value.

Example 11 An account paying 6% interest compounded quarterly is opened with $500. When will the account contain $800?

Solution The amount after t years is given by

$$A = 500\left(1 + \frac{0.06}{4}\right)^{4t} = 500(1 + 0.015)^{4t}$$

When the amount is 800,

$$800 = 500(1 + 0.015)^{4t}$$

$$\frac{800}{500} = 1.6 = (1 + 0.015)^{4t}$$

Now use Appendix Table 1. Look down the column for 1.5% interest until the number 1.6 occurs and then look in the leftmost column to find the number n of payment periods. In this case, the number 1.610324 is found in the table and $n = 32$. Then $4t = 32$, so $t = 8$. The account will contain $800 (actually $805.16) after eight years.

Example 12 If $100 is deposited in an account paying 8% compounded semiannually, how long will it take the money to double?

Solution This leads to the equation

$$200 = 100\left(1 + \frac{0.08}{2}\right)^{2t}$$

$$2 = (1.04)^{2t}$$

Using Appendix Table 1 again gives $n = 18$. Since $2t = 18$, $t = 9$. The money will double after nine years. In fact, any amount of money put into this account will double in nine years.

∎

Exercises 5-1

Compute the interest using the simple interest formula.
1. $5000 at 9% for 21 months.
2. $3000 at 12% for 4 months.
3. $600 at 18% for 6 months.
4. $1200 at 8% for 1 month.
5. $1000 at 7.5% for 2 months.
6. $10,500 at 11% for 9 months.

Compute the amount.
7. $750 at 6% compounded quarterly after 2 years.
8. $900 at 8% compounded semiannually after 5 years.
9. $1200 at 5% compounded annually after 8 years.
10. $800 at 10% compounded quarterly after 2.5 years.
11. $1000 at 7% compounded quarterly after 6 years.
12. $400 at 5% compounded quarterly after 8 years.

Determine the effective interest rate.
13. 5% compounded quarterly.
14. 15% compounded monthly.
15. 10% compounded semiannually.
16. 12% compounded quarterly.
17. How long will it take money to double at 8% compounded quarterly?
18. How long will it take money to quadruple at 6% compounded annually?
19. How long will it take money to triple at 12% compounded quarterly?
20. How long will it take money to double at 4% compounded annually?
21. A bond sells for $500 and will be redeemed for $540 in six months. If Juanita Hernandez buys the bond, what rate of simple interest will she earn on her investment?
22. Society Saving and Loan advertises a $5000 certificate of deposit that pays $5400 after one year. What is the simple interest rate on the certificate?
23. Ralph Williams will come into a large inheritance in two months, and he plans to use part of it to buy a car that sells for $8000. He learns that the price of the car will soon rise to $9000. Instead of waiting and paying the higher price, he borrows $8000 at 15% simple interest for two months and uses this money to buy the car. How much interest does he

pay when the loan is repaid in two months? Overall, how much does he save by buying the car now?

24. Electric rates are rising by 8% each year. An average monthly electric bill is now $30. What will an average monthly electric bill be in ten years?

25. The population of a country is about 20 million and is increasing by 4% each year. What will the population be after 20 years? After 50 years?

26. The cost of houses in a large suburb has been increasing by 15% annually. If this trend continues, how much will a house worth $50,000 today be worth in ten years?

27. John Ellsworth will retire and sell his home in two years. Meanwhile, he wants to invest $30,000 in the stock market. He decides to borrow $30,000 from a bank and repay the loan and the interest as soon as he sells his house. Northern State Bank offers to lend him the money at 12% compounded annually. Central National offers the loan at 11.75% compounded quarterly. Which bank offers him the better deal? How much does he save by using that bank instead of the other one?

28. Suppose the price of lumber increases by 12% each year. A certain size board costs $5. After how many years will it cost $167?

Section 5-2 Annuities

Suppose you get tired of being short of cash every Christmas and open a Christmas Club account. On the fifteenth of each month from January to November, you deposit the same amount into the account. On December 15, you close the account by withdrawing the whole amount you deposited along with all the interest that has accumulated. Your sequence of payments is an annuity. Payroll savings plans and retirement fund contributions also are likely to be annuities.

An *annuity* is a sequence of equal payments made at equal time intervals. The time interval between payments is the *payment period*. The *term* of the annuity is the time from the beginning of the first payment period to the end of the last payment period. The annuity *matures*, or reaches *maturity*, at the end of the last payment period. The *amount*, or *future value*, of an annuity is the sum at maturity of all the payments plus all the interest earned.

5-2 ANNUITIES

> **Definition.** An *ordinary annuity* is an annuity in which payments are made and interest is compounded at the *end* of each payment period.

Right now, let's assume that annuities are ordinary unless stated otherwise. Diagram 1 shows how the payments and interest accumulate for an annuity with a term of four payment periods. Here R is the amount of each payment, and i is the interest rate per payment period. The amount of this annuity is

$$R + R(1 + i) + R(1 + i)^2 + R(1 + i)^3$$

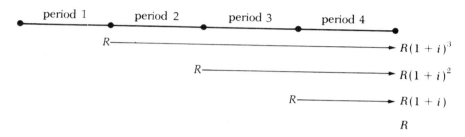

Diagram 1.

For n payment periods, the amount is

$$A = R + R(1 + r) + R(1 + r)^2 + \cdots + R(1 + r)^{n-1}$$

Now let's derive a formula for the amount of an ordinary annuity. To do so, we first consider the mathematical subject of geometric series.

> A geometric series is a sum of the form
> $$b + bc + bc^2 + bc^3 + \cdots + bc^{n-1}$$

Note that there are n terms in the series. The first term is b, and each successive term is obtained by multiplying the previous term by c. Let S be the sum of the series.

$$S = b + bc + bc^2 + bc^3 + \cdots + bc^{n-1}$$

If $c = 1$, each term in the series is b, so $S = bn$. If $c \neq 1$, multiply by c to obtain

$$cS = bc + bc^2 + bc^3 + bc^4 + \cdots + bc^{n-1} + bc^n$$

Recall $S = b + bc + bc^2 + bc^3 + bc^4 + \cdots + bc^{n-1}$, so subtracting gives

$$cS - S = bc^n - b$$
$$(c - 1)S = b(c^n - 1)$$

Thus, if $c \neq 1$, the sum of the geometric series

$$b + bc + bc^2 + bc^3 + \cdots + bc^{n-1}$$

is given by

$$\boxed{S = \frac{b(c^n - 1)}{c - 1}} \quad \text{Sum of a geometric series}$$

Observe that the amount A of an annuity whose term is n payment periods is the sum of a geometric series with n terms, $b = R$, and $c = 1 + i$. Thus the formula for the amount of an ordinary annuity is

$$\boxed{A = \frac{R\left[(1 + i)^n - 1\right]}{i}} \quad \text{Amount of an ordinary annuity}$$

When using this formula, be sure to remember that i is the *interest rate per payment period*. The payment period is not always one year.

Example 1 To save for retirement, Henry and Elsie Kraus deposit $1000 in their savings account each December 31. The account pays 6% interest compounded annually. How much money will be in the account when they retire after 30 years?

Solution The question asks for the amount of an annuity with a term of 30 payment periods, 6% interest per payment period, and payments of $1000. Thus $n = 30$, $i = 6\% = 0.06$, and $R = \$1000$. The amount is

$$A = \frac{1000(1.06^{30} - 1)}{0.06}$$

The numerical value of A can now be computed using a calculator. Alternatively, you can use Appendix Table 3 with $i = 6\%$ and $n = 30$ to obtain

$$\frac{1.06^{30} - 1}{0.06} = 79.058184$$

Then

$$A = 1000 \frac{(1.06)^{30} - 1}{0.06} = 1000 \cdot 79.058184 = 79058.184$$

Since one cent is the smallest unit of money, A is rounded off to the nearest penny. Thus, when they retire, they will have $79,058.18 in the account. Of this money, $30,000 is the sum of their deposits and the remaining $49,058.18 is interest.

Example 2 Alan and Laura Page are saving money for the down payment on a house. At the end of each month, they put $100 in an account that pays 9% interest compounded monthly. How much will the account contain after four years?

Solution This is an annuity with a term of 48 payment periods, $\frac{3}{4}$% interest per payment period, and payments of $100. The amount is

$$A = \frac{100(1.0075^{48} - 1)}{0.0075} = \$5752.07$$

After four years, $5752.07 will be in the account.

Sinking fund

Sometimes it is important to have a definite amount of money available at some definite time in the future. One way to do this is to establish an annuity that matures when the money is needed. Such an annuity is called a *sinking fund*. Suppose the sinking fund is required to contain the amount A after n payment periods and that the interest rate per payment period is i. If R is the amount of each payment, then

$$A = \frac{R[(1+i)^n - 1]}{i}$$

so that

$$R = \frac{Ai}{(1+i)^n - 1} \quad \text{Payments into sinking fund}$$

(*Note:* Appendix Table 4 is useful in many problems involving sinking funds.)

Example 3 The Outland Steel Company needs to replace a furnace in eight years and needs to have $6,000,000 available then. A sinking fund is established with an annual interest rate of 10% compounded quarterly. What are the quarterly payments into the fund?

Solution The quarterly interest rate i is $10\%/4 = 2.5\% = 0.025$ and there are $8 \cdot 4 = 32$ payment periods. Thus

$$R = \frac{(6{,}000{,}000)(0.025)}{1.025^{32} - 1} = \$124{,}608$$

The Outland Steel Company must deposit $124,608 into the sinking fund each quarter. The total amount Outland will put into the fund is $3,987,456; the rest of the six million is accrued interest.

Example 4 Mr. and Mrs. Ernesto Garcia want to save money for their new baby's college education. On each of her birthdays, starting when she is one year old, and ending on her eighteenth birthday, they deposit the same amount into an account paying 8% compounded annually. At maturity (on her eighteenth birthday), the account must contain $20,000. How much is each deposit?

Solution

$$R = \frac{(20{,}000)(0.08)}{1.08^{18} - 1} = \$534.04$$

Each deposit is $534.04.

Example 5 Joan Harris is 50 years old today, and plans to retire on her sixty-fifth birthday. She wants to have $40,000 in her retirement account then. The account pays interest at the rate of 8% compounded monthly. How much should she put into the account each month?

Solution Here $n = 15 \cdot 12 = 180$ and $i = 8\%/12 = 0.0066667$. Then

$$R = \frac{(40{,}000)(0.0066667)}{(1.0066667)^{180} - 1} = \$115.59$$

She must put $115.59 into the account each month. ∎

Annuity due

We next discuss annuities other than ordinary annuities.

> An *annuity due* is an annuity in which payments are made at the *beginning* of each payment period.

The amount A of an annuity due with n payments of R each and interest rate per payment period i is as follows.

$$A = \frac{R\left[(1+i)^{n+1} - 1\right]}{i} - R \quad \text{Amount of an annuity due}$$

Example 6 What is the amount of an annuity due with ten annual payments of $150 each at 9% interest compounded annually?

Solution In the formula, $R = 150$, $i = 0.09$, and $n = 10$.

$$A = \frac{150(1.09^{11} - 1)}{0.09} - 150 = \$2484.04$$

The amount of the annuity due is $2484.04.

Example 7 Wilbur Smith opens a Christmas Club account at a bank. On the fifteenth of each month, starting in January, he puts $25 in his account. The interest rate is 6% compounded monthly. On November 15, he makes his last payment, and on December 15 withdraws all the money in the account. How much does he get?

Solution This is an annuity due with 11 payment periods, so $n = 11$. Since the annual interest rate is 6%, the monthly interest rate is 6%/12 = 0.5%. Thus $i = 0.005$. Then

$$A = \frac{25(1.005)^{12} - 1)}{0.005} - 25 = \$283.39$$

Mr. Smith receives $283.39.

■

Exercises 5-2

In all of the following, every annuity is an ordinary annuity unless stated otherwise.

What is the value at maturity of an annuity with annual payments of $600 and 7% interest compounded annually, if the term of the annuity is

1. Five years?
2. Ten years?
3. Twenty years?
4. Forty years?

What is the value at maturity of an annuity with quarterly payments of $200 and 10% interest compounded quarterly if the term of the annuity is

5. Two years?
6. Five years?
7. Ten years?
8. Twenty years?

What is the value at maturity of an annuity with monthly payments of $10 and 6% interest compounded monthly if the term of the annuity is

9. One year?
10. Five years?
11. Twenty years?
12. Forty years?

What is the value at maturity of an annuity with annual payments of $1000 and 8% interest compounded annually if the term of the annuity is

13. Ten years?
14. Twenty years?
15. Forty years?
16. Fifty years?

A 40-year annuity requires annual payments of $500. What is the value of this annuity at maturity if interest is compounded annually at the rate of

17. 6%?
18. 7%?
19. 8%?
20. 10%?

A school savings plan requires monthly payments of $10. What is the value of this account at maturity (after 12 payments) if interest is compounded monthly at the nominal rate of

21. 6%?
22. 9%?
23. 12%?
24. 15%?

A sinking fund is required to contain $100,000 after five years. Deposits are made annually, and interest is compounded annually. How much is de-

posited each time if the interest rate is

25. 5%?   **26.** 6%?   **27.** 8%?   ◆ **28.** 10%?

A sinking fund is required to contain $250,000 after ten years. Deposits are made quarterly. How much is deposited each time, if interest is compounded quarterly at the rate of

29. 5%?   ◆ **30.** 9%?   **31.** 12%?   **32.** 16%?

A sinking fund is required to contain $10,000 after six years. Interest is paid at the rate of 8%. How much is deposited each time, if the deposits are made and interest is compounded

33. Annually?   **34.** Semiannually?
35. Quarterly?   ◆ **36.** Monthly?

An annuity due with monthly payments of $20 pays 9% interest compounded monthly. What is the amount of this annuity if its term is

37. One year?   **38.** Two years?
39. Four years?   ◆ **40.** Five years?

An annuity due with a term of three years requires monthly payments of $10. Interest is compounded monthly. What is the amount of this annuity, if the interest rate is

41. 9%?   **42.** 12%?   **43.** 15%?   **44.** 18%?

An annuity due with a term of ten years requires monthly payments of $25. Interest is compounded monthly. What is the amount of this annuity, if the interest rate is

◆ **45.** 6%?   ◆ **46.** 8%?   ◆ **47.** 10%?   ◆ **48.** 12%?

49. Joe Tew is saving money for the down payment on a Datsun. Every month he puts $50 in an account that pays 9% interest compounded monthly. How much is in the account after three years?

50. Sara Wadsworth wants to go on a vacation to Europe in a year. The vacation will cost $1800. How much must she deposit each month in an account paying 9% interest compounded monthly in order to have enough saved for the vacation?

The ABCD Company will need to build a new factory in the future to replace its current facility, which is becoming obsolete. To begin construction of the new factory, the company needs to have $10,000,000 in cash. A *sinking fund is established. There will be annual payments, and the interest rate is 9% compounded annually. How much must be put into the sinking fund each year, if the construction of the new factory is to begin in*

51. Five years?
52. Ten years?
53. Fifteen years?
54. Twenty years?

Henrietta Dove wants to accumulate $8000 for a down payment on a house. To do this, she makes monthly payments into an account that pays 9% interest compounded monthly. How much must her monthly payments be, if the $8000 is to be available in

55. One year?
56. 18 months?
57. Two years?
58. Four years?

(*Note:* This exercise refers to Example 5 of Section 5-1.)

59. The town of Hamilton starts a sinking fund to pay off its bonds. How much must the annual payments be if the interest rate on the sinking fund is 7% compounded annually?

Exercises 60, 61, and 62 involve annuities due.

60. Jim Moore participates in a payroll savings plan at work. When he gets paid on the first of the month, $50 is deducted from his paycheck and put into a savings account that pays 6% interest compounded monthly. How much is in the account at the end of $2\frac{1}{2}$ years?

61. To help pay for her grandchild's education, Mary Bisbee puts $500 into an account on the day the child is born and on each of the child's first 17 birthdays. If the account pays 9% interest compounded annually, how much is in the account on her grandchild's eighteenth birthday?

62. Repeat exercise 61 using an interest rate of 12%.

*63. Joe Klugman is 50 years old today and is planning his future. He plans to work full time for ten more years. During this time, he will deposit $1000 each quarter into a retirement account that pays 12% interest compounded quarterly. He will then be semiretired for five years. During this time, the money in his retirement account will remain there and accumulate interest, but he will make no further contributions.

   a. How much will be in Joe's retirement account when he retires at age 65?

   b. Answer part (a) if the interest rate is 10% instead of 12%. How much less would Joe have at the lower interest rate?

c. Assume a 12% interest rate. Joe rethinks his plans. When he retires, he needs $5000 each quarter to live as he desires. He wants the interest on his account to provide this amount so that he can avoid touching the principal. For this to be possible, how large must his quarterly deposits be?

Section 5-3 Present Value and Amortization

Jill Cranston recently bought a $600 television set at Appliance House for $200 down, and two years of easy monthly payments. Her payments form an annuity. In Section 5-2, we learned how to determine the amount of an annuity at the end of its term.

However, for its annual audit, Appliance House needs to be able to assess the value of this annuity now. In this section, we learn how to compute the present value of annuities. Let's begin by studying the present value of a single deposit made in the future.

> The *present value* of a given amount to be received in the future, given a specific interest rate, is its worth now.

For example, given a 6% interest rate, the present value of $10.60 one year from now is $10.00. This is true, because, if $10.00 is deposited now at 6% interest, it will become $10.60 after one year. Suppose P is the present value of an amount A to be received at time t. Then if the amount P is deposited in an account now, the amount in the account, including interest, at time t will be A.

If i is the interest rate per compounding period, the compound interest formula says that

$$A = P(1 + i)^n$$

after n compounding periods have elapsed. Solve this relation for P to obtain the formula for present value.

$$P = A(1 + i)^{-n} = \frac{A}{(1 + i)^n} \quad \text{Present value}$$

(*Note:* Appendix Table 2 is useful in many problems involving present value.)

Example 1 What is the present value of $2000 two years from now if interest is 12% compounded quarterly?

Solution Since interest is compounded four times each year, $i = 12\%/4 = 3\%$ and $n = 2 \times 4 = 8$. Then

$$P = \frac{2000}{(1.03)^8} = \$1578.82$$

Example 2 What is the present value of $10,000 twenty years from now if interest is compounded annually at the rate of 9%?

Solution
$$A = \frac{10,000}{(1.09)^{20}} = \$1784.31$$

Example 3 Helen wants to deposit enough now in bonds paying 11% compounded annually so that her new-born grandson will have $10,000 on his eighteenth birthday. How much must she deposit?

Solution The problem asks for the present value of $10,000 eighteen years from now at 11% interest.

$$P = \frac{10,000}{(1.11)^{18}} = \$1528.22$$

She must deposit $1528.22. In 18 years, the account will earn $8471.78 interest. ∎

Now we will derive a formula for the present value of an ordinary annuity. Suppose the interest rate per payment period is i, the amount of each payment is R, and the term of the annuity is n payment periods. Suppose A is the amount of the annuity at the end of its term and P is the present value of A. Then

$$P = A(1 + i)^{-n}$$

by the formula for present value. By the formula for the amount of an annuity,

$$A = \frac{R[(1+i)^n - 1]}{i}$$

Substitute this value for A in the formula for P to obtain

$$P = \frac{R[(1+i)^n - 1]}{i}(1+i)^{-n} = \frac{R[1-(1+i)^{-n}]}{i}$$

The formula for the present value of an annuity is

$$P = \frac{R[1-(1+i)^{-n}]}{i} \quad \text{Present value of an annuity}$$

(*Note:* Appendix Table 5 is useful in determining the present value of an annuity.)

Example 4 At retirement, Joe Torelli wants to have a savings account from which he can draw $900 monthly for 20 years, at which time the account will be depleted. If his money draws 8% interest compounded monthly how much must he have in the account at retirement?

Solution The problem asks for the present value of an annuity paying $900 monthly for 20 years. The interest rate per month is $8\%/12 = \frac{2}{3}\%$ and there are $20 \cdot 12 = 240$ payment periods.

$$P = \frac{900(1 - 1.0066667^{-240})}{0.0066667} = \$107{,}599$$

He must have $107,599 in the account at retirement.

Example 5 A Million Dollar Winner in a lottery receives $50,000 at the end of each year for 20 years. What is the present value of the amount won, if 7.75% interest compounded annually is available?

Solution

$$P = 50{,}000 \frac{(1 - 1.0775^{-20})}{0.0775} = \$500{,}176$$

The amount won has present value $500,176. The winner should be called a Half-Million Dollar Winner. ∎

Amortization

One extremely important use of annuities is in paying off loans. A loan is amortized if both the principal and the interest are paid off by a sequence of equal payments at equal time intervals, that is, by an annuity. In this case, the principal on the loan is equal to the present value of the annuity.

The amount of each payment is denoted R in the formula for annuities. Solving the formula for present value of an annuity for R gives the formula for amortization

$$R = \frac{Pi}{1 - (1 + i)^{-n}} \quad \text{Amortization}$$

(*Note:* Appendix Table 6 is useful in amortization problems.)

Example 6 Mr. and Mrs. McDonald buy a house for $60,000 with a $20,000 down payment. They take out a 15-year mortgage at 10% interest compounded monthly for the remaining $40,000. How much is their monthly payment?

Solution The monthly interest rate is $10\%/12 = 0.0083333$. There are $15 \cdot 12 = 180$ monthly payments, so that $n = 180$.

$$R = \frac{(40{,}000)(0.0083333)}{1 - (1.0083333)^{-180}} = \$429.84$$

The monthly payment is $429.84. If the mortgage were for 20 years, $n = 240$ and the monthly payment would be

$$R = \frac{(40{,}000)(0.0083333)}{1 - (1.0083333)^{-240}} = \$386.01$$

If the mortgage were for 30 years, $n = 360$ and the mortgage payment would be

$$R = \frac{(40{,}000)(0.0083333)}{1 - (1.0083333)^{-360}} = \$351.03$$

5-3 PRESENT VALUE AND AMORTIZATION

Example 7 A car dealer offers a specially equipped Antelope for $6000, with $1000 down and four years to pay off the rest at 15% interest compounded monthly. How much are the monthly payments?

Solution The amount to be amortized is $5000. There are 48 payment periods, and the interest per payment period is 15%/12 = 1.25%.

$$R = \frac{(5000)(0.0125)}{1 - 1.0125^{-48}} = \$139.16$$

The monthly payments are $139.16.

Example 8 Ralph Heimlich buys the Antelope according to the terms of Example 7. How much interest does he pay?

Solution Ralph makes 48 payments of $139.16 each, for a total of $6679.68. Since he has borrowed $5000, he pays 6679.68 − 5000 = $1679.68 interest.

Exercises 5-3

Assume all annuities are ordinary.

What is the present value of $50,000 ten years from now, if interest is 12% compounded

 1. Annually? 2. Semiannually?
3. Quarterly?   4. Monthly?

What is the present value of $12,000 four years from now if interest is compounded annually at the rate of

5. 4%?   6. 6%?   7. 8%?   8. 10%?

How much must be deposited now in an account paying 10% interest compounded semiannually in order to have $12,500 in

9. One year?   10. Two years?
11. Five years?   12. Ten years?

How much must be deposited now in an account paying 7% interest compounded quarterly in order to have $1000 after

13. One year?   14. Two years?
15. Five years?   16. Eight years?

What is the present value of an annuity paying $250 annually for ten years if interest is compounded annually at the rate of

17. 6%? 18. 8%? 19. 10%? 20. 20%?

What is the present value of an annuity paying $5000 annually for 40 years, if interest is compounded annually and the interest rate is

21. 4%? 22. 6%? 23. 7%? 24. 8%?

What is the present value of an annuity paying $10 monthly for two years if interest is compounded monthly at the annual rate of

25. 9%? 26. 12%? 27. 15%? 28. 18%?

What is the present value of an annuity paying $4000 annually for 40 years if interest is compounded annually and the interest rate is

29. 5%? 30. 6%? 31. 7%? 32. 8%?

A $10,000 loan is to be paid off with four years of monthly payments. How much is each payment, if interest is compounded monthly and the nominal interest rate is

33. 9%? 34. 12%? 35. 15%? 36. 18%?

A $5000 loan is to be paid off with 20 monthly payments. How much is each payment, if interest is compounded monthly and the nominal interest rate is

37. 9%? 38. 12%? 39. 15%? 40. 18%?

A five-year home improvement loan of $3000 requires monthly payments. How much is each payment, if interest is compounded monthly and the interest rate is

41. 10%? 43. 12%? 43. 14.5%? 44. 16.2%?

A three-year car loan of $4500 requires monthly payments. How much is each payment, if interest is compounded monthly and the interest rate is

45. 9%? 46. 12%? 47. 15%? 48. 18%?

49. John Henry, age 50, just sold his business for $200,000. He wants to save some of this money for retirement at age 62 and to spend the rest now. He wants to have $250,000 in his account when he retires. How much can he spend now, if the interest rate is 10% compounded annually?

50. Joanna and John Kerchoff have a new daughter. To provide for her

education, they want to deposit enough money now into an account in the baby's name so that she will have $20,000 on her eighteenth birthday. How much must they deposit, if the account pays 8% interest compounded annually?

51. Willie Hayes is a fullback for a football team. To avoid paying extremely high taxes now and not having much income after retiring from football, his agent negotiates a deferred payment plan for him. For his services this year, Willie will receive $50,000 now. In addition, on each of the first ten anniversaries of this contract, Willie will receive $30,000. If interest is compounded annually at the rate of 10%, what lump sum payment now would be equivalent to his salary?

52. A stereo system sells for $500 down and three years of monthly payments of $25. The interest rate is 15% compounded monthly.
   a. What is the price of the stereo system?
   b. What is the total amount of interest paid?

53. Helen O'Neill purchases a home for $80,000 with a $20,000 down payment. How much are her monthly mortgage payments, if the interest rate is 9.5% compounded monthly and the mortgage is for
   a. 15 years?
   b. 20 years?
   c. 30 years?

54. Wilbur Jones purchases a home for $75,000 with a $25,000 down payment. He takes out a 30-year mortgage. How much are his monthly payments, if interest is compounded monthly at the rate of
   a. 7%?
   b. 8%?
   c. 9%?
   d. 10%?

55. Max Mocha decides to buy a new sports car. He finds the car he wants at both Sam Miller Sports Cars and at Square Deal Cars. Sam Miller offers to sell the car for $9000. This is to include $2000 down and 36 monthly payments to be figured using an annual interest rate of 15%. At Square Deal, the car costs $8800 with a $2000 down payment and 36 monthly payments figured using an annual interest rate of 18%. Where should Max buy the car? Both dealers compound interest

monthly.

*56. Laura Wilcox is starting college and takes out a student loan. Each September 1, for four years, she borrows $1200 at 3% interest compounded annually. Laura graduates as planned in June of her fourth year as a student and gets a full-time job. Beginning the next September, the interest rate on the loan increases to 9% compounded monthly. Laura is expected to pay the loan off with 48 monthly payments, with the first payment due at the end of September. How much is each payment?

*57. Mr. Pugnace has little confidence in his son and daughter-in-law. When their first child Matilda is born, he worries about her future and decides to provide for her by selling some of his stock and establishing a trust fund for her. Interest on the trust fund is at the rate of 10% compounded annually. Mr. Pugnace decides that Matilda should receive $5000 on her eighteenth birthday and on each birthday thereafter, concluding with her thirtieth birthday. The last payment to Matilda is to exhaust the trust fund. How much must Mr. Pugnace deposit in the trust fund? (*Hint:* Treat the sequence of payments to Matilda as the difference of two annuities. This annuity is called a *deferred annuity* because the payments to Matilda begin more than one payment period after the money is deposited in the account.)

58. Repeat exercise 57 if the interest rate is
   a. 6%?
   b. 12%?

Section 5-4 Amortization Tables

Two years ago, Wanda Johnson bought a new car for $8000 with $1500 down and a loan at 16.8% annual interest, compounded monthly, for the balance. The loan is to be paid off with 36 equal monthly payments. Because of the high interest rate, Wanda is thinking of paying off the balance of the loan all at once. How much is the balance on the loan?

To determine the balance, first compute her monthly payment. Note that the monthly interest rate i is $16.8\%/12 = 1.4\% = .014$ and that her loan is for $6500.

$$R = \frac{(6500)(0.14)}{1 - (1.014)^{-36}} = \$231.10$$

Now we'll compute the balance on the loan in two different ways.

1. *Method A.* Wanda still has to make 12 payments of $231.10 each. These payments form an annuity. The present value of this annuity is

$$P = \frac{231.10[1 - 1.014^{-12}]}{0.014} = \$2536.50$$

Wanda still owes $2536.50 on her car.

2. *Method B.* Wanda has already made 24 payments of $231.10 each. These payments form an annuity. By the formula for present value of an annuity, the value of this annuity at the time she bought the car is

$$P = \frac{231.10[1 - 1.014^{-24}]}{0.014} = \$4683.24$$

Thus, she has still not paid off $6500.00 - 4683.24 = \$1816.76$ of the principal. This remaining principal has been earning interest for two years and its value now is

$$(1816.76) \cdot (1.014)^{24} = \$2536.35$$

The results of methods A and B differ by 15¢. Both methods are correct in theory, so how can they give different results?

The answer to this question is that the amount of each mortgage payment is not exactly the amount given by the amortization formula. The reason for this is that the smallest unit of money is the penny, and therefore the payment amount given by the amortization formula must be rounded off to the nearest cent. This rounding off causes methods A and B to give slightly different results.

In business, amounts of money must be accurate to the penny.

To correctly compute the balance on Wanda's loan, and also to answer other questions about her mortgage, we must construct an *amortization*

TABLE 1 Amortization Table

Period	Outstanding principal at start of period	Payment	Interest	Principal paid	Outstanding principal at end of period
1	6500.00	231.10	91.00	140.10	6359.90
2	6359.90	231.10	89.04	142.06	6217.84
3	6217.84	231.10	87.05	144.05	6073.79
4	6073.79	231.10	85.03	146.07	5927.72
5	5927.72	231.19	82.99	148.11	5779.61
6	5779.61	231.10	80.91	150.19	5629.42
7	5629.42	231.10	78.81	152.29	5477.13
8	5477.13	231.10	76.68	154.42	5322.71
9	5322.71	231.10	74.52	156.58	5166.13
10	5166.13	231.10	72.33	158.77	5007.36
11	5007.36	231.10	70.10	161.00	4846.36
12	4846.36	231.10	67.85	163.25	4683.11
13	4683.11	231.10	65.56	165.54	4517.57
14	4517.57	231.10	63.25	167.85	4349.72
15	4349.72	231.10	60.90	170.20	4179.52
16	4179.52	231.10	58.51	172.59	4006.93
17	4006.93	231.10	56.10	175.00	3831.93
18	3831.93	231.10	53.65	177.45	3654.48
19	3654.48	231.10	51.16	179.94	3474.54
20	3474.54	231.10	48.64	182.46	3292.08
21	3292.08	231.10	46.09	185.01	3107.07
22	3107.07	231.10	43.50	187.60	2919.47
23	2919.47	231.10	40.87	190.23	2729.24
24	2729.24	231.10	38.21	192.89	2536.35
25	2536.35	231.10	35.51	195.59	2340.76
26	2340.76	231.10	32.77	198.33	2142.43
27	2142.43	231.10	29.99	201.11	1941.32
28	1941.32	231.10	27.18	203.92	1737.40
29	1737.40	231.10	24.32	206.78	1530.62
30	1530.62	231.10	21.43	209.67	1320.95
31	1320.95	231.10	18.49	212.61	1108.34
32	1108.34	231.10	15.52	215.58	892.76
33	892.76	231.10	12.50	218.60	674.16
34	674.16	231.10	9.44	221.66	452.50
35	452.50	231.10	6.34	224.76	227.74
36	227.74	230.93	3.19	227.74	0.00

table. See Table 1. An amortization table tells you how much of each payment is interest and how much goes towards principal, and how much is still owed after each payment.

To compute the first row of the amortization table, note that the entire $6500 borrowed is outstanding since no payments have been made yet. The interest on this amount is given by the simple interest formula as

$$(6500.00) \cdot (0.014) \cdot (1) = \$91.00$$

because the interest rate per payment period is 1.4% = 0.014. Since 231.10 was paid,

$$231.10 - 91.00 = \$140.10$$

is applied towards the principal, and the amount still owed is

$$6500.00 - 140.10 = \$6359.90$$

To compute the second row of the amortization table, repeat the same process. Note that the outstanding principal at the start of this period equals the outstanding principal at the end of the previous period, in this case $6359.90. The other rows of the table, except for the last row, are computed in a similar way.

To compute the last row, note that the outstanding principal at the end of this last payment period must be exactly 0, because the loan is exactly paid off. Therefore the outstanding principal at the start of this period, $227.74 in this case, must equal the principal paid. The interest paid during this payment period is

$$(227.74) \cdot (0.014) \cdot (1) = \$3.19$$

Then the payment must be

$$227.74 + 3.19 = \$230.93$$

This last payment is $0.17 less than each of the other payments. *The last payment on a mortgage is usually slightly different from the other payments.* However, sometimes the last payment equals the other payments. Note that this last payment is not computed with the amortization formula. An amortization table must be constructed in order to determine the last payment accurately.

Let's return to the original question: how much is the balance on Wanda's loan after 24 payments? The amount in question is the outstanding principal at the end of 24 payment periods. From Table 1, this amount is $2536.35. Note that method B gave the same amount. Method B is very accurate but will not always give exactly the correct amount. Method A is less accurate because the last payment differs from all the other payments.

Once the amortization table has been constructed, most questions about the loan can be answered easily.

Example 1 Wanda made her first payment in July, and during that year made a total of six payments. How much interest did she pay that year?

Solution Add the interest for the first six payment periods to obtain

$$91.00 + 89.04 + 87.05 + 85.03 + 82.99 + 80.91 = \$516.02$$

Wanda paid $516.02 in interest that year.

Example 2 How many payments did Wanda make until half of the principal was paid off?

Solution When half of the principal is paid off, half (or slightly less) of the principal remains outstanding. Since half of the principal is $3250, look down the last column of Table 1 until you see a number less than $3250. The first such number is $3107.07 in period 21. Therefore Wanda made 21 payments until half of the principal was paid off.

Exercises 5-4

Do exercises 1–8 without constructing amortization tables.

Bob and Sue Brown have a four-year $7000 car loan at 18% annual interest.

1. How much is their monthly payment?
2. How much do they still owe on the car after making 30 payments?
3. How much do they still owe on the car after making 40 payments?
4. How much of their forty-first payment is interest?

Ron and Joan Jambrowski buy a house for $130,000 with $30,000 down. They take out a 30-year mortgage for the balance at 15% annual interest.

5. How much is their monthly payment?
6. How much do they still owe on the house after making 100 payments?
7. How much of their 101st payment is interest?
8. How much of their 101st payment is used to reduce the outstanding principal?

5-4 AMORTIZATION TABLES

The following table is an amortization table for a four-year $10,000 loan at 16% interest compounded quarterly, with quarterly payments. Use this table for exercises 9–16.

Period	Outstanding principal at start of period	Payment	Interest	Principal paid	Outstanding principal at end of period
1	10000.00	858.20	400.00	458.20	9541.80
2	9541.80	858.20	381.67	476.53	9065.27
3	9065.27	858.20	362.61	495.59	8569.68
4	8569.68	858.20	342.79	515.41	8054.27
5	8054.27	858.20	322.17	53.03	7518.24
6	7518.24	858.20	300.73	557.47	6960.77
7	6960.77	858.20	278.43	579.77	6381.00
8	6381.00	858.20	255.24	602.96	5778.04
9	5778.04	858.20	231.12	627.08	5150.96
10	5150.96	858.20	206.04	652.16	4498.80
11	4498.80	858.20	179.95	678.25	3820.55
12	3820.55	858.20	152.82	705.38	3115.17
13	3115.17	858.20	124.61	733.59	2381.58
14	2381.58	858.20	95.26	762.94	1618.64
15	1618.64	858.20	64.75	793.45	825.19
16	825.19	858.20	33.01	825.19	0.00

9. How much is the first payment?

10. How much is the last payment?

11. How much of the third payment is interest?

12. How much of the eighth payment is interest?

13. How much of the eleventh payment is used to reduce the outstanding principal?

14. How much of the fourteenth payment is used to reduce the outstanding principal?

15. How much interest is paid during the first year of the loan?

16. How much interest is paid during the last year of the loan?

17. Construct an amortization table for a five-year loan of $2500 with annual payments and an annual interest rate of 14%.

18. Construct an amortization table for a six-year loan of $3000 with annual payments and an annual interest rate of 12%.

19. Construct an amortization table for a one-year loan of $1000 with monthly payments at an annual interest rate of 15.6%.

20. Construct an amortization table for a one-year loan of $1500 with monthly payments at an annual interest rate of 17.4%.

Chapter 5 Review

Compute the interest using the simple interest formula.
1. $500 at 11% for six months.
2. $8000 at 9% for two months.

3. Compute the interest on a loan of $2000 at 8% compounded quarterly after seven years.

4. Compute the interest on a loan of $250 at 18% compounded monthly after eight months.

5. What is the effective interest rate on a loan at 12% compounded monthly?

6. How long will it take money to triple at 10% compounded semiannually?

7. What is the value at maturity of an annuity with 30 monthly payments of $50 and an interest rate of 9% compounded monthly?

8. What is the value at maturity of an annuity with 25 annual payments of $400 and an interest rate of 6% compounded annually?

9. A sinking fund must contain $250,000 in five years. Deposits are made quarterly into an account paying 8% interest compounded quarterly. How much is deposited each time?

10. A sinking fund must contain $2,000,000 after eight years. Deposits are made semiannually into an account paying 8% interest compounded semiannually. How much must be deposited each time?

11. What is the present value of $35,000 four years from now at 8% interest compounded annually?

12. How much must be deposited now into an account paying 12% interest compounded monthly in order to have $500 in the account in two years?

13. What is the present value of an annuity paying $800 annually for 40 years if the interest rate is 6% compounded annually?

14. What is the present value of an annuity paying $200 monthly for four years if the interest rate is 15% compounded monthly?

15. How much are the monthly payments on a two-year car loan if $5000 is

financed at an annual interest rate of 18% and interest is compounded monthly?

16. How much are the quarterly payments on a nine-year $14,000 loan if interest is 8% compounded quarterly?

17. Central State Bank offers to lend Johannes Wolfgang $2000 for one year at 14% simple interest. How much interest would Mr. Wolfgang pay if he took the loan?

18. Citizens National Savings and Loan offers to lend Mr. Wolfgang the $2000 at 12% compounded monthly. How much interest would there be on this loan?

19. Wanda Bartlett deposits $90 into an account at the end of each month. The account pays 9% interest compounded monthly. How much is in the account after $3\frac{1}{2}$ years? When does the amount in the account reach $5000?

20. The Southern Stream Railroad needs $5,000,000 in ten years to replace a bridge. A sinking fund is established to pay for the bridge replacement. How much must be the annual deposit into the fund, if the interest rate is 8% compounded annually?

21. Henry and Elizabeth Wang have just won first prize in a promotional contest sponsored by a furniture company. They can have $2500 now as their prize, or they can have $1000 now and $150 a year for 20 years, starting a year from now. If 9% interest compounded annually is available, which option should they take?

22. Esther Polyansky buys $1500 of furniture with 20% down and 18 monthly payments, with 15% annual interest compounded monthly. How much is each payment? How much does she still owe on the furniture after making half of the payments? How much interest does she pay?

23. Mr. and Mrs. Abramowitz buy a house with a 30-year $80,000 mortgage at 12% annual interest compounded monthly. How much is each monthly payment? How much do they still owe on the house after making payments for 20 years? How much of their first payment in the twenty-first year is interest and how much goes to paying off the principal?

Probability

You usually think of mathematics as dealing with certainty. A given number either is or is not a solution of a specific equation. In a linear programming problem where cost is to be minimized, the minimum possible cost is $128 or it is not.

Mathematics also deals with uncertainty. The following statements deal with uncertainty but can also be treated mathematically.

1. There is a 40% chance of rain today.
2. We have a 70% chance of getting that account.
3. The probability of a nuclear reactor accident this year is .03.
4. This test will detect 98% of all tuberculosis cases.

Probability theory is a branch of mathematics that treats such statements.

Section 6-1 An Introduction to Probability

In probability theory, one studies the *likelihood* or *probability* of the various outcomes of an experiment. Later in this section two basic ways of computing probabilities are discussed. First, we need some terminology. If a die is thrown, there are six possible *outcomes*, since 1, 2, 3, 4, 5, or 6 dots will show.

These six outcomes together form the *sample space* of the experiment. If I have bet $10 that an odd number of dots show, my concern is not with the specific outcome but with the *event* that 1, 3, or 5 dots show.

> **Definition.** The *sample space* of an experiment is the set of all possible outcomes of the experiment. An *event* is any set of outcomes, that is, any subset of the sample space.

Consider an experiment with a finite sample space S. Each outcome O_i will have a probability p_i. The numbers p_i are nonnegative and their sum is 1. If E is any event, the *probability of E*, written $p(E)$, is the sum of the probabilities of the outcomes contained in E.

Example 1 Alice, Conrad, Joe, and Marcia are playing a game. $P(\text{Alice wins}) = .4$ $P(\text{Conrad wins}) = .1$, $P(\text{Joe wins}) = .3$, and $P(\text{Marcia wins}) = .2$. The experiment is the playing of the game. The sample space is {Alice wins, Conrad wins, Joe wins, Marcia wins}.

PROBLEM A What is the probability that Conrad or Marcia wins?

Solution The probability that Conrad or Marcia wins is .1 (the probability that Conrad wins) + .2 (the probability that Marcia wins) = .3.

PROBLEM B What is the probability that a girl wins?

Solution Alice and Marcia are the girls in the game. The probability that a girl wins is $P(\text{Alice wins}) + P(\text{Marcia wins}) = .4 + .2 = .6$.

PROBLEM C What is the probability that Joe loses?

Solution Joe loses is the same event as Alice, Conrad, or Marcia wins and has probability $.4 + .1 + .2 = .7$.

PROBLEM D If the game is played 1000 times, how many times do you expect Joe to lose?

Solution Since the probability of Joe losing a game is .7, we would expect Joe to lose about $(.7)(1000) = 700$ times. However, it is *possible* that Joe would lose every game, or that he would win every game.

Example 2 A die is loaded so that the probability of showing n is proportional to n.

PROBLEM A What is the probability of each member of the sample space?

Solution The sample space is $\{1, 2, 3, 4, 5, 6\}$. If $P(1) = x$, then $P(2) = 2x$, $P(3) = 3x$, $P(4) = 4x$, $P(5) = 5x$, and $P(6) = 6x$. Since the sum of the probabilities of all members of the sample space is 1, then $1 = x + 2x + 3x + 4x + 5x + 6x = 21x$, so that $x = 1/21$. Thus,

$$P(1) = \frac{1}{21} \quad P(2) = \frac{2}{21} \quad P(3) = \frac{3}{21}$$

$$P(4) = \frac{4}{21} \quad P(5) = \frac{5}{21} \quad P(6) = \frac{6}{21}$$

PROBLEM B What is the probability that the die shows an even number?

Solution The even numbers in the sample space are 2, 4 and 6. Thus, the probability that the die shows an even number is

$$P\{2, 4, 6\} = \frac{2}{21} + \frac{4}{21} + \frac{6}{21} = \frac{12}{21} = \frac{4}{7}$$

PROBLEM C What is the probability that the die shows a number less than 1? less than 7? less than 5?

Solution The probability that the die shows a number less than 1 is $P(\emptyset) = 0$—this event cannot occur because 1 is the smallest number on the die. The probability that the die shows a number less than 7 is $P\{1, 2, 3, 4, 5, 6\} = P(S) = 1$—this event is certain, since every number on the die is less than 7. The probability that the die shows a number less than 5 is $P\{1, 2, 3, 4\} = 1/21 + 2/21 + 3/21 + 4/21 = 10/21$. If the die is tossed 210 times, we would expect it to show a number less than 5 about $(10/21)(210) = 100$ times.

■

If an event E has probability p and the experiment is repeated m times, we expect E to occur pm times. However, it is possible for E to occur anywhere from 0 to m times. If m is very large, however, it is very likely that E will occur about pm times.

Since the empty set $\emptyset$ contains no outcomes, $P(\emptyset) = 0$. An event has probability 0 if it is impossible for the event to occur. $P(S) = 1$, since all outcomes are in S. An event has probability 1 if the event is certain to occur. For any event $E, 0 \leq p(E) \leq 1$.

Empirical probability

How are the probabilities of the different outcomes of an experiment determined? One way is called *empirical* or *experimental probability*. If an experiment has been repeated n times, and E has occurred r times, then $P(E)$ is assumed to be r/n. For this procedure to be reasonable, n must be a large number.

Example 3 Last year, 50,000 sales were made at a store. 20,000 sales were for over $10. The probability that the next sale will be for over $10 is $20{,}000/50{,}000 = .4$.

Example 4 A factory produces radios. Last year, 240,000 out of the 250,000 radios produced passed inspection. The probability that a radio will pass inspection is $240{,}000/250{,}000 = .96$.

Equally likely outcomes

Now let's consider an alternate way to assign probabilities. First, a bit of notation is needed. If T is any set, then

$$n(T) \text{ is the number of members of } T.$$

Thus, $n\{a, b, c\} = 3$ and $n\{4, 6, 8, 15, 23\} = 5$. Now assume that there are m possible outcomes of an experiment and that these outcomes are all equally likely. Then each outcome has probability $1/m$. If E is any event, then

$$P(E) = \frac{n(E)}{n(S)} \qquad \text{Equally likely outcomes}$$

This formula is used throughout Chapters 6 and 7 to assign probabilities. Later on we will study several kinds of counting techniques in order to learn how to compute $n(E)$ and $n(S)$.

Example 5 A card is drawn at random from a standard deck of cards. What is the probability that the card is a spade? Is an ace? Is a spade or an ace?

Solution The phrase "*at random*" means that any card is as likely to be drawn as any other card. Since there are 52 cards and 13 of the cards are spades, the probability that the card is a spade is $13/52 = 1/4$. The probability that the card is an ace is $4/52 = 1/13$, since there are 4 aces. The probability that the card is a spade or an ace is $16/52$—do not count the ace of spades twice.

Example 6 A digit is selected at random. What is the probability that the digit is odd? Is greater than 5?

Solution There are ten digits: $0, 1, 2, 3, 4, 5, 6, 7, 8,$ and 9. Five of the digits are odd, so that the probability that a digit selected at random is odd is $5/10 = 1/2$. The probability that a digit selected at random is greater than 5 is $4/10$, since the digits that exceed 5 are $6, 7, 8,$ and 9.

Example 7 A die is tossed. The die is not loaded, so that all faces are equally likely to show. What is the probability that 4 or 5 shows? That 2 does not show?

Solution The sample space is $\{1, 2, 3, 4, 5, 6\}$ and contains 6 equally likely outcomes. Thus, the probability that 4 or 5 shows is $2/6$. The probability that 2 does not show is $5/6$, since there are 5 outcomes besides 2.

Example 8 A jar contains 6 green marbles, 8 red marbles, and 12 yellow marbles. A marble is selected at random from the jar. What is the probability that the marble is red?

Solution The sample space is the set of marbles in the jar. The probability that the marble is red is $8/26$, since the sample space has $6 + 8 + 12 = 26$ members and 8 of them are red.

Exercises 6-1

The sample space of an experiment has seven members, named $A, B, C, D, E, F,$ and G. $P(A) = .1, P(B) = .05, P(C) = .2, P(D) = .15, P(E) = .25, P(F) = .18,$ and $P(G) = .07$. What is the probability of each of the following events?

1. $\{A, B, C\}$
2. $\{F, G\}$
3. $\{A, B, C, D, E, F, G\}$
4. $\{A, E, F\}$
5. $\{E, B\}$
6. $\{A, C, D, F\}$
7. $\{\emptyset\}$
8. $\{G, A, D, D\}$

A card is drawn at random from a deck. What is the probability that the card is

9. A heart?
10. A club or a diamond?
11. A seven?
12. An ace?
13. A three or a jack?
14. A five, six, or seven?
15. A five, six, or heart?
16. An eight and a club?

A jar contains 5 red marbles, 7 green marbles, 10 yellow marbles, and 18 pink marbles. A marble is selected at random. What is the probability that the marble is

17. Pink?
18. Green?
19. Yellow or red?
20. Pink and red?
21. Not green?
22. Red or purple?
23. Orange?
24. Not blue?

Use the same jar as in exercises 17–24. After the marble is selected, it is replaced and the jar is shaken. This procedure is performed 200 times. About how many times do you expect the marble to be

25. Green?
26. Yellow?
27. Pink or yellow?
28. Not red?

29. A lottery awards 1000 prizes. 100,000 tickets are sold. If Max has one ticket, what is the probability that he will win a prize? If Koko has 23,000 tickets, about how many prizes should she expect to win?

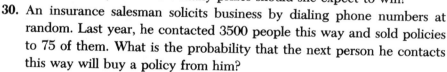
30. An insurance salesman solicits business by dialing phone numbers at random. Last year, he contacted 3500 people this way and sold policies to 75 of them. What is the probability that the next person he contacts this way will buy a policy from him?

A new drug is being tested on people with colds. The drug is given to 1200 people and a placebo is given to 1200 other people.

31. The drug relieves symptoms in 900 of the people who take it. What is the probability that the drug will relieve symptoms in someone with a cold?

32. The drug produces unwanted side effects in 500 people. What is the probability that the drug will have side effects in someone with a cold?

33. The placebo relieves symptoms in 600 of the people who take it. What is the probability that the placebo will relieve symptoms in someone with a cold?

34. The placebo produces unwanted side effects in 200 people. What is the

probability that the placebo will have side effects in someone with a cold?

The Granite Insurance Company has 1200 female employees and 1000 male employees. As part of a program to improve morale, an employee to be selected at random will be given an all-expenses-paid vacation for two in Sarasota.

35. What is the probability that a woman will be selected?
36. What is the probability that the company president will be selected?
37. Mr. and Mrs. Steven Marshall both work for Granite. What is the probability that they will get the vacation?
38. The company's 100 executives (all male) are ruled ineligible. Now what is the probability that the person selected is female?

39. In Yorkville, there are normally 120,000 dollar bills in circulation on a given day. This morning, a counterfeiter added an additional 5000 bills, all counterfeit, to the supply. When I receive a dollar bill as part of my change at a store, what is the probability that it is counterfeit?

40. The Healthy Bread Bakery bakes 3000 one-pound loaves of whole wheat bread each day. Because of various factors, 300 loaves each day weigh at least 17 oz. If I buy one loaf of whole wheat bread at Healthy Bread, what is the probability it weighs at least 17 oz?

A sociologist doing communications research selected 5000 households at random and surveyed them to find the relationship between household income and the number of radios in the home. The results of the survey are shown in the following table.

	Number of Radios in the Home					
Annual Income	0	1	2	3	4	More than 4
Less than $5,000	20	35	42	15	12	0
$5,000–$9,999	32	48	160	90	23	15
$10,000–14,999	25	223	473	268	47	30
$15,000–19,999	20	472	523	305	112	93
$20,000–29,999	5	106	305	410	231	115
$30,000–39,999	2	68	95	116	158	140
More than $40,000	3	12	25	32	41	53

Compute the empirical probability that:

41. A household has no radios.
42. A household has three or more radios.
43. A household with income of at least $15,000 a year has a radio.
44. A household with income of at least $20,000 a year has more than one radio.
45. A household with income in the $10,000–$14,999 range has two or fewer radios.
46. A household with income in the $15,000–$29,999 range has more than one radio.

Section 6-2 Cartesian Products

In Section 6-1, we discussed how the formula $P(E) = n(E)/n(S)$ is used to compute probabilities when all outcomes are equally likely. In order to use this formula, it is important to be able to count the number of members in a set. In this section, we study an operation on sets called cartesian product and learn how to compute the number of members in a cartesian product.

> **Definition.** If A and B are sets, the *cartesian product* $A \times B$ of A with B is $\{(a, b) \mid a \in A \text{ and } b \in B\}$.

The cartesian product $A \times B$ is a set whose members are ordered pairs. The first member of each ordered pair is from A and the second member of each ordered pair is from B. If $A = \{a, b, c\}$ and $B = \{1, 2, 3, 4\}$, then $A \times B = \{(a,1), (a,2), (a,3), (a,4), (b,1), (b,2), (b,3), (b,4), (c,1), (c,2), (c,3), (c,4)\}$. In this example, $n(A) = 3$, $n(B) = 4$, and $n(A \times B) = 12$. In general,

$$n(A \times B) = n(A) \cdot n(B) \qquad \text{Number of elements in a cartesian product}$$

The cartesian product of three sets can be defined in a similar way and is a set of ordered triples. The formula $n(A \times B \times C) = n(A) \cdot n(B) \cdot n(C)$ holds. The cartesian product of m sets is a set of ordered m-tuples. For any m sets, the number of members of the cartesian product is the product of the number of members of each of the m sets.

Example 1 Henry has five pairs of pants and 15 shirts. In how many ways can he select a pants-shirt outfit?

Solution The set of pants-shirt outfits is the cartesian product of the set of pairs of pants with the set of shirts. There are $5 \cdot 15 = 75$ ways.

Example 2 Suppose Henry has six ties. In how many ways can he select a pants-shirt-tie outfit?

Solution The set of pants-shirt-tie outfits is the cartesian product of the set of pairs of pants, the set of shirts, and the set of ties. He has $5 \cdot 15 \cdot 6 = 450$ possible pants-shirt-tie outfits.

Example 3 What is the probability that a three-letter word will contain only consonants?

Solution For our purpose, a *word* is any sequence of letters, so that *cat* and *zqw* are both words. We consider *a*, *e*, *i*, *o*, and *u* to be vowels and the 21 other letters to be consonants. We can think of a three-letter word as an ordered triple of letters. There are $26 \cdot 26 \cdot 26 = 26^3$ different three-letter words. There are $21 \cdot 21 \cdot 21$ three-letter words with only consonants. The probability that a three-letter word contains only consonants is $21^3/26^3 = .53$.

Example 4 What is the probability that a three-letter word will begin with two consonants and end with a vowel?

Solution This set of words can be thought of as {consonants} × {consonants} × {vowels} and therefore has $21 \cdot 21 \cdot 5$ members. The answer is $(21 \cdot 21 \cdot 5)/26^3 = .13$.
∎

When doing problems in probability, we do not usually use the phrase "cartesian product." To do the last example, one would usually say that there are 21 possibilities for the first letter, 21 possibilities for the second letter, and 5 possibilities for the third letter. The number of three-letter words beginning with 2 consonants and ending with a vowel is $21 \cdot 21 \cdot 5$.

Example 5 The employees at the National Armature Factory have been subjected to a speed-up. The assembly-line workers as well as the foremen object. At a union meeting, they decide to send a delegation of one assembly-line worker and one foreman to discuss the issue with management. The delegation is to be selected at random from those who volunteer. Fifty assembly-line workers volunteer as well as ten foremen. Among the volunteers are the five Stuart brothers: Alan, Frank, and Jesse, who work in the assembly line, and Bill and

Ken who are foremen.

PROBLEM A What is the probability that the delegation consists of two of the Stuart brothers?

Solution The worker on the delegation must be selected from among the 50 workers who volunteer. The foreman on the delegation must be selected from among the ten foremen who volunteer. Thus, there are $50 \cdot 10 = 500$ possible delegations.

If the delegation is to consist of two Stuart brothers, there are three possibilities for the worker and two possibilities for the foreman. This gives a total of $3 \cdot 2 = 6$ possibilities.

The probability that the delegation consists of two of the Stuart brothers is $6/500 = .012$.

PROBLEM B What is the probability that the delegation contains none of the Stuart brothers?

Solution If the delegation contains none of the Stuart brothers, then there are $50 - 3 = 47$ possible workers and $10 - 2 = 8$ possible foremen. The number of possible delegations is thus $47 \cdot 8 = 376$.

The probability that the delegation contains none of the Stuart brothers is $376/500 = .752$.

Exercises 6-2

List the members of the following cartesian products.

1. $\{x, z\} \times \{a, b, x\}$
2. $\{p, q, r\} \times \{p, q, 5\}$
3. $\{1, 2\} \times \{a, c\} \times \{a, 1, c\}$
4. $\{7\} \times \{a, b, c\} \times \{4, 5, 6\}$

5. A store sells ten colors of enamel paint and 12 colors of flat paint. In how many ways can someone select one color of enamel and one color of flat paint at this store?

6. An ice cream parlor has 25 flavors of ice cream, ten flavors of syrup, and four kinds of nuts. How many kinds of sundae can be made? A sundae consists of ice cream (two scoops of the same flavor), one flavor of syrup, one kind of nut, whipped cream, and a cherry on top.

7. What is the probability that a three-letter word starts with a q?

8. A three-letter word is selected at random. What is the probability that

the word is either dog or cat?

9. What is the probability that a four-letter word contains only vowels?

10. What is the probability that a four-letter word starts with two vowels and ends with two consonants?

A numeral is any sequence of digits.

11. What is the probability that a three-digit numeral contains only odd digits?

12. What is the probability that a three-digit numeral ends with a 9?

13. What is the probability that the sum of the digits in a three-digit numeral is 26?

14. What is the probability that a four-digit numeral begins with a 7?

15. What is the probability that a three-digit numeral does not contain a 5?

16. What is the probability that a three-digit numeral does not contain any of the digits 2, 3, 4, or 8?

17. What is the probability that a four-digit numeral contains no digits except 3, 6, and 9?

18. What is the probability that a four-digit numeral consists of the same digit repeated four times?

19. A refrigerator comes in five sizes, three colors, and four models. How many kinds of refrigerator can be ordered?

20. A meal at a restaurant consists of salad, main course, dessert, and beverage. There are four kinds of salad, 16 main courses, 10 kinds of dessert, and six beverages. A waiter forgets to take the order but delivers Mr. Hansen a meal anyway. What is the probability that Mr. Hansen gets what he would have ordered?

21. A contractor builds garages. She has available three basic styles and five exterior colors. The garage can be economy grade, good grade, or deluxe. How many types of garage are possible?

22. How many license plate numbers are possible, if a license plate "number" consists of two letters followed by four digits?

23. A radio station name is a sequence of four letters. The first letter must be K or W. If a radio station selects its name at random, what is the

probability that its name is *KISS* or *WISH*?

24. Sew and Seam Fabric normally carries 16 shades of brown double knit and 12 shades of yellow. All of these fabrics are equally popular. Today, they are out of six shades of brown and five shades of yellow. I want to buy my favorite shade of brown as well as my favorite shade of yellow. What is the probability that Sew and Seam has what I want?

25. A political pollster classifies potential voters according to sex, party (Democrat, Republican, or Independent), age (18–30, 31–40, 41–50, 51–65, over 65) and residence (rural, suburban, urban). How many classifications are there?

26. An epidemiologist studying influenza classifies people according to age (0–4, 5–25, 26–45, 46–60, 61–70, over 70) sex, general level of health (extremely poor, poor, fair, good, very good), and whether or not the person had flu vaccine that year. How many classifications are there?

*27. If a set S has n members, how many distinct subsets does S have?

*28. S is a set with n elements, and $x \in S$. A subset of S is selected at random. What is the probability that the subset contains x?

29. Determine $n(A \times B \times C)$ if $n(A) = 143$, $n(B) = 247$, and $n(C) = 86$.

30. Determine $n(A \times B \times C \times D)$ if $n(A) = 74$, $n(B) = 97$, $n(C) = 168$, and $n(D) = 47$.

Section 6-3 Set Operations and Counting

In Chapter 1, we studied the set operations of union, intersection, and complementation. Here these operations will be related to counting and then to probability.

Let's begin with a very simple case—the union of disjoint sets.

If A and B are disjoint, then
$n(A \cup B) = n(A) + n(B)$ Number of elements in a union of disjoint sets
If $C_1, C_2, \ldots, C_m$ are disjoint sets, then
$n(C_1 \cup C_2 \cup \cdots \cup C_m) = n(C_1) + n(C_2) + \cdots + n(C_m)$

Example 1 If A and B are disjoint, with $n(A) = 17$ and $n(B) = 12$, then $n(A \cup B) = n(A) + n(B) = 17 + 12 = 29$.

Example 2 If A, B, and C form a disjoint collection of sets, with $n(A) = 3$, $n(B) = 5$, and $n(A \cup B \cup C) = 18$, determine $n(C)$.

Solution Use the formula $n(C_1 \cup C_2 \cup C_3) = n(C_1) + n(C_2) + n(C_3)$ with $C_1 = A$, $C_2 = B$, and $C_3 = C$ to obtain $n(A \cup B \cup C) = n(A) + n(B) + n(C)$. Then $18 = 3 + 5 + n(C)$, so that $n(C) = 10$.
■

Now consider the very special and important case of a set and its complement. If A is any set, A and A' are disjoint and $A \cup A' = U$. In the formula $n(A \cup B) = n(A) + n(B)$, put $B = A'$. Then $n(U) = n(A) + n(A')$. Solve this equation for $n(A')$ to obtain the formula for the number of elements in the complement of a set

$$n(A') = n(U) - n(A) \qquad \text{Number of elements in the complement}$$

Example 3 If $n(U) = 510$ and $n(A) = 470$, what is $n(A')$?

Solution $n(A') = n(U) - n(A) = 510 - 470 = 40$.

Venn diagrams

We now consider $n(A \cup B)$ when A and B are not disjoint. Venn diagrams are useful here.

Venn diagrams are drawings that make it easy to visualize operations on sets. In Diagram 1, the entire rectangle represents the universal set U. The inside of the left circle represents a set A and the inside of the right circle represents a set B. The two circles divide the rectangle into four regions, labeled with Roman numerals. These four regions can be thought of as disjoint. Regions I and II together form A, so that regions III and IV form A'. Regions II and III form B, so that regions I and IV form B'. Each of the four regions can be described in terms of A and B. For example, region I is part of A but is not part of B, so that region I is $A \cap B'$. Here is a table of regions and descriptions.

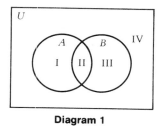
Diagram 1

Region(s)	Description
I	$A \cap B'$
II	$A \cap B$
III	$A' \cap B$
IV	$A' \cap B'$
I, II, III	$A \cup B$
I, II, IV	$A \cup B'$
II, III, IV	$A' \cup B$
I, III, IV	$A' \cup B'$

Let's derive a formula for $n(A \cup B)$. If we take $n(A) + n(B)$, we have counted every member of regions I and III once but have counted every member of region II twice. Then $n(A) + n(B) - n(A \cap B)$ counts every member of regions I, II, and III once. But $A \cup B$ is regions I, II, and III.

$$n(A \cup B) = n(A) + n(B) - n(A \cap B)$$
Number of elements in a union

This formula is also frequently written in the form

$$n(A) + n(B) = n(A \cup B) + n(A \cap B)$$

This formula can be solved for any one of the four numbers in it, if the other three are known.

Example 4 If $n(A) = 15$, $n(B) = 8$, and $n(A \cap B) = 6$, find $n(A \cup B)$.

Solution Use the formula $n(A \cup B) = n(A) + n(B) - n(A \cap B)$. Then $n(A \cup B) = 15 + 8 - 6 = 17$.

Example 5 If $n(A) = 12$, $n(B) = 15$, and $n(A \cup B) = 23$, find $n(A \cap B)$.

Solution Use the formula $n(A) + n(B) = n(A \cup B) + n(A \cap B)$. Then $12 + 15 = 23 + n(A \cap B)$, so that $n(A \cap B) = 4$.

Example 6 If $n(A) = 15$, $n(B) = 20$, and $n(A \cup B) = 35$, find $n(A \cap B)$.

Solution Use the formula $n(A) + n(B) = n(A \cup B) + n(A \cap B)$. Then $15 + 20 = 35 + n(A \cap B)$, so that $n(A \cap B) = 0$. Then $A \cap B = \emptyset$, so that A and B are disjoint.
∎

Some problems involving the number of elements in a set are most easily solved with a Venn diagram. To do these problems, first draw a Venn diagram and find the number of members of each of the four regions.

Example 7 Find $n(A \cup B')$ if $n(A) = 25$, $n(A \cap B) = 11$, $n(B) = 32$, and $n(U) = 65$.

Solution Draw a Venn diagram. (See Diagram 2.) Since $A \cap B$ is region II, and $n(A \cap B) = 11$, region II has 11 members. Write the number 11 in region II. Now A consists of regions I and II. Since A has 25 members and region II has 11 members, region I has $25 - 11 = 14$ members. Now B consists of regions II and III. Since B has 32 members and region II has 11 members, region III has $32 - 11 = 21$ members. Now regions I, II, and III together have $14 + 11 + 21 = 46$ members. Since U has 65 members and consists of regions I, II, III, and IV, region IV has $65 - 46 = 19$ members. We have now found the number of members of each region in the Venn diagram.

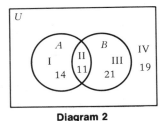

Diagram 2

Since $A \cup B'$ consists of regions I, II, and IV, $n(A \cup B') = 14 + 11 + 19 = 44$.

Example 8 In a survey of 100 business executives, it was found that 75 read the *Wall Street Journal*, 60 read the *New York Times*, and 10 read neither newspaper.

PROBLEM A How many read the *New York Times* and the *Wall Street Journal*?

PROBLEM B How many read the *New York Times* or the *Wall Street Journal*?

PROBLEM C How many read only the *Wall Street Journal*?

Solution Let U be the set of those surveyed, let W be the set of those surveyed who read the *Wall Street Journal*, and let N be the set of those surveyed who read the *New York Times*. We are given $n(U) = 100$, $n(W) = 75$, $n(N) = 60$, and $n(W' \cap N') = 10$.

Draw a Venn diagram. (See Diagram 3.) Since $W' \cap N'$ is region IV, and $n(W' \cap N') = 10$, region IV has 10 members. Now regions I and II together form N, and $n(N) = 60$. Use the equation

$$n(\text{I}) + n(\text{II}) + n(\text{III}) + n(\text{IV}) = n(U)$$

to obtain $60 + n(\text{III}) + 10 = 100$, so that $n(\text{III}) = 30$.

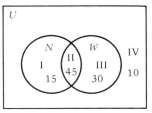

Diagram 3

Now W consists of regions II and III and has 75 members, so that $n(\text{II}) + 30 = 75$. Then $n(\text{II}) = 45$. Since N consists of regions I and II, and N has 60 members, $n(\text{I}) + 45 = 60$. Then $n(\text{I}) = 15$. The Venn diagram is now filled in.

PROBLEM A A person reads the *New York Times and* the *Wall Street Journal* precisely when the person is a member of both N (the set of those who read the *New York Times*) and W (the set of those who read the *Wall Street Journal*), that is, when the person is a member of $N \cap W$. Thus, $n(\text{II}) = 45$ people read both newspapers.

PROBLEM B A person reads the *New York Times or* the *Wall Street Journal* precisely when the person is a member of either N or W, that is, when the person is a member of $N \cup W$. Thus $n(\text{I}) + n(\text{II}) + n(\text{III}) = 15 + 45 + 30 = 90$ people read one (or both) of these newspapers.

PROBLEM C A person reads only the *Wall Street Journal* precisely when the person is a member of both W and N', that is, of $W \cap N'$. Thus, $N(\text{III}) = 30$ people read only the *Wall Street Journal*.

Exercises 6-3

Determine which regions in Diagram 1 form the given set.

1. $A' \cup (A \cap B)$
2. $(A' \cup B) \cap (A \cup B')$
3. $(A \cup U) \cap B$
4. $(A \cap \emptyset) \cap B$

5. $A' \cup (A' \cap B)$
6. $A \cap (A' \cap B)$
7. $(B' \cap U) \cap (A \cup B)'$
8. $(A' \cup B') \cap (A \cup B)$

Given $n(U) = 20, n(E) = 8, n(F) = 3, n(G) = 5,$ and $E, F,$ and G are disjoint, determine

9. $n(E \cap G)$
10. $n(E \cup F \cup G)$
11. $n(F \cup G)$
12. $n(E \cap U)$
13. $n(E \cup F \cup G \cup U)$
14. $n((E \cup F)')$

15. Determine the number of members of the indicated set, given that $n(A) = 15, n(U) = 35, n(B) = 12,$ and $n(A \cap B) = 4$. Use a Venn diagram.
  a. $n(A')$
  b. $n(B')$
  c. $n((A \cap B)')$
  d. $n(A \cup B)$
  e. $n((A \cup B)')$
  f. $n(B \cap U)$

16. Determine the number of members of the indicated set, given that $n(V) = 23, n(W') = 40, n(V \cap W) = 12,$ and $n(U) = 70$. Use a Venn diagram.
  a. $n(V')$
  b. $n(W)$
  c. $n(V' \cap W')$
  d. $n((V \cap W')')$
  e. $n((V' \cup W)')$
  f. $n(V \cap (V' \cup W))$

17. Determine the number of members of the indicated set, given $n(P) = 30, n(Q) = 20, n(P \cup Q) = 35,$ and $n(U) = 60$. Use a Venn diagram.
  a. $n(P \cap Q)$
  b. $n((P \cap Q)')$
  c. $n(P' \cap Q')$
  d. $n(P' \cup Q')$
  e. $n(Q')$
  f. $n((P' \cap Q')')$

18. A survey of 100 people is taken to find their news sources. Forty of the people watch the news on television each day and 35 read the newspaper. Twenty do both. How many of these 100 people
  a. Read the newspaper or watch the news on television each day?
  b. Neither read the newspaper nor watch the news on television each day?

19. Human blood can contain the A antigen alone (type A blood), the B antigen alone (type B blood), both antigens (type AB blood), or neither antigen (type O blood). In an expeditionary force of 80 soldiers, 15 have type AB blood, 55 have the B antigen, and 20 have type O blood.

a. How many have type A blood?
b. How many have the A antigen?
c. How many have type B blood?
d. How many have exactly one of these two antigens?

20. Crossroads Datsun has 40 cars in stock with air conditioning and 100 cars with 4-cylinder engines. It has 80 cars with 6-cylinder engines but no air conditioning and 30 cars with 6-cylinder engines and air conditioning. All of the cars have either 4-cylinder or 6-cylinder engines.
   a. How many have 6-cylinder engines?
   b. How many 4-cylinder cars have air conditioning?
   c. How many cars does Crossroads Datsun have?
   d. How many cars have 6 cylinders or air conditioning but not both?

21. The manager at Quickburger requires that 500 hamburgers be on hand when the lunch rush begins at 11:45. Four hundred of these hamburgers must have ketchup on them, 300 have pickle, and 75 are plain.
   a. How many hamburgers have ketchup and pickle?
   b. How many hamburgers have ketchup without pickle?
   c. How many hamburgers have pickle without ketchup?
   d. How many hamburgers have ketchup or pickle?

A market research company interviews 100 people in depth and obtains the following information:
   54 *are willing to pay a premium for brand-name canned fruit.*
   64 *are willing to pay a premium for brand-name paper towels.*
   49 *are willing to pay a premium for brand-name toothpaste.*
   17 *are willing to pay premiums for brand names on canned fruit, paper towels, and toothpaste.*
   10 *are willing to pay premiums for brand names on canned fruits and toothpaste but not on paper towels.*
   56 *are willing to pay premiums for brand names on at least two of canned fruit, paper towels, and toothpaste.*
   12 *are willing to pay a premium for brand-name canned fruit but not for brand-name paper towels or toothpaste.*

How many of the people interviewed

*22. Are unwilling to pay a premium for a brand name on any of the three items discussed?

*23. Are willing to pay a premium for brand-name toothpaste but not for brand-name canned fruit?

*24. Are willing to pay a premium for a brand name on exactly one of the items discussed?

*25. Are willing to pay a premium for brand-name paper towels only?

*26. Are willing to pay a premium for brand-name canned fruit and toothpaste?

*27. Are willing to pay a premium for a brand name on at least one of the items discussed?

Section 6-4 Complements, Unions, and Intersections.

If E is any event, the complement E' of E is the event that E does not occur. To obtain a formula for $P(E')$, divide the equation $n(E) + n(E') = n(S)$ by $n(S)$.

$$\frac{n(E)}{n(S)} + \frac{n(E')}{n(S)} = 1$$

If each possible outcome is equally likely, this equation becomes $P(E) + P(E') = 1$. Then

$$\boxed{P(E') = 1 - P(E) \qquad \text{Probability of the complement}}$$

The equation $P(E') = 1 - P(E)$ is valid throughout probability theory. Sometimes it is very difficult to compute the probability of an event directly, but easy to compute the probability of its complement, as in Example 3.

Example 1 Determine $P(E')$ if $P(E) = .3$.

Solution $P(E') = 1 - P(E) = 1 - .3 = .7$.

Example 2 Determine $P(A \cap B)$ if $P(A' \cup B') = .2$.

Solution Use a Venn diagram to verify that $(A' \cup B')' = (A \cap B)$. Then

$$P(A \cap B) = 1 - P(A' \cup B') = 1 - .2 = .8$$

Example 3 A coin is tossed six times. What is the probability of at least one head?

Solution This is difficult to compute directly. However, the complement of "at least one head" is "no heads," which is the same as "all tails." There is only one way to get all tails, and there are $2^6 = 64$ possible outcomes. Thus $P(\text{all tails}) = 1/64$ and $P(\text{at least one head}) = 1 - 1/64 = 63/64$.

∎

If A and B are events, $A \cap B$ is the event that both A occurs and B occurs. $A \cup B$ is the event that either A or B occurs. Thus the notions of union and intersection are important in probability. We now derive a formula involving the probabilities of unions and intersections. In Section 6-3, the formula $n(A) + n(B) = n(A \cup B) + n(A \cap B)$ was derived. Assume all possible outcomes are equally likely and divide by $n(S)$ to obtain

$$P(A) + P(B) = P(A \cup B) + P(A \cap B)$$
Probability of the union and intersection

This equation is valid throughout probability theory. This equation can be used to find any one of the four numbers in it if the other three are known.

Example 4 Determine $P(A \cap B)$ if $P(A) = .2$, $P(B) = .6$, and $P(A \cup B) = .7$.

Solution Use the formula $P(A) + P(B) = P(A \cup B) + P(A \cap B)$ to obtain $.2 + .6 = .7 + P(A \cap B)$, so that $P(A \cap B) = .1$.

Example 5 Harris Construction needs to sublet two electrical jobs, one on the Hilltop Highrise apartment building, and one on Euclid Beach Mall. John Walker, the owner of Walker Electrical Company, is bidding for both jobs. He estimates that he has a .6 chance of getting the job at Hilltop, a .3 chance of getting the job at the mall, and a .2 chance of getting both jobs. What is the probability that John gets at least one job?

Solution Let H be the event that John gets the job at Hilltop Highrise and let E be the event that he gets the job at Euclid Beach Mall.

Now $H \cap E$ is the event that he gets both jobs, and $H \cup E$ is the event that he gets at least one job. Now $P(H) = .6$, $P(E) = .3$, and $P(H \cap E) = .2$, so that

$$.6 + .3 = P(H \cup E) + .2$$

Then $P(H \cup E) = .7$, so Walker Electrical Company has a .7 chance of getting at least one job.

∎

An important special case of the last formula occurs if A and B are disjoint. The $A \cap B = \emptyset$, so that $P(A \cap B) = 0$ and

$$P(A \cup B) = P(A) + P(B) \quad \text{Probability of a disjoint union}$$

This generalizes to any collection of disjoint sets. If $A_1, A_2, \ldots, A_k$ is a disjoint collection of sets, then $P(A_1 \cup A_2 \cup \cdots \cup A_k) = P(A_1) + P(A_2) + \cdots + P(A_k)$.

Example 6 Alice, Betty, David, and Frank are playing a game. One person must win the game. The probability that Alice wins is .3, the probability that Betty wins is .25, and the probability that David wins is .28.

PROBLEM A What is the probability that Alice or Betty wins?

Solution The probability that Alice or Betty wins is $.3 + .25 = .55$, since the event {Alice, Betty} is the union of the two disjoint events {Alice} and {Betty}.

PROBLEM B What is the probability that Frank wins?

Solution The complement of {Frank} is {Alice, Betty, David} = {Alice} $\cup$ {Betty} $\cup$ {David}. The three sets in this union are disjoint, so that $P\{\text{Alice, Betty, David}\} = .3 + .25 + .28 = .83$. The probability that Frank wins is $1 - .83 = .17$. ∎

Some problems can be done using the formulas of this section, but can also be done much more easily using Venn diagrams.

Example 7 If $P(A \cup B) = .7$, $P(B) = .5$, and $P(A' \cap B) = .4$, what is $P(A \cap B')$?

Solution Draw the Venn diagram.

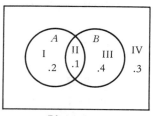

Diagram 1

$A' \cap B$ is region III in Diagram 1, so $P(\text{III}) = .4$. $A \cup B$ is regions I, II, and III in the diagram. Region IV is $(A \cup B)'$. Thus $P(\text{IV}) = 1 - P(A \cup B) = 1 - .7 = .3$. B is regions II and III in the diagram, so that $.5 = P(\text{II}) + .4$, and $P(\text{II}) = .1$. Since $P(\text{I}) + P(\text{II}) + P(\text{III}) + P(\text{IV}) = 1$, $P(\text{I}) = .2$

$A \cap B'$ is region I, so that $P(A \cap B') = .2$

Once the Venn diagram is complete and the probabilities of regions I, II, III, and IV have been filled in, the probability of any region can be found easily.

Example 8 If $P(A') = .4$, $P(B') = .6$, and $P(A \cap B) = .2$, find $P(A \cup B)$, $P(A' \cap B)$, and $P(A' \cup B)$.

Solution Draw the Venn diagram. See Diagram 2.

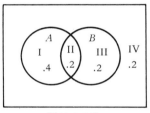

Diagram 2

Since $A \cap B$ is region II, $P(\text{II}) = .2$. Since A' is regions III and IV, $A \cap B$ and A' are disjoint. However, $(A \cap B) \cup A'$ is regions II, III, IV, or (region I)'. Therefore,

$$P(\text{region I}) = 1 - P[(A \cap B) \cup A']$$
$$= 1 - [P(A \cap B) + P(A')] = 1 - [.2 + .4] = .4.$$

Since B' is regions I and IV, $.6 = .4 + P(\text{IV})$, so that $P(\text{IV}) = .2$. Since $P(S) = 1$, $P(\text{III}) = .2$.

Note that we first completed the Venn diagram by finding the probability of each of the four regions. Now let's answer the questions. $A \cup B$ is regions I, II, and III, so that $P(A \cup B) = .8$. $A' \cap B$ is region III, so that $P(A' \cap B) = .2$. $A' \cup B$ is regions II, III, and IV, so that $P(A' \cup B) = .6$.

Example 9 Market Research Associates, Inc. claims that 72% of all families in Brandon Heights use fluoride toothpaste, 28% use dental floss, and 20% use both fluoride toothpaste and dental floss. Crown Products is selling a new product, Gem Toothcleaner, which will be of interest only to people who use neither fluoride toothpaste nor dental floss.

Jim Smith works for Crown Products. His assignment is to go door-to-door in Brandon Heights and try to convince people to use Gem Toothcleaner. When he knocks at a door, he first determines whether the people use either fluoride toothpaste or dental floss. If they do, he leaves. This takes a total of three minutes, including time to walk to the next house. If they do not, he gives them a free sample and a promotional pitch. This takes an extra five minutes. How many houses do you expect him to visit in his eight-hour workday? How many free samples do you expect him to give away?

Solution First draw a Venn diagram. The inside of circle F represents users of fluoride toothpaste. The inside of circle D represents users of dental floss.

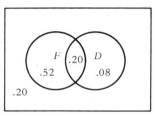

Diagram 3

Diagram 3 shows that the probability that a family uses either fluoride toothpaste or dental floss $[P(D \cup F)]$ is .80 and the probability that a family uses neither $[P(D' \cap F')]$ is .20.

Now let x be the number of houses he visits in a day. Then he visits about $.8x$ houses in which either fluoride toothpaste or dental floss is used and spends three minutes at each of them, for a total of $2.4x$ minutes. He also visits about $.2x$ houses in which he makes his promotional pitch and spends 8 minutes at each of them, for a total of $1.6x$ minutes.

Now his total time spent is $2.4x + 1.6x = 4x$ minutes. He works 8 hours = 480 minutes. Thus,

$$4x = 480$$
$$x = 480/4 = 120$$

Jim Smith is expected to visit about 120 houses each day. He is expected to give away about $(0.2) \cdot (120) = 24$ free samples each day.

Exercises 6-4

1. What is the probability that a three-letter word contains at least one letter a?

2. What is the probability that a four-letter word contains at least one vowel?

3. What is the probability that a five-digit number has at least one odd digit?

4. What is the probability that a four-digit number contains at least one 7?

5. If A, B, C, and D are disjoint sets, $P(A) = .1$, $P(B) = .07$, $P(C) = .3$, and $P(D) = .41$, determine $P(A \cup B \cup C \cup D)$ and $P[(A \cup C \cup D)']$.

6. If K, L, and M are disjoint sets, $P(K) = .5$, $P(L) = .2$, and $P(M) = .16$, determine $P(K \cup L \cup M)$ and $P[(K \cup M)']$.

7. If $P(A \cup B) = .72$, $P(B) = .30$, and $P(A \cap B) = .10$, determine $P(A \cap B')$, $P(A \cup B')$, and $P(A' \cup B)$.

8. If $P(A) = .7$, $P(A \cap B') = .5$, and $P(A \cup B) = .95$, determine $P(A' \cap B')$, $P(A' \cap B)$, and $P(A' \cup B')$.

9. If $P(A) = .5$, $P(B) = .46$, and $P(A' \cap B') = .23$, determine $P(A \cap B)$, $P(A' \cap B)$, and $P(A' \cup B)$.

10. If $P(C \cap D') = .25$, $P(C' \cap D) = .12$, and $P(D') = .80$, determine $P(C)$, $P(D)$, and $P(C \cup D)$.

11. If $P(C \cap D) = .23$, $P(C \cup D) = .60$, and $P(C' \cap D) = .07$, determine $P(C)$, $P(D)$, and $P(C' \cup D)$.

12. If $P(C') = .27$, $P(D') = .41$, and $P(C \cap D) = .5$, determine $P(C)$, $P(D)$, and $P(C \cup D)$.

13. When Katie Robinson shops for clothes, she pays with either Master-Card (probability .23), Visa (probability .26), check (probability .38), or cash (probability .13). The next time she goes shopping, what is the probability that:
    a. She pays with a credit card?
    b. She pays with a credit card or a check?
    c. She does not pay with Visa?
    d. She does not pay with cash?

14. Roger Kuhn is a very indecisive voter. The probability that he will not vote on election day is .4, the probability that he will vote and write in his own name for governor is .2, and the probability that he will vote for an independent for governor is .1. If he does none of the above, he will vote for either the Democratic or the Republican candidate for governor with equal likelihood. What is the probability that

a. Roger votes on election day?
b. Roger votes for the Democratic candidate for governor?
c. Roger votes either for himself or an independent for governor?
d. Roger votes for someone other than himself for governor?

15. A survey of corporation presidents shows that 60% own a vacation home, 50% own a yacht, and 30% own neither. What is the probability that a corporation president owns a yacht but not a vacation home?

16. In National Motors cars, the engine fails within the warranty period 3% of the time, and the transmission fails 2% of the time. Both fail within the warranty period 1% of the time. If you buy a new National Motors car, what is the probability that neither the engine nor the transmission will fail during the warranty period?

A survey of top business executives by an economics forecasting company reveals that:

55% *expect high inflation next year.*
48% *expect a recession next year.*
21% *expect both high inflation and a recession next year.*
16% *expect high inflation and an oil shortage but no recession next year.*
6% *expect an oil shortage and a recession without high inflation next year.*
13% *expect an oil shortage next year without high inflation or a recession.*
44% *expect an oil shortage next year.*

An executive is selected from those surveyed and is interviewed in depth by a weekly news magazine. What is the probability that this executive expects

*17. Neither high inflation, a recession, nor an oil shortage next year?

*18. High inflation without either a recession or an oil shortage next year?

*19. High inflation, an oil shortage, and a recession next year?

*20. A recession without either high inflation or an oil shortage next year?

*21. A recession and high inflation without an oil shortage next year?

*22. An oil shortage and high inflation next year?

Section 6-5 Independent Events

If two events do not influence the probability of each other, the events are said to be *independent*. If either event influences the probability of the other, the events are said to be *dependent*.

Successive tosses of a coin, or throws of a die, are independent. Suppose two cards are selected from a deck in sequence without replacement. Let E_1 be that the first card is a club and E_2 be that the second card is a club. E_1 and E_2 are dependent, since $P(E_2) = 12/51$ if E_1 has occurred and $P(E_2) = 13/51$ if E_1 has not occurred. The reason why E_1 and E_2 are dependent is that the condition of the deck when the second card is drawn is different if the first card was a club than if the first card was not a club. This difference in the condition of the deck changes the probability of E_2.

If the two cards are drawn with replacement, E_1 and E_2 are independent. In this case, the second card is drawn from a whole deck of 52 cards.

The precise definition of independent is as follows.

Definition. Two events E_1 and E_2 are *independent* if and only if $P(E_1 \cap E_2) = P(E_1) \cdot P(E_2)$. E_1 and E_2 are *dependent* if they are not independent.

Example 1 If $P(E_1) = .3$, $P(E_2) = .8$, and $P(E_1 \cap E_2) = .2$, are E_1 and E_2 independent?

Solution $P(E_1) \cdot P(E_2) = (.3) \cdot (.8) = .24 \neq P(E_1 \cap E_2)$. E_1 and E_2 are dependent.

Example 2 If $P(E_1) = .4$, $P(E_2) = .15$, and $P(E_1 \cup E_2) = .49$, are E_1 and E_2 independent?

Solution We first need to determine $P(E_1 \cap E_2)$ by using the formula $P(E_1) + P(E_2) = P(E_1 \cup E_2) + P(E_1 \cap E_2)$. This gives $.4 + .15 = .49 + P(E_1 \cap E_2)$, so that $P(E_1 \cap E_2) = .06$.

Now $P(E_1) \cdot P(E_2) = (.4) \cdot (.15) = .06 = P(E_1 \cap E_2)$. Therefore E_1 and E_2 are independent.

Example 3 Frank Strasek has just had two job interviews. He estimates that the probability of a job offer from Kovatch Industries is .6 and from Dwyer Manufacturing is .7. What is the probability that he will have offers from both Kovatch and Dwyer? What is the probability that he will have exactly one job offer?

Solution First, let K be the event of a job offer from Kovatch Industries and let D be the event of a job offer from Dwyer Manufacturing. It is reasonable to assume that K and D are independent. Then the probability that he will have offers from both Kovatch and Dwyer is $P(K \cap D) = P(K) \cdot P(D) = (.6) \cdot (.7) = .42$.

The probability that Frank receives an offer from Kovatch only is $P(K) - P(K \cap D) = .6 - .42 = .18$. The probability that Frank receives an offer from Dwyer only is $.7 - .42 = .28$. Thus the probability that Frank receives exactly one offer is $.18 + .28 = .46$. You should draw a Venn diagram for this problem, and use the diagram to answer this last question. ∎

If E_1 and E_2 are independent, E_1 and E_2' are independent, E_1' and E_2 are independent, and E_1' and E_2' are independent.

Independence can be defined for more than two events, but the definition is very complicated and is not necessary here. We will, however, use the notion of independence of more than two events as follows:

> If $\{E_1, E_2, \ldots, E_m\}$ is a set of independent events,
> $$P(E_1 \cap E_2 \cap \cdots \cap E_m) = [P(E_1)][P(E_2)] \cdots [P(E_m)]$$

Example 4 A coin is tossed ten times. What is the probability that it lands heads each time?

Solution The ten events {heads on toss 1, heads on toss 2, ..., heads on toss 10} are independent, each has probability $1/2$, and their intersection is the event heads each time. Thus, P(heads each time) is $(1/2)^{10} = 1/2^{10}$. Similarly, the event HHTTHTTHTT, or any given sequence of ten heads and tails, has probability $(1/2)^{10} = 1/1024$.

Example 5 A three-letter word is picked at random. What is the probability that the first and third letters are consonants and the second letter is a q?

Solution Let E_1 be the event that the first letter is a consonant, E_2 that the second letter is a q, and E_3 that the third letter is a consonant. We are asked to find $P(E_1 \cap E_2 \cap E_3)$. Since E_1, E_2, and E_3 are independent $P(E_1 \cap E_2 \cap E_3) = P(E_1) \cdot P(E_2) \cdot P(E_3) = (21/26) \cdot (1/26) \cdot (21/26) = .025$.

Example 6 A machine has four component subassemblies. Each subassembly has probability .02 of failure each time it is used. Assume that the performance of each subassembly is independent of the performance of the other subassemblies, and that all four subassemblies must function properly in order for the machine to function properly. What is the probability that the machine functions properly?

Solution The probability that a subassembly functions properly is $1 -$ (probability of subassembly failure) $= 1 - .02 = .98$. The probability that the machine functions properly is the probability that all four subassemblies function properly. Since the performances of the subassemblies are independent of each other, this is $(.98)(.98)(.98)(.98) = .98^4 = .92$.

■

Sometimes one cannot tell by looking whether certain events are independent and must use the definition to decide.

Example 7 Let a die be loaded so that P(die shows n) is proportional to n. (See Example 2 in Section 6-1.) The die is thrown once. Let $E_1 = \{3,4\}$, $E_2 = \{2,4,6\}$, and $E_3 = \{4,5\}$. There is no reason to suspect off hand that any two of these events are independent. However, one pair of these events *is* independent! Without doing any computation, try out your intuition by guessing which pair of events is independent.

To check independence, first compute that $P(E_1) = 7/21$, $P(E_2) = 12/21$, and $P(E_3) = 9/21$. Then test each pair of events.

$$P(E_1 \cap E_2) = P(\{4\}) = \frac{4}{21}$$
$$= P(E_1) \cdot P(E_2) \quad \text{so that } E_1 \text{ and } E_2 \text{ are independent!}$$

$$P(E_2 \cap E_3) = P(\{4\}) = \frac{4}{21}$$
$$\neq P(E_2) \cdot P(E_3) \quad \text{so that } E_2 \text{ and } E_3 \text{ are dependent.}$$

$$P(E_1 \cap E_3) = P(\{4\}) = \frac{4}{21}$$
$$\neq P(E_1) \cdot P(E_3) \quad \text{so that } E_1 \text{ and } E_3 \text{ are dependent.}$$

Exercises 6-5

1. If $P(A) = .2$, $P(B) = .3$, and $P(A \cap B) = .1$, are A and B independent?
2. If $P(A) = .3$, $P(B) = .35$, and $P(A \cap B) = .105$, are A and B independent?
3. If $P(A) = .4$, $P(B') = .7$, and $P(A \cap B) = .12$, are A and B independent?

4. If $P(A') = .6$, $P(B) = .4$, and $P(A \cap B) = .24$, are A and B independent?
5. If $P(E) = .3$, $P(F) = .8$, and $P(E \cup F) = .95$, are E and F independent?
6. If $P(E) = .35$, $P(F) = .6$, and $P(E \cup F) = .8$, are E and F independent?
7. If C and D are independent, $P(C) = .2$, and $P(D) = .5$, what are $P(C \cap D)$, $P(C \cup D)$, and $P(C \cap D')$?
8. If E and F are independent, $P(E) = .6$ and $P(F) = .3$, what are $P(E \cap F)$, $P(E' \cap F')$, $P(E \cup F)$, and $P(E' \cap F)$?
9. A coin is tossed five times. What is the probability that it lands tails each time?
10. A fair die is tossed four times. What is the probability that it shows 6 each time? What is the probability that it shows 2 or 3 each time?
11. What is the probability that a three-letter word contains only vowels? only consonants?
12. On a multiple choice test, there are five choices for each question. What is the probability of getting all eight questions right by guessing? What is the probability of getting all wrong by guessing?

A card is selected from a deck and replaced. The deck is shuffled. A second card is then drawn. What is the probability that

13. Both cards are diamonds?
14. Neither card is a diamond?
15. The first card is a seven and the second card is a nine?
16. The first card is a ten and the second card is a spade?
17. Both cards are eights?
18. The first card is the three of hearts and the second card is a three?

A jar contains 8 red balls, 7 green balls, and 5 orange balls. A ball is drawn from the jar and replaced. The jar is shaken, and a second ball is drawn. What is the probability that

19. Both balls are red?
20. The first ball is red and the second ball is green?
21. Neither ball is orange?
22. The first ball is orange and the second ball is yellow?

A seed packet contains seeds for yellow, orange, pink, and red cosmos. The color refers to the color of the flower. The percentages of seeds are: 20% yellow, 25% orange, 40% pink, and 15% red. The probability that a seed will germinate is 70% and is independent of color. Assume that any seed that germinates will result in flowers. One seed is selected and planted. What is the probability that

23. A yellow plant will result?
24. A pink or red plant will result?
25. No orange flowers will result?
26. Flowers will result, but they will not be red?

Suppose two seeds are planted. Since an enormous number of seeds is in a package, removing the first seed will not change the percentages for germination and color of the remaining seeds. We may thus assume that the outcomes from the two seeds are independent events. What is the probability that

27. Both seeds germinate?
28. Neither seed germinates?
29. Both seeds produce orange flowers?
30. Both seeds produce pink or red flowers?

31. A certain flu vaccine gives immunity to flu to 80% of those who take it. Experimental evidence shows that the success or failure of the vaccine on a person is independent of its success or failure on others in the person's family. Jeffrey and Olive Olbinsky and their children Ruth, Lou, and Paula all are vaccinated.
    a. What is the probability that the whole Olbinsky family is immunized to flu?
    b. What is the probability that at least one of the Olbinsky's is immunized to flu?
32. A survey gives the following information: 60 people who drink heavily have liver ailments and 160 do not; 20 people who do not drink heavily have liver ailments and 510 do not.
    a. What is the probability that a person in the survey has a liver ailment?
    b. For those in the survey, are drinking heavily and having a liver ailment independent events?

Section 6-6 Conditional Probability

In Section 6-5, independent events, that is, events that do not affect each other's probability, were discussed. In this section, we study dependent events, that is, events that do affect each other's probability, and learn how the occurrence of one event can change the probability of another event.

A jar contains ten red marbles, four green marbles, and six yellow marbles. A marble is selected at random from the jar. The probability that the marble is green is 4/20.

Suppose now that a marble is selected at random from the jar and the marble is known to be not red. What is the probability that the marble is green?

Knowing that the marble is not red changes the sample space. The sample space now has ten members: four green marbles and six yellow marbles. Thus the probability that the marble is green is 4/10.

It rains here about one day in three. Therefore the probability that it will rain tomorrow is 1/3. However, the weather bureau makes some observations and measurements and based on them says that the probability of rain tomorrow is 70%.

In both of the preceding examples, the probability of an event changed when certain information was obtained.

Example 1 A survey of women voters in Wilsonville is tabulated in the following table.

	Family Income			
	0–$9999	$10,000–$19,999	$20,000–$29,999	$30,000 or more
Democrat	470	1320	520	130
Independent	50	210	240	110
Republican	120	700	550	280

PROBLEM A What is the probability that a person voted Republican?

Solution There are 4700 voters tabulated in the survey, and 120 + 700 + 550 + 280 = 1650 of them voted Republican. The probability that a person voted Republican is 1650/4700 = .35.

PROBLEM B What is the probability that a person voted Republican, if her family income is $24,600?

Solution Her family income puts her in the $20,000–$29,999 range. There are 520 + 240 + 550 = 1310 voters in this income range and 550 of them voted Republican. The probability that she voted Republican, given that her family income is $24,600, is 550/1310 = .42.

PROBLEM C What is the probability that a voter's family income is below $20,000?

Solution A total of $470 + 50 + 120 + 1320 + 210 + 700 = 2870$ voters had family income below \$20,000. Since there are a total of 4700 voters, the probability that a voter's family income is below \$20,000 is $2870/4700 = .61$.

PROBLEM D What is the probability that a voter's family income is below \$20,000, if she voted Democratic?

Solution A total of $470 + 1320 + 520 + 130 = 2440$ voters voted Democratic. Of these, $470 + 1320 = 1790$ had family incomes below \$20,000. The probability that her family income is below \$20,000, given that she voted Democratic, is $1790/2440 = .73$.

■

A coin is tossed three times. The probability of heads on the first toss is $1/2$. Suppose now it is known that heads occurred at least twice. There are now four possible outcomes: HHH, HHT, HTH, and THH. These four outcomes are equally likely, and heads occurs on the first toss in three of them. Thus, the probability of heads on the first toss is now $3/4$.

Here is another way to think of this. Let E_1 be the event $\{HHH, HHT, HTH, HTT\}$, that is, heads on the first toss. Let $E_2 = \{HHH, HHT, HTH, THH\}$, that is, at least two heads. Given that E_2 occurred, E_2 is certain and E_2' is impossible. Thus, $E_1 \cap E_2 = \{HHH, HHT, HTH\}$ now has the same probability as E_1, since the outcomes in $E_1 \cap E_2'$ cannot occur now. To find the new probability of E_1, divide the old probability of $E_1 \cap E_2$ by the old probability of E_2. This gives $(3/8) \div (4/8) = 3/4$. What we have done is ignore outcomes in E_2', and divide by $P(E_2)$ to "change the scale" in order to reflect that E_2 is now the entire sample space.

Definition. Let E_1 and E_2 be events with $P(E_2) \neq 0$. The *conditional probability of E_1 given E_2* [written $P(E_1 | E_2)$] is defined by

$$P(E_1 | E_2) = \frac{P(E_1 \cap E_2)}{P(E_2)}$$

We require that $P(E_2) \neq 0$ in order to avoid division by 0. If $P(E_2) = 0$, then $P(E_1 | E_2)$ is undefined for any event E_1.

Example 2 Let $P(A) = .5$, $P(B) = .7$, and $P(A \cap B) = .4$.

PROBLEM A Find $P(A|B)$.

Solution $P(A|B) = \dfrac{P(A \cap B)}{P(B)} = \dfrac{.4}{.7} = \dfrac{4}{7}$.

PROBLEM B Find $P(B|A)$.

Solution $P(B|A) = \dfrac{P(B \cap A)}{P(A)} = \dfrac{.4}{.5} = \dfrac{4}{5}$.

PROBLEM C Find $P(A|A \cup B)$.

Solution At this stage, we need a Venn diagram. See Diagram 1.

$$P(A|A \cup B) = \dfrac{P[A \cap (A \cup B)]}{P(A \cup B)}$$

However, $A \cap (A \cup B) = A$. Thus,

$$P(A|A \cup B) = \dfrac{P(A)}{P(A \cup B)} = \dfrac{.5}{.8} = \dfrac{5}{8}$$

PROBLEM D Find $P(A' \cap B|B)$.

Solution First notice that $(A' \cap B) \cap B = A' \cap B$. Then

$$P(A' \cap B|B) = \dfrac{P[(A' \cap B) \cap B]}{P(B)}$$
$$= \dfrac{P(A' \cap B)}{P(B)}$$
$$= \dfrac{.3}{.7} = \dfrac{3}{7}$$

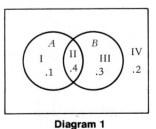

Diagram 1

Example 3 The relationship between scheduled and actual arrival times for flights by Kitty Hawk Airline is contained in the following table.

Amount of time early or late	6 or more min. early	Between 5 min. early and 5 min. late	6–15 min. late	16–30 min. late	31–60 min. late	More than 60 min. late
Probability	.12	.43	.20	.05	.12	.08

PROBLEM A What is the probability that a flight is at least 31 minutes late, given that it is at least 6 minutes late?

Solution Let E_1 be the event that the flight is at least 31 minutes late and E_2 be the event that the flight is at least 6 minutes late. The problem asks for $P(E_1|E_2)$.

$$P(E_1|E_2) = \frac{P(E_1 \cap E_2)}{P(E_2)} = \frac{.12 + .08}{.20 + .05 + .12 + .08} = \frac{.20}{.45} = \frac{20}{45} = \frac{4}{9}$$

PROBLEM B What is the probability that a flight scheduled to arrive at 2:00 arrives by 2:30, given that the flight is at least 6 minutes late?

Solution Let E_1 be the event that the flight is at most 30 minutes late and E_2 be the event that the flight is at least 6 minutes late. We are asked for $P(E_1|E_2)$. Note that $E_1 \cap E_2$ is the event that the flight is between 6 and 30 minutes late.

$$P(E_1|E_2) = \frac{P(E_1 \cap E_2)}{P(E_2)} = \frac{.20 + .05}{.20 + .05 + .12 + .08} = \frac{.25}{.45} = \frac{25}{45} = \frac{5}{9}$$

∎

Now let's find some conditional probabilities. Assume $P(E) \neq 0$. $P(E'|E) = P(E' \cap E)/P(E) = P(\emptyset)/P(E) = 0/P(E) = 0$. Thus,

$$P(E'|E) = 0$$

This makes sense, because if E occurs, then it is impossible for E' to occur. Now assume $P(E \cap F) \neq 0$. Then

$$P(E|E \cap F) = P(E \cap E \cap F)/P(E \cap F) = P(E \cap F)/P(E \cap F) = 1$$
$$P(E|E \cap F) = 1$$

If $E \cap F$ occurs, then both E and F must occur, so that E occurs with certainty.

If S is the sample space, then $P(E|S) = P(E \cap S)/P(S) = P(E)/1 = P(E)$.

$$P(E|S) = P(E)$$

This occurs because being given S contains no new information—all possible outcomes are in S.

The formula defining conditional probability can be written $P(A \cap B) = P(A|B)P(B)$. Recall that A and B are independent precisely when $P(A \cap B) = P(A)P(B)$. Thus, if $P(B) \neq 0$, A and B are independent precisely when $P(A|B) = P(A)$, that is, precisely when the occurrence of B has no effect on the probability of A.

We can also write the formula for conditional probability as

$$P(A \cap B) = P(A)P(B|A)$$

This formula can be used to compute certain probabilities.

Example 3 Two cards are dealt from a deck. What is the probability that the first is a heart and the second is a spade?

Solution Let A be that the first card is a heart and B be that the second card is a spade. We are asked to find $P(A \cap B)$. $P(A) = 13/52 = 1/4$. If A has occurred the deck contains 51 cards, 13 of which are hearts. Thus, $P(B|A) = 13/51$. Then $P(A \cap B) = (1/4) \cdot (13/51) = 13/204 = .064$.

Example 4 Lester Long, an account executive for Ludlow Advertising, is trying to get the account of Swisshelm Chocolate. If he gets the account, he will try to get Swisshelm to spend $300,000 on advertising instead of the $200,000 they are planning on. He estimates the probability of getting the account at .7 and the probability of getting Swisshelm to spend $300,000, if he gets the account, at .4. What is the probability Lester gets a $300,000 account with Swisshelm?

Solution Let A be the event that Lester gets the account and B the event that Swisshelm spends $300,000 on advertising. We are given $P(A) = .7$ and

$P(B|A) = .4$ and are asked for $P(A \cap B)$. Now $P(A \cap B) = P(A)P(B|A)$
$= (.7)(.4) = .28$.

The probability that Lester gets a \$300,000 account with Swisshelm is .28.

Example 5 The procedure can also be used for longer sequences of events. Four cards are dealt from a deck. What is the probability that the first two are kings and the last two are aces?

Solution The answer is $(4/52) \cdot (3/51) \cdot (4/50) \cdot (3/49)$. The probability that the first card is a king is $4/52$. The probability that the second card is a king, given that the first card is a king, is $3/51$. The probability that the third card is an ace, given that the first two cards are kings, is $4/50$. The probability that the fourth card is an ace, given that the first two cards are kings and the third card is an ace, is $3/49$.

Exercises 6-6

A card is drawn from a deck. What is the probability that the card is
1. A spade, given that the card is not a club?
2. A ten, given that the card is not an ace or king?
3. A five, given that the card is a diamond?
4. A diamond, given that the card is a five?
5. A jack, given that the card is a jack, queen, or king?
6. A diamond, given that the card is a diamond or a heart?

7. Find $P(A|B)$ if $P(A) = .5$, $P(B) = .6$, and $P(A \cap B) = .2$.
8. Find $P(E|F)$ if $P(F) = .5$ and $P(E \cap F) = .3$.
9. Given $P(A) = .4$, $P(B) = .7$, and $P(A \cup B) = .8$, find $P(A|B)$, $P(B|A)$, $P(A'|A' \cup B)$, and $P(A|A \cup B')$.
10. Given $P(X) = .7$, $P(Y) = .5$, and $P(X \cap Y) = .4$, find $P(X|X \cup Y)$, $P(Y|X)$, $P(X|Y')$, and $P(X \cap Y|X \cup Y')$.
11. Given $P(E \cup F) = .8$, $P(E \cup F') = .7$, and $P(E \cap F) = .4$, find $P(E|F)$, $P(F|E)$, $P(F'|E \cup F)$, and $P(E \cup F|F')$.
12. Given $P(G \cap H) = .3$, $P(G \cup H') = .6$, and $P(G) = .55$, determine $P(G|H)$, $P(H|G)$, $P(H'|G)$, and $P(H'|G')$.

A jar contains three green balls, eight red balls, and nine blue balls.
13. Four balls are selected in sequence without replacement. What is the probability that all four balls are blue?

14. Three balls are selected in sequence without replacement. What is the probability that the balls are red, green, and red in that order?
15. Three balls are selected in sequence with replacement. What is the probability that all three balls are green?
16. Four balls are selected in sequence without replacement. What is the probability that all four balls are green?
17. Four balls are selected in sequence without replacement. What is the probability that the first ball is the only red one selected?
18. Three balls are selected in sequence with replacement. What is the probability that the first ball is the only green ball selected?
19. One ball is selected. What is the probability the ball is green, if the ball is not blue?
20. One ball is selected. What is the probability the ball is red, if the ball is red or green?
21. Use the table in Example 1. What is the probability that a voter is an Independent, given that her family income is less than $30,000?
22. Use the table in Example 1. What is the probability that a voter's family income is $10,000 or more, if she is not a Republican?

The following table shows the probability that a newly married couple wants a certain number of children

Number of Children	0	1	2	3	4	5	More than 5
Probability	.30	.13	.21	.15	.12	.06	.03

23. What is the probability that a newly married couple wants two or more children, given that they do not want to remain childless?
24. What is the probability that a newly married couple wants more than five children, given that they want at least four children?
25. What is the probability that a newly married couple wants fewer than four children, given that they want at least one child?
26. The State Licensing Bureau has found that 80% of the applicants for real estate agent licenses pass the test on the first attempt. Of those who fail, 60% of those who take the test a second time pass it then. An individual is determined to take the test until he passes. What is the probability that two attempts are enough?

Insurance tables concerning drinking and driving give the following information:

	Percent of drivers	Probability of a serious accident within one year for an individual
Nondrinker	27%	.002
Light drinker	41%	.003
Moderate drinker	23%	.008
Heavy drinker	6%	.014
Alcoholic	3%	.126

27. What is the probability that a driver is a moderate drinker and will have a serious accident within one year?

28. What is the probability that a driver will have a serious accident within one year?

29. What is the probability that a driver, who is a moderate drinker or a heavy drinker, will not have an accident within one year?

30. What is the probability that a driver, who is a heavy drinker or an alcoholic, is an alcoholic?

Section 6-7 Bayes' Theorem

In Section 6, we studied conditional probability and learned how the occurrence of one event can influence the probability of another event. In this section, we will learn how the occurrence of an event F can influence the probability of an earlier event E. This involves expressing $P(E|F)$ in terms of $P(F|E)$. Let's begin with an example.

Example 1 Urn I contains 5 red balls and 15 green balls. Urn II contains 20 red balls and 10 green balls. One of these urns is selected at random, and a ball is drawn from the urn. The ball is red. What is the probability that urn II was selected?

Solution This problem in conditional probability asks for $P(\text{II}|\text{red})$. Without the information that the ball is red, $P(\text{II})$ is $1/2$. However, this probability must be modified in the light of extra information.

Now $P(\text{II and red}) = P(\text{II}) \cdot P(\text{red}|\text{II}) = (1/2) \cdot (20/30) = 1/3$. A red ball could have come from Urn I or Urn II. Thus, $P(\text{red}) = P(\text{I and red}) + P(\text{II and red}) = P(\text{I}) \cdot P(\text{red}|\text{I}) + P(\text{II}) \cdot P(\text{red}|\text{II}) = (1/2) \cdot (5/20) + (1/2) \cdot (20/30) = 11/24$. Finally, $P(\text{II}|\text{red}) = P(\text{II and red})/P(\text{red}) = (1/3) \div (11/24) = 8/11$.

∎

The procedure we have just used is called Bayes' Theorem. We solved the problem using the equation:

$$P(\text{II} \mid \text{red}) = \frac{P(\text{II}) \cdot P(\text{red} \mid \text{II})}{P(\text{I}) \cdot P(\text{red} \mid \text{I}) + P(\text{II}) \cdot P(\text{red} \mid \text{II})}$$

> **Bayes' Theorem.** Suppose $\{E_1, E_2, \ldots, E_m\}$ is a disjoint set of events and $P(E_1) + P(E_2) + \cdots + P(E_m) = 1$. If A is any event and $P(A) \neq 0$, then
>
> $$P(E_1 \mid A) = \frac{P(E_1) \cdot P(A \mid E_1)}{P(E_1) \cdot P(A \mid E_1) + P(E_2) \cdot P(A \mid E_2) + \cdots + P(E_m) \cdot P(A \mid E_m)}$$

In Example 1, we used Bayes' Theorem for the case $m = 2$. To derive Bayes' Theorem, first note that

$$P(E_1 \mid A) = \frac{P(E_1 \cap A)}{P(A)}$$

Now we will show that the numerator of the fraction on the right side of Bayes' equation is $P(E_1 \cap A)$ and the denominator is $P(A)$.

$$P(E_1) \cdot P(A \mid E_1) = P(E_1) \cdot \frac{P(A \cap E_1)}{P(E_1)} = P(A \cap E_1) = P(E_1 \cap A)$$

This shows that the numerator is $P(E_1 \cap A)$. Next consider the denominator. The last equation is valid for any of the sets E_i, where $i = 1, 2, \ldots, m$.

$$P(E_i) \cdot P(A \mid E_i) = P(E_i \cap A)$$

Note that $\{E_1, E_2, \ldots, E_m\}$ is a disjoint collection of events. Therefore, $\{A \cap E_1, A \cap E_2, \ldots, A \cap E_m\}$ is a disjoint collection of events.

Since $P(E_1) + P(E_2) + \cdots + P(E_m) = 1$ and these sets are disjoint, $P(E_1 \cup E_2 \cup \cdots \cup E_m) = P(E_1) + P(E_2) + \cdots + P(E_m) = 1$. Thus $E_1 \cup E_2 \cup \cdots \cup E_m = S$ since S is the only event with probability 1. Then

$$P(E_1) \cdot P(A \mid E_1) + P(E_2) \cdot P(A \mid E_2) + \cdots + P(E_m) \cdot P(A \mid E_m)$$
$$= P(E_1 \cap A) + P(E_2 \cap A) + \cdots + P(E_m \cap A)$$

$$= P[(E_1 \cap A) \cup (E_2 \cap A) \cup \cdots \cup (E_m \cap A)]$$
$$= P[(E_1 \cup E_2 \cup \cdots \cup E_m) \cap A]$$
$$= P(S \cap A)$$
$$= P(A)$$

Example 2 Suppose $\{E_1, E_2, E_3, E_4\}$ is a disjoint set of events, $P(E_1) = .3$, $P(E_2) = .2$, $P(E_3) = .1$, $P(E_4) = .4$, $P(A|E_1) = .7$, $P(A|E_2) = .3$, $P(A|E_3) = .4$, and $P(A|E_4) = 0$. Determine $P(E_1|A)$.

Solution
$$P(E_1|A) = \frac{(.3)(.7)}{(.3) \cdot (.7) + (.2) \cdot (.3) + (.1) \cdot (.4) + (.4) \cdot (0)}$$
$$= \frac{.21}{.31} = \frac{21}{31} = 0.68.$$

Example 3 Susan Roth, personnel director for Iris Cosmetics, is dissatisfied with her current hiring practices. Of the people she has hired, 70% are good workers and the remaining 30% are bad workers. She gives her employees a test and finds that 80% of the good workers and 40% of the bad workers pass. Assume she adds the test to her hiring procedure, so that in order to be hired, a person must meet the previous requirements and also pass the test. What percent of the people she how hires will be good workers?

Solution In solving this problem, we must assume that on the job experience did not change a person's chance of passing the test. Let G be the event that a worker is good, B that a worker is bad, and T that a worker passes the test. We are given $P(G) = .7$, $P(B) = .3$, $P(T|G) = .8$, and $P(T|B) = .4$ and are asked for $P(G|T)$. By Bayes' Theorem,

$$P(G|T) = \frac{P(G) \cdot P(T|G)}{P(G) \cdot P(T|G) + P(B) \cdot P(T|B)}$$
$$= \frac{(.7)(.8)}{(.7)(.8) + (.3)(.4)} = \frac{.56}{.68} = \frac{56}{68} = .82$$

Thus 82% of the people she now hires will be good workers.

Example 4 The following table contains data concerning the mortgages held by Hillcrest Bank.

Down Payment (in %)	10%	20%	25%	30%
Percentage of mortgages with this down payment	45	25	20	10
Estimated probability of default	.05	.02	.01	.01

If a default occurs, what is the probability that it is on a mortgage with a 10% down payment?

Solution Let D denote default. Then, by Bayes' Theorem,

$P(10\%|D) =$

$$\frac{P(10\%) \cdot P(D|10\%)}{P(10\%) \cdot P(D|10\%) + P(20\%) \cdot P(D|20\%) + P(30\%) \cdot P(D|30\%) + P(40\%) \cdot P(D|40\%)}$$

$$= \frac{(.45) \cdot (.05)}{(.45) \cdot (.05) + (.25) \cdot (.02) + (.20) \cdot (.01) + (.10) \cdot (.01)}$$

$$= \frac{.0225}{.0305} = .74$$

If a default occurs, the probability that it is on a mortgage with a 10% down payment is .74.

Exercises 6-7

1. Given $P(E_1) = .5$, $P(E_2) = .4$, $P(E_3) = .1$, $P(A|E_1) = .6$, $P(A|E_2) = .8$, and $P(A|E_3) = .2$, determine $P(E_1|A)$, $P(E_2|A)$, and $P(E_3|A)$.

2. Given $P(E_1) = .2$, $P(E_2) = .5$, and $P(E_3) = .3$, $P(B|E_1) = 1$, $P(B|E_2) = .1$, and $P(B|E_3) = 0$, determine $P(E_1|B)$, $P(E_2|B)$, and $P(E_3|B)$.

3. Given $P(E_1) = .1$, $P(E_2) = .1$, $P(E_3) = .3$, $P(E_4) = .5$, $P(A|E_1) = .7$, $P(A|E_2) = .3$, $P(A|E_3) = .8$, and $P(A|E_4) = .3$, determine $P(E_1|A)$, $P(E_2|A)$, $P(E_3|A)$, and $P(E_4|A)$.

4. Given $P(E_1) = .2$, $P(E_2) = .2$, $P(E_3) = .5$, $P(E_4) = .1$, $P(C|E_1) = .6$, $P(C|E_2) = 0$, $P(C|E_3) = 1$, and $P(C|E_4) = .2$, determine $P(E_1|C)$, $P(E_2|C)$, $P(E_3|C)$, and $P(E_4|C)$.

5. A bottle contains 15 fair coins and 1 two-headed coin. A coin is picked at random from the bottle and is tossed 5 times. The coin shows heads each time. What is the probability that the coin is two-headed?

6. Do exercise 5 if the bottle contains 5 fair coins and 5 two-headed coins.

Urn I contains 5 red balls and 5 green balls. Urn II contains 1000 red balls. Urn III contains 200 red balls and 300 green balls. An urn is selected at random and a ball is selected from the urn. Determine the following:

7. $P(\text{I} \mid \text{red})$
8. $P(\text{I} \mid \text{green})$
9. $P(\text{II} \mid \text{red})$
10. $P(\text{II} \mid \text{green})$
11. $P(\text{III} \mid \text{red})$
12. $P(\text{III} \mid \text{green})$
13. $P(\text{red} \mid \text{III})$
14. $P(\text{green} \mid \text{III})$
15. $P(\text{green} \mid \text{II})$

16. A factory made ten batches of faucets yesterday. Nine of the batches were good. In a good batch, experience has shown that 99% of the faucets work. One batch was spoiled because of a broken machine, so that none of the faucets in this batch works. I select a batch at random and choose a faucet from it.
    a. If this faucet works, what is the probability that I have chosen a good batch?
    b. If this faucet does not work, what is the probability that I have chosen a bad batch?

17. Insurance company statistics show that a new driver who has completed a driver education course has probability .85 of an accident-free first year of driving. Without the course, the probability is .75. In Appleton, 70% of all new drivers have taken the driver education course.
    a. What is the probability that a person with an accident-free first year of driving took the driver education course?
    b. What is the probability that a person with an accident in the first year of driving took the driver education course?

18. A medical researcher has devised a quick and cheap test for strep throat. To evaluate the test, she performs the test on 5000 adults with sore throats. She also tests each adult with the traditional test of taking a throat culture and finds that 500 of them have strep throat. Her test indicates strep in 96% of those who have strep and in 3% of those who do not.
    a. If the test indicates strep in a person, what is the probability that the person has strep?
    b. If the test indicates that a person does not have strep, what is the

probability that the person has strep?

19. Abco has three plants, called Alpha, Beta, and Gamma, which make the motors for its fans. The large plant, Alpha, makes 50% of the motors and only 1% of Alpha's motors are defective. Beta makes 35% of the motors and only 2% of Beta's motors are defective. Gamma has poor management and a bad assembly line. Gamma makes 15% of the motors, and 10% of Gamma motors are defective.
   a. If a motor selected at random is good, what is the probability it came from Beta?
   b. If a motor selected at random is defective, what is the probability it came from Gamma?
   c. A motor is selected at random. Compute the probability that it is bad.
   d. A motor selected at random is bad. What is the probability it came from Alpha or Beta?

20. The probability of mechanical failure on an assembly line is .02 each day. The probability that a mechanical failure will cause an accident is .05. The probability of a human error is .01 on each day and the probability that a human error will cause an accident is .03. Assume only one accident can occur in a day.
   a. What is the probability that there will be an accident on a given day?
   b. If there is an accident, what is the probability it was caused by human error?

21. The university's records show that only 40% of the students who register for calculus pass the course. To improve the pass rate, the university gives a screening test to determine which students need remedial work. For one year, the test is given to students on the first day of the calculus class. Data is collected and shows that 90% of the students who pass calculus passed the screening test. However, only 20% of those who do not pass calculus passed the screening test.
   a. If an incoming student passes the screening test, what is the probability that he will pass calculus?
   b. If an incoming student fails the screening test, what is the probability that he will pass calculus?

*22. The university now institutes a remedial course. This course is required of all students who fail the screening test. The records show that 60% of all students who fail the screening test pass the remedial course and take calculus. If a student passes the remedial course and registers for calculus, he has a .8 chance of passing. Assume this system has been in effect for several years now, and that enrollments have remained constant.

a. What is the probability that an entering student who would have taken calculus under the old system (no screening test) takes calculus at some time under the new system?

b. What is the pass rate in calculus now?

23. Nelson Pharmaceuticals has employed 637 salespeople over the last five years. Of these, 28 have generally exceeded their sales quota, 87 have met their quota, and the remainder have been short of their quota. A management consulting firm offers Nelson a test which is supposedly passed by 87.4% of pharmaceutical salespeople who exceed quota, 37.2% of those who meet quota, and 11.9% of those who fall short of quota. Assume the claims for the test are true, and Nelson from now on only hires salespeople who meet all previous criteria for hiring and also pass the test. What is the probability that a salesperson who is hired under the new procedures will

a. Exceed quota?
b. Meet quota?
c. Be short of quota?

Chapter 6 Review

1. An audit of a firm's billing procedures revealed that errors were made on 1500 of the 300,000 bills that were checked. What is the probability that a bill selected at random has an error? Is correct?

2. An optical store carries 20 styles of eyeglass frame. The frames come in five colors and seven sizes. How many kinds of frame are possible?

Suppose A and B are sets, $P(A) = .4$, and $P(B) = .25$. Determine each of the following:

3. $P(A')$
4. $P(A \cup B)$, if A and B are disjoint
5. $P(A \cup B)$, if $P(A \cap B) = .2$
6. $P(A|B)$, if $P(A \cap B) = .2$
7. $P(A \cup B)$, if A and B are independent events

Happy Edgerton sells cosmetics door to door. Last year, she knocked on 8000 doors and made 1600 sales. What is the probability that she will make

8. No sales at the first three doors she knocks on today?
9. At least one sale at the first three doors she knocks on today?
10. A sale at each of the first three doors she knocks on today?
11. If $P(X) = .3$, $P(Y) = .7$, and $P(X \cap Y) = .20$, are X and Y independent?
12. If $P(Q) = .4$, $P(R) = .35$, and $P(Q' \cap R') = .39$, are Q and R independent?
13. Determine $P(C|D)$, if $P(C) = .5$, $P(D) = .3$, and $P(C \cap D) = .2$.
14. Determine $P(A'|A \cup B)$, if $P(A' \cap B') = .2$, $P(A) = .5$, and $P(A \cap B) = .1$.

A card is drawn at random from a deck. What is the probability that

15. The card is a ten?
16. The card is a ten or a diamond?
17. The card is a nine, given that the card is not a three or a four?
18. The card is a nine, given that the card is a nine, a ten, or a diamond?

Two cards are drawn at random from a deck. Find the probability of each of the following events, if

    a. The first card is replaced and the deck is shuffled before the second card is drawn.
    b. The second card is drawn before the first card is replaced.

19. Both cards are hearts.
20. Neither card is a heart.
21. At least one of the cards drawn is a heart.
22. The five of clubs is drawn both times.
23. The second card is a spade, given that the first card is the ten of diamonds.
24. Both cards are the same suit.
25. The cards are of different suits.

26. Determine $P(E_1|A)$ if $P(E_1) = .2$, $P(E_2) = .7$, $P(E_3) = .1$, $P(A|E_1) = .5$, $P(A|E_2) = .3$, and $P(A|E_3) = .8$.
27. Two mechanics, Ed and Ray, run an auto repair shop. Ed words on 2/3 of the cars, and Ray works on the other 1/3. Ed's work is satisfactory only 70% of the time because he works too fast. Ray's work is satisfactory 95% of the time.

Susanna Shapiro takes her car into the shop for repairs. If the work is satisfactory, what is the probability that Ed did it? If the work is unsatisfactory, what is the probability that Ray did it?

7

More Topics in Probability

In this chapter, we continue the study of probability by discussing counting techniques involving permutations and combinations.

Section 7-1 Permutations

We begin by learning how to count the number of ways in which objects can be *permuted*, that is, arranged in order.

In how many ways can three people form a line? If their initials are A, B, and C, there are six possible ways: ABC, ACB, BAC, BCA, CAB, and CBA. The answer can be obtained by saying there are three slots __ __ __ to fill. There are three ways of filling the first slot, since A or B or C can go there. There are two ways of filling the second slot, since one person is already in line. There is only one way to fill the third slot, since only one person is not yet in line. Thus, there are $3 \cdot 2 \cdot 1 = 6$ ways of filling all three slots.

If six people had to stand in line, there would be $6 \cdot 5 \cdot 4 \cdot 3 \cdot 2 \cdot 1 = 720$ possible orders.

> **Definition.** If n is a counting number, n *factorial* (written $n!$) is the product of the counting numbers from 1 to n.
> $$n! = 1 \cdot 2 \cdot 3 \cdot \cdots \cdot (n-1) \cdot n = n \cdot (n-1) \cdot \cdots \cdot 3 \cdot 2 \cdot 1$$

Two special definitions will be very convenient later on. These are: $0! = 1$ and $1! = 1$.

> **Definition.** A *permutation* of n objects is an arrangement of the n objects in order. There are $n!$ permutations of n objects.

Thus n objects can be arranged in order in $n!$ different ways. Note that $n!$ becomes large very quickly. Some values of $n!$ are:

$0! = 1$ $8! = 40{,}320$
$1! = 1$ $9! = 362{,}880$
$2! = 2$ $10! = 3{,}628{,}800$
$3! = 6$ $15! = 1{,}307{,}674{,}368{,}000 = 1.3 \times 10^{12}$
$4! = 24$ $20! = 2.4 \times 10^{18}$
$5! = 120$ $25! = 1.6 \times 10^{24}$
$6! = 720$ $70! = 1.2 \times 10^{100}$
$7! = 5040$ $100! = 9.3 \times 10^{157}$

Example 1 Compute $27!/26!$.

Solution Although $27!$ is a very large number, $27!/26!$ is easy to compute, and in fact, $27!/26! = 27$. The reason is that $27!$ is obtained by multiplying $26!$ by 27. In general,

$$(n+1)! = (n+1) \cdot n!$$

Similarly, $(n+2)! = (n+2)(n+1)n!$. Thus $52!/50! = 52 \cdot 51 = 2652$ and $12!/14! = 1/(13 \cdot 14) = 1/182$. Also, $8!/5! = 8 \cdot 7 \cdot 6 = 336$ and $9!/5! = 9 \cdot 8 \cdot 7 \cdot 6 = 3024$.

Example 2 There are ten rooms in a house. The owner buys ten paintings and wants to put one painting in each room. In how many ways can this be done?

Solution This can be done in $10!$ ways. To see this, suppose the rooms are numbered 1–10. Then the problem asks in how many ways can the paintings be numbered from 1–10. If each painting is put in the room with the same number, different numberings of the paintings will lead to different arrangements of the paintings among the rooms.

Example 3 At a party, each of the 20 guests hangs his or her coat in a closet. At the end of the party, each guest grabs a coat at random. What is the probability that each person gets the right coat?

Solution There is only one way for each guest to take the right coat. There are 20! possible ways in which each guest can take one coat. Thus, the probability that each person has the right coat is $1/20! = 4.1 \times 10^{-19}$.

Example 4 A child has six blocks. Each block has one of the letters A, E, G, N, O, and R on it, and different blocks have different letters. If the six blocks are lined up at random, the probability that they will spell ORANGE is $1/6! = 1/720$.

Example 5 Nine men decide to form a baseball team. There are 9! different ways in which they can fill the nine positions on the team. There are also 9! possible batting orders.
■

Now we consider the case in which there are n objects, but only some of them are arranged in order.

A club with 25 members is to select a president, treasurer, and secretary. (No person may hold more than one office.) This can be done in $25 \cdot 24 \cdot 23 = 13{,}800$ ways. There are 25 possible presidents. After the president is selected, there are 24 remaining members from which to choose the treasurer. Then there are 23 remaining members from which to pick the secretary. The number $25 \cdot 24 \cdot 23$ may also be written as $25!/22!$.

> **Definition.** A *permutation of n objects taken r at a time* is an ordering of r objects selected from among the n objects.
> **Definition.** $P(n, r)$ is the number of permutations of n objects taken r at a time.

The preceding example involved permutations of 25 objects taken three at a time. The number of such permutations is $P(25, 3) = 25 \cdot 24 \cdot 23 = 25!/22!$. The general formula for $P(n, r)$ is

$$P(n, r) = n(n-1) \cdot \cdots \cdot (n - r + 1) = \frac{n!}{(n-r)!}$$

Permutations of n objects taken r at a time

The middle expression in this formula is used to actually compute $P(n,r)$. Note that the middle expression is the product of r different numbers. The last expression is convenient only for algebraic manipulation.

Example 6 Simplify $P(n+1,3) - P(n,2)$.

Solution
$$P(n+1,3) - P(n,2)$$
$$= (n+1)n(n-1) - n(n-1)$$
$$= n(n-1)(n+1-1)$$
$$= n^2(n-1)$$
$$= n^3 - n^2$$

Example 7 Seven people try out for a basketball team. In how many different ways can the five positions be filled?

Solution The five positions can be filled in $P(7,5) = 7 \cdot 6 \cdot 5 \cdot 4 \cdot 3 = 2520$ ways.

Example 8 Three cards are dealt from a deck. What is the probability that all three are spades?

Solution The probability that all are spades is

$$P(13,3)/P(52,3) = (13 \cdot 12 \cdot 11)/(52 \cdot 51 \cdot 50) = .013.$$

Note that there are $P(13,3)$ ways of dealing 3 spades and $P(52,3)$ ways of dealing three cards.

Example 9 Forty men and sixty women apply for jobs at a company. If ten people are hired at random for ten jobs, the probability that 10 women are hired is $P(60,10)/P(100,10) = .004$.

■

Exercises 7-1

Compute each of the following.

1. $5!$
2. $7!$
3. $\dfrac{10!}{8!}$
4. $\dfrac{20!}{19!}$
5. $\dfrac{6!}{7!}$
6. $\dfrac{11!}{13!}$
7. $P(15,2)$
8. $P(n+1,2)$

9. $\dfrac{(2n)!}{(2n-2)!}$

10. $P(30, 1)$

11. $\dfrac{(2n+3)!}{(2n+2)!}$

12. $P(2n, 2)$

13. In how many ways can eight people seat themselves at a table with eight chairs?
14. In how many ways can five lanes be assigned to the five contestants in a race?
15. How many three-digit numerals contain three different digits?
16. How many five-letter words contain five different letters?
17. At a banquet, seven main courses are served, one at a time. In how many orders can the main courses be served?
18. A four-digit numeral is selected at random. What is the probability that the numeral contains four different digits?
19. A movie comes on five reels. The projectionist messes them up and does not know what order to show them in. However, he knows which reel comes last because one reel is not full. He guesses an order to show the other reels in. What is the probability that he guesses correctly?
20. Four cards are dealt from a deck. What is the probability that all are aces?
21. Three cards are dealt from a deck. What is the probability that all are spades, given that none is hearts?
22. Three cards are dealt from a deck. What is the probability that all three are fives or sixes?

A jar contains five red balls, ten green balls, and ten yellow balls. Three balls are selected in sequence without replacement.

23. In how many ways can the three balls be selected?
24. What is the probability that all three balls are green?
25. What is the probability that none of the three balls is green?
26. What is the probability that the first ball is red and the next two balls are yellow?
27. What is the probability that all three balls are red, given that none is yellow?
28. What is the probability that all three balls are yellow, given that the first ball is yellow?
29. A four-digit numeral contains four different digits. What is the probability that all four digits are even?
30. A seven-letter word contains seven different letters. What is the proba-

bility that all seven letters are vowels? Are consonants?

Six diplomats must form a receiving line at a party celebrating the signing of a peace treaty. Two of the diplomats are from the host country, and the others represent countries A, B, C, and D. How many orders are possible for the receiving line if

*31. A diplomat from the host country must be first in line?

*32. The two diplomats from the host country may not be next to each other?

*33. The two diplomats from the host country must both be in the middle of the line?

*34. The diplomat from country A must be ahead of the diplomat from country B?

Section 7-2 Combinations and Hypergeometric Probability

In the previous section we discussed permutations. There order was critical. Here we discuss the selection of elements of a set without regard to order.

Suppose seven people apply for jobs as cashiers at a new branch of Stop and Shop, and three jobs are available. In how many ways can Ms. Goldstein, the manager, decide whom to hire?

Ms. Goldstein can choose the best, second best, and third best applicants in $P(7,3) = 7 \cdot 6 \cdot 5 = 210$ ways. However, different choices can lead to the same three people being hired. For example, the choices

$$
\begin{array}{ll}
ABC & BCA \\
ACB & CAB \\
BAC & CBA
\end{array}
$$

all lead to A, B, and C being hired. These six arrangements are the ways of arranging A, B, and C in order, that is, the $3! = 6$ permutations of A, B, and C.

Therefore, in computing $P(7,3)$, we have counted each possible way of selecting three applicants $3!$ times. There are consequently $P(7,3)/3! = 35$ ways in which Ms. Goldstein can decide whom to hire.

Definition. A subset with r elements of a set with n elements is a *combination* of n objects taken r at a time.

$$C(n, r) = \frac{P(n, r)}{r!} = \frac{n!}{(n-r)!r!}$$

Ms. Goldstein has selected a combination of seven objects taken three at a time. There are $C(7,3)$ such combinations. By reasoning similar to that in the preceding example, one can show that $C(n,r)$ is the number of combinations of n objects taken r at a time. $C(n,r)$ is always a counting number because the combinations of n objects taken r at a time can always be listed and counted.

Examples 1. $C(8,5) = \dfrac{8!}{3!5!} = \dfrac{8 \cdot 7 \cdot 6}{3!} = \dfrac{8 \cdot 7 \cdot 6}{3 \cdot 2 \cdot 1} = 56$

2. $C(10,6) = \dfrac{10!}{4!6!} = \dfrac{10 \cdot 9 \cdot 8 \cdot 7}{4!} = \dfrac{10 \cdot 9 \cdot 8 \cdot 7}{4 \cdot 3 \cdot 2 \cdot 1} = 210$

∎

For certain values of r, $C(n,r)$ is easy to compute. For example, $C(n,n)$ is the number of ways of selecting n objects from n—the only way to do this is to select all the objects. Thus $C(n,n) = 1$. Similarly, $C(n,0) = 1$, because there is only one way to select zero objects—don't take anything.

$$\boxed{C(n,n) = 1 \qquad C(n,0) = 1}$$

There are n ways to select one object from n. There also are n ways to select $n-1$ objects from n, since the selection of $n-1$ objects can be made by deciding which one object to leave behind.

$$\boxed{C(n,1) = n \qquad C(n,n-1) = n}$$

Suppose a set has n members and r of them are selected. Then $n-r$ members are not selected. Thus every time a subset with r members is formed, a set with $n-r$ members is formed. Similarly, every time a set with $n-r$ members is formed, a set (its complement) with r members is formed. Therefore,

$$\boxed{C(n,r) = C(n, n-r)}$$

Example 3 A basketball coach can choose five starters from among eight players in $C(8,5) = 56$ ways.

7-2 COMBINATIONS AND HYPERGEOMETRIC PROBABILITY

Example 4 A committee of ten can choose a subcommittee of three in $C(10,3) = 120$ ways.

Example 5 A poker hand contains five cards. What is the probability that a poker hand contains all spades?

Solution The probability that a poker hand contains all spades is $C(13,5)/C(52,5) = .00050$, since $C(13,5)$ is the number of ways of selecting five cards from the 13 spades, and $C(52,5)$ is the number of ways of selecting five cards from the 52 cards in the deck.

Example 6 A bridge hand contains 13 cards. What is the probability that a bridge hand contains all spades?

Solution The probability that a bridge hand contains all spades is $1/C(52,13) = 1.6 \times 10^{-12}$, since there is only one way to select all 13 spades.

Example 7 What is the probability that a poker hand contains no fours, fives, or sixes?

Solution The probability that a poker hand contains no fours, fives, or sixes is $C(40,5)/C(52,5) = .25$, since there are 40 cards in the deck other than fours, fives, and sixes.

Example 8 What is the probability that a poker hand contains two jacks and three aces?

Solution Since there are four jacks in a deck of cards, there are $C(4,2) = 6$ ways of selecting two jacks. Since there are four aces, there are $C(4,3) = 4$ ways of selecting three aces. Since two jacks and three aces are five cards, they fill up the whole poker hand. The probability that a poker hand contains two jacks and three aces is $C(4,2) \cdot C(4,3)/C(52,5) = 24/C(52,5) = 9.2 \times 10^{-6}$.

Example 9 What is the probability that a bridge hand contains exactly seven spades?

Solution There are $C(13,7)$ ways of selecting seven cards from the 13 spades in a deck. Since a bridge hand has 13 cards, six more cards must be selected from among the $52 - 13 = 39$ cards that are not spades; this can be done in $C(39,6)$ ways. There are $C(13,7) \cdot C(39,6)$ bridge hands containing exactly seven spades. The probability that a bridge hand contains exactly seven spades is $C(13,7) \cdot C(39,6)/C(52,13) = .0088$.

Example 10 An urn contains five red balls, seven green balls, and eight yellow balls. A set of six balls is selected from the urn. What is the probability that one ball in the set is red, three are green, and two are yellow?

Solution The probability that one ball in the set is red, three are green, and two are yellow is $C(5,1) \cdot C(7,3) \cdot C(8,2)/C(20,6) = .13$.

Hypergeometric probability

Examples 8, 9, and 10 demonstrate *hypergeometric probability*. Suppose a set has n members, n_1 of one kind and n_2 of a second kind, so that $n = n_1 + n_2$. The probability that a subset with r members has r_1 of the first kind and r_2 of the second kind ($r = r_1 + r_2$) is $C(n_1, r_1) \cdot C(n_2, r_2)/C(n, r)$. This formula generalizes, as shown in the preceding urn problem, to allow for three or more kinds of elements in the set. The reason for multiplying the numbers in the numerator is that the union of any set of r_1 members of the first kind with any set of r_2 members of the second kind is a subset with r members.

Example 11 Certain valves for a boiler come in batches of 1000. When a batch arrives at Dalhousie Plumbing, a sample of ten valves is selected at random for testing. If there is more than one defective valve in the sample, the batch is rejected. Otherwise the batch is accepted.

Suppose there are 200 defective valves in the batch. What is the probability that the batch is accepted?

Solution The number of possible samples is $C(1000, 10)$. A batch is accepted if the sample contains no defective valves or one defective valve.

If a sample contains no defective valves, its ten members must be chosen from among the $1000 - 200 = 800$ good valves. Thus there are $C(800, 10)$ samples with no defective valves. The probability that a sample contains no defective valves is

$$\frac{C(800, 10)}{C(1000, 10)} = .106$$

If a sample contains one defective valve, it contains nine good valves. The one defective valve must be chosen from among the 200 defective valves, and the nine good valves must be chosen from among the 800 good valves. Thus there are $C(200, 1) \cdot C(800, 9)$ samples with exactly one defective valve. The probability that a sample contains exactly one defective valve is

$$\frac{C(200,1) \cdot C(800,9)}{C(1000,10)} = .268$$

The probability that the sample contains no defective valves or one defective valve, and that the batch is therefore accepted, is $.106 + .268 = .374$.

■

Exercises 7-2

1. In how many ways can Bill choose three flavors of ice cream for a banana split, if 31 flavors are available?
2. In how many ways can a teacher decide which 10 of 25 problems to assign?
3. In how many ways can John decide which four main courses to order at a Chinese restaurant, if there are 150 main courses listed on the menu?
4. In how many ways can a child select two toys, if 20 toys are available?
5. A shipment of 20 smoke detectors contains six that are defective. Five are selected at random by a customer. What is the probability that the customer has exactly two defective smoke detectors?
6. A box contains 48 flashbulbs, of which 40 are good. If Sue buys 12 flashbulbs, what is the probability that she has exactly two defective ones?
7. What is the probability that a poker hand contains exactly two kings?
8. What is the probability that a poker hand contains exactly two deuces and exactly one ace?
9. What is the probability that a poker hand contains exactly three queens, if the hand contains no aces?
10. What is the probability that a poker hand contains exactly two sevens, if the hand contains the ten of diamonds?
11. Three cards are selected from a deck. What is the probability that the two of clubs is selected?
12. Two cards are selected from a deck. What is the probability that both are fours?
13. What is the probability that a bridge hand contains four clubs and three of each other suit?
14. What is the probability that a bridge hand contains exactly five clubs and two diamonds?

15. The City Council has 15 members. Of these, nine are liberals and six are conservatives. A delegation of five members is chosen at random to attend a convention. What is the probability that a majority of the delegation is liberal?

16. A box of a baker's dozen (13) doughnuts contains four stale ones.
    a. If I grab three doughnuts, what is the probability that all three are fresh?
    b. If I grab two doughnuts, what is the probability that one is fresh and one is stale?
    c. If I discard two stale doughnuts and then pick three, what is the probability that these three are all fresh?
    d. If I pick six doughnuts, what is the probability that I get four fresh ones and two stale ones?
    e. If I pick six doughnuts, what is the probability that all are stale?
    f. I eat four fresh doughnuts and then offer the doughnuts to my boss. He picks three. What is the probability he gets two fresh ones and one stale one?

17. A jar contains seven pink marbles, ten green marbles, and eight yellow marbles. A set of nine marbles is selected from the jar.
    a. What is the probability that three of the marbles are pink and the other six are yellow?
    b. What is the probability that three marbles of each color are selected?
    c. What is the probability that exactly four of the marbles are green, given that none is yellow?
    d. What is the probability that all nine marbles are green?

18. Ten men and 15 women are employed in a store. The store manager is required to pick four employees at random. These employees, together with the manager, form a committee to find and end any sex discrimination in the store. Of the four employees who are chosen,
    a. What is the probability that all are men?
    b. What is the probability that all are women?
    c. What is the probability that all are men, if Henry Jacobs is chosen but Sheila Sontag is not chosen?
    d. What is the probability that two are women and two are men?
    e. What is the probability that exactly three are women?
    f. What is the probability that exactly three are women, if Sam Traylor and Charley Jones are not chosen?

19. A batch of 20 dishwashers is to be tested for acceptance. Four of the dishwashers are selected at random and are thoroughly inspected. The batch is accepted if at least three of the four dishwashers in the sample

pass inspection. What is the probability that the batch is accepted, if the batch contains
a. No defective dishwashers?
b. One defective dishwasher?
c. Two defective dishwashers?
d. Three defective dishwashers?
e. Four defective dishwashers?

The following exercises involve distinguishable permutations. In each of these exercises, assume a child has spelled out the given word with blocks. Two words can be distinguished only if the letters are in a different order. If the blocks spell ABCA and the two As are exchanged, the new word is still ABCA and cannot be distinguished from the old word. However ABCA and, for example, ABAC can be distinguished because they do not have all the letters in the same order. How many distinguishable permutations are there of

20. ABB?
22. AAABB?
24. AABBC?
*26. AAABBCD?

21. AABB?
23. ABBC?
*25. AAAABBBCC?
*27. MISSISSIPPI?

Section 7-3 Binomial Probability

There are many circumstances in which an experiment with two possible outcomes is performed, under identical conditions, many times. For example, a person doing market research may ask shoppers whether they prefer regular or menthol flavored toothpaste, or a public health worker may test samples from a city's water supply to see if they are fit to drink. It is sometimes important to know how likely it is for a particular outcome to occur a certain number of times. This information is provided by what is called binomial probability.

Example 1 A die is tossed five times. What is the probability that the die lands on 6 exactly twice?

Solution The probability that the die lands on 6 on the first two tosses and does not land on 6 afterwards is

$$\frac{1}{6} \cdot \frac{1}{6} \cdot \frac{5}{6} \cdot \frac{5}{6} \cdot \frac{5}{6}$$

since the outcome on each toss is independent of the outcomes on the other tosses. The die could land on 6 on any two tosses, however, and there are $C(5,2)$ ways of selecting two tosses from the five tosses. Thus the probability the die lands on 6 exactly twice is $C(5,2) \cdot (1/6)^2 (5/6)^3 = .16$.

∎

Example 1 demonstrates *binomial probability*. Binomial probability occurs in this situation: An experiment has two outcomes E and E'. The experiment is repeated n times. The probability that E occurs exactly r times is

$$\boxed{C(n,r)P(E)^r P(E')^{n-r}} \quad \text{Binomial probability formula}$$

Note that if E occurs exactly r times, then E' occurs exactly $n - r$ times. Recall that $p(E') = 1 - p(E)$. If $p(E)$ is denoted by p, then $p(E') = 1 - p$. The binomial probability formula can also be written

$$C(n,r)p^r(1-p)^{n-r}$$

Example 2 A fair coin is tossed 100 times.

PROBLEM A What is the probability of exactly 95 heads?

Solution Use the binomial probability formula with $n = 100$, $r = 95$, $P(E) = 1/2$, and $P(E') = 1/2$. The probability of exactly 95 heads is $C(100,95)(1/2)^{95}(1/2)^5 = 5.94 \times 10^{-23}$.

PROBLEM B What is the probability of at least 98 heads?

Solution The probability of at least 98 heads is

$$C(100,98)\left(\frac{1}{2}\right)^{98}\left(\frac{1}{2}\right)^2 + C(100,99)\left(\frac{1}{2}\right)^{99}\left(\frac{1}{2}\right) + \left(\frac{1}{2}\right)^{100}$$
$$= (4950 + 100 + 1)/2^{100} = 5051/2^{100} = 3.98 \times 10^{-27},$$

since the event "at least 98 heads" is the union of the disjoint events 98 heads, 99 heads, and 100 heads.

Example 3 A certain experiment has two outcomes. E is one of the outcomes, and

$P(E) = .17$. Repetitions of the experiment are independent of each other.

PROBLEM A What is the probability that E occurs exactly twice in eight trials?

Solution The probability that E occurs exactly twice in eight trials is $C(8, 2)(.17)^2(.83)^6 = .26$.

PROBLEM B What is the probability that E occurs at most twice in eight trials?

Solution The probability that E occurs at most twice in eight trials is $C(8, 2)(.17)^2(.83)^6 + C(8, 1)(.17)(.83)^7 + .83^8 = .86$ since the event "E occurs at most twice" is the union of the disjoint events E occurs twice, E occurs once, and E does not occur.

PROBLEM C What is the probability that E occurs at least once in eight trials?

Solution The probability that E occurs at least once in eight trials is $1 - (.83)^8 = .77$, because the events "E occurs at least once" and "E does not occur" are complements of each other.
■

In Boston, 80% of the people love dogs and 20% hate dogs. Twelve people are selected at random in Boston. The probability that exactly nine people in the sample love dogs is $C(12, 9)(.8)^9(.2)^3 = .24$. This is computed using binomial probability. Note that the population of Boston is not given. This problem really involves hypergeometric probability, but when the size of the sample is very small compared to the size of the population, binomial probability is a very good approximation to hypergeometric probability. Binomial probability is then frequently used instead of hypergeometric probability, because the numerical computations needed are much simpler.

Example 4 Five tellers usually work in a bank branch office each day. If fewer than four show up, the remaining ones cannot handle all the customers fast enough. This leads to many mistakes by the tellers and losses for the bank. As a result of experience, current policy is to close the office if fewer than four tellers show up.

Exactly one week ago, an executive from the bank's central office visited the branch. The next day, he came down with influenza. Thus all five tellers were exposed. For this particular kind of influenza, 70% of those who are exposed develop the disease. The incubation time is eight days. What is the probability that the branch office will open tomorrow?

Solution Let E be the event that an individual comes down with flu. Then $P(E) = .7$, so that $P(E') = 1 - .7 = .3$. The branch will open tomorrow if at least four tellers show up, that is, if at most one teller has the flu. The probability that the branch opens tomorrow is

$$(.3)^5 + C(5,1)(.7)^1(.3)^4 = .03$$

Exercises 7-3

1. On a multiple-choice test, there are five possible answers for each question. There are ten questions on the test, all of which are scored equally. Joe Cool has not studied at all, so he guesses. What is the probability that
   a. His grade is 60%?
   b. He gets a grade of 90% or higher?
   c. His grade is 75%?
   d. His grade is 70% or 80%?

2. A jar contains five red balls, eight green balls, and 12 yellow balls. Four balls are selected in sequence with replacement. What is the probability that, of the four balls selected,
   a. Exactly two are red?
   b. Exactly three are green?
   c. At most one is red?
   d. At least two are yellow?
   e. None is green?
   f. At most three are yellow?

3. A saleswoman usually makes a sale to 20% of the people she calls on. What is the probability she makes a sale on Wednesday, if she calls on five people that day?

4. Of all the trowels made by U.S. Gadget Corp. 90% are good and the remaining 10% are defective. A sample of ten trowels is selected. What is the probability that nine of the trowels in the sample are good and 1 is defective?

5. In Bradford Heights 60% of the voters favor a school levy. Twenty voters

are selected at random, and each is asked his or her opinion on the levy. What is the probability that, of the 20 voters sampled,
a. All favor the levy?
b. None favors the levy?
c. Exactly half favor the levy?
d. Exactly 14 favor the levy?
e. At least two favor the levy?
f. At least 17 favor the levy?
g. At most three favor the levy?
h. At most one favors the levy?

6. Currently one-third of all marriages in the United States end in divorce. The local records' office recorded 12 marriages today.
a. How many of these marriages do you expect will end in divorce?
b. What is the probability that exactly four of these marriages will end in divorce?

7. In Gorham, 40% of the school children will have no cavities within a year, without the use of toothpaste. If a child uses McBrite Toothpaste, the child has a 60% chance of no cavities within a year.
a. If ten children in Gorham use McBrite Toothpaste for a year, what is the probability that at least nine of them will have no cavities?
b. McBrite arranges to have 200 school children divided into 20 groups of 10 each. What is the probability that in at least one group of 10, at least 9 will have no cavities? In this case, McBrite will advertise that, "In one scientific study of McBrite Toothpaste, 90% of the school children who used McBrite had no cavities in a year."

8. A new drug is being tried on 12 people with colds. The drug will cure the cold within a day in 70% of all people. Otherwise the drug is ineffective and the cold lasts at least three days. What is the probability that
a. None of the people is cured by the drug?
b. Exactly eight of the people are cured by the drug?
c. Exactly nine of the people are cured by the drug?
d. At least two of the people are cured by the drug?

In the discussion in this section about Boston, $C(12,9)(.8)^9(.2)^3$ was computed to approximate $C(.8N, 9) \cdot C(.2N, 3)/C(N, 12)$, where N is the population of Boston.

♦ 9. Show that this approximation is "good" (correct to two decimal places) even if $N = 500$. The approximation is even better if N is increased.

♦ 10. How good is this approximation if $N = 100$?

Section 7-4 More Examples

In this section, we look at some problems that seem similar to one another but are solved using different methods. The examples in this section provide a review of most of the counting techniques you have studied.

Example 1 A jar contains six blue balls, five green balls, and nine orange balls.

PROBLEM A A sequence of four balls is selected with replacement. What is the probability that the first two balls are blue and the last two balls are orange?

Solution The probability that the first two balls are blue and the last two balls are orange is $(6/20) \cdot (6/20) \cdot (9/20) \cdot (9/20) = .018$. Since each ball drawn is replaced before the next ball is drawn, the contents of the jar are the same at each draw. The draws are independent of each other.

PROBLEM B A sequence of four balls is selected without replacement. What is the probability that the first two balls are blue and the last two balls are orange?

Solution The probability that the first two balls are blue and the last two are orange is $(6/20) \cdot (5/19) \cdot (9/18) \cdot (8/17) = .019$. The probability that the first ball is blue is $6/20$. The probability that the second ball is blue, given that the first ball is blue, is $5/19$, since when the second ball is drawn, only 19 balls remain in the jar, of which five are blue. The probability that the third ball is orange, given that the first two balls are blue, is $9/18$. The probability that the fourth ball is orange, given that the first two balls are blue and the third ball is orange, is $8/17$—under the given conditions, 17 balls, eight of them orange, are in the jar when the fourth ball is drawn. Conditional probability is involved here.

PROBLEM C A set of four balls is selected from the jar. What is the probability that two of the balls are blue and two are orange?

Solution The probability that two are blue and two are orange is $C(6,2) \cdot C(9,2)/C(20,4) = .111$—this is hypergeometric probability.

PROBLEM D A set of four balls is selected from the jar. What is the probability that two of the balls are blue and two are orange, given that none is green?

Solution The probability that two are blue and two are orange given that none is green is $C(6,2) \cdot C(9,2)/C(15/4) = .396$—this is conditional probability and hypergeometric probability.

PROBLEM E A sequence of four balls is selected without replacement. What is the probability that the third ball drawn is green?

Solution This problem *seems* very complicated, since the selection of the first two balls changes the contents of the jar. However, there are originally 20 balls in the jar. Each ball is as likely to be selected third as any other ball. Thus, the probability that any given ball is selected third is $1/20$. Since five balls are green, the probability that a green ball is selected third is $5/20$.

Example 2 Ten people get on a elevator on the ground floor. Besides the ground floor, the building has floors 1 through 12.

PROBLEM A What is the probability that everyone gets off on the seventh floor?

Solution Each person has probability $1/12$ of getting off on the seventh floor (or any other particular floor) since there are 12 floors and all floors are equally likely. The probability that everyone gets off on the seventh floor is $(1/12)^{10} = 1.62 \times 10^{-11}$, since the selections of floors are independent events.

PROBLEM B What is the probability that everyone gets off on the same floor?

Solution The probability that everyone gets off on a specific floor (such as the seventh floor) is $(1/12)^{10}$ by problem A. The probability that everyone gets off on the same floor is $12 \cdot (1/12)^{10} = 1/12^9 = 1.94 \times 10^{-10}$, since it is possible for everyone to get off on any one of the 12 floors.

PROBLEM C What is the probability that exactly four people get off on the fifth floor?

Solution The probability that exactly four people get off on the fifth floor is $C(10,4) \cdot (1/12)^4 \cdot (11/12)^6 = .00601$. This is binomial probability. There is one "experiment" for each person. The experiment has two possible outcomes: the person gets off on the fifth floor (probability $= 1/12$) and the person does not get off on the fifth floor (probability $11/12$).

PROBLEM D What is the probability that four people get off on the fifth floor and six people get off on the ninth floor?

Solution The probability that four people get off on the fifth floor and six people get off on the ninth floor is $C(10,4)/12^{10} = 3.39 \times 10^{-9}$. $C(10,4)$ is the number of ways of selecting four people from ten and is thus the number of ways in which four people can get off on the fifth floor. After four people get off on the fifth floor, six people are left in the elevator and therefore there is only one way that six people can get off on the ninth floor. The number of ways in which ten people can get off the elevator is 12^{10}.

PROBLEM E What is the probability that the ten people get off on ten different floors?

Solution The probability that the ten people get off on ten different floors is $P(12,10)/12^{10} = .00387$, since the first (in alphabetical order, for example) person can choose any of 12 floors, the second person can choose any of the remaining 11 floors, and so on.

Example 3 A coin is tossed 101 times. What is the probability that it shows heads a majority of the times?

Solution A majority of 101 is 51 or more. This problem could be done using binomial probability, but one term would have to be computed for each counting number from 51 to 101. To do the problem simply, note that since the number of tosses is odd, there must be a majority of heads or a majority of tails. Since heads and tails are equally likely on each toss, a majority of heads and a majority of tails are equally likely. If two events are equally likely and precisely one of them must occur, then each has probability $1/2$. Therefore, the probability of a majority of heads in 101 coin tosses is $1/2$.

Example 4 A coin is tossed 100 times. What is the probability that it shows heads a majority of the time?

Solution If the coin is tossed 100 times, a tie (50 heads and 50 tails) is possible and has probability $C(100,50) \cdot (1/2)^{50} \cdot (1/2)^{50} = C(100,50)/2^{100} = .0796$ by the binomial probability formula. The probability of not having a tie (that is, of a majority of heads or a majority of tails) is $1 - C(100,50)/2^{100}$. Since a majority of heads and a majority of tails are equally likely, the probability of a majority of heads is $[1 - C(100,50)/2^{100}]/2 = 1/2 - C(100,50)/2^{101} = .460$.

Example 5 There are ten boys and 15 girls in a second-grade class.

PROBLEM A The teacher calls on three children at random. What is the probability that Henry is called?

Solution Since there are 25 children and three are called, the probability that Henry is called is 3/25.

PROBLEM B The teacher calls on two boys and four girls. What is the probability that Henry is called?

Solution If the teacher calls on two boys and four girls, the probability that Henry is called is 2/10. There are ten boys and two are called, so Henry has two chances out of ten of being called; the number of girls being called does not matter.

PROBLEM C The teacher calls on seven children. What is the probability that three boys and four girls are selected?

Solution If the teacher calls on seven children, the probability that three boys and four girls are called is $C(10,3) \cdot C(15,4)/C(25,7) = .34$—this is hypergeometric probability.

PROBLEM D The teacher calls on four children. What is the probability that Cindy is called but Clark is not called?

Solution If the teacher calls on four children, the probability that Cindy is called but Clark is not called is $C(23,3)/C(25,4) = 7/50$. There are $C(25,4)$ ways of selecting four children from the class. If Cindy is called but Clark is not called, there are 23 other children and three of them must be selected—this can be done in $C(23,3)$ ways.

Exercises 7-4

What is the probability that a poker hand contains

1. The six of clubs and exactly one diamond?
2. Exactly two queens, given that the hand contains exactly one king?
3. Three or more clubs?
4. At least one spade?
5. At least two clubs, given that it contains no diamonds?

6. At least one card in each suit?

Twenty people attend a meeting at which three door prizes are awarded in sequence. Winners are determined by drawing stubs from a can. No one may win more than one prize at the meeting. What is the probability that

7. Daisy wins a prize?
8. Dagwood and Herb both win prizes?
9. Dagwood wins a prize and that later on Herb wins a prize?
10. Neither Dagwood nor Blondie wins a prize?
11. Herb wins third prize?
12. Dagwood wins a prize, given that Mr. Dithers wins the first prize?

Suppose that a meeting is held once each month for a year. What is the probability that

13. Dagwood and Blondie together win at least one first prize?
14. Herb wins at most two prizes?

A jar contains seven green balls, 18 red balls, and five orange balls.

15. Five balls are selected at once. What is the probability that two are green, two are red, and one is orange?
16. Six balls are selected at once. What is the probability that at least one is red?
17. Six balls are selected in sequence with replacement. What is the probability that at least one is red?
18. Six balls are selected in sequence with replacement. What is the probability that the last two balls are orange?
19. Five balls are selected in sequence without replacement. If the first two balls selected are green, what is the probability that the last three balls are orange?
20. Four balls are selected in sequence without replacement. What is the probability that all four balls have the same color?
21. Five balls are selected at once. What is the probability that exactly two of the balls are red, given that exactly one is orange?
22. Eight balls are selected in sequence with replacement. What is the probability that exactly three of these balls are green?

Chapter 7 Review

Compute each of the following

1. $\dfrac{15!}{13!}$ 2. $P(20, 18)$ 3. $P(n, 3)$

4. $C(16,3)$

5. $\dfrac{(2n+4)!}{(2n+2)!}$

6. In how many ways can six co-pilots be assigned to six pilots?

7. Four people write a book together. The order of their names on the title page is to be determined by drawing slips out of a hat. What is the probability that their names appear in alphabetical order?

8. Fifteen people apply for positions as plant managers for National Steel. Managers are needed at each of five plants. In how many ways can these five positions be filled?

9. In how many ways can a supervisor of 20 people select three of them for promotion?

10. Ten men and 15 women apply for jobs at Halle's. All the applicants are qualified, so that five of them are chosen at random to be hired. What is the probability that two men and three women are hired?

Of the customers of Wolf Construction Company, 80% are satisfied with the work performed for them by Wolf. Ten customers of Wolf are selected at random and interviewed. Of these ten customers, what is the probability that

11. All are satisfied?

12. None is satisfied?

13. At least one is satisfied?

14. Exactly four are satisfied?

15–18. Suppose that Wolf Construction had only 50 customers last month, 40 of them were satisfied, and 10 of them are selected at random. Answer exercises 11–14 under these conditions.

Applications of Probability to Decision Making

Suppose you operate the concession stands at a baseball stadium and must decide how much food to order for tomorrow's game. You will be concerned about the weather, and especially about the probability of rain. In the first two sections of this chapter, the use of probability theory in making decisions such as this is discussed.

When the outcome of a decision will be determined by chance, it is obvious that probabilities must be involved in making the decision. How about when the outcome is determined by someone else's decision as well as yours?

Game theory is the study of competitive decision making and is the topic of the last two sections of this chapter. Surprisingly enough, probability is involved in finding optimum strategies for many games.

Section 8-1 Expected Value and Odds

Many business decisions must be made in the face of uncertainty. For example, suppose some grain, which may be contaminated, is for sale at $2 a bushel. The grain is worthless if it is contaminated, and is worth $5 a bushel otherwise. Should you buy the grain now, before the tests for contamination are performed?

8-1 EXPECTED VALUE AND ODDS

The current value of the grain must be some sort of weighted average of the value if it is contaminated, and the value if it is not. In this section, the weighted average, which is called the *expected value* of the grain, is considered.

The concept of expected value originated in gambling, and thus we begin by considering a lottery. Suppose there are 100 tickets in a lottery. The first prize is $50, second prize is $25, and third prize is $10. What is a ticket worth?

This question cannot be answered absolutely. Economists have given such problems much thought, and they have formulated various theories. The actual value of a ticket depends on an individual's preferences and is different for different people.

We can answer some related questions. How much should a ticket sell for if the amount paid for all the lottery tickets is to equal the amount paid out in prizes? If I keep on playing the lottery, at what ticket price should I expect to break even in the long run?

The lottery pays out $85. If 100 tickets sell for $85, so that each ticket costs 85¢, the amount paid for tickets equals the amount of prize money awarded. At this price, if I play the lottery, I should expect to break even in the long run.

> **Definition.** If an experiment has outcomes $O_1, O_2, \ldots, O_m$ and outcome O_i has a value or payoff of $V(O_i)$, the *expected value* of the experiment is
> $$P(O_1)V(O_1) + P(O_2)V(O_2) + \cdots + P(O_m)V(O_m)$$

Example 1 What is the expected value of a ticket in the lottery?

Solution In the preceding lottery, there are four outcomes: O_1 is winning the first prize, O_2 is winning the second prize, O_3 is winning the third prize, and O_4 is not winning. $P(O_1) = P(O_2) = P(O_3) = 1/100$ and $P(O_4) = 97/100$. $V(O_1) = \$50$, $V(O_2) = \$25$, $V(O_3) = \$10$, and $V(O_4) = 0$.

The expected value is $(1/100) \cdot 50 + (1/100) \cdot 25 + (1/100) \cdot 10 + (97/100) \cdot 0 = .85$. Thus, if the price of each ticket equals the expected value of the ticket, I should expect to break even in the long run. If tickets sell for $1.00 each, I should expect to lose, on the average $1.00 - \$.85 = \$.15$ for each ticket I buy.

Example 2 A fair die is tossed. The payoff is $1 for each dot that shows. Determine the expected value.

Solution The expected value is

$$\frac{1}{6}(1) + \left(\frac{1}{6}\right)(2) + \left(\frac{1}{6}\right)(3) + \left(\frac{1}{6}\right)(4) + \left(\frac{1}{6}\right)(5) + \left(\frac{1}{6}\right)(6) = \$3.50$$

Example 3 Bill pays $1 for the privilege of selecting a number on a roulette wheel. There are 38 numbers on the roulette wheel. If his number comes up Bill receives $36. What is the expected value? If Bill plays 1000 times, how much should he expect to win or lose?

Solution There are two outcomes: win and lose. The outcome win has probability 1/38 and value $35. The outcome lose has probability 37/38 and value −$1. The expected value is

$$\frac{1}{38}(35) + \frac{37}{38}(-1) = \frac{-2}{38} - \$0.0526$$

If Bill plays 1000 times, he should expect to lose $1000(2/38) = \$52.63$. ■

The concept of expected value was originally invented and used by gamblers to determine whether the rules and procedures of a game favored them or not, that is, whether they should expect to win or lose. *The expected value is a weighted average of the values of the possible outcomes of a game or an experiment.* As we observe in Example 2, the expected value need not be one of the possible outcomes.

If a game or an experiment has expected value E and is repeated n times, the expected value for the n trials is nE. If n is a large number, it is very likely that the actual gain or loss after playing a game n times will be close to the expected value. Gambling casinos base their operation on this principle. The games in a casino are so arranged (see Example 3) that the player has a negative expected value and the casino has a positive expected value.

The concept of expected value is used to make business decisions, as in Examples 4 and 5, even when the experiment only occurs once. The notion of expected value is of especially great importance in the insurance industry in the determination of insurance rates.

Example 4 Susan is wondering whether or not to buy stock in C, D, and E, Inc. The stock is selling at $35. There are rumors of a merger. She thinks that the rumors have a 1/3 chance of being true, and that if the rumors are true, the stock will rise to $50. If the rumors are false, she thinks the stock will fall to $25. What is the expected profit or loss on a share?

Solution There are two possible outcomes: true and false. $P(\text{true}) = 1/3$, so that $P(\text{false}) = 1 - 1/3 = 2/3$. There is a profit of $15 per share if the rumors are true and a loss of $10 per share if the rumors are false. The expected value is $(1/3)(15) + (2/3)(-10) = -\$1.67$. Thus, there is an expected loss of $1.67 on each share, so Susan should not buy the stock.

Example 5 Two major construction jobs in Canton are open for bids. Harry Sanborn, owner of Sanborn Construction, must decide whether to bid on one job or both. He has probability .8 of getting the school renovation job and would make a profit of $500,000 there. He has probability .3 of getting the road resurfacing job and would make $800,000 profit there. His other choice is to bid on both jobs. However, if he did this and received both contracts, he would be unable to fulfill them properly. Paying overtime would reduce his total profit by $300,000 and finishing late would cost him an estimated $400,000 in loss of goodwill and in penalties. His actual profit would therefore be $600,000. Which job, or jobs, should he bid on?

Solution Harry has three choices: bid on the school job only, bid on the road job only, or bid on both jobs. We need to compute the expected value of each choice.

If he bids on the school job only, there are two possible outcomes: he gets the job or he does not. The expected value is

$$(.8)(\$500{,}000) + (1 - .8)(\$0) = \$400{,}000$$

Similarly, if he bids on the road job alone, the expected value is

$$(.3)(\$800{,}000) + (1 - .3)(\$0) = \$240{,}000$$

If Harry bids on both jobs, there are four possible outcomes:

Outcome	Both jobs	School job only	Road job only	Neither job
Probability	.24	.56	.06	.14
Value	$600,000	$500,000	$800,000	$0

To obtain these probabilities, assume that obtaining the school job and obtaining the road job are independent events. Then the probability of obtaining both jobs is $(.8)(.3) = .24$. You could draw a Venn diagram to determine the other probabilities.

The expected value of bidding on both jobs is:

$$(.24)(\$600{,}000) + (.56)(\$500{,}000) + (.06)(\$800{,}000) + (.14)(\$0) = \$472{,}000$$

The expected value is greatest if Harry bids on both jobs, and therefore this is his best strategy.

Odds

Let's return to gambling and discuss odds and bets. Suppose that the odds on Snowshoe are 3:2 in the fourth race. If I bet on Snowshoe and he wins, I receive \$2; if he loses, I pay \$3. The odds on Snowshoe could also be expressed as 30:20, 6:4, or as any other pair of positive numbers with quotient 1.5.

> **Definition.** *A bet is fair if the expected value of the outcome is* 0.

If a bet is fair, *the odds on E are* $P(E):P(E')$ (or any other pair of positive numbers with the same quotient). This is true because the expected value with these odds is $P(E) \cdot P(E') + P(E')(-P(E)) = 0$. In the previous equation, $P(E')$ is the payoff if E occurs and $-P(E)$ is the payoff if E' occurs.

> $P(E):P(E')$ Odds on E for a fair bet

Example 6 The odds on Snowshoe are 3:2. If the bet is fair and E is the event Snowshoe wins, determine $P(E)$.

Solution $P(E)/P(E') = 3/2$. But $P(E') = 1 - P(E)$. Then

$$\frac{P(E)}{1 - P(E)} = \frac{3}{2}$$

$$2P(E) = 3[1 - P(E)] = 3 - 3P(E)$$

$$5P(E) = 3$$

$$P(E) = \frac{3}{5} = .6$$

The probability that Snowshoe wins is .6.

Example 7 The probability that the Browns beat the Bengals is .7. What are the odds for a fair bet on the Browns? On the Bengals?

Solution The probability that the Browns lose is $1 - .7 = .3$. We are neglecting the possibility of a tie. The odds for a fair bet on the Browns are $.7:.3 = 7:3$. If I bet \$70 on the Browns and they win, I make \$30.

The odds for a fair bet on the Bengals are $.3:.7$. In general, the odds on E' are $P(E'):P(E)$. Thus, if the odds on E are $b:c$, the odds on E' are $c:b$.

Exercises 8-1

Compute the expected value.
1. $P(O_1) = .3$, $P(O_2) = .5$, $P(O_3) = .2$, $V(O_1) = 7$, $V(O_2) = 0$, $V(O_3) = 12$
2. $P(A) = .6$, $P(B) = .1$, $P(C) = .3$, $V(A) = -8$, $V(B) = -10$, $V(C) = -5$
3. $P(X) = .25$, $P(Y) = .35$, $P(Z) = .2$, $P(Q) = .2$, $V(X) = 3$, $V(Y) = 2$, $V(Z) = 5$, $V(Q) = -15$
4. $P(X) = .1$, $P(Y) = .2$, $P(Z) = .3$, $P(Q) = .4$, $V(X) = -9$, $V(Y) = 8$, $V(Z) = 10$, $V(Q) = -6$
5. A fair die is tossed. The payoff is n^2 if n dots show.
6. Receipts at a concert will be \$15,000 if it snows, and \$30,000 otherwise. The probability of snow is 40%.
7. A coin is tossed. The payoff is $-\$1$ for heads and \$2 for tails.
8. A fair die is tossed. The payoff is \$5 if 1, 2, 3, or 4 shows, $-\$10$ if 5 shows, and $-\$20$ if 6 shows.
9. How much should I expect to win or lose if I play the game in exercise 7 ten times? 500 times?
10. How much should I expect to win or lose, if I play the game in exercise 8 20 times? 100 times?
11. The probability of snow today is 80%. What are the odds on snow today? If I make a fair bet of \$20 on snow and win, how much do I win?
12. A card is picked at random from a deck. What are the odds that it is a diamond, if the bet is fair? If I bet \$25 that the card is a diamond, how much do I profit if I win the bet?
13. A coin is tossed. What are the odds on heads?
14. A number from 1 to 10 is picked at random. What are the odds that the number is a multiple of 3?
15. If the odds for a fair bet on E are 5:7, what is $P(E)$?
16. If the odds for a fair bet on H are 9:5, what is $P(H)$?
17. On Friday afternoon, Al's Car Wash is very busy. One car can get

washed every 90 seconds. What is the expected waiting time when you arrive?

Number of cars in line	0	1	2	3	4	5	6	7	8	9	10	11	12
Probability	0	.01	.01	.02	.04	.07	.12	.15	.18	.20	.12	.05	.03

18. Given in the following table are the probabilities for the number of live puppies in a litter for a certain kind of dog. Each puppy is worth $200. What is the expected number of puppies in a litter? What is the expected value of a litter?

Number	0	1	2	3	4	5	6
Probability	.20	.07	.13	.36	.15	.08	.01

19. Grabco Corporation estimates that the death or incapacitation of its president within the next year would cost the company $5,000,000 and takes out a one-year insurance policy on her for that amount. The policy costs $30,000. The probability of collecting on the policy, based on the president's age and health and on insurance company statistics, is .005. What is the expected value of this insurance to Grabco? To the insurance company?

20. The annual premium on a certain auto insurance policy is $83. The probability of collecting on it is .03, and the average amount collected is $2400. What is the expected value of the policy to the customer? To the insurance company?

21. An insurance company has written 10,000 policies of $100,000, 7000 policies of $50,000, 4000 policies of $25,000, and 8000 policies of $10,000 for people who are now 60 years old. The probability that a person who is 60 years old will die within a year is .025. How much should the company expect to pay out on these policies within the next year?

To prevent a flu epidemic, all 10,000 soldiers at an army base are innoculated with a new vaccine. The vaccine is so effective that 95% of those innoculated do not get flu and 3% get only a mild case of flu. The other 2% get a severe case of flu. A soldier with a mild case of flu does not need to be put in the infirmary but cannot work for one week. A soldier with a severe case of flu must be put in the infirmary for a week and cannot work for two weeks.

The base commander estimates that the loss of a week's work by a soldier costs the army $180 and that keeping a soldier in the infirmary for a week costs the army an additional $250.

22. How many soldiers are expected to come down with a mild case of flu? With a severe case?
23. How many man-weeks of work are expected to be lost because of the epidemic?
24. How much is the epidemic expected to cost the army because of infirmary costs and lost work?

Section 8-2 Decision Theory

Most business decisions are made in the face of uncertainty. In many cases, an executive's objective is to maximize profit. (Other objectives would be to maximize sales, minimize costs, maximize market share, and so on.) Since there is uncertainty, no one can ever be sure how much profit will result from a given decision. Thus, one must instead try to maximize the expected value of profit.

Example 1 Some otherwise worthless land in the desert is suspected of having oil beneath it. If there is oil beneath the land, the land is worth $10,000,000. Geologists for the Texxon Oil Company estimate that the probability of finding oil beneath the land is 30%.

Marcy Bosco, an executive for Texxon, must decide whether or not to buy the land. The price is $4,000,000. One other choice is possible for Ms. Bosco. For $500,000, she can purchase an option to buy the land in six months. During these six months, she could have an exploratory oil well dug on the land at a cost of $1,000,000 and would then know for sure whether or not there was oil. At the end of six months, she could buy the land for $4,000,000 if there was oil. Should she buy the land, not buy the land, or purchase the option?

Solution In Table 1, the value of each decision, if there is oil and if there is no oil, and the expected value of each decision are shown. The expected value is greatest if Texxon takes the option to purchase, and so this is the correct decision for Ms. Bosco to make.

TABLE 1

	Buy	Do Not Buy	Option to Purchase
Value if Oil	$6,000,000	0	$4,500,000
Value if No Oil	−$4,000,000	0	−$1,500,000
Expected Value	−$1,000,000	0	$ 300,000

Here is how the numbers in Table 1 are computed. If Texxon buys the land, there is a profit of $10,000,000 − $4,000,000 = $6,000,000 if there is oil. If there is no oil, the purchase price of $4,000,000 is lost. If Texxon does not buy the land, there is nothing ventured and nothing gained. The option to purchase and the exploratory well together cost $1,500,000. This amount is lost if there is no oil. If there is oil, the complete cost of the land to Texxon is this amount plus the $4,000,000 purchase price, for a total of $5,500,000. Thus, in this case, the profit is $10,000,000 − $5,500,000 = $4,500,000. The expected values are computed using a probability of .3 for finding oil and of $1 − .3 = .7$ for not finding oil.

Example 2 John Kocab owns an apricot orchard. During the apricot season, he takes apricots to a nearby produce market and sells them for $1000 a truckload. At the end of the day, any unsold apricots must be discarded. He can also sell apricots to a canning company for $700 a truckload. The probability p_n that John can sell n truckloads at the market on a particular day is given in Table 2.

TABLE 2

n	0	1	2	3	4	5
p_n	.1	.1	.2	.3	.2	.1

Suppose John has five truckloads of apricots available today. How many apricots should he take to market if his objective is to maximize expected revenue?

Solution Table 3 lists the expected revenue for each of John's choices. The expected revenue is greatest if John takes two truckloads of apricots to market.

TABLE 3

Number of Truckloads Taken to Market	0	1	2	3	4	5
Expected Revenue	3500	3700	3800	3700	3300	2700

Now let's compute some of the numbers in Table 3. If John takes none of the apricots to market, he will sell five truckloads for $700 each to the canning company for a total of $3500. If John takes five truckloads to market, the expected number that he will sell is 2.7—this is the expected value of n in Table 2. At $1000 a truckload, the expected revenue is $2700.

Suppose John takes three truckloads to market. In this case, he sells two truckloads to the cannery for a total of $1400. Now we must determine John's expected revenue at the market. In Table 4, q_m is the probability that John will be able to sell m truckloads of apricots at the market if he takes three truckloads there.

TABLE 4

m	0	1	2	3
q_m	.1	.1	.2	.6

Table 4 differs from Table 2 because John cannot now sell four or five truckloads of apricots at the market, since he is only taking three truckloads there. The probability that John will now sell three truckloads at the market is the sum of the numbers p_3, p_4, and p_5 from Table 2. At the market, the expected number of truckloads that John would sell is 2.3 for an expected revenue of $2300. The number 2.3 is obtained by taking the expected value of m in Table 4. John's total expected revenue is thus $1400 + $2300 = $3700.

Exercises 8-2

What is the correct decision in Example 1 if the geologists change their estimate of the probability of finding oil beneath the land to

1. $\frac{2}{3}$?
2. $\frac{1}{5}$?

3. In Example 2, compute the expected revenue when one, two, or four truckloads of apricots are taken to market.

Repeat Example 2 if Table 2 is replaced by

4.

n	0	1	2	3	4	5
p_n	0	.1	.1	.3	.3	.2

5.

n	0	1	2	3	4	5
p_n	.1	.1	.1	.2	.3	.2

6. The Economy Car Rental Company rents cars for $30 a day. The daily

cost per car to the company is $12. Thus the company makes $18 profit each time it rents a car for a day but loses $12 each time a car is unrented for a day. Each day the company has five customers with probability .3, six customers with probability .5, and seven customers with probability .2. How many cars should the company keep on hand in order to maximize expected profit? (*Hint*: Consider the expected value of keeping five, six, or seven cars on hand.)

7. Bill Bailey, the owner of the Bailey Machine Tool Company, must decide whether or not to expand the business. Without expansion, the business will make $7,000,000 profit next year if the economy expands and $2,000,000 if there is a recession. With expansion, the company will make $12,000,000 if the economy expands but will lose $2,000,000 if there is a recession. What decision should Mr. Bailey make, if the probability of a recession is .4?

Ruth Balaton has $10,000 to invest in stocks. She can buy blue-chip stocks, stock in small but established companies, or highly speculative stocks. The table shows how much each kind of stock would be worth after a year under different economic conditions.

	Expansion	Stagnation	Recession
Blue-Chip	$12,000	$11,000	$ 9,000
Small, Established Companies	14,000	10,000	8,000
Speculative	18,000	9,000	3,000

8. Ruth reads an article in *Fortune* which says that there is a probability of .1 of the economy expanding next year, .2 of stagnation, and .7 of recession. If she believes this article and wants to maximize the expected value of her stock a year from now, in which stocks should she invest?

9. Ruth also reads an article in the *Wall Street Journal* which says that there is a probability of .4 of the economy expanding next year, .3 of stagnation, and .3 of recession. In which stocks should she invest if she believes this article?

10. Ruth then reads an article in *Business Week* which says there is a probability of .6 of the economy expanding next year, .3 of stagnation, and .1 of recession. In which stocks should she invest if she believes this article?

11. Ray Rufus has written a successful novel. A speculator has offered him $200,000 for the movie and television rights to the book.

A movie company has expressed interest in the novel and has asked Ray to write some suggestions for adapting the novel to the screen. Ray estimates that doing this would cost him $30,000, mostly in foregone income for his time. If the movie company buys the screen rights to the novel, they would pay $500,000.

If a movie is made from the novel, there is a 20% chance that a television series would also be made from the novel and that Ray would receive $250,000 for the television rights.

Should Ray do business with the speculator, if he estimates a probability of .4 that the movie company will buy screen rights to the novel?

 12. A contractor is allowed to bid for one of two jobs. The profit would be $3,000,000 on job A and $4,000,000 on job B. The cost of preparing the bid is $100,000 for job A and $150,000 for job B. The probability of getting job A is estimated to be .4. The probability of getting job B is estimated to be only .3 because of more competition. Which job should the contractor bid on?

Section 8-3 Game Theory

So far, we have considered decisions whose outcome depended on chance, and have tried to maximize the expected value of the outcome. Now we will study decision making when there is competition.

The study of competitive decision making is called *game theory*. Game theory has applications to business, politics, warfare, recreational games, and many other aspects of human life.

All games studied in this book will be *two-person zero-sum games*. *Two-person* means that there are two players. *Zero-sum* means that one player's winnings are the other player's losses. In a zero-sum game, the player's interests are completely opposed so that there can never be any cooperation.

A game can be completely described by a matrix, called the *payoff matrix*, such as

$$
\begin{array}{c}
\ \ \ \ \ \ \ \ \ \ \ \ \ \ B \\
\ \ \ \ 1\ \ \ \ \ \ 2\ \ \ \ \ \ 3 \\
A\ \begin{array}{c}1\\2\\3\end{array}\left[\begin{array}{rrr}4 & 3 & 5 \\ 7 & -2 & -6 \\ -3 & -1 & 4\end{array}\right]
\end{array}
$$

In this game, the players are A and B. Player A has three possible strategies—he can choose row 1, row 2, or row 3. Player B also has three possible strategies—he can choose column 1, column 2, or column 3. The payoff is at the intersection of the row and column selected.

Suppose A chooses row 2 and B chooses column 1. Then the *payoff* is 7. This means that B gives A $7. If A chooses row 3 and B chooses column 2, the payoff is -1. A *negative payoff* means that A pays B, so in this case A pays B $1.

What is the best strategy for each player? Suppose each player seeks the biggest possible payoff at first. Then A plays row 2 each time and B plays column 3.

At this point, A is losing heavily and switches to row 1, since this is his best counter to column 3. B now switches to column 2, since this does best against row 1.

At this point, neither player could gain if he alone switched strategies. Thus A would continue to play row 1 and B to play column 2. The pair of strategies $(1, 2)$ is called a *saddle point*. The outcome of these strategies, in this case 3, is called the *value* of the game.

> **Definition.** If the entry in position i, j in the payoff matrix of a game is the smallest number in its row and the largest number in its column, the ordered pair (i, j) is called a *saddle point* of the game and the entry in position i, j is called the *value* of the game.

Example 1 Determine all saddle points and the value of the game with matrix

PROBLEM A $\begin{bmatrix} 4 & -3 & -10 & -2 \\ 1 & 0 & 3 & -1 \end{bmatrix}$

Solution Find the smallest number in each row and decide whether or not it is the largest number in its column. The smallest number in row 1 is -10, but -10 is not the largest number in its column. The smallest number in row 2 is -1, and this is the largest number in column 4.

Thus $(2, 4)$ is a saddle point and the value of the game is -1.

PROBLEM B $\begin{bmatrix} 6 & 4 & 4 \\ 19 & 2 & -3 \\ -1 & -1 & 0 \end{bmatrix}$

Solution The 4's in positions 1, 2 and 1, 3 are the smallest entries in row 1. Each 4 is the largest entry in its column. Thus $(1, 2)$ and $(1, 3)$ are both saddle points, and the value of the game is 4. You should verify that there are no other saddle points.

Whenever a game has more than one saddle point, all the saddle points have the same entry.

PROBLEM C $\begin{bmatrix} 2 & -8\checkmark & -3 \\ -6\checkmark & 5 & -1 \\ 1 & 4 & 0\checkmark \end{bmatrix}$

Solution The smallest number in each row has a $\checkmark$ next to it. Of these numbers, only the 0 in position 3, 3 is the largest entry in its column. Thus (3, 3) is a saddle point, and the value of the game is 0.

If the value of a game is 0, neither player is favored and the game is *fair*.

∎

When a game has a saddle point, the *optimum strategy* for each player is to always play a row or column that contains a saddle point. If one player plays his optimum strategy and the other does not, the other player will not do as well as he could. If both players use optimum strategies, the payoff will be the same each time the game is played. For this reason, the game is said to be *strictly determined*.

> **Definition.** A game is *strictly determined* if and only if it has a saddle point.

Example 2 Two companies, American Wicker and National Wicker, are the sole suppliers of wicker furniture. Currently American has 32% of the market and National has 68%. Each company can produce the same quality furniture next year, or can lower or raise the quality of its furniture. The matrix

$$\text{American} \begin{array}{c} \\ \text{Lower} \\ \text{Same} \\ \text{Raise} \end{array} \overset{\begin{array}{ccc} & \text{National} & \\ \text{Lower} & \text{Same} & \text{Raise} \end{array}}{\begin{bmatrix} -1 & -5 & 10 \\ 3 & 7 & 2 \\ 4 & 6 & 5 \end{bmatrix}}$$

shows the payoffs from the various strategies. Here the payoffs are measured in percents of the market. What should each company do, and what will be their market shares next year?

Solution Position $(3, 1)$ is a saddle point because 4 is the smallest entry in row 3 and is the largest entry in column 1. Thus American Wicker should produce higher quality furniture next year and National Wicker should produce lower quality furniture. American will increase its market share by 4% to 36% and National's share will drop by 4% to 64%. ■

Note that in this example the entries in the matrix represent percents of the market for wicker furniture. The payoffs and value of a game do not have to be measured in money.

Example 3 Determine all saddle points and the value of the game with matrix

$$A \begin{matrix} & B \\ & \begin{bmatrix} -5 & 4 \\ 3 & -2 \end{bmatrix} \end{matrix}$$

Solution This game has no saddle points and so is not strictly determined. Suppose A plays row 1. B will soon start playing column 1. At this point, A will be losing $5 each time the game is played and will start playing row 2 in order to win $3 each time. B will now switch to column 2, and then A will switch to row 1. The players will keep switching strategies and trying to outsmart each other.

It is clear that in this game each player should mix his strategies. In the next section, we will learn the best way for the players to do so.

Exercises 8-3

In the following game, determine the payoff when the indicated strategies are used.

$$\begin{bmatrix} 3 & -4 & 6 & -2 \\ -2 & 7 & 1 & 4 \\ 1 & -5 & 0 & -3 \end{bmatrix}$$

1. $(2, 4)$ 2. $(3, 1)$
3. $(1, 3)$ 4. $(2, 2)$

Determine whether or not each game is strictly determined. Find the saddle point (or points), if any, and value of each strictly determined game.

5. $\begin{bmatrix} 14 & -1 \\ -8 & 9 \end{bmatrix}$ 6. $\begin{bmatrix} 12 & 7 \\ -8 & 15 \end{bmatrix}$

7. $\begin{bmatrix} 6 & 2 & 3 \\ -1 & 1 & 4 \end{bmatrix}$ 8. $\begin{bmatrix} 3 & -6 \\ -9 & -6 \\ -2 & -4 \end{bmatrix}$

9. $\begin{bmatrix} 4 & -1 & 7 \\ 2 & 0 & 3 \\ 6 & -4 & -3 \end{bmatrix}$ 10. $\begin{bmatrix} 3 & -7 & 4 \\ -2 & -1 & 5 \\ 4 & 5 & 3 \end{bmatrix}$

11. $\begin{bmatrix} 3 & -1 & 7 & -1 \\ 2 & -2 & 0 & -4 \\ 5 & -1 & 1 & -1 \end{bmatrix}$ 12. $\begin{bmatrix} 3 & -1 & 0 & -3 \\ 2 & 4 & 1 & 5 \\ -4 & 5 & -2 & 7 \end{bmatrix}$

13. General Alexander commands an army that is about to invade a neighboring country. He can attack either north or south of a mountain range, or can attack both places at once. General Khan commands the defending forces. Besides the border defense forces, General Khan has a reserve division that can be stationed in either the north or the south.

    The entries in the following matrix represent the number of square miles of territory General Alexander will gain in the first week of fighting. Note that this number may be negative, because General Khan can counterattack, and General Alexander is very weak in any location he does not attack. What are the best strategies for the two generals if their only concern is control of territory?

		General Khan	
		North	South
	North	0	120
General Alexander	South	20	−30
	Both	50	100

14. General Manufacturing Corporation (GMC) and the United Workers Union (UWU) are negotiating next year's contract. They have agreed on all matters except pay raise. Since they cannot agree, the matter is

subject to binding arbitration. The procedure is as follows: each side proposes a raise to the arbitrator, and the arbitrator selects one of the proposals.

This arbitrator is in fact in favor of a raise of $19, so that he will select whichever proposal is closer to $19. GMC is considering proposing a low raise of $10 weekly, a moderate raise of $18, and a high raise of $27. UWU is considering a low raise of $12, a moderate raise of $21, and a high raise of $30. What are the optimum strategies? If both players use optimum strategies, what will the raise be?

*15. Suppose each payoff in a game is doubled. What effect will this have on
   a. Whether or not the game is strictly determined?
   b. Optimum strategies and the value of the game if the game is strictly determined?

*16. Answer the questions in exercise 15 if, instead of doubling, a number k is added to each payoff.

*17. Suppose each payoff in a strictly determined game is multiplied by -1. Must the new game be strictly determined?

Section 8-4   Games with Mixed Strategies

In Section 8-3, games with saddle points were discussed. In such a game, the optimum strategy for each player is to always choose a row or a column containing a saddle point. Such a strategy is called a *pure strategy* because the same row or column may always be chosen. The game is said to be strictly determined since the payoff is always the same.

In this section, games without saddle points are studied. Such a game was discussed in Example 3 of Section 8-3 and had matrix

$$\begin{array}{c} & B \\ A & \begin{bmatrix} -5 & 4 \\ 3 & -2 \end{bmatrix} \end{array}$$

As we saw then, a player who always plays the same row or column will be outplayed by the opponent. In this game, each player should use a *mixed strategy*. This means that the player should mix the possible choices in a certain proportion. Now let's determine the proportion.

Suppose A plays row 1 with probability x. Then A must play row 2 with probability $1 - x$. If B plays column 1, A loses 5 with probability x and wins 3 with probability $1 - x$. The expected value E_1 of the game to A is

$$E_1 = -5(x) + 3(1-x) = -5x + 3 - 3x = 3 - 8x$$

as long as B plays column 1.

Similarly, if B plays column 2, the expected value E_2 of the game to A is

$$E_2 = 4(x) - 2(1-x) = 4x - 2 + 2x = 6x - 2$$

Now assume that A wishes to do as well as possible regardless of what B does. In Diagram 1, the line segments representing E_1 and E_2 meet at the point $(5/14, 2/14)$. If $x = 5/14$, the expected value of the game to A is $2/14$ no matter how B plays. If $x < 5/14$, the expected value of the game to A is less than $2/14$ if B plays column 2. If $x > 5/14$, the expected value of this game to A is less than $2/14$ if B plays column 1. Thus A should choose the value $5/14$ for x.

The optimum strategy for A is to select row 1 with probability $5/14$ and row 2 with probability $9/14$. A should mix these rows at random so that there is no pattern to the selection. This prevents B from outguessing A. A could do this random mixing in many possible ways. For example, A could place five red balls and nine green balls in a jar. Each time the game is played, A would shake the jar and select a ball without looking. A would play row 1 if the ball were red and row 2 if the ball were green. Note that the selected ball must be replaced before the game is repeated in order to have the correct proportion of colors in the jar each time.

What is the optimum strategy for B? Suppose B plays column 1 with probability y and column 2 with probability $1 - y$. If A plays row 1, the expected value E_1 of the game to A is

$$E_1 = -5y + 4(1-y) = -5y + 4 - 4y = 4 - 9y$$

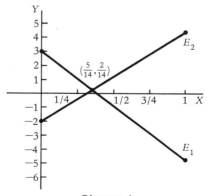

Diagram 1

If A plays row 2, the expected value E_2 is

$$E_2 = 3y - 2(1-y) = 3y - 2 + 2y = 5y - 2$$

By considerations similar to those earlier in this section, E_1 and E_2 should be equated to give the equation

$$4 - 9y = 5y - 2$$

The solution of this equation is $y = 6/14$. Thus, B should play column 1 with probability 6/14 and column 2 with probability 8/14. If B plays this optimum strategy, the expected value of the game to A is 2/14, no matter which row A plays.

Note that in this game the expected value of the game to A is 2/14, provided that either player uses his optimum strategy. If either player uses his optimum strategy, the expected value of the game does not change no matter what the other player does.

> **Definition.** Assume a game is not strictly determined and that at least one player uses his optimum strategy. The *value* of the game is the expected value of the payoff to the player who selects rows.

Example 1 Regis Walton, a developer, is building a strip shopping center. Two ice cream chains, Dairysweet and Sundae's 75 Flavors, want to have stores there. Locations are available in the middle and at the north end. The total amount of business for the two stores will be the same no matter where they are located, but the percent market shares for each store will vary and are shown in the matrix

$$\text{Dairysweet} \begin{array}{c} \\ \text{Middle} \\ \text{North End} \end{array} \overset{\begin{array}{cc} \text{Sundae's} & \\ \text{Middle} & \text{North End} \end{array}}{\begin{bmatrix} 60 & 55 \\ 50 & 60 \end{bmatrix}}$$

Where should Dairysweet locate? What is Dairysweet's expected market share?

Solution First note that there are no saddle points, since the smallest number in a row of the payoff matrix is never the largest number in its column. Thus Dairysweet must use a mixed strategy. Assume Dairysweet chooses the middle with probability x. Equate the expected market shares if Sundae's selects the middle and the north end to obtain the equation.

$$60x + 50(1 - x) = 55x + 60(1 - x)$$
$$60x + 50 - 50x = 55x + 60 - 60x$$
$$10x + 50 = -5x + 60$$
$$15x = 10$$
$$x = \frac{2}{3}$$

Dairysweet should locate in the middle with probability 2/3 and in the north end with probability 1/3. To obtain Dairysweet's expected market share of 56 $\frac{2}{3}$%, substitute 2/3 for x in either side of the first equation in this example.

Dominated strategies

So far in this section, only games with 2×2 matrices have been considered. When a game has a larger matrix and is not strictly determined, finding optimum strategies usually involves setting up a linear programming problem and solving it by the simplex method. Such problems are not considered in this book. Sometimes, however, a game with a larger matrix can be reduced to one with a 2×2 matrix by eliminating certain strategies that would never be used. Consider the game

$$A \begin{array}{c} \\ 1 \\ 2 \\ 3 \\ 4 \end{array} \begin{array}{c} B \\ \begin{array}{ccc} 1 & 2 & 3 \end{array} \\ \left[\begin{array}{ccc} 2 & -5 & 4 \\ 1 & 2 & -3 \\ 6 & 3 & -2 \\ 0 & -6 & 4 \end{array} \right] \end{array}$$

No matter what strategy B chooses, A is always better off with row 3 than with row 2, because each entry in row 3 is larger than the corresponding entry in row 2. Thus A would never choose row 2. Since each entry in row 1 is greater than or equal to the corresponding entry in row 4, A will also never choose row 4. Eliminating these two rows gives the matrix

$$\begin{array}{c} & \quad\quad B \\ & \quad 1 \quad\ 2 \quad\ 3 \\ A \begin{array}{c} 1 \\ 3 \end{array} & \left[\begin{array}{rrr} 2 & -5 & 4 \\ 6 & 3 & -2 \end{array}\right] \end{array}$$

Now note that B will never choose column 1, since column 2 is better for him no matter what A does. Eliminate column 1 to obtain the matrix

$$\begin{array}{c} & \quad\quad B \\ & \quad 2 \quad\ 3 \\ A \begin{array}{c} 1 \\ 3 \end{array} & \left[\begin{array}{rr} -5 & 4 \\ 3 & -2 \end{array}\right] \end{array}$$

The game with this particular 2×2 matrix was studied earlier in this section. The optimum strategy for A is to play *row 1* with probability $5/14$ and *row 3* with probability $9/14$. A should never play rows 2 or 4. B should play *column 2* with probability $6/14$, *column 3* with probability $8/14$, and never play columns 1 or 4. The value of the game is the same as the value of this 2×2 game and is $2/14$.

Note that originally column 2 was not always better for B than column 1. However, B can deduce that A will never play rows 2 or 4 and should eliminate them from consideration. Then B can decide never to select column 1. In solving this game, we have eliminated *dominated strategies*.

Definition. A row r^* is *dominated* by a row r if every entry of r^* is less than or equal to the corresponding entry in r.

A column c^* is *dominated* by a column c if every entry of c^* is greater than or equal to the corresponding entry in c.

Example 2 Eliminate all dominated strategies in the game with matrix

$$\left[\begin{array}{rrrr} 3 & 1 & 4 & 2 \\ -1 & 3 & 1 & 0 \\ 2 & 4 & 2 & 3 \\ 5 & 2 & 6 & 4 \end{array}\right]$$

Solution First note that every entry in row 4 is larger than the corresponding entry in row 1 so that row 4 dominates row 1. Similarly, row 3 dominates row 2.

Eliminate rows 1 and 2 to obtain the matrix

$$\begin{array}{c} \\ 3 \\ 4 \end{array}\begin{array}{c} \begin{array}{cccc} 1 & 2 & 3 & 4 \end{array} \\ \left[\begin{array}{cccc} 2 & 4 & 2 & 3 \\ 5 & 2 & 6 & 4 \end{array}\right] \end{array}$$

Note that every entry in column 1 is less than or equal to the corresponding entry in column 3, so that column 1 dominates column 3. (Recall that B is losing the amount shown in the matrix, so that B wants to play the column with the smaller numbers.) Delete column 3 to obtain the matrix

$$\begin{array}{c} \\ 3 \\ 4 \end{array}\begin{array}{c} \begin{array}{ccc} 1 & 2 & 4 \end{array} \\ \left[\begin{array}{ccc} 2 & 4 & 3 \\ 5 & 2 & 4 \end{array}\right] \end{array}$$

This last matrix does not contain any dominated strategies.

Exercises 8-4

Let A be the row player and B be the column player. The following games are not strictly determined. In each game, find optimum strategies for the players and the value of the game.

1. $\begin{bmatrix} 2 & 1 \\ -4 & 5 \end{bmatrix}$ 2. $\begin{bmatrix} -1 & 3 \\ 0 & -2 \end{bmatrix}$

3. $\begin{bmatrix} -5 & 7 \\ 2 & 1 \end{bmatrix}$ 4. $\begin{bmatrix} -6 & -2 \\ -4 & -5 \end{bmatrix}$

Let A be the row player and B be the column player. Some of the following games are strictly determined and some are not. Find optimum strategies and the value for each game. Any of the games that are not strictly determined can be reduced to a game with a 2×2 matrix by eliminating dominated strategies.

5. $\begin{bmatrix} -1 & -3 & 2 \\ 2 & -2 & 0 \\ 0 & -1 & 1 \end{bmatrix}$ 6. $\begin{bmatrix} -2 & 3 & 4 \\ 1 & 0 & 2 \\ 0 & -2 & -1 \end{bmatrix}$

7. $\begin{bmatrix} 10 & -12 \\ 6 & 8 \\ 4 & 5 \end{bmatrix}$ 8. $\begin{bmatrix} 5 & 2 & -4 \\ -1 & -3 & 1 \end{bmatrix}$

9. $\begin{bmatrix} -2 & 6 & 1 & -4 \\ 4 & 3 & 0 & 6 \\ 3 & 5 & 2 & 8 \\ 7 & -3 & -2 & 3 \end{bmatrix}$ 10. $\begin{bmatrix} 4 & 2 & 5 & 1 \\ 3 & 1 & 5 & 0 \\ 6 & 3 & 7 & 4 \\ 0 & 5 & 1 & 6 \end{bmatrix}$

11. $\begin{bmatrix} 3 & -1 & 2 & 5 \\ 4 & 8 & 6 & 9 \\ 6 & 4 & 3 & 5 \\ 7 & 5 & 4 & 6 \end{bmatrix}$ 12. $\begin{bmatrix} 7 & 10 & 5 & 12 \\ 14 & 8 & 3 & 8 \\ 6 & 15 & 1 & 11 \\ 8 & 9 & 4 & 7 \end{bmatrix}$

13. John Allgood is taking a cruise on the Mississippi River. A gentleman named Maverick convinces him to play a game of matching coins. Each player simultaneously reveals a dime. If the coins match (both are heads or both are tails), Maverick wins Allgood's coin, otherwise, Allgood wins. What are the optimum strategies and what is the value of the game?

14. After a while, Maverick suggests raising the stakes and also changing the game a little to make it more interesting. From now on, when the coins match, Maverick wins $2. When Allgood chooses heads and Maverick chooses tails, Allgood wins only $1. However, when Allgood chooses tails and Maverick chooses heads, Allgood wins $3. Determine optimum strategies and the value of the game.

15. Gnome Manufacturing has 90% of the market in teak rocking chairs and produces a superior product than its only competitor, Oriental Hardwood. Oriental is planning a major advertising campaign, and can use either radio or television. To counter this, Gnome will also advertise on either radio or television. If both companies choose the same medium, Gnome will gain 2% of the market. If Gnome chooses radio and Oriental chooses television, Oriental will gain 3%. If the opposite choices are made, Oriental will gain 5%. Determine optimum strategies and the value of the game.

16. At a NATO war game, the green army is attacking and the blue army is defending. The blue army has two forts, both minimally garrisoned, and mobile troops that can be stationed to defend either fort, but not both. The green army can attack either fort, but cannot attack both. The green army would gain two points for capturing the north fort and seven points for capturing the south fort. These points are lost by the blue army. If a fort is attacked, it will be captured unless the blue

army's mobile troops are there. Determine optimum strategies and the value of the game.

*17. Suppose a game is not strictly determined and that each payoff is doubled. What effect will this have on optimum strategies and on the value of the game?

*18. Repeat exercise 17 if, instead of doubling the payoffs, a number k is added to each payoff.

Chapter 8 Review

1. Compute the expected value of an experiment with outcomes A, B, and C if $P(A) = .2$, $P(B) = .7$, $P(C) = .1$, $V(A) = 10$, $V(B) = 3$, and $V(C) = 6$.

2. Compute the expected value of a toss of a fair die if the payoff is $n^2 - n + 1$ when n dots show.

3. The probability that the Dodgers will beat the Giants in tonight's baseball game is .6. What are the odds for a fair bet on the Dodgers?

4. Swifty is a 2:5 underdog in a race. What is the probability that Swifty will win the race if the given odds are for a fair bet?

5. The annual premium on a homeowner's policy is $180. The probability of collecting on it within a year is .01, and the average amount collected is $15,000. What is the expected value of the policy to the homeowner? To the insurance company?

6. Joanna James is a land speculator. She is considering two deals but only has enough funds available to accept one. Some land adjoining a proposed highway is available for $2,000,000. If the highway is built, the land will be worth $5,000,000 if it is between interchanges (probability .3) and $8,000,000 if it is located at an interchange (probability .1). Otherwise the land is worth $500,000.

   The other deal concerns land near a proposed shopping mall. The land is available for $2,500,000. If the mall is built, the land will be worth $5,000,000. Otherwise, the land is worth $1,500,000. Conservation groups are opposing the mall and have an even chance of preventing its construction.

   Which deal should Ms. James take?

Find optimum strategies and the value of each game.

7. $\begin{bmatrix} 3 & 4 \\ 5 & -2 \end{bmatrix}$

8. $\begin{bmatrix} -2 & 3 \\ 1 & -1 \end{bmatrix}$

9. $\begin{bmatrix} 0 & 5 & -3 \\ -2 & 1 & -1 \\ 1 & 2 & 4 \end{bmatrix}$ 10. $\begin{bmatrix} -2 & 2 & -3 \\ -3 & 1 & -5 \\ 1 & -2 & 0 \end{bmatrix}$

11. $\begin{bmatrix} -2 & 0 & -1 & -3 \\ -4 & 1 & -3 & -1 \\ 1 & 2 & 1 & -2 \end{bmatrix}$ 12. $\begin{bmatrix} -1 & -3 & 4 \\ -4 & 3 & 1 \\ -2 & 5 & -4 \\ 0 & 2 & 3 \end{bmatrix}$

13. The Democratic candidate for president is a better debater than the Republican. The candidates will have a series of debates and must determine their strategies. In each debate, a candidate can emphasize either domestic matters or foreign affairs. The matrix

$$\text{Democrat} \begin{array}{c} \\ \text{Domestic} \\ \text{Foreign} \end{array} \overset{\begin{array}{cc} \text{Republican} \\ \text{Domestic} \quad \text{Foreign} \end{array}}{\begin{bmatrix} 5 & -2 \\ -3 & 3 \end{bmatrix}}$$

describes the change in the percentage of the votes received by the candidates as a result of one debate. Determine optimum strategies. Which candidates is expected to gain by these debates?

Functions and Graphs

In business as in many other endeavors, it is often important to be able to describe the relationship between two or more quantities. For example, the relationship between total sales and the number of defective items returned is of vital interest to any quality control manager (and also to consumers).

The main subject of this chapter is functions. A function is a mathematical way of expressing how one quantity determines another. We begin by studying the concept of function, and then discuss different kinds of functions and their graphs.

Section 9-1   The Concept of Function

Norton Shipping is discussing the purchase of some freighters from the Portland Shipyard. Portland could make up to four freighters next year and offers to make one freighter for $12,000,000, two for $22,000,000, three for $32,000,000, or four for $41,000,000.

Portland's offer could be expressed by listing the following *ordered pairs*

(1, 12,000,000)   (2, 22,000,000)   (3, 32,000,000)   (4, 41,000,000)

A *relation* is a set of ordered pairs. If (c, d) is an ordered pair, c is called the *first component* and d is called the *second component*. The set of first components is called the *domain* of the relation. The set of second components is called the *range* of the relation.

Portland's offer can be expressed as a relation. The domain of this relation is $\{1, 2, 3, 4\}$ and the range is $\{12{,}000{,}000, 22{,}000{,}000, 32{,}000{,}000, 41{,}000{,}000\}$. This relation has the special property that the second component of each ordered pair is completely determined by the first component.

> **Definition.** A *function f* is a relation with this property: If two ordered pairs in f have the same first component, they have the same second component.

In symbols, this special property is written:

$$\text{If } (b, c) \in f \text{ and } (b, d) \in f, \text{ then } c = d$$

Example 1 $\{(1,5), (2,3), (3,8), (4,5), (5,9)\}$ is a function. The domain is $\{1, 2, 3, 4, 5\}$ and the range is $\{3, 5, 8, 9\}$.

Example 2 $\{(a, m), (e, v), (g, 5), (v, g)\}$ is a function. The domain is $\{a, e, g, v\}$ and the range is $\{g, m, v, 5\}$.

Example 3 $\{(a, q), (b, q), (b, h), (c, d)\}$ is not a function because the pairs (b, q) and (b, h) have the same first member but different second members. The domain of this relation is $\{a, b, c\}$ and the range is $\{d, h, q\}$.

∎

A function can often be described by a formula. For example, instead of writing $k = \{(x, y) \mid y = 3x - 2\}$, one can write $y = k(x) = 3x - 2$. Then x is called the independent variable. Since the value of y depends on and is in fact determined by x, y is called the *dependent variable*.

Most of the functions of concern to us will have as domain an infinite set of real numbers. Since we cannot write infinitely many ordered pairs, such functions are described by formulas. Some examples of such functions are $f(x) = 5x + 3$, $g(x) = \sqrt{x}$, and $h(x) = 1/x$.

The domain of a function defined by a formula may be specified. Otherwise, the domain is the set of all real numbers that can be substituted into the formula to obtain a real number.

Examples Find the domain of each function.

Example 4 $f(w) = \sqrt{9 - w^2}$. Since the square root of a negative number is not defined, we must have $9 - w^2 \geq 0$. Solve this inequality with a sign graph to obtain $-3 \leq w \leq 3$.

Example 5 $g(y) = (4 - y^2)^{1/3}$. Since the cube root of any real number is defined, the domain is the set of all real numbers.

Example 6 $h(t) = 1/(t^2 + 4t + 3)$. Since division by 0 is undefined, the domain excludes the solutions of $t^2 + 4t + 3 = 0$, which are -3 and -1. The domain is $\{t \mid t \neq -3 \text{ and } t \neq -1\}$.
■

When we write $f(x) = x^2 + 7x + 5$, the rule for computing $f(x)$ determines f. The variable x can be replaced by any other variable and the same function results. Thus, f is also given by the formulas $f(v) = v^2 + 7v + 5$ and $f(q) = q^2 + 7q + 5$. The *value* of $f(x)$ at a number t is obtained by substituting t for x in the formula that defines f. Thus,

$$f(0) = 0^2 + 7 \cdot 0 + 5 = 5$$
$$f(8) = 8^2 + 7 \cdot 8 + 5 = 125$$
$$f(-t) = (-t)^2 + 7(-t) + 5 = t^2 - 7t + 5$$
$$f(4q) = (4q)^2 + 7(4q) + 5 = 16q^2 + 28q + 5$$
$$f(y - 1) = (y - 1)^2 + 7(y - 1) + 5 = y^2 + 5y - 1$$

When does a formula define y as a function of x? An expression like $y = g(x)$ always does, but consider $2x + 3y = 6$ or $x^2 + y^2 = 25$. Such an equation will define y as a function of x if it can be solved for y and gives a unique value of y. If the equation gives two or more values of y for one value of x, it does not define y as a function of x.

Now consider $2x + 3y = 6$. This can be solved to give $y = 2 - 2x/3$. For each number x, one value of y is determined. Therefore $2x + 3y = 6$ defines y as a function of x.

Consider $x^2 + y^2 = 25$. Solving this equation for y gives $y = \sqrt{25 - x^2}$ or $y = -\sqrt{25 - x^2}$. If $x = 3$, for example, y could be either 4 or -4. The equation $x^2 + y^2 = 25$ does not define y as a function of x.

340   CHAPTER 9   FUNCTIONS AND GRAPHS

A formula provides a concise algebraic description of a function. Graphs are often very useful in visualizing functions. If the domain and range of a function are sets of real numbers, the *graph* of the function is the set of points in the coordinate plane that correspond to the ordered pairs in the function. Later in this chapter, the graphs of several kinds of functions are studied.

When is a subset S of the plane the graph of a function f? Recall that if $(b, c) \in f$ and $(b, d) \in f$, then $c = d$. Geometrically this says that two different points of S must have different first coordinates. This is equivalent to saying that no vertical line meets S in more than one point. Diagrams 1, 2, and 4 show graphs of functions. The circle shown in Diagram 3 is not the graph of a function.

Cost, revenue, and profit functions

Now let's study several specific functions that are of importance in business. Suppose the Royal Wax Company produces elaborate hand-made candles. Let q denote the number of candles produced.

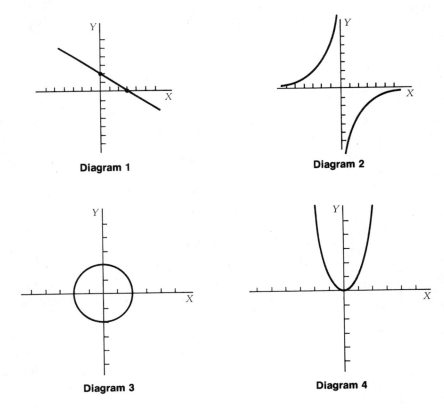

Diagram 1

Diagram 2

Diagram 3

Diagram 4

The *cost function* $C(q)$ is the cost of producing q candles. Two types of costs are incurred in producing these candles. Overhead expenses, called *fixed costs*, are incurred by the Royal Wax Company even if nothing is produced. Some examples of fixed costs are rent on the candle factory, insurance, and licensing fees. The number $C(0)$ represents fixed costs. The number $C(q) - C(0)$ is called the *variable cost* of producing q candles and represents actual production costs. Since

$$C(q) = [C(q) - C(0)] + C(0)$$

cost is the sum of variable cost and fixed costs. The *average cost function* is defined by $AC(q) = C(q)/q$.

Example 7 The cost function for Royal Wax is $C(q) = 1500 + 10q$.

PROBLEM A Determine fixed costs.

Solution The fixed costs are $C(0) = \$1500$.

PROBLEM B Determine the cost of producing 1000 candles and their average cost.

Solution The cost of producing 1000 candles is $C(1000) = \$11,500$. The average cost of these candles is $AC(1000) = C(1000)/1000 = \$11.50$.

The owners of Royal Wax would rather hear about revenue than about costs. The *revenue function* $R(q)$ is the amount received by the company from the sale of the first q candles. The *average revenue function* is defined by $AR(q) = R(q)/q$.

Example 8 The revenue function for Royal Wax is $R(q) = 20q$. Determine the revenue from the sale of the first 1000 candles and the average revenue from them.

Solution The revenue from the sale of the first 1000 candles is $R(1000) = \$20,000$. The average revenue from these candles is $AR(1000) = R(1000)/1000 = \$20$.

Revenue is the amount of money coming into the business and costs are the amount leaving. The difference, $R(q) - C(q)$, represents profit. The *profit function* is defined by

$$P(q) = R(q) - C(q) \quad \text{Profit function}$$

For Royal Wax, the profit function is

$$P(q) = (20q) - (1500 + 10q) = 20q - 1500 - 10q$$
$$= 10q - 1500$$

The *average profit function* represents the average profit on each candle sold. This function is denoted $AP(q)$ and is defined by $AP(q) = P(q)/q$. Another formula for average profit is $AP(q) = AR(q) - AC(q)$.

Example 9 Determine the profit from the sale of 1000 candles and the average profit.

Solution The profit is $P(1000) = \$8500$. The average profit is $P(1000)/1000 = \$8.50$. ∎

The manager of Royal Wax is very much concerned with whether the company shows a profit or a loss. If $P(q) < 0$, there is a loss. This happens if

$$10q - 1500 < 0$$
$$10q < 1500$$
$$q < 150$$

Thus, if fewer than 150 candles are made and sold, there is a loss. The company makes a profit if $P(q) > 0$. This happens if

$$10q - 1500 > 0$$
$$10q > 1500$$
$$q > 150$$

Royal Wax makes a profit if more than 150 candles are sold.

What happens if exactly 150 candles are made and sold? Then there is neither a profit nor a loss, so that Royal Wax breaks even. Note that $P(150) = 0$.

Definition. A *breakeven point* is a solution of the equation $P(q) = 0$.

Since $P(q) = R(q) - C(q)$, the equation $P(q) = 0$ has the same solutions as the equation $R(q) = C(q)$. Therefore a breakeven point is a value of q at which revenue equals costs. Royal Wax has only one breakeven point. This happens fairly often, but it is also common to have two breakeven points.

Supply and demand functions

Let's proceed now to study two important functions in economics: the supply and demand functions.

The *supply function* $s(q)$ gives the price at which producers supply the quantity q of an item to the market. The *demand function* $d(q)$ gives the price at which consumers demand q of the item. Usually $s(q)$ increases and $d(q)$ decreases as q gets larger. The number q^* for which $s(q^*) = d(q^*)$ is called the *equilibrium quantity*. The common value p^* of $s(q^*)$ and $d(q^*)$ is called the *equilibrium price*, and the point (q^*, p^*) is called the *equilibrium point*. In certain market situations, q^* items will be traded at the price p^*.

Example 10 Suppose $s(q) = 100 + 2q$ and $d(q) = 150 - 3q$.

PROBLEM A At what price are 20 items supplied?

Solution Twenty items are supplied at the price $s(20) = 100 + 2(20) = \$140$.

PROBLEM B At what price are 20 items demanded?

Solution Twenty items are demanded at the price $d(20) = 150 - 3(20) = \$90$.

PROBLEM C Determine the equilibrium price.

Solution Equate supply with demand to obtain the equation $100 + 2q = 150 - 3q$. The solution q^*, in this case 10, is the equilibrium quantity. To obtain the equilibrium price, substitute 10 for q in either the supply function or the demand function and obtain \$120. ∎

Exercises 9-1

Determine which of the following are functions. State the domain and range of each.

1. $\{(-1, 4), (4, -1), (-1, 4), (3, 0)\}$
2. $\{(3, 2), (2, 4), (4, 2), (3, 4)\}$

3. $\{(6,8),(7,6),(7,6),(4,2),(4,3)\}$
4. $\{(1,5),(4,2),(3,2),(2,1),(5,6)\}$
5. $\{(0,0),(1,0),(2,2),(3,1)\}$
6. $\{(8,0),(7,0),(6,0),(5,0)\}$

Determine the domain of each function.

7. $h(s) = \dfrac{1}{\sqrt{s}}$
8. $k(t) = \sqrt{2t-3}$
9. $j(v) = v^2 + \dfrac{1}{(v-2)}$
10. $g(w) = \dfrac{1}{w} + \dfrac{1}{(w^2-4)}$
11. $f(y) = \sqrt{4-y^2}$
12. $j(y) = \sqrt{y^2-25}$
13. $f(t) = t^2 - 6t + 3$
14. $m(s) = \sqrt{s^2-5s+6}$
15. $j(t) = \dfrac{1}{\sqrt{t^2+4t-5}}$
16. $f(x) = 3x^2 + \dfrac{5}{x^2-8x+15}$

Let $f(v) = v^2 + 2v + 3$, $g(v) = 4v - 5$, and $h(v) = 1/v$. Compute

17. $f(h(v))$
18. $h(f(v))$
19. $g(h(v))$
20. $h(g(v))$
21. $g(f(v))$
22. $f(g(v))$
23. $f(y)$
24. $g(w)$
25. $f(3x)$
26. $f(5t)$
27. $g(g(t))$
28. $h(h(w))$
29. $g(2v+1)$
30. $g(3-5v)$
31. $h(2x-7)$
32. $h(4-3x^2)$
33. $f(f(1))$
34. $f(f(-1))$
35. $f(g(2))$
36. $f\left(h\left(\dfrac{1}{2}\right)\right)$

Determine whether or not the equation defines y as a function of x. If it does, solve for y.

37. $2x + 3y = 24$
38. $4x - 5y = 20$
39. $x^2 - y^2 = 1$
40. $x^2 + 4x + 3y - 12 = 0$
41. $x^2 y = 9$
42. $xy = 4$
43. $xy^2 = 4$
44. $(x+3)(y-2) = 8$
45. $x^3 - y^3 = 9$
46. $x^3 + y^3 = 1$
47. $(x+y)^3 = x^2$
48. $6x^4 + 3x - 2y^3 + 12 = 0$
49. $(x-y)^2 = 4$
50. $6x^3 - 4x + 5 - 2y^2 = 0$

51. Let $C(q) = 1000 + 12q$ and $R(q) = 16q$. Determine each of the following:

 a. Fixed costs
 b. $C(50)$
 c. $C(400)$
 d. $AC(400)$
 e. $R(80)$
 f. $AR(80)$
 g. $AR(100)$
 h. $P(100)$
 i. $AP(100)$
 j. $P(1000)$
 k. $AP(1000)$
 l. Breakeven point

Answer the same questions as in exercise 51.
52. $C(q) = 800 + 10q$, $\quad R(q) = 14q$
53. $C(q) = 120{,}000 + 200q$, $\quad P(q) = q^2 + 100q - 120{,}000$
54. $R(q) = q^2 + 500q$, $\quad P(q) = q^2 + 300q - 400{,}000$

55. Let $s(q) = 200 + 3q$ and $d(q) = 500 - 7q$. Determine each of the following:

 a. $s(100)$
 b. $d(40)$
 c. Equilibrium point

Answer the same questions as in exercise 55.
56. $s(q) = 400 + 5q$, $\quad d(q) = 900 - 20q$
57. $s(q) = q^2 + 100q$, $\quad d(q) = 8000 - 10q$
58. $s(q) = q^2 + 80q$, $\quad d(q) = 900$
59. $s(q) = q^2 + 10q$, $\quad d(q) = 1000 - 20q$
60. $s(q) = 625 + 20q$, $\quad d(q) = 2000 - 10q - q^2$

61. Selena Foster is considering purchasing business stationery from Smythe Printing. Smythe charges $15.00 for setting up the presses plus $.03 for each printed sheet of stationery.
 a. Write a formula expressing the cost c as a function of the number n of sheets of stationery.
 b. How much would 500 sheets of stationery cost?
 c. How much stationery could Ms. Foster buy for $60.00?
 d. If Ms. Foster buys 100 sheets, what is the average cost per sheet?
 e. If Ms. Foster buys 1000 sheets, what is the average cost per sheet?
 f. What is the approximate cost per sheet if Ms. Foster places a very large order with Smythe Printers?

62. An operator-assisted phone call from Des Moines to Sioux City during business hours costs $1.10 for up to three minutes. For each additional

minute or part of a minute, there is an additional charge of $.19.
  a. How much does a two-minute and five-second call cost?
  b. How much does a four-minute call cost?
  c. How much does a five-minute and 30-second call cost?
  *d. Draw the graph of cost as a function of time, for times from one to ten minutes.

Section 9-2 Lines and Linear Functions

We begin our study of specific kinds of functions with linear functions. Linear functions have many applications. Cost, revenue, and profit functions, for instance, are often linear.

> **Definition.** A *linear function* is a function of the form
> $$f(x) = ax + b$$

Here a and b are constants. We will show in this section that the graph of a linear function is a line. To do so, lines must be studied.

A line is determined by two points. A line can also be determined by specifying one point along with the steepness of the line. The steepness of a line is conveniently described by a number called the *slope*. The concept of slope is illustrated in Diagram 1.

> **Definition.** Let L be a line that is not vertical. If $P_1(x_1, y_1)$ and $P_2(x_2, y_2)$ are any two points on L, the *slope* of L is given by
> $$m = \frac{y_2 - y_1}{x_2 - x_1}$$

If a line is vertical with equation $x = c$, suppose (c, y_1) and (c, y_2) are any two different points on it. Substituting into the formula for slope would give $(y_2 - y_1)/(c - c) = (y_2 - y_1)/0$, which involves division by 0. Consequently *the slope of a vertical line is undefined.*

The slope of a line L is usually denoted by the letter m. The value of m does not depend on which two points on the line are used to compute the slope. To see this, note that in Diagram 2 the triangles P_1BP_2 and P_3CP_4 are similar and thus the ratios of corresponding parts are equal. In particular, the ratios of rise to run are equal.

9-2 LINES AND LINEAR FUNCTIONS

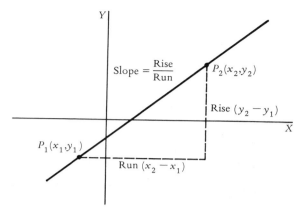

Diagram 1

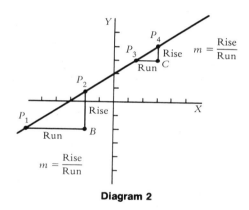

Diagram 2

Examples 1. The slope of the line through $(3, 4)$ and $(2, 7)$ is
$$\frac{7-4}{2-3} = -3.$$

2. The slope of the line through $(0, -3)$ and $(3, -6)$ is
$$\frac{-6-(-3)}{3-0} = -1.$$

3. The slope of the line through $(-3, -5)$ and $(4, 7)$ is
$$\frac{7-(-5)}{4-(-3)} = \frac{12}{7}.$$

4. The slope of the line through $(3, 5)$ and $(3, 2)$ is

$$\frac{2 - 5}{3 - 3},$$

which is undefined. This line is vertical and has equation $x = 3$.

∎

A line with positive slope is described as rising. The lines in Diagram 3 are all rising. A line with negative slope is described as falling. The lines in Diagram 4 are all falling.

If L is a horizontal line with equation $y = c$, suppose (x_1, c) and (x_2, c) are any two different points on L. Then

$$m = \frac{c - c}{x_2 - x_1} = \frac{0}{x_2 - x_1} = 0.$$

The slope of a horizontal line is 0.

A horizontal line is the graph of a special kind of function called a constant function.

> **Definition.** A *constant function* is a function whose range contains only one member.

A function is a constant function precisely when there is a number c such that $f(x) = c$ whenever x is in the domain of f.

Now let's use the notion of slope to find the equation of a line. Let L be the line through the point (x_1, y_1) with slope m. If (x, y) is any other point

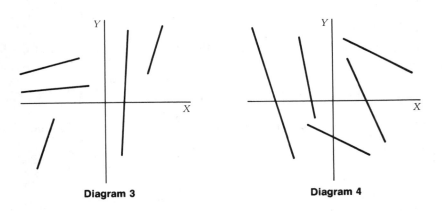

Diagram 3   Diagram 4

of L, then $(y - y_1)/(x - x_1) = m$. Multiplying by $x - x_1$ gives the following equation.

$$y - y_1 = m(x - x_1) \quad \text{Point-slope equation of a line}$$

This is called the equation of L in *point-slope form*.

Examples
5. The equation of the line through $(3, 5)$ with slope 6 is $y - 5 = 6(x - 3)$.
6. The equation of the line through $(-2, 8)$ with slope -4 is $y - 8 = -4[x - (-2)] = -4(x + 2)$.
7. The equation of the line through $(0, 0)$ with slope 2 is $y - 0 = 2(x - 0)$, or $y = 2x$.
8. The equation of the line through $(4, 7)$ with slope 0 is $y - 7 = 0(x - 4)$, or $y - 7 = 0$. This line is horizontal.

■

Let us find the equation of the line through $(2, 4)$ and $(5, 13)$. $m = (13 - 4)/(5 - 2) = 3$. The equation is $y - 4 = 3(x - 2)$. The equation could also be written as $y - 13 = 3(x - 5)$. These two equations describe the same line, and both equations reduce to $y = 3x - 2$.

Example 9 Find the equation of the line through $(-2, 4)$ and $(7, 3)$.

Solution

$$m = \frac{3 - 4}{7 - (-2)} = \frac{-1}{9} \quad \text{Find the slope.}$$

$$y - 4 = \frac{-1}{9}[x - (-2)] \quad \text{Use the point-slope formula.}$$

$$= \left(\frac{-1}{9}\right)(x + 2)$$

Example 10 Find the equation of the line through $(-5, -3)$ and $(-3, -7)$.

Solution

$$m = \frac{-7 - (-3)}{-3 - (-5)} = \frac{-7 + 3}{-3 + 5} = \frac{-4}{2} = -2 \quad \text{Find the slope.}$$

$$y - (-3) = -2[x - (-5)] \quad \text{Use the point-slope formula.}$$

$$y + 3 = -2(x + 5) \qquad \text{Simplify.}$$

∎

A line has many different equations in point-slope form—one for each point on the line. It would be convenient to be able to specify one particular equation for each line. We will now do this, but first some definitions are needed.

> **Definition.** Let $y = f(x)$ be a function. The number a is an *x-intercept* of the graph of f if $f(a) = 0$. The number $f(0)$ is the *y-intercept* of the graph of f.

If a is an x-intercept, $(a, 0) \in f$ and the graph of f intersects the x-axis at $(a, 0)$. The graph of f intersects the y-axis at $(0, f(0))$.

A nonvertical line always has one y-intercept. A line with nonzero slope always has one x-intercept.

The equation of the line through (x_1, y_1) with slope m has equation $y - y_1 = m(x - x_1)$, or $y = mx + (y_1 - mx_1)$. If $x = 0$, then $y = y_1 - mx_1$, so that $y_1 - mx_1$ is the y-intercept of the line. Let $b = y_1 - mx_1$, so that b is the y-intercept of the line. Then

> $y = mx + b$ Slope-intercept equation of a line

This is called the *slope-intercept form* of the equation of the line. In this form, y is written as a function of x. If L is a nonvertical line, there is exactly one way of writing the equation of L in slope-intercept form.

Example 11 The equation of the line with slope 3 and y-intercept 5 is $y = 3x + 5$.

Example 12 The equation of the line with slope -2 and y-intercept 0 is $y = -2x + 0$ or $y = -2x$.

Example 13 Find the equation in slope-intercept form of the line through $(2, 7)$ with slope 4.

Solution

$$y - 7 = 4(x - 2) \qquad \text{Use the point-slope formula.}$$

$$y - 7 = 4x - 8 \qquad \text{Solve the equation for } y.$$

$$y = 4x - 1$$

Example 14 Find the equation in slope-intercept form of the line through $(3, 6)$ and $(5, 16)$.

Solution

$$m = \frac{16 - 6}{5 - 3} = \frac{10}{2} = 5 \quad \text{Find the slope}$$

$$y - 6 = 5(x - 3) \quad \text{Use the point-slope formula.}$$

$$y - 6 = 5x - 15 \quad \text{Solve the equation for } y.$$

$$y = 5x - 9$$

Example 15 Find the equation in slope-intercept form of the line with x-intercept 6 and y-intercept 4.

Solution The line contains the points $(6, 0)$ and $(0, 4)$.

$$m = \frac{4 - 0}{0 - 6} = \frac{4}{-6} = -\frac{2}{3} \quad \text{Find the slope.}$$

$$y = -\frac{2}{3}x + 4 \quad \text{Use the slope-intercept formula.}$$

Example 16 In one production run, Easton Appliances can produce 100 telephones for $2300 or 250 telephones for $5300. Write the cost function, assuming it is linear.

Solution The problem asks for the equation of the line through the points $(100, 2300)$ and $(250, 5300)$. The slope of this line is

$$m = \frac{5300 - 2300}{250 - 100} = \frac{3000}{150} = 20$$

Let n denote the number of telephones and C denote cost. Then use the point-slope formula to obtain

$$C - 2300 = 20(n - 100)$$

Now solve this equation for C.

$$C - 2300 = 20n - 2000$$
$$C = 20n + 300$$

■

Now let's return to the relationship between linear functions and lines in order to demonstrate that the graph of any linear function is a line. First, any nonvertical line has an equation of the form $y = mx + b$ and is thus the graph of a linear function. If $f(x) = mx + b$ is a linear function, the graph of f is the line with slope m and y-intercept b.

In Chapter 2, it was stated that the graph of the *general linear equation* $Ax + By = C$ is a line. (In this equation, A and B cannot both be 0.) To see this, observe that if $B = 0$, this equation is $Ax = C$ or $x = C/A$. The graph in this case is a vertical line. If $B \neq 0$, the equation can be solved for y to give

$$y = \frac{-A}{B}x + \frac{C}{B}$$

which is the equation of a line.

Example 17 Find the slope and y-intercept of the line with equation $3x + 4y = 24$.

Solution Solve the equation for y.

$$4y = -3x + 24$$
$$y = \frac{-3}{4}x + 6$$

The last equation is in slope-intercept form. Thus the line has slope $-3/4$ and y-intercept 6.

■

Exercises 9-2

Compute the slope of the line through each pair of points.

1. $(1, 3), (5, 9)$
2. $(2, 4), (4, 8)$
3. $(3, 6), (-3, 6)$
4. $(-3, -10), (-5, -12)$
5. $(-4, 6), (4, -6)$
6. $(2, 1), (-3, 1)$
7. $(4, 5), (5, 5)$
8. $(2, 1), (1, 8)$
9. $(2, 3), (2, 6)$
10. $(-1, 1), (0, 0)$
11. $(4, 1), (-3, 2)$
12. $(-1, 4), (-1, -4)$

Write the equation of the line through the given point with the given slope in (a) point-slope, (b) slope-intercept, and (c) general linear form. Graph each line.

13. $(2, 8)$ $m = -1$
14. $(8, 4)$ $m = 3$
15. $(-1, 5)$ $m = 3$
16. $(2, -6)$ $m = -4$
17. $(2, 4)$ $m = 0$
18. $(5, 3)$ $m = 1$
19. $(-3, -6)$ $m = 2$
20. $(-2, -1)$ $m = -4$
21. $(3, 7)$ $m = -2$
22. $(4, 5)$ $m = 0$

Write the equation of the line through the given pair of points in (a) point-slope, (b) slope-intercept, and (c) general linear form. Graph each line.

23. $(1, 4), (3, 10)$
24. $(2, 7), (4, 11)$
25. $(-1, 1), (1, -1)$
26. $(3, 5), (-2, 5)$
27. $(0, 0), (4, 12)$
28. $(5, 0), (0, 10)$
29. $(4, 0), (0, -8)$
30. $(4, 1), (0, 0)$
31. $(-3, 3), (0, 6)$
32. $(1, 1), (3, -3)$

Put each equation into slope-intercept form and determine the slope and the y-intercept.

33. $2x + 3y = 6$
34. $4x - 5y = 20$
35. $3x - 15y = 15$
36. $2x + y = 0$
37. $3x = 6$
38. $5y = 10$
39. $y + 2 = 5(x - 1)$
40. $y - 3 = -2(x + 3)$
41. $y - 7 = -4(x + 2)$
42. $y + 2 = 4(x - 1)$

43. The Eastern Gas Company charges $4.00 a month for service plus $2.50 for each thousand cubic feet of gas used. Express the monthly bill b as a function of the amount x of gas used. Here one unit of gas is 1000 cu ft.

44. John Newman is a salesman for Drork Enterprises. He earns $120 a week plus a 3% commission. Express his weekly income I as a function of his weekly sales S.

45. Southern Maple can produce 20 tables for $3500. To produce 50 tables, the cost is $8000. Write the cost function, assuming it is linear.

46. In one production run, Belkey Electronics can produce 40 control units for $1600 or 60 for $2300. Write the cost function, assuming it is linear.

47. Evergreen Enterprises was founded in 1970 and had sales of $5 million that year. Since then, sales have increased linearly and have reached $12 million in 1979. How much will sales be in 1982, if this trend continues?

The pollution level in Los Angeles was 40 at 6 A.M. and rose linearly during the day until it reached 208 at 8 P.M.

48. Express the pollution level P as a function of the time t, measured in hours, since 6 A.M. Be sure to state the domain of this function.
49. At 11 A.M., the air pollution level reached the unhealthy range. What was the air pollution level then?
50. At 7:20 P.M., the air pollution level reached the very unhealthy range. What was the air pollution level then?

The air pollution level rose overnight to a high of 260 and then declined until it reached 214 at 6 A.M. An air pollution alert was called and, as a result, pollution declined linearly during the day until 8 P.M., at which time the pollution level was 88.

51. Express the pollution level P as a function of the time t, measured in hours, since 6 A.M.
52. When did the air pollution level fall below the very unhealthy range?
53. When did the air pollution level fall below the unhealthy range?

The straight-line, or linear, method is allowed by the Internal Revenue Service to be used for depreciating business property. With this method, a capital asset depreciates by the same amount each year from the time it is purchased until it is no longer useful. The total amount of depreciation within this period, called the depreciable value V, is the original cost of the asset minus its scrap value. (Scrap value is the value of the asset when it is no longer useful.) The undepreciated balance B is the depreciable value minus the amount of depreciation already taken. Use linear depreciation in exercises 54–59.

A computer costs $4,000,000 and has a useful life of eight years, at which time its scrap value is $400,000.

54. Determine the depreciable value V.
55. How much does the computer depreciate each year of its useful life?
56. Write a formula expressing the undepreciated balance B as a function of time t. Here t is measured in years and $0 \leq t \leq 8$.

A car used for business purposes costs $8000, has a useful life of four years, and has scrap value $2000.

57. Determine the depreciable value V.

58. How much does the car depreciate each year of its useful life?

59. Express B as a function of time t, where $0 \leq t \leq 4$.

Certain assets depreciate more rapidly when they are new than when they are older. For such assets, the Internal Revenue Service permits depreciation by the sum-of-the-year's-digits method. With this method, the depreciation D_j in year j of an asset with a useful life of n years is given by:

$$D_j = \frac{n - j + 1}{n(n + 1)} \cdot 2V$$

Suppose the car from exercises 57–59 is depreciated by the sum-of-the-year's-digits method. How much does the car depreciate during

60. The first year?

61. The second year?

62. The third year?

63. The fourth year?

64. Its useful life?

Section 9-3 Quadratic Functions

The formula for a linear function involves a polynomial of degree 1 or a constant. If the polynomial were of degree 2, the function would be called quadratic.

> **Definition.** A *quadratic function* is a function of the form $f(x) = ax^2 + bx + c$, with $a \neq 0$.

The graph of a quadratic function is a curve called a *parabola*. A parabola is symmetric about a line called the *axis*. The axis meets the parabola at a point called the *vertex*. In Diagram 1, note that the vertex is the lowest point on the parabola.

Parabolas have useful properties. For example, light rays parallel to the axis of a parabola all pass through one point, called the *focus*, after being reflected by the parabola. (See Diagram 2.) This greatly concentrates the light. For this reason, parabolas are used in telescopes and radar receivers. If a point source of light is placed at the focus of a parabola, the rays of light are all parallel to the axis after they are reflected by the parabola. For this reason, parabolas are used in flashlights and automobile headlights.

To graph a quadratic function $f(x) = ax^2 + bx + c$, find the intercepts and plot a few points. It is also necessary to find the vertex of the parabola,

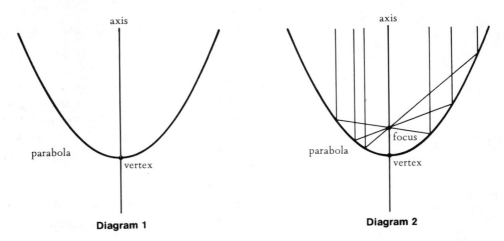

Diagram 1

Diagram 2

since the vertex is either the lowest point (if $a > 0$), or the highest point (if $a < 0$) of the graph. Now we will develop a procedure for finding the vertex. The formula for $f(x)$ can be manipulated into the form:

$$f(x) = a\left(x + \frac{b}{2a}\right)^2 + \frac{4ac - b^2}{4a}$$

You should verify this by simplifying the last formula. Now suppose $a > 0$. Since

$$\left(x + \frac{b}{2a}\right)^2$$

is nonnegative, f takes its minimum value when $x + [b/(2a)] = 0$, that is, when $x = -b/(2a)$. Thus this value for x must be the first coordinate of the vertex and of all points on the axis.

$$\boxed{x = \frac{-b}{2a} \quad \text{Axis of a parabola}}$$

Now we will draw the graph of $f(x) = x^2 + 4x + 3$. First find the intercepts. Since $f(0) = 3$, the y-intercept is 3. Since $x^2 + 4x + 3$ factors as $(x + 1)(x + 3)$, the x-intercepts are -1 and -3. Next find the vertex. Since the axis has equation $x = -b/(2a) = -4/(2 \cdot 1) = -2$, the vertex is the point $(-2, f(-2)) = (-2, -1)$. Plot a few more points and then fill in a smooth curve (see Diagram 3). Some more examples of graphing quadratic functions follow.

9-3 QUADRATIC FUNCTIONS 357

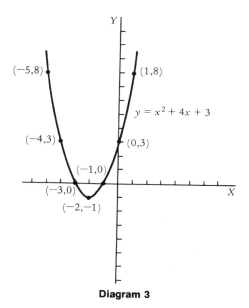

Diagram 3

Example 1 Draw the graph of $f(x) = 4x^2 + 4x + 3$.

Solution The y-intercept is $f(0) = 3$. The axis has equation $x = -b/(2a) = -1/2$, and $f(-1/2) = 2$. Since the vertex is the lowest point on this parabola, there are no x-intercepts. We could also determine this by trying to solve the equation $4x^2 + 4x + 3 = 0$ using the quadratic formula. Plot a few points and fill in the parabola. See Diagram 4.

Example 2 Draw the graph of $f(x) = -x^2 + 3x - 2$.

Solution The y-intercept is $f(0) = -2$. To find the x-intercepts, solve the equation $-x^2 + 3x - 2 = 0$ to obtain $x = 1$ and $x = 2$. Since the coefficient of x^2 is negative, this parabola "opens down" and its vertex is its highest point. The vertex is $(3/2, 1/4)$. The axis can be obtained by using the formula $x = -b/(2a)$. The axis can also be obtained by using symmetry—the axis of a parabola passes midway between the intercepts. Plot some more points and fill in the curve. See Diagram 5.

Example 3 South Side Enterprises has been manufacturing 500 dishwashers a day and selling them for $400 each. There has been very little profit, however, and an outside consultant is called in to study their operation. She finds that the cost function is $C(q) = q^2 - 440q + 165{,}000$, provided at least 250 dishwashers are made daily. What advice does she give them?

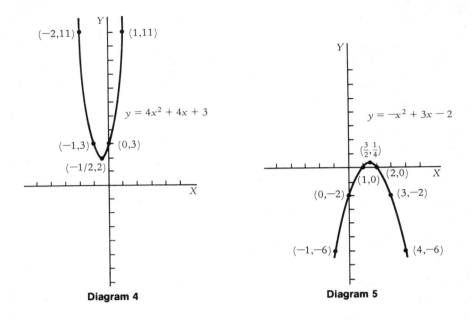

Diagram 4

Diagram 5

Solution The profit function is

$$P(q) = R(q) - C(q) = (400q) - (q^2 - 440q + 165{,}000)$$
$$= 400q - q^2 + 440q - 165{,}000$$
$$= -q^2 + 840q - 165{,}000$$

The graph of the profit function is a downward opening parabola. This parabola has its highest point when

$$q = \frac{-840}{2(-1)} = 420$$

Thus profit will be greatest if production is 420 dishwashers a day. Her advice was to cut production from 500 dishwashers daily to 420. This would result in profit increasing from \$5000 [$= P(500)$] to \$11,400 [$= P(420)$] daily.

Example 4 The market changes so that the selling price for a dishwasher is \$350. What should South Side Enterprises do now?

Solution The revenue function is now given by $R(q) = 350q$ and the profit function is

$$P(q) = 350q - (q^2 - 440q + 165{,}000)$$
$$= 350q - q^2 + 440q - 165{,}000$$
$$= -q^2 + 790q - 165{,}000$$

Profit is maximum when

$$q = \frac{-790}{2(-1)} = 395$$

Since $P(395) = -\$8975$, the best South Side can do is lose \$8975 daily. South Side should leave the dishwasher business unless the price of dishwashers is expected to rise.

Exercises 9-3

Draw the graph of each quadratic function. In the graph, label the vertex and all intercepts.

1. $y = x^2$
2. $y = x^2 - 1$
3. $y = x^2 - 2x + 1$
4. $y = x^2 + 2x + 1$
5. $y = -x^2$
6. $y = 4 - x^2$
7. $y = -x^2 + 2x$
8. $y = -x^2 - x$
9. $y = x^2 - x - 2$
10. $y = x^2 + x - 2$
11. $y = 2x^2 - 3x - 2$
12. $y = 2x^2 + x - 1$

13. The profit function for Wedgewood Tables is $P(q) = 160q - q^2 - 3000$, where q is the number of tables made in a week. What is the maximum possible weekly profit?

14. The profit function for Bill's Hand-made Shoes is $P(q) = 60q - 2q^2 - 200$, where q is the number of pairs of shoes made each week. What is the maximum possible weekly profit?

15. The revenue function for portraits from Mary's Art Studio is $R(q) = 150q - 3q^2$. Mary has fixed costs of \$500 and additional costs of \$30 per portrait.
    a. Determine the average revenue function.
    b. How many portraits must be sold to maximize revenue?
    c. Determine the cost function.
    d. Determine the profit function.
    e. What is the maximum possible profit?
    f. How many portraits must be sold to maximize profit?

16. The daily revenue function for hamburgers at Big Burger is $R(q) = 1.50q - 0.001q^2$. Big Burger allocates $100 of its daily fixed costs to hamburgers. The variable cost per hamburger is 60¢.
    a. Determine the average revenue function.
    b. How many hamburgers must be sold each day in order to maximize revenue?
    c. Determine the cost function.
    d. Determine the profit function.
    e. What is the maximum possible profit?
    f. How many hamburgers must be sold each day in order to maximize profit?

17. The demand function for vacuum cleaners is $d(q) = 400 - 5q - q^2/10$ and the supply function is $s(q) = q^2/20 + 3q + 25$. Find the equilibrium point.

18. The demand function for electric lawnmowers is $d(q) = 239 - 0.0002q^2 - 0.03q$ and the supply function is $s(q) = 50 + 0.1q$. Find the equilibrium point.

19. Transworld Travel Agency is organizing a tour of Europe for up to 200 people. If the tour is filled, each person will pay $2000. Otherwise the price will increase by $20 for each vacancy. What is the maximum possible revenue for Transworld from the tour?

20. A theater with 400 seats is running a special performance for a charity. Tickets are sold for $5 each. However, the theater will receive only $1 per ticket if all the tickets are sold. This amount will be increased by one cent for each unsold ticket. What is the maximum amount the theater can possibly receive?

Find the minimum value taken by each function.

*21. $f(w) = w^4 - 4w^2 + 12$
*22. $g(v) = v^4 + 6v^2 - 15$
*23. Determine which quadratic functions satisfy the relation $f(x + y) = f(x) + f(y)$ for all numbers x and y.
*24. Determine which quadratic functions satisfy the relation $f(kx) = k^2 f(x)$ for all numbers x provided $k \neq 0$.

Find the minimum value taken by each function.

25. $f(x) = 2.37x^2 - 9.41x + 12.17$
26. $f(x) = 135x^2 - 87x - 493$

27. $f(x) = 23.2x^2 + 11.5x - 118.4$
 28. $f(x) = 1.43x^2 - 0.27x + 8.97$

Section 9-4 Arithmetic and Composition of Functions

In the two previous sections, we have studied specific types of functions. Now it is time to study ways of combining functions. We have already combined the cost function $C(q)$ and the function q via division to obtain the average cost function. The profit function $P(q)$ has been obtained by subtracting the cost function from the revenue function. Two functions may be combined by any of the operations of arithmetic.

> Let f and g be functions, and let c denote a constant.
> The *sum* $f + g$ is defined by $(f + g)(x) = f(x) + g(x)$.
> The *difference* $f - g$ is defined by $(f - g)(x) = f(x) - g(x)$.
> The *function* cf is defined by $(cf)(x) = c \cdot f(x)$.
> The *product* fg is defined by $(fg)(x) = f(x)g(x)$.
> The *quotient* f/g is defined by $(f/g)(x) = f(x)/g(x)$. (Note that if $g(r) = 0$, then r is not in the domain of f/g because division by 0 is not defined.)

The following examples illustrate the arithmetic of functions.

Example 1 Let $f(x) = x^2 + 3x$ and $g(x) = 2x - 5$.

a. $(f + g)(x) = (x^2 + 3x) + (2x - 5) = x^2 + 5x - 5$
b. $(f - g)(x) = (x^2 + 3x) - (2x - 5) = x^2 + x + 5$
c. $(3x)(x) = 3(x^2 + 3x) = 3x^2 + 9x$
d. $(fg)(x) = (x^2 + 3x)(2x - 5) = 2x^3 + x^2 - 15x$
e. $\left(\dfrac{f}{g}\right)(x) = \dfrac{x^2 + 3x}{2x - 5}$
f. $(2f - 5g)(x) = 2(x^2 + 3x) - 5(2x - 5) = 2x^2 - 4x + 25$

Example 2 Wilma's Nursery grows and sells dogwoods. To grow q dogwoods until they are tall enough for sale requires $2q + 0.001q^2$ hours of labor and entails other expenses of $q - 0.001q^2 + 200$. Labor costs $10 an hour. The dogwoods can be sold for $25 each. Determine the profit function.

Solution The cost of labor is the product of the hourly rate with the number of hours and is

$$10(2q + 0.001q^2) = 20q + 0.01q^2$$

The total cost is the sum of labor costs and other costs and is

$$C(q) = (20q + 0.01q^2) + (q - 0.001q^2 + 200) = 21q + 0.009q^2 + 200$$

The revenue function is

$$R(q) = 25q$$

Profit is revenue minus cost, so that

$$P(q) = 25q - (21q + 0.009q^2 + 200)$$
$$= 25q - 21q - 0.009q^2 - 200$$
$$= 4q - 0.009q^2 - 200$$

■

Composition of functions

Now suppose that Bouncy's Beanbags has the following pricing formula. The retail price R of a beanbag is \$5 plus 1.1 times the wholesale cost W of the beanbag. Beanbags, Inc., the supplier, charges Bouncy's a \$15 fee for processing an order for Resto beanbags plus \$25 for each beanbag. Express R as a function of the number q of Resto beanbags ordered by Bouncy's.

First, R is given as a function of W by the formula

$$R = 5 + 1.1W$$

Now W is a function of q. The cost of q beanbags is $15 + 25q$. The wholesale cost of each beanbag is the average cost of the q beanbags, so that

$$W = \frac{15 + 25q}{q} = \frac{15}{q} + 25$$

Note that R is a function of W and W is a function of q. R is said to be a *composite function*. To express R directly as a function of q, substitute

$15/q + 25$ for W in the formula for R.

$$R = 5 + 1.1\left(\frac{15}{q} + 25\right)$$

$$= 5 + \frac{16.5}{q} + 27.5 = 32.5 + \frac{16.5}{q}$$

> The *composition* $f \circ g$ of f with g is defined by $(f \circ g)(x) = f(g(x))$. The domain of $f \circ g$ is $\{x \mid x \in \text{domain } g \text{ and } g(x) \in \text{domain } f\}$.

Example 3 Let $f(x) = x^2$, $g(x) = 2x + 3$, $h(x) = 1/x$, and $j(x) = \sqrt{x-9}$.
  a. $(f \circ g)(x) = f(g(x)) = f(2x + 3) = (2x + 3)^2 = 4x^2 + 12x + 9$
  b. $(g \circ f)(x) = g(x^2) = 2x^2 + 3$ (Note that the functions $f \circ g$ and $g \circ f$ are not equal.)
  c. $(h \circ f)(x) = h(x^2) = \dfrac{1}{x^2}$
  d. $(j \circ g)(x) = j(g(x)) = j(2x + 3) = \sqrt{2x - 6}$ The domain of $j \circ g$ is $x \geq 3$.
  e. $(g \circ j)(x) = g(\sqrt{x-9}) = 2\sqrt{x-9} + 3$ The domain of $g \circ j$ is $x \geq 9$.
  f. $(j \circ f)(x) = j(x^2) = \sqrt{x^2 - 9}$ The domain of $j \circ f$ is $x \leq -3$ or $x \geq 3$.
  g. $(f \circ j)(x) = f(\sqrt{x-9}) = x - 9$ The domain of $f \circ j$ is $x \geq 9$, even though any number can be substituted into the formula for $f \circ j$.

■

Change in a function

In calculus the concept of the change in a function will be used many times. Since this concept involves functions, let's introduce it now.

> **Definition.** The *change in f from a to b* (denoted $f(x)\vert_a^b$) is $f(b) - f(a)$.

If the temperature is 53° at 9:00 A.M. and 70° at 2:00 P.M. the change in temperature between 9 and 2 is $70° - 53° = 17°$.

Example 4 If $f(x) = 5x + 4$, the change in f from 3 to 9 is $f(9) - f(3) = 49 - 19 = 30$.

Example 5 $(x^2 + 3x)\vert_{-1}^2 = (4 + 6) - (1 - 3) = 10 - (-2) = 12$.

Example 6 If $g(x) = x^2$, then $g(2x)\big|_1^5 = g(10) - g(2) = 100 - 4 = 96$, and $g(x-2)\big|_1^5 = g(3) - g(-1) = 9 - 1 = 8$.

∎

Exercises 9-4

Let $f(w) = 4w + 2$, $g(w) = 5 - w$, $h(w) = w^2 + 2w$, and $i(w) = w^{1/2}$. Compute each of the following:

1. $f + g$
2. $g - 2h$
3. $3f + 12g$
4. fg
5. hi
6. $\dfrac{h}{i}$
7. $\dfrac{g}{i}$
8. $2h - f$
9. $f \circ g$
10. $g \circ f$
11. $f \circ h$
12. $h \circ f$
13. $f \circ i$
14. $i \circ f$
15. $g \circ i$
16. $i \circ g$
17. $f(x)\big|_0^5$
18. $g(s)\big|_1^4$
19. $h(t)\big|_2^3$
20. $i(w)\big|_4^{25}$
21. $(f - g)(x)\big|_1^7$
22. $(g + h)(t)\big|_0^2$
23. $[i(w) + 2]\big|_1^9$
24. $(fg)(t)\big|_0^1$

25. Juanita's Ceramics makes vases. The material for each vase costs $2.50. If q vases are made in a week, the labor cost per vase is $10.00 - 0.025q$. Weekly overhead is $300. The revenue function is $R(q) = 40q - 0.3q^2$.
    a. Write the cost function.
    b. Write the average cost function.
    c. Write the profit function.
    d. What is the maximum possible profit?
    e. What is the average profit per vase when profit is maximum?
    f. If production is currently 30 vases per week and is increased to 40, how much does profit change?
    g. If production is changed from 40 vases weekly to 50, how much does profit change?
    h. If production is changed from 50 vases weekly to 70, how much does profit change?

In exercises 26 and 27, determine the change in (a) costs, (b) revenue, and (c) profit.

26. Farm Dairy increases production and sales from 400 pints of yogurt a day to 600. The cost function is $C(q) = 50 + 0.40q$ and yogurt sells for 90¢ a pint.

27. The Whole Grain Bakery increases production and sales from 500

loaves of bread a day to 600. The cost function is $C(q) = 100 + 0.45q$ and bread sells for 80¢ a loaf.

*28. Find all quadratic functions f that satisfy the equation $f(x) = (f \circ g)(x)$, where $g(x) = 2 - x$.

*29. Find all linear functions that satisfy the relation $(f \circ f)(x) = x$.

*30. Show that $f(x)|_a^a = 0$. Is it true if $f(x)|_a^b = 0$, then $a = b$?

*31. Show that $f(x)|_a^b = -f(x)|_b^a$. Is it true that if $f(x)|_a^b = -f(x)|_c^d$, then $a = d$ and $b = c$?

Section 9-5 Inverse of a Function

Joe Kleinman has compacted clay soil in his back yard and decides to improve his land by adding topsoil. He calls the Greenplant Garden Center and finds that topsoil costs 5¢ a pound plus a $40 delivery charge. He wants to know how much topsoil he could get for various amounts of money.

Now the cost function is given by $C(q) = 0.05q + 40$, where q is the number of pounds of topsoil. Joe wants to know q in terms of $C(q)$ so he solves this equation for q to obtain

$$q = 20C(q) - 800$$

Joe has expressed quantity as a function of cost. Since cost is now the independent variable, let us call cost c instead of $C(q)$. Then

$$q = 20c - 800$$

What Joe has done is to find the *inverse* of the cost function. The inverse of a function is the relationship that expresses the independent variable in terms of the dependent variable.

Not all functions have inverse functions. For example, if $f(x) = x^2$, x could be either -5 or 5 when $f = 25$. A function is able to have an inverse function when it is one-to-one.

Definition. The function f is one-to-one if $(b, j) \in f$ and $(c, j) \in f$ imply $b = c$. If f is defined by a formula, this condition can be stated:
$$f(b) \neq f(c) \quad \text{if } b \neq c$$

Definition. Let f be a one-to-one function. The function f^{-1} is called the *inverse of* f and is defined by $f^{-1} = \{(b, a) | (a, b) \in f\}$.

If $(b, j) \in f^{-1}$ and $(b, k) \in f^{-1}$, then $(j, b) \in f$ and $(k, b) \in f$. Since f is one-to-one, $j = k$. This shows that f^{-1} is a function. If f is not one-to-one, f does not have an inverse function.

The domain of f^{-1} is the range of f. The range of f^{-1} is the domain of f.

Example 1 If $f = \{(1, 3), (2, 5), (3, 9), (4, 2)\}$, then $f^{-1} = \{(3, 1), (5, 2), (9, 3), (2, 4)\}$.

Example 2 If $g = \{(1, 5), (2, -3), (4, 5), (7, 1)\}$, then g is not one-to-one, because $g(1) = g(4) = 5$. Thus g does not have an inverse function.

∎

If the function $y = f(x)$ is defined by a formula, f will have an inverse if the equation $y = f(x)$ can be solved for x and gives only one value for x.

Example 3 Find the inverse of the function $y = f(x) = 3x + 5$.

Solution Solve the formula for x in terms of y.

$$y = 3x + 5$$
$$y - 5 = 3x$$
$$\frac{y-5}{3} = x \quad \text{so} \quad x = \frac{y-5}{3}$$

Thus $f^{-1}(y) = (y - 5)/3$. It is customary to write f and f^{-1} as functions of the same variable. Thus the formula for f^{-1} can be written

$$f^{-1}(x) = \frac{x-5}{3}$$

Example 4 Find the inverse of the function $w = v^2 + 4v + 3$.

Solution Solve this equation for v. First write the formula as $v^2 + 4v + (3 - w) = 0$.

$$v = \frac{-4 \pm \sqrt{4^2 - 4(1)(3-w)}}{2(1)} \qquad \text{Use the quadratic formula.}$$

$$= \frac{-4 \pm \sqrt{4 + 4w}}{2} = -2 \pm \sqrt{1 + w}$$

Since there can be two values of v for one value of w, the function is

not one-to-one and has no inverse.
■

Several useful facts about the relation between a function and its inverse are contained in the following theorems, which are proved in more advanced courses.

Theorem. If f is a one-to-one function,
1. f^{-1} is a one-to-one function, and $(f^{-1})^{-1} = f$.
2. $(f^{-1} \circ f)(x) = x$ whenever $x \in$ domain f.
3. $(f \circ f^{-1})(x) = x$ whenever $x \in$ domain f^{-1}.

Theorem. If f and g are functions, $g = f^{-1}$ if and only if $(g \circ f)(x) = x$ for all x in domain f and $(f \circ g)(x) = x$ for all x in domain g.

If g is the inverse of f, f is the inverse of g. Thus we can say that f and g are inverses of each other.

Example 5 Are $g(x) = x^3 + 1$ and $f(x) = x^{1/3} - 1$ inverses of each other?

Solution Since $(f \circ g)(x) = f(x^3 + 1) = (x^3 + 1)^{1/3} - 1 \neq x$, the answer is no.

Example 6 Are $g(t) = t^2$ and $h(t) = \sqrt{t}$ inverses of each other?

Solution $(g \circ h)(t) = g(\sqrt{t}) = t$. However, $(h \circ g)(t) = h(t^2) = |t|$. In particular, $(h \circ g)(-2) = 2$. Thus g and h are not inverses of each other.

Example 7 Are $g(t) = t^3 - 7$ and $h(t) = (t + 7)^{1/3}$ inverses of each other?

Solution
$$(g \circ h)(t) = g(h(t)) = g\big[(t+7)^{1/3}\big]$$
$$= \big[(t+7)^{1/3}\big]^3 - 7 = (t+7) - 7 = t$$

and
$$(h \circ g)(t) = h(g(t)) = h(t^3 - 7)$$
$$= (t^3 - 7 + 7)^{1/3} = (t^3)^{1/3} = t$$

Thus g and h are inverses of each other.
■

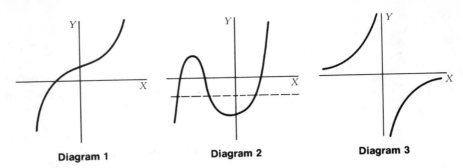

Diagram 1 Diagram 2 Diagram 3

We can also use graphs to decide if a function has an inverse. This is useful, because some functions are given by complicated formulas, and solving the equation $y = f(x)$ for x may be difficult or even impossible. A function f is one-to-one if and only if no horizontal line meets the graph of f in more than one point. By this test, the functions in Diagrams 1 and 3 have inverses but the function in Diagram 2 has no inverse.

By the graphical test, any linear function, other than a constant function, has an inverse. No quadratic function has an inverse.

Sometimes, a function has no inverse, but can have an inverse if its domain is made smaller. For instance, the function $f(x) = x^2$ has no inverse. Let $g(x) = x^2$, domain $g = \{x \mid x \geq 0\}$. The function g has an inverse: $g^{-1}(x) = \sqrt{x}$. Here we've restricted the domain of f to a set on which f is one-to-one; the resulting function g has an inverse.

If a function h has an inverse, the graph of h^{-1} can be obtained very easily from the graph of h. The points (a, b) and (b, a) are mirror images of each other in a mirror placed on the line $y = x$. The graph of h^{-1} is obtained by reflecting the graph of h in the line $y = x$. See Diagram 4.

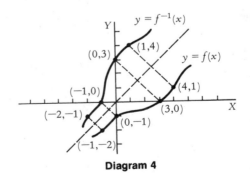

Diagram 4

Exercises 9-5

Determine whether or not each of the following functions has an inverse. If a function has an inverse, write either the inverse function or a formula for the inverse function.

1. $\{(1,4),(1,4),(2,3),(3,2),(4,4)\}$
2. $\{(2,3),(1,3),(3,1)\}$
3. $\{(1,5),(-3,6),(-4,7),(0,0)\}$
4. $\{(0,0),(4,1),(-3,-6),(-2,1)\}$
5. $\{(0,0),(1,1),(5,5),(7,9)\}$
6. $\{(1,2),(2,3),(3,5),(5,4)\}$
7. $2x + 3y = 6$
8. $3x - 4y = 12$
9. $y = 2x + 3$
10. $y = 2x - 5$
11. $\dfrac{x}{4} - \dfrac{y}{5} = 20$
12. $\dfrac{x}{3} + \dfrac{y}{4} = 1$
13. $y = x^2 - x$
14. $y = 2x^2 + x$
15. $y = x^3$
16. $y = x^3 + 2$
17. $y = \dfrac{1}{x}$
18. $y = \dfrac{1}{x^2}$
19. $y = x^{2/3}$
20. $y = x^{4/3}$
21. $y = x^{-1/3} + 4$
22. $y = (x + 4)^{-1/3}$
23. $y = x^{3/2}$
24. $y = x^{5/2}$

Is the function equal to its inverse?

25. $y = \dfrac{2}{x}$
26. $w = \dfrac{-3}{v}$
27. $t = s - 2$
28. $t = 2 - s$
29. $w = 4 - v$
30. $w = 4 - 3v$

31. The equation $P = 0.04S - 50{,}000$ expresses monthly profit at Quigby's as a function of sales. Management wants to know what sales levels are needed in order to have various amounts of profit. Find a formula to aid them.

32. The demand function for Ace record changers is $d(q) = 400 - 0.05q$. Find the inverse function, and use it to determine how many record changers would be sold at the price of $325.

33. The demand function for milk shakes at Elm Center Dairy is $d(q) = 1.00 - 0.0002q$. Find the inverse function, and use it to determine how many milk shakes will be sold if the price is 80¢.

34. The supply function for a new antibiotic is $s(q) = 6.00 + 0.001q$. Here q is the number of bottles of pills. Find the inverse function, and use it to determine how many bottles of pills will be supplied at the price of $7.50 per bottle.

35. John Ito earns $8 an hour as a mechanic and receives time-and-a-half for overtime. He works a 40-hour week in addition to occasional overtime. Express his weekly earnings W as a function f of the number H of hours of overtime he puts in. Also express H as a function g of W.

What is the relationship between f and g?

Crickets chirp more often in warmer temperatures than in cooler temperatures. The number n of times a cricket chirps per minute is expressed as a function of the temperature t (in Fahrenheit degrees) by the equation

$$n = f(t) = 4t - 160$$

36. Express t as $f^{-1}(n)$.
37. If a thermometer is not available, the temperature can be estimated by counting a cricket's chirps. What is the temperature if a cricket chirps 120 times a minute?

Assume f and g are functions and each has all real numbers as domain and range. Assume f and g both have inverses. Write a formula for the inverse of each of the following functions.

*38. $f \circ f$ *39. $5f$
*40. $f \circ g$ *41. $g \circ f$
*42. $f^{-1} \circ g$

Section 9-6   The Exponential Function

We now study an important class of functions called exponential functions. These functions have numerous applications to situations involving growth and decay.

For example, the amount of money in an account earning continuously compounded interest and the mass of a sample of radioactive material are both given by exponential functions of time.

> **Definition.** Let b be any positive number except 1. The *exponential function with base b* is the function $f(x) = b^x$.

The graph of $y = 2^x$ is shown in Diagram 1. Note that the domain consists of all real numbers and the range consists of all positive numbers. As x increases, 2^x also increases. If $b > 1$, the graph of $y = b^x$ resembles, but is not identical to, the graph of $y = 2^x$.

In Diagram 2, the graph of $y = (1/2)^x$ is shown. Note that $y = (1/2)^x$ is the same as $y = 2^{-x}$. The domain is all real numbers and the range is all positive numbers. However, $(1/2)^x$ gets smaller as x gets bigger. If $0 < b < 1$,

9-6 THE EXPONENTIAL FUNCTION

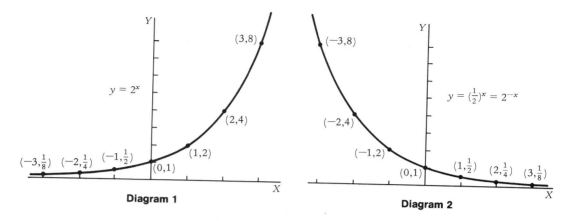

Diagram 1

Diagram 2

the graph of $y = b^x$ resembles, but is not identical to, the graph of $y = (1/2)^x$.

The two graphs just discussed are congruent to each other. If you reflect either of the graphs about the y-axis, you obtain the other graph.

We will now look at some specific examples of growth and decay involving the exponential functions 2^x and 2^{-x}.

Example 1 The population of the new city of Mountainview is growing rapidly and is expected to grow according to the formula $P(t) = 10{,}000 \cdot 2^{t/8}$. What is the population expected to be after 24 years?

Solution $P(24) = 10{,}000 \cdot 2^{24/8} = 10{,}000 \cdot 2^3 = 80{,}000$. The population is expected to be 80,000 after 24 years.

Example 2 The half-life of the radioactive isotope carbon 14 is 5730 years. This means that half of any sample of carbon 14 decays to other substances within 5730 years. Thus, if an amount M_0 is present initially, the amount $M(t)$ left after t years is given by the formula

$$M(t) = M_0 2^{-t/5730}$$

If 500 grams are present initially, how much is left after 2865 years?

Solution

$$M(2865) = 500 \cdot 2^{-2865/5730} = 500 \cdot 2^{-1/2} = 354$$

After 2865 years, 354 grams of carbon 14 are left.

The data of an ancient object can be determined by measuring the amount of carbon 14 in it. How this is done will be discussed in the next section. ∎

There are infinitely many different exponential functions. The exponential function with *base e* is of great importance since it is used most often. The number *e* will be defined later in this section and is an irrational number. The value of *e* is approximately 2.7.

> **Definition.** The *exponential function* is the function $f(x) = e^x$. This function is also denoted $f(x) = \exp(x)$.

The exponential function is convenient for several reasons. We will see in calculus that many formulas are much simpler for $y = e^x$ than for $y = b^x$ with $b \neq e$. For various technical reasons, calculators and computers can evaluate e^x much more easily than b^x. In fact, these machines usually use the exponential function to evaluate b^x.

The graphs of $y = \exp(x)$ and $y = \exp(-x)$ are shown in Diagrams 3 and 4.

To define *e*, let's consider compound interest and continuously compounded interest. In Section 5-1, it was shown that if $1 is left in an account paying $r\%$ interest compounded n times annually, the amount $A(n)$ in the account after one year is given by

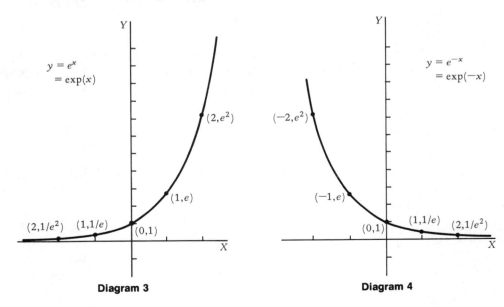

Diagram 3   Diagram 4

$$A(n) = \left(1 + \frac{r}{n}\right)^n$$

As n gets bigger, interest is compounded more frequently and $A(n)$ increases. $A(n)$ does not increase without limit however, and in fact it can be shown that as n keeps increasing, the difference between $\exp(r)$ and $A(n)$ gets smaller and smaller. This difference can be made as small as desired by making n big enough. This is symbolized by writing

$$\lim_{n \to \infty} \left(1 + \frac{r}{n}\right)^n = \exp(r)$$

The statement is read: "The limit as n approaches infinity of $[1 + (r/n)]^n$ equals $\exp(r)$."

If $r = 1$, then $\exp(r) = \exp(1) = e^1 = e$ and the definition of e is obtained

Definition. The number e is given by the formula
$$e = \lim_{n \to \infty} \left(1 + \frac{1}{n}\right)^n$$

Continuously compounded interest

Recall from Section 5-1 that interest is *compounded continuously* if it is compounded at every instant. The formula for continuously compounded interest is:

$$A = P \exp(rt) \quad \text{Continuously compounded interest}$$

In this formula, A is the amount, P is the principal, r is the nominal interest rate, and t is time measured in years. The following examples involve continuously compounded interest.

Example 3 An account pays 8% interest compounded continuously. If the account contains $200 now, how much will it contain after three years?

Solution $A = 200 \cdot \exp[(0.08)(3)] = 200 \cdot \exp(0.24) = \$254.25$. The account will contain $254.25 after three years.

Example 4 An account containing $500 pays 9% interest compounded continuously. How much is in the account after 21 months?

Solution First, 21 months equals $21/12 = 1.75$ years. Then
$$A = 500 \cdot \exp[(0.09)(1.75)] = 500 \cdot \exp(0.1575) = \$585.29$$
The account contains $585.29 after 21 months. ■

In Section 1-3, algebraic manipulations involving exponents were discussed. You should review that section now. For convenience, some properties of exponents are listed here.

1. $e^v e^w = e^{v+w}$
2. $\dfrac{e^v}{e^w} = e^{v-w}$
3. $e^{-v} = \dfrac{1}{e^v}$
4. $(e^v)^w = e^{vw}$

Exercises 9-6

Determine each of the following using a calculator.

1. e
2. e^2
3. e^{-20}
4. $e^{1.5}$
5. $e^{2.3}$
6. e^{47}
7. e^{35}
8. e^{-12}

Express each of the following in the form exp( ).

9. $\dfrac{e^3(e^5)^2}{e^{-4}}$
10. $\left(\dfrac{e^2}{e^4}\right) \div \left(\dfrac{e^3}{e^{-2}}\right)^2$
11. $e^{x+y} \cdot e^{x-y}$
12. $(e^{-x})^2 \cdot \exp(-x^2)$
13. $\dfrac{\exp(x+3w)}{\exp(x-2w)}$
14. $\dfrac{\exp[(x+1)^2]}{[\exp(x+1)]^2}$
15. $(\exp 5)^x \cdot \exp(x^3)$
16. $\dfrac{\exp(1)}{\exp(-x^2)}$

Graph the following functions by plotting a few carefully chosen points and connecting them.

17. $f(x) = \exp\left(\dfrac{x}{2}\right)$

18. $f(x) = e - \exp(x)$

19. $f(x) = \exp\left(\dfrac{x}{3}\right) - e^2$

20. $f(x) = \exp\left(\dfrac{x}{2}\right) - e$

21. A bacterial culture is being grown for use in research on antibiotics. The mass of the culture in grams after t hours is given by the formula

$$m(t) = 0.0001 \exp\left(\dfrac{t}{4}\right), \quad 0 \le t \le 48$$

What is the mass of the culture after two hours? After 24 hours?

22. A bacterial culture weighs 10.0 grams. An antibiotic is added to the culture and the weight of the culture after t hours is

$$m(t) = 10.0 \exp\left(\dfrac{-t}{5}\right), \quad 0 \le t \le 20$$

What is the weight of the culture after one hour? After eight hours?

If interest is compounded continuously at the nominal rate r, the effective interest rate r_E is given by the formula

$$r_E = \exp(r) - 1$$

23. What is the effective interest rate for 8% interest compounded continuously? For 12%?

24. Bill McGee is going to open a savings account and expects to leave the money there for two years. Which bank should he use, if Security Federal offers 9% compounded annually and East River Bank offers 8.75% compounded continuously?

25. A typical experienced worker on TIK's assembly line can process 1000 components a day. Experimentation has shown that on day t at the job, a new worker can usually process $1000(1 - e^{-0.2t})$ components.
    a. How many components can a new worker usually process on the fifth day at the job?
    b. How many components can a new worker usually process on the twelfth day at the job?
    c. After 40 days of work, should a new worker be considered experienced?

26. The population P of a country is increasing according to the formula

$$P = (3.4 \times 10^6) \times \exp\left[0.02(Y - 1982)\right]$$

where Y is the year. What will the country's population be in the year 2000? In the year 2050?

A rumor is initially known to the proportion P_0 of the citizens of Chelm and spreads according to the formula

$$P(t) = \frac{P_0 \exp(0.02t)}{1 + P_0[\exp(0.02t) - 1]}$$

Here t is the number of hours since the rumor starts and $P(t)$ is the proportion of the citizens of Chelm who have heard the rumor after t hours. What proportion of the citizens of Chelm have heard the rumor after one day if P_0 equals

  27. 0.01?   28. 0.02?
  29. 0.05?   30. 0.10?

Section 9-7 The Natural Logarithm Function

The function $y = e^x$ has as its domain all real numbers, as its range $\{x \mid x > 0\}$, and is one-to-one. Thus the exponential function has an inverse. The *natural logarithm function* is the inverse of the exponential function. The natural logarithm of x is written as $\ln x$.

The function $y = \ln x$ has as its domain $\{x \mid x > 0\}$ and as its range the set of all real numbers. In particular, $\ln 0$ is undefined, and $\ln x$ is undefined if $x < 0$. The graph of $y = \ln x$ is obtained by reflecting the graph of $y = e^x$ about the line $y = x$. See Diagram 1.

From the graph, we see that $\ln x < 0$ if $0 < x < 1$, $\ln x = 0$ if $x = 1$, and $\ln x > 0$ if $x > 1$, and that $\ln b > \ln c$ if $b > c > 0$.

Since the natural logarithm function and the exponential function are inverses of each other, the following identities hold.

$$\ln\left[\exp(x)\right] = x \quad x \text{ any real number}$$
$$\exp\left[\ln(x)\right] = x \quad x > 0$$

The function ln has the following properties:

1. $\ln(bc) = \ln b + \ln c$

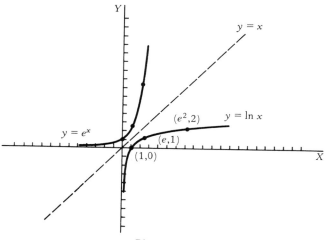

Diagram 1

2. $\ln\left(\dfrac{b}{c}\right) = \ln b - \ln c$

3. $\ln\left(\dfrac{1}{c}\right) = -\ln c$

4. $\ln(b^r) = r \ln b$

To derive property 1, apply the exponential function to both sides of the equation. Now

$$\exp\left[\ln(bc)\right] = bc \quad \text{because ln and exp are inverses of each other}$$

and

$$\exp\left[\ln b + \ln c\right] = \exp(\ln b) \cdot \exp(\ln c)$$
$$= bc$$

Recall that the exponential function is one-to-one. Since the exponential function gives the same value when applied to $\ln(bc)$ and to $\ln b + \ln c$, these quantities must be equal. The other properties can be derived in similar ways. The following examples illustrate these properties.

Examples
1. $\ln 10 + \ln 2 = \ln(10 \cdot 2) = \ln 20$
2. $\ln 24 - \ln 6 = \ln\left(\dfrac{24}{6}\right) = \ln 4$
3. $\ln\left(\dfrac{17}{e^2}\right) = \ln 17 - \ln(e^2) = \ln(17) - 2$

4. $\ln(5e) = \ln 5 + \ln e = \ln(5) + 1$

5. $\ln x \big|_3^{15} = \ln 15 - \ln 3 = \ln 5$

6. $\ln 8 = \ln(2^3) = 3 \ln 2$

7. $\ln\left(\dfrac{1}{\sqrt{3}}\right) = \ln(3^{-1/2}) = -\left(\dfrac{1}{2}\right)\ln 3$

8. $4 \ln x \big|_2^6 = 4 \ln 6 - 4 \ln 2 = 4(\ln 6 - \ln 2)$
$= 4 \ln 3 = \ln(3^4) = \ln 81$

9. $\ln\left[\dfrac{(x+1)^2}{(x-3)}\right] = \ln[(x+1)^2] - \ln(x-3) = 2\ln(x+1) - \ln(x-3)$

10. $\left(\dfrac{1}{3}\right)\ln\left[\dfrac{(4x-3)^6}{(2x+1)^3}\right] = \dfrac{1}{3}\ln[(4x-3)^6] - \dfrac{1}{3}\ln[(2x+1)^3]$
$= 2\ln(4x-3) - \ln(2x+1)$

11. $\left(\dfrac{1}{2}\right)\ln x \big|_3^{48} = \left(\dfrac{1}{2}\right)\ln 48 - \left(\dfrac{1}{2}\right)\ln 3 = \left(\dfrac{1}{2}\right)(\ln 48 - \ln 3)$
$= \left(\dfrac{1}{2}\right)\ln 16 = \ln(16^{1/2}) = \ln 4$

12. $[\ln(x+3) - 2\ln(x-4)]\big|_5^{10} = (\ln 13 - 2 \ln 6) - (\ln 8 - 2 \ln 1)$
$= \ln 13 - \ln 36 - \ln 8$
$= \ln\left[\dfrac{13}{(36 \cdot 8)}\right] = \ln\left(\dfrac{13}{288}\right)$

13. $\exp(2 \ln 3) = \exp(\ln 9) = 9$

■

The natural logarithm function is used in finding the inverse of composite functions that involve the exponential function. For example, to find the inverse of the function $y = h(t) = 3\exp(t+4)$, solve for t as follows:

$$y = 3\exp(t+4)$$

$$\dfrac{y}{3} = \exp(t+4)$$

$$\ln\left(\dfrac{y}{3}\right) = \ln[\exp(t+4)] = t + 4$$

$$\ln\left(\dfrac{y}{3}\right) - 4 = t$$

$$t = \ln\left(\dfrac{y}{3}\right) - 4$$

Thus $h^{-1}(y) = \ln(y/3) - 4$ and $h^{-1}(t) = \ln(t/3) - 4$.

Similarly the exponential function is used in finding the inverse of a composite function that involves ln.

Example 14 Find the inverse of $w = g(v) = 2\ln(5v + 1)$.

Solution

$$w = 2\ln(5v + 1) \quad \text{Solve this equation for } v.$$

$$\frac{w}{2} = \ln(5v + 1)$$

$$\exp\left(\frac{w}{2}\right) = \exp\left[\ln(5v + 1)\right] = 5v + 1$$

$$\exp\left(\frac{w}{2}\right) - 1 = 5v$$

$$v = \frac{\left[\exp\left(\frac{w}{2}\right) - 1\right]}{5}$$

Thus

$$g^{-1}(v) = \frac{\left[\exp\left(\frac{v}{2}\right) - 1\right]}{5}$$

■

We now study applications of the natural logarithm function. The first application is to compute b^c. The formula to be used is

$$b^c = \left(e^{\ln b}\right)^c = e^{c \ln b}$$

With this formula, any positive number to any power can be computed using the functions exp and ln. This is in fact how most computers and calculators evaluate b^c. If your calculator does not have a special button for b^c, you can compute this quantity using the functions exp and ln.

Example 15 $10^{4.3} = \exp(4.3 \ln 10) = \exp((4.3) \cdot (2.3026)) = \exp(9.90112)$
$= 2.00 \times 10^4$

Example 16 $2^{100} = \exp(100 \ln 2) = \exp(100 \cdot (0.69315)) = \exp(69.315)$
$= 1.27 \times 10^{30}$

■

The next application is to *carbon 14 dating*. The ratio of radioactive carbon 14 to nonradioactive carbon 12 is almost constant in the atmosphere and in living things. In nonliving things, such as firewood and paper, the carbon 14 gradually decays to nitrogen while the carbon 12 remains carbon 12. To date an object, scientists measure the amount of carbon 12 and of carbon 14 in it and determine what proportion P of the original carbon 14 is there. P obeys the equation

$$P = 2^{-t/5730}$$

where t is the age of the object. For example, suppose an ancient parchment is found and has 60% of its original carbon 14. Then

$$0.6 = 2^{-t/5730}$$

$$\ln 0.6 = \frac{-t}{5730} \ln 2 \qquad \text{Take the ln of both sides.}$$

$$t = \frac{-5730 \ln 0.6}{\ln 2}$$

$$= \frac{-5730(-0.51083)}{0.69315} = 4223$$

The parchment is about 4200 years old.

Example 17 Some ashes are found in a pot at an ancient campsite. The ashes contain 10% of their original carbon 14. How long ago was the campsite used?

Solution The proportion P of carbon 14 remaining satisfies the equation $P = 2^{-t/5730}$. Thus

$$0.1 = 2^{-t/5730}$$

$$\ln 0.1 = \frac{-t}{5730} \ln 2 \qquad \text{Take the ln of both sides.}$$

$$t = \frac{-5730 \ln 0.1}{\ln 2}$$

$$= \frac{-5730(-2.3026)}{0.69315} = 19035$$

The campsite is about 19,000 years old.
∎

This last application involved exponential decay. The natural logarithm has applications to many growth and decay processes. For example, suppose $100 is deposited in an account paying 11% interest compounded continuously. When will the account contain $350?

To solve this problem, use the formula for continuously compounded interest to obtain the equation

$$350 = 100 \exp(0.11t)$$

Now solve this equation for t.

$$\frac{350}{100} = 3.5 = \exp(0.11t)$$

$$\ln 3.5 = \ln[\exp(0.11t)] = 0.11t$$

$$t = \frac{\ln 3.5}{0.11} = \frac{1.2528}{0.11} = 11.4$$

The account will contain $350 after 11.4 years.

Example 18 An account paying 8% interest compounded continuously is opened with a deposit of $250. When will the account contain $800?

Solution Use the formula for continuously compounded interest to obtain $800 = 250 \exp(0.08t)$.

$$\frac{800}{250} = \exp(0.08t)$$

$$\ln\left(\frac{800}{250}\right) = \ln[\exp(0.08t)] \qquad \text{Take the ln of both sides.}$$

$$\ln\left(\frac{800}{250}\right) = 0.08t \qquad \text{The inverse of the function exp is ln.}$$

$$\frac{1}{0.08}\ln\left(\frac{800}{250}\right) = t$$

$$t = \frac{1}{0.08}(1.1632) = 14.5$$

The account will contain $800 after about 14.5 years.
■

Exercises 9-7

Express each of the following as the natural logarithm of one number or expression.

1. $\ln 12 - \ln 3$
2. $\ln 15 + \ln 4$
3. $\ln\left(\dfrac{2}{3}\right) + \ln 5$
4. $\ln\left(\dfrac{3}{7}\right) - \ln 9$
5. $4 \ln 2$
6. $-\left(\dfrac{1}{2}\right) \ln 36$
7. $2 \ln 3 - 3 \ln 5$
8. $3 \ln 4 - \left(\dfrac{1}{2}\right) \ln 49$
9. $\ln(x+1) - \ln(x-2)$
10. $\ln(x+3) - \ln(2x+1)$
11. $2 \ln x - 3 \ln y$
12. $3 \ln b + \left(\dfrac{4}{3}\right) \ln c$
13. $2 \ln x - \ln y - 4 \ln z$
14. $3 \ln v + 2 \ln w - 8 \ln v$
15. $2 + \ln 5$
16. $-3 + \ln 8$
17. $\ln x \,|_5^{35}$
18. $\ln(x+3)\,|_0^5$
19. $[\ln(2x+1) + \ln(x+5)]\,|_0^5$
20. $[2\ln(x+1) - \ln(x-1)]\,|_3^9$

Determine each of the following using a calculator.

21. $\ln 2$
22. $\ln 10$
23. $\ln(0.00001)$
24. $\ln(7^{4.3})$
25. $\ln(10^{25})$
26. $\ln(8^{-60})$
27. $3^{3/2}$
28. $\left(\dfrac{1}{2}\right)^{10.5}$
29. 3^{70}
30. $40^{2/3}$
31. $20^{1/4}$
32. $(0.0037)^{1.5}$

Determine the inverse of each function.

33. $\ln(x-1)$
34. $2\ln(x+1)$
35. $4\ln(5-2x)$
36. $3\ln(x+6)$
37. $-\exp(x+3)$
38. $\exp(5-2x)$
39. $4\exp(2x+1)$
40. $-\exp(4-8x)$

Graph each of the following functions by plotting a few carefully chosen points and connecting them.

41. $f(x) = \ln(2x)$
42. $f(x) = \ln(1+x)$
43. $f(x) = \ln(x-1)$
44. $f(x) = 1 - \ln x$

45. An accident contaminates a nuclear reactor. The radiation level inside the reactor dome is currently 1000 units and has a half-life of 400 years.

The reactor can be entered for repairs only after the radiation level has fallen to 50 units. How long will this take?

46. A scroll is found and is believed to contain a message from Julius Caesar to Rome. The scroll contains 85% of its original carbon 14. Could Caesar have ever used this scroll?

47. The mass of a bacterial sample is increasing according to the formula $m(t) = 50 \exp(0.03t)$. Here mass is measured in grams and t in hours. When will the sample weigh 200 g?

48. The population of a city is now 350,000 and is doubling every eight years. How long will it take for the population to reach a million?

49. Bill McGraw deposits $3000 in an account paying 9% interest compounded continuously. How long will it be until the account contains $10,000 for the down payment on a house?

50. The supply function for hand carved walnut music boxes is $s(q) = 40 + \ln(100 + q)$.

   a. At what price would 900 of these music boxes be supplied?
   b. How much would the price rise if 1500 were supplied instead of 900?

Derive each of the following formulas.

*51. $\ln\left(\dfrac{b}{c}\right) = \ln b - \ln c$

*52. $\ln\left(\dfrac{1}{c}\right) = -\ln c$

*53. $\ln(b^r) = r \ln b$

*54. $\ln x \big|_a^b = \ln\left(\dfrac{1}{x}\right)\bigg|_b^a$

Section 9-8 The Common Logarithm Function

The exponential function with base 10, $y = 10^x$, is one-to-one. This function therefore has an inverse, called the *common logarithm function*. The common logarithm of x is written as $\log x$. The graphs of $y = 10^x$ and $y = \log x$ are shown in Diagram 1.

Since the functions $y = \log x$ and $y = 10^x$ are inverses of each other, the following identities hold.

$$\log(10^x) = x \qquad x \text{ any real number}$$
$$10^{\log x} = x \qquad x > 0$$

For example,

$$\log(10^2) = 2 \qquad \log(10^{-3}) = -3$$
$$10^{\log 4} = 4 \qquad 10^{\log 0.037} = 0.037$$

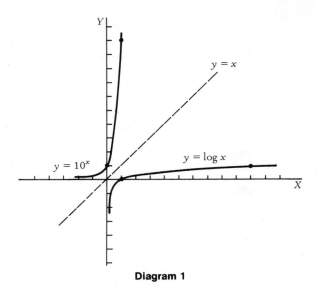

Diagram 1

The function log is closely related to the natural logarithm function. To derive the relationship, apply the function ln to both sides of the equation $10^{\log x} = x$ to obtain

$$\ln\left(10^{\log x}\right) = \ln x$$

$$\log x \cdot \ln 10 = \ln x \qquad \text{and divide by } \ln 10 \text{ to obtain}$$

$$\log x = \frac{\ln x}{\ln 10}$$

Using this equation, the function log can easily be shown to have the following properties, which are similar to properties of the function ln.

1. $\log(bc) = \log b + \log c$
2. $\log\left(\dfrac{b}{c}\right) = \log b - \log c$
3. $\log\left(\dfrac{1}{c}\right) = -\log c$
4. $\log(b^r) = r \log b$

The following examples illustrate these properties.

Examples 1. $\log 5 + \log 7 = \log(5 \cdot 7) = \log 35$

2. $\log 75 - \log 3 = \log\left(\dfrac{75}{3}\right) = \log 25$

3. $\log\left(\dfrac{1}{4}\right) = -\log 4$

4. $\log(2^{1/3}) = \dfrac{1}{3}\log 2$

5. $\log(10x^2) = \log 10 + \log(x^2) = 1 + 2\log x$ ∎

Until the advent of calculators, the common logarithm function was often used to make arithmetical computations easier. Although this is no longer done, there still are places where the function log is used. An example is in determining the loudness of a sound.

If L is the loudness with which a sound is perceived and I is the physical intensity of the sound, psychologists have shown that

$$L = 10\log\left(\dfrac{I}{I_0}\right)$$

where I_0 is a constant and is the lowest intensity sound that can be heard. Here I and I_0 are measured in watts per square meter and L is measured in *decibels*.

Example 5 A whisper has a loudness of 10 decibels and an ordinary conversation has a loudness of 40 decibels. What is the ratio of their sound intensities?

Solution Let I_w be the intensity of the whisper and I_c be the intensity of the conversation. Then

$$40 = 10\log\left(\dfrac{I_c}{I_0}\right)$$

$$10 = 10\log\left(\dfrac{I_w}{I_0}\right)$$

Subtract the second equation from the first to obtain

$$30 = 10\log\left(\dfrac{I_c}{I_0}\right) - 10\log\left(\dfrac{I_w}{I_0}\right)$$

$$= 10\left[\log\left(\dfrac{I_c}{I_0}\right) - \log\left(\dfrac{I_w}{I_0}\right)\right]$$

$$= 10\log\left(\frac{I_c}{I_0} \div \frac{I_w}{I_0}\right)$$

$$= 10\log\left(\frac{I_c}{I_0} \cdot \frac{I_0}{I_w}\right)$$

$$= 10\log\left(\frac{I_c}{I_w}\right)$$

Since

$$30 = 10\log\left(\frac{I_c}{I_w}\right)$$

then

$$3 = \log\left(\frac{I_c}{I_w}\right)$$

Apply the function 10^x to both sides. Then

$$10^3 = 10^{\log(I_c/I_w)}$$

$$1000 = \frac{I_c}{I_w}$$

Thus the conversation has 1000 times the sound intensity of the whisper.
∎

Exercises 9-8

Express each of the following as the common logarithm of one number or expression.

1. $\log 5 + \log 23$
2. $\log 9 + \log 7$
3. $\log\left(\frac{3}{4}\right) - \log 6$
4. $\log\left(\frac{5}{7}\right) - \log 21$
5. $3\log 4$
6. $\frac{1}{2}\log 64$
7. $\log(x^3 y) - \log(xy^2)$
8. $\log(4xy) + \log\left(\frac{6x}{y}\right)$

The rustle of leaves in a forest has a loudness of 10 decibels, a quiet automobile 40 decibels, a telephone conversation 75 decibels, and thunder 120 decibels. Determine the ratio of the sound intensities of

9. Thunder and a telephone conversation.
10. Thunder and a quiet automobile.
11. Thunder and the rustle of leaves.
12. A telephone conversation and a quiet automobile.
13. A telephone conversation and the rustle of leaves.
14. A quiet automobile and the rustle of leaves.

Acidity or alkalinity is described by chemists using a number called the pH, defined by

$$\text{pH} = -\log[\text{H}^+]$$

Here $[\text{H}^+]$ is the concentration, in moles/liter, of the hydrogen ion H^+. A pH below 7 corresponds to an acid, while a pH greater than 7 corresponds to an alkali or base. Determine the pH of each of the following.

15. Distilled water, $[\text{H}^+] = 10^{-7}$. (Distilled water is said to be neutral.)
16. Human blood, $[\text{H}^+] = 4 \times 10^{-8}$.
17. Acid rain, $[\text{H}^+] = 2.1 \times 10^{-5}$.
18. Sea water, $[\text{H}^+] = 6 \times 10^{-9}$.
19. Oranges, $[\text{H}^+] = 0.0003$.
20. Limes, $[\text{H}^+] = 0.013$.

Derive each of the following formulas.

*21. $\log(bc) = \log b + \log c$
*22. $\log\left(\dfrac{b}{c}\right) = \log b - \log c$
*23. $\log\left(\dfrac{1}{c}\right) = -\log c$
*24. $\log(b^r) = r \log b$

Chapter 9 Review

1. Is $\{(3,5),(4,6),(6,3),(7,2),(5,6)\}$ a function? What are its domain and range?
2. Determine the domain of $f(x) = \sqrt{5 - 3x}$.
3. Determine the domain of $g(w) = \dfrac{w^2 - 1}{w^2 - 9}$.
4. Let $f(w) = w^2 - 3$ and $g(w) = 4w + 1$. Compute
   a. $f(g(w))$
   b. $g(f(w))$
   c. $f(1 + x^2)$
   d. $g(3 - v)$

5. Does the equation $x^2y - 3x = 5y$ define y as a function of x? If so, solve for y.
6. Does the equation $xy + 3y^2 = 15$ define y as a function of x? If so, solve for y.
7. Does the equation $x^3 + y^3 = 27$ define y as a function of x? If so, solve for y.
8. Write the equation of the line through $(2, -3)$ with slope 7 in
   a. Point-slope form   b. Slope-intercept form
9. Write the equation of the line through $(9, 2)$ and $(6, 8)$ in
   a. Point-slope form   b. Slope-intercept form
10. Determine the slope and y-intercept of the line with equation $5x - 2y = 40$.
11. Draw the graph of $y = x^2 + 4x - 5$. Label the vertex and all intercepts.
12. Draw the graph of $y = -x^2 + 3x + 4$. Label the vertex and all intercepts.

Let $f(v) = 3v + 2$ and $g(v) = v^2 + 8v$. Compute each of the following.

13. $(f + g)(v)$
14. $(2f - 4g)(v)$
15. $(fg)(v)$
16. $(f \circ g)(v)$
17. $(g \circ f)(v)$
18. $\left(\dfrac{f}{g}\right)(v)$
19. $g(v)\big|_2^3$
20. $f^{-1}(v)$

Determine whether or not each function has an inverse. When there is an inverse function, write a formula for it.

21. $3x - 5y = 30$
22. $4x^2 + y = 8$
23. $y = 2\ln(x + 3)$
24. $y = \exp(2x + 19)$

Express each of the following in the form $\exp(\ \ )$.

25. $\dfrac{e^{3x-2}}{e^{4-x}} \cdot e^x$
26. $\dfrac{\exp(x^2 + 3) \cdot \exp(x)}{\exp(3x + 3)}$

Express each of the following in the form $\ln(\ \ )$.

27. $\ln 36 - \ln 3$
28. $4\ln 2 - \dfrac{2}{3}\ln 8$
29. $\ln(x^2 - 9) - \ln(x^2 - 6x + 9)$
30. $0.5\ln(3x + 4)\big|_0^7$

Express each of the following in the form $\log(\ \ )$.

31. $\log 72 - 3\log 2$
32. $2\log(xy^3) - 3\log(x^{1/2}y)$

9-8 THE COMMON LOGARITHM FUNCTION

Nate's Chocolate Chips sells 15 different kinds of chocolate chip cookies wholesale in cartons of 144 cookies. Nate's has fixed costs of $5000 a week. In addition, making a carton of cookies costs Nate's $15. The average revenue function for Nate's is $AR(q) = 26 - (q/500)$. Here q is the number of cartons of cookies. Determine each of the following.

33. $C(q)$

34. $AC(q)$

35. $R(q)$

36. $P(q)$

37. $AP(q)$

38. Breakeven point

39. Maximum possible profit

40. q when profit is maximum

41. The change in profit if production is changed from 1000 cartons to 2000 cartons a week.

42. The change in revenue if production is changed from 1200 cartons a week to 2400.

43. Wilfred James used 500 kWh of electricity in August and had an electric bill of $34.50. In September, he used 460 kWh, and his bill was $32.10. Express his electric bill B as a function of the amount of electricity A he uses. Assume this function is linear.

44. Express the amount of electricity Mr. James uses as a function of his electric bill. What is the relation between this function and the function in exercise 43.

45. Julie Horvath has $2500 invested in an account that pays 10% interest compounded continuously. How much will the account contain after 3.5 years? How long will it take the $2500 to double? To triple?

10

Differential Calculus

Pre-calculus mathematics is concerned with static situations; it considers how high production levels are at a certain time, or how much an account contains after two years. Calculus is concerned with dynamic situations, such as how fast production levels are increasing, or how rapidly interest is accruing.

Calculus is traditionally divided into two parts. The first part, *differential calculus*, is concerned with the rate of change, called the *derivative*, of a function. The derivative has many applications, and is extremely useful in optimization, that is, in making quantities as large as possible (for example, profit) or as small as possible (for example, average cost).

The second part of calculus, called *integral calculus*, is concerned with an operation called *integration*. Integral calculus is used to find areas and probabilities, and to solve equations involving derivatives. Integration is also used to determine a function whose rate of change (derivative) is known, for example, to determine the profit function from the marginal profit function. Thus, there is a fundamental connection between the two parts of calculus.

Section 10-1 Limits and Continuity

Since the central notion of both parts of calculus is that of *limit*, we begin calculus by studying limits.

Suppose $g(x) = (x^2 - 9)/(x - 3)$. Then $g(3)$ is not defined, because substituting 3 for x in the formula for g would lead to division by 0. How does $g(x)$ behave when x is near 3? Table 1 provides the basis for a good guess.

TABLE 1

x	3.1	2.9	3.01	2.99	3.001	2.999	3.0001	2.9999
$g(x)$	6.1	5.9	6.01	5.99	6.001	5.999	6.0001	5.9999

Table 1 seems to indicate that $g(x)$ is close to 6 when x is close to 3. In addition, $g(x)$ gets closer and closer to 6 as x gets closer and closer to 3. We therefore say that the limit as x approaches 3 of $g(x)$ is 6.

> **Definition.** The *limit* as x approaches b of $f(x)$ is L if $f(x)$ can be made arbitrarily close to L by requiring x to be close to but not equal to b. This is written
> $$\lim_{x \to b} f(x) = L$$

Note that nothing is said about $f(b)$—indeed, $f(b)$ may even be undefined. The definition given here of limit is not precise. A more precise definition of limit is studied in advanced mathematics courses.

Let's return to the function g. Table 1 only provides a basis for guessing that $\lim_{x \to 3} g(x) = 6$. You do not know from the table whether or not $g(2.9997)$ is close to 6. To avoid guessing, algebraically simplify the formula for g when $x \neq 3$.

$$g(x) = \frac{x^2 - 9}{x - 3} = \frac{(x-3)(x+3)}{x-3} = x + 3, \quad x \neq 3$$

Now if x is close to 3 but $x \neq 3$, the formula $g(x) = x + 3$ is valid and $g(x)$ is close to 6. As x gets closer and closer but not equal to 3, $g(x)$ gets closer and closer to 6. Therefore the statement $\lim_{x \to 3} g(x) = 6$ is valid.

Many different kinds of limits are studied in mathematics. We will be concerned mainly with limits of the form $\lim_{x \to b} p(x)/q(x)$, where p and q are polynomials and $q(b) = 0$. If $p(b) \neq 0$, we are trying to evaluate $p(b)/0$. This is impossible, and the limit does not exist. If $p(b) = 0$, factor $x - b$ out of p and q and proceed further.

Now let's compute

$$\lim_{x \to 2} \frac{x^2 - 5x + 6}{x^2 - 6x + 8}$$

Note that the numerator and denominator both take the value 0 when $x = 2$. This suggests factoring and cancelling out a factor of $x - 2$.

$$\frac{x^2 - 5x + 6}{x^2 - 6x + 8} = \frac{(x-2)(x-3)}{(x-2)(x-4)} = \frac{x-3}{x-4}, \quad x \neq 2$$

As long as $x \neq 2$, our original quotient is equal to $(x-3)/(x-4)$. As x gets closer and closer to 2 but $x \neq 2$, $(x-3)/(x-4)$ gets closer and closer to $(2-3)/(2-4) = -1/-2 = 1/2$. Therefore,

$$\lim_{x \to 2} \frac{x^2 - 5x + 6}{x^2 - 6x + 8} = \frac{1}{2}$$

Example 1 Compute $\lim_{x \to 4} \dfrac{x^2 - 7x + 12}{x^2 - 8x + 16}$.

Solution The numerator and denominator both take the value 0 when $x = 4$. Simplify the function.

$$\frac{x^2 - 7x + 12}{x^2 - 8x + 16} = \frac{(x-3)(x-4)}{(x-4)^2} = \frac{x-3}{x-4}$$

If x gets closer and closer to 4, the numerator of $(x-3)/(x-4)$ gets closer and closer to 1 and the denominator gets closer and closer to 0. The absolute value of this fraction gets bigger and bigger without bound, and so the limit does not exist.

Continuity

The concept of *continuity* is very closely related to the concept of limit. A function f is *continuous* at a number a if, whenever x is close to a, $f(x)$ is close to $f(a)$. We all use the notion of continuity. For example, let's say that Swiss cheese is $2.79 a pound. You go to the store and ask for a one-pound chunk. The clerk offers to cut you a chunk, but cannot guarantee that it will weigh exactly one pound. You say to go ahead, because *if the weight is close to one pound, the price is close to $2.79.* The precise definition of continuity involves the notion of limit.

> **Definition.** A function f is *continuous at the number b* if b is in the domain of f and $\lim_{x \to b} f(x) = f(b)$.
>
> **Definition.** A function f is *continuous* on an interval if f is continuous at every number in the interval.

Most of the functions we have considered in this book, in particular polynomials, exp, and ln, are continuous. Many more functions are also continuous. It can be shown that the sum, difference, product, quotient (whenever the denominator is not zero), and composition of continuous functions are continuous.

The continuity of a function is closely related to the appearance of its graph. If a function is continuous and its domain is an interval or all real numbers, the graph of the function can be drawn without removing the pencil from the paper. Conversely, if the graph of a function can be drawn without removing the pencil from the paper, the function is continuous.

Many limit statements are valid because the function involved is continuous. For example, since the exponential function is continuous, $\lim_{x \to 5} e^x = e^5$. Similarly

$$\lim_{x \to 7} x^2 = 7^2 = 49$$

$$\lim_{x \to 4} (3 + \ln x) = 3 + \ln 4$$

$$\lim_{x \to 10} \frac{x^2 + 3x + 5}{x^2 - 4x} = \frac{10^2 + 3 \cdot 10 + 5}{10^2 - 4 \cdot 10} = \frac{135}{60}$$

$$\lim_{x \to 2} \left(5e^{3x} + \frac{1}{x}\right) = 5e^6 + \frac{1}{2}$$

Example 2 Determine $\lim_{x \to -2} g(x)$, where $g(x) = \dfrac{x^2 + x - 2}{x^3 + 6x^2 + 8x}$.

Solution Substituting -2 for x in the formula for g gives $0/0$. Now

$$g(x) = \frac{x^2 + x - 2}{x^3 + 6x^2 + 8x} = \frac{(x + 2)(x - 1)}{x(x^2 + 6x + 8)}$$

$$= \frac{\cancel{(x+2)}(x-1)}{x\cancel{(x+2)}(x+4)} = \frac{x - 1}{x(x + 4)} \quad \text{provided } x \neq -2.$$

Then

$$\lim_{x \to -2} g(x) = \lim_{x \to -2} \frac{x-1}{x(x+4)} \qquad \text{Substitute } -2 \text{ for } x \text{ since } \frac{x-1}{x(x+4)} \text{ is continuous at } -2.$$

$$= \frac{-2-1}{(-2)(-2+4)}$$

$$= \frac{-3}{-4} = \frac{3}{4}$$

Example 3 Evaluate $\lim_{h \to 0} \frac{g(x+h) - g(x)}{h}$, where $g(x) = 7x + 9$.

Solution Recall that $g(x+h)$ is computed by substituting $x+h$ for x in the formula for g. Thus,

$$g(x+h) = 7(x+h) + 9$$

Now

$$\lim_{h \to 0} \frac{g(x+h) - g(x)}{h} = \lim_{h \to 0} \frac{[7(x+h) + 9] - [7x + 9]}{h}$$

$$= \lim_{h \to 0} \frac{\cancel{7x} + 7h + \cancel{9} - \cancel{7x} - \cancel{9}}{h}$$

$$= \lim_{h \to 0} \frac{7h}{h} = \lim_{h \to 0} 7 \qquad \text{Since } h \neq 0.$$

$$= 7 \qquad \text{Since the constant function 7 is continuous.}$$

Example 4 Let $k(x) = 4x^2$. Evaluate $\lim_{h \to 0} \frac{k(x+h) - k(x)}{h}$.

Solution $\lim_{h \to 0} \frac{k(x+h) - k(x)}{h} = \lim_{h \to 0} \frac{4(x+h)^2 - 4x^2}{h}$

$$= \lim_{h \to 0} \frac{\cancel{4x^2} + 8hx + 4h^2 - \cancel{4x^2}}{h} = \lim_{h \to 0} \frac{h(8x + 4h)}{h}$$

$$= \lim_{h \to 0} (8x + 4h) = 8x$$

Treat x as a constant; then $8x + 4h$ is a continuous function of h.

■

Not all functions are continuous. The postage function $P(x)$ is an example of a discontinuous function. Here $P(x)$ is the amount of postage required on a letter of weight x ounces. The graph of $P(x)$ is shown in Diagram 1. In this diagram, a small, empty circle means the point is excluded from the graph. A small, filled-in circle means the point is part of the graph. Note that the function P is discontinuous at every counting number n. If a letter weighs even the tiniest amount more than n ounces, it requires one more stamp than a letter that weighs exactly n ounces. Other discontinuous functions are shown in Diagrams 2 and 3.

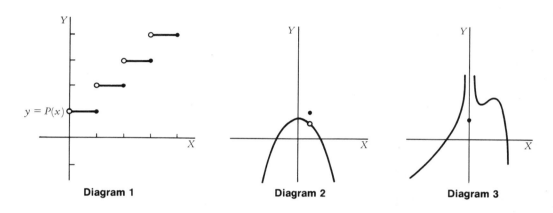

Diagram 1   **Diagram 2**   **Diagram 3**

Exercises 10-1

Evaluate the following limits.

1. $\lim_{x \to 2} (x^3 - 4x^2)$
2. $\lim_{x \to 3} (e^x + 5)$
3. $\lim_{x \to e} (3 + \ln x)$
4. $\lim_{x \to -2} (4x + 5)$
5. $\lim_{x \to 0} e^{5x+4}$
6. $\lim_{x \to 5} (x^2 - 3x + 2)$
7. $\lim_{x \to 1} \ln x$
8. $\lim_{x \to 0} e^x$
9. $\lim_{x \to 4} \dfrac{x^2 - 5x + 6}{x^2 + 5x - 6}$
10. $\lim_{x \to 2} \dfrac{x^2 - 4x + 4}{x + 2}$
11. $\lim_{x \to 1} \dfrac{x^2 - 1}{x^2 + 6x - 7}$
12. $\lim_{x \to 3} \dfrac{x^2 + 6x + 9}{x^2 + 5x + 6}$
13. $\lim_{x \to -1} \dfrac{x^2 + x}{(x + 1)^2}$
14. $\lim_{x \to 4} \dfrac{x^2 - 9x + 20}{x^2 + 9x - 52}$
15. $\lim_{x \to 5} \dfrac{(x + 5)^2(x - 7)}{(x - 5)(x + 5)^2}$

16. $\lim\limits_{x \to -4} \dfrac{(x+4)(x-4)(x+3)}{(x+7)(x+4)(x-3)}$

17. $\lim\limits_{x \to 1} \dfrac{(x^2 - 3x + 2)^2}{(x^2 - 1)^2}$

18. $\lim\limits_{x \to 0} \dfrac{x^3 - 5x^2}{x^4 - 4x^2}$

*19. $\lim\limits_{x \to 4} \dfrac{|x^2 - 16|}{x - 4}$

*20. $\lim\limits_{x \to 3} \dfrac{(x-3)^2}{|x-3|}$

*21. $\lim\limits_{x \to 16} \dfrac{\sqrt{x} - 4}{x - 16}$

*22. $\lim\limits_{x \to 25} \dfrac{(5 - \sqrt{x})^2}{25 - x}$

*23. $\lim\limits_{x \to 0} \left(\dfrac{x}{|x|} - \dfrac{|x|}{x} \right)$

*24. $\lim\limits_{x \to 0} \left(\dfrac{x^3}{|x|} - \dfrac{|x|^3}{x} \right)$

In each of the following exercises, construct a table similar to Table 1 in this section and use it to guess the value of the limit.

◈ 25. $\lim\limits_{x \to 0} x \ln |x|$

◈ 26. $\lim\limits_{x \to 0} x^{1/3} \ln |x|$

◈ 27. $\lim\limits_{x \to 1} \dfrac{x - 1}{\ln x}$

◈ 28. $\lim\limits_{x \to 5} \dfrac{\ln x - \ln 5}{x - 5}$

◈ 29. $\lim\limits_{x \to 4} \dfrac{x - 4}{\ln x - \ln 4}$

◈ 30. $\lim\limits_{x \to 0} \dfrac{e^x - 1}{x}$

◈ 31. $\lim\limits_{x \to 0} \dfrac{e^{3x} - 1}{x}$

◈ 32. $\lim\limits_{x \to 0} \dfrac{e^{5x} - 1}{x}$

Section 10-2 The Derivative

The *derivative* of a function is its instantaneous rate of change. To introduce the derivative, we must first discuss the concept of rate of change. Let's start with a familiar situation—the motion of a car on a straight road.

Suppose a car is stopped at a traffic light and at time $t = 0$, the light turns green. The car then accelerates until it reaches a speed of 60 mph or 88 ft/sec. (Recall that *speed* is the rate at which distance changes with time.) Suppose the position s of the car while it is accelerating is given as a function of time t by

$$s(t) = 2t^2, \qquad 0 \le t \le 22$$

Here s is measured in feet and t in seconds.

At $t = 5$, the car has moved $s(5) = 50$ ft. At time $t = 9$, the car has moved $s(9) = 162$ ft. During the 4 sec from $t = 5$ to $t = 9$, the car has moved $162 - 50 = 112$ ft. The *average speed* of the car during this time interval is the distance traveled divided by the length of the interval, that is, $112/4 = 28$ ft/sec.

> **Definition.** Let $f(x)$ be a function. The *average rate of change* of f between x_1 and x_2 is
> $$\frac{f(x_2) - f(x_1)}{x_2 - x_1}$$

What is the speed of the car when $t = 5$ sec? One way to attempt to answer this question is to look at the average speed of the car over various time intervals beginning at 5. See Table 1.

TABLE 1

Time Interval: 5 to	9	8	7	6	5.5	5.1	5.01	5.001
Final Position	162	128	98	72	60.5	52.02	50.2002	50.020002
Change in Position	112	78	48	22	10.5	2.02	0.2002	0.020002
Length of Time Interval	4	3	2	1	0.5	0.1	0.01	0.001
Average Speed	28	26	24	22	21	20.2	20.02	20.002

The average speeds appear to be approaching a limit of 20 ft/sec. To verify this, let us compute the average speed of the car over the time interval from 5 to $5 + h$. This average speed is

$$\frac{s(5+h) - s(5)}{(5+h) - 5} = \frac{2(5+h)^2 - 2(5^2)}{5+h-5}$$

$$= \frac{2(25 + 10h + h^2) - 50}{h} = \frac{50 + 20h + 2h^2 - 50}{h}$$

$$= \frac{20h + 2h^2}{h} = \frac{h(20 + 2h)}{h} = 20 + 2h$$

As h approaches 0, the average speed approaches 20, since

$$\lim_{h \to 0} (20 + 2h) = 20$$

The limit of the average speed is called the *instantaneous speed* and is the *derivative* of the position function. We have just computed the value of the derivative $s'(t)$ at $t = 5$.

> **Definition.** Let $f(x)$ be a function. The *derivative* f' is the function defined by
> $$f'(x) = \lim_{h \to 0} \frac{f(x+h) - f(x)}{h}$$
> The function f is said to be *differentiable* at a number x if x is in the domain of f', that is, if the above limit exists.

Other notations for the derivative are

$$\frac{df}{dx} \quad \text{and} \quad \frac{d}{dx} f(x)$$

These notations hide the fact that the derivative is a function. If $y = f(x)$, the derivative is sometimes denoted y' or $y'(x)$ or dy/dx.

The derivative f' is a function defined by a formula. The domain of f' thus consists of those numbers x in the domain of f for which

$$\lim_{h \to 0} \frac{f(x+h) - f(x)}{h}$$

exists.

Now we will use the definition to compute some derivatives.

Example 1 Let $f(x) = x^2$. Compute $f'(x)$.

Solution First compute $\dfrac{f(x+h) - f(x)}{h}$ in simplified form.

$$\frac{f(x+h) - f(x)}{h} = \frac{(x+h)^2 - x^2}{h}$$
$$= \frac{x^2 + 2hx + h^2 - x^2}{h} = \frac{2hx + h^2}{h} = 2x + h$$

Then $f'(x) = \lim\limits_{h \to 0} (2x + h) = 2x$.

Example 2 Let $g(t) = 1/t$. Compute $g'(t)$.

Solution

$$\frac{g(t+h)-g(t)}{h} = \frac{\frac{1}{t+h}-\frac{1}{t}}{h}$$

$$= \frac{\frac{1}{t+h}-\frac{1}{t}}{h} \cdot \frac{t(t+h)}{t(t+h)} \qquad \text{Multiply by } [t(t+h)]/[t(t+h)] \text{ to avoid a fraction in the numerator of another fraction.}$$

$$= \frac{t-(t+h)}{ht(t+h)} = \frac{t-t-h}{ht(t+h)}$$

$$= \frac{-h}{ht(t+h)} = \frac{-1}{t(t+h)}$$

Then

$$g'(t) = \lim_{h \to 0} \frac{-1}{t(t+h)} = \frac{-1}{t^2}$$

Example 3 Let $k(w) = \sqrt{w}$. Compute $k'(w)$.

Solution

$$\frac{k(w+h)-k(w)}{h} = \frac{\sqrt{w+h}-\sqrt{w}}{h}$$

$$= \frac{\sqrt{w+h}-\sqrt{w}}{h} \cdot \frac{\sqrt{w+h}+\sqrt{w}}{\sqrt{w+h}+\sqrt{w}} \qquad \text{Rationalize the numerator.}$$

$$= \frac{w+h-w}{h(\sqrt{w+h}+\sqrt{w})}$$

$$= \frac{h}{h(\sqrt{w+h}+\sqrt{w})}$$

$$= \frac{1}{\sqrt{w+h}+\sqrt{w}}$$

Then

$$k'(w) = \lim_{h \to 0} \frac{1}{\sqrt{w+h}+\sqrt{w}} = \frac{1}{2\sqrt{w}}$$

Example 4 Let $j(z) = z^{1/3}$. Compute $j'(0)$.

Solution
$$\frac{j(0+h) - j(0)}{h} = \frac{h^{1/3} - 0^{1/3}}{h} = \frac{1}{h^{2/3}}$$

As $h \to 0$, $h^{2/3}$ also approaches 0, so that the quotient $1/h^{2/3}$ gets bigger and bigger without bound. Therefore, $\lim_{h \to 0} 1/h^{2/3}$ does not exist, so that $j'(0)$ is not defined.

Example 5 Let $f(x) = c$. Compute $f'(x)$.

Solution
$$\frac{f(x+h) - f(x)}{h} = \frac{c - c}{h} = \frac{0}{h} = 0$$

Then $f'(x) = \lim_{h \to 0} 0 = 0$.

■

We have just derived a formula for the derivative of any constant function.

> The derivative of a constant function is 0.

Tangent lines

So far, the derivative of a function has been interpreted as its *instantaneous rate of change*. Now we will interpret the derivative geometrically in terms of tangent lines.

What is meant by the tangent line to a curve? In geometry, a tangent line to a circle is a line that meets the circle at only one point. (See Diagram 1.) This definition does not generally work for other curves. In Diagram 2, the line L is tangent to the curve but meets it more than once, and the line M meets the curve only once but is obviously not a tangent line.

To define the tangent line to a curve, we will use a limiting process. In Diagram 3, fix the point $A(x, f(x))$. If $P(x + h, f(x + h))$ is any other point on the graph of f, the line L_P through A and P is called a *secant line* to the graph of f.

Suppose there is a "tangent line" L to the graph of f at A. If P is near A, then L_P should be "near" L. Since L_P and L both contain A, this means

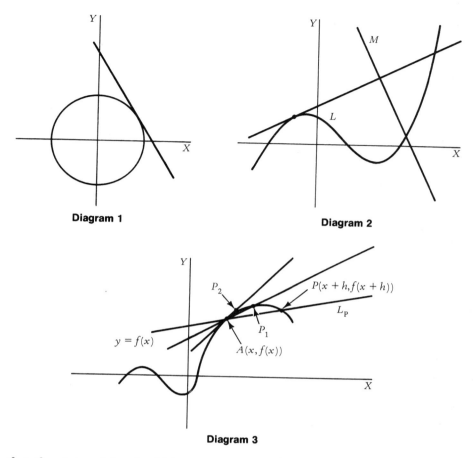

Diagram 1

Diagram 2

Diagram 3

that the slope of L_P should be close to the slope of L. As P gets closer and closer to A, the slope of L_P should get closer and closer to the slope of L. By the slope formula, the slope of L_P is given by $[f(x+h) - f(x)]/h$. Therefore, the slope of L should be given by $\lim_{h \to 0} [f(x+h) - f(x)]/h$, that is, by $f'(x)$.

Definition. If the function f is differentiable at x_0, the *tangent line* to the graph of f at $(x_0, f(x_0))$ is the line through the point $(x_0, f(x_0))$ with slope $f'(x_0)$.

Thus the tangent line is the limit of secant lines. This new definition of tangent line gives the same tangent line to a circle as the definition in geometry.

Now we will use the derivatives already computed in Examples 1, 2, and 3 to find the equations of some tangent lines.

Example 5 Find the equation of the tangent line to $f(x) = x^2$ at $x = 3$.

Solution If $f(x) = x^2$, then $f'(x) = 2x$ and $f'(3) = 2 \cdot 3 = 6$. The tangent line to $f(x) = x^2$ at $x = 3$ has slope 6 and passes through the point $(3, f(3)) = (3, 9)$. The equation of this tangent line is $y - 9 = 6(x - 3)$ in point-slope form and $y = 6x - 9$ in slope-intercept form.

Example 6 Find the equation of the tangent line to $g(t) = 1/t$ at $t = 1/2$.

Solution If $g(t) = 1/t$, then $g'(t) = -1/t^2$ and $g'(1/2) = -4$. The tangent line to $g(t) = 1/t$ at $t = 1/2$ has slope -4 and passes through $(1/2, g(1/2)) = (1/2, 2)$. Its equation is $y - 2 = -4(x - 1/2)$ or $y = -4x + 4$.

Example 7 Find the equation of the tangent line to $k(w) = \sqrt{w}$ at $w = 16$.

Solution If $k(w) = \sqrt{w}$, then $k'(w) = 1/(2\sqrt{w})$ and $k'(16) = 1/8$. The tangent line to $k(w) = \sqrt{w}$ at $w = 16$ has slope $1/8$ and passes through $(16, 4)$. Its equation is $y - 4 = (x - 16)/8$ or $y = (x/8) + 2$.

Exercises 10-2

Compute the derivative of each of the following functions. Evaluate the derivative at 16.

1. $f(x) = 2x + 3$
2. $g(x) = 5 - 3x$
3. $f(t) = 4t + 8$
4. $k(t) = 6t - 7$
5. $h(s) = 4s^2$
6. $f(x) = x^2 + 4$
7. $j(v) = 6\sqrt{v}$
8. $g(j) = \dfrac{4}{j}$

Using the derivatives computed in this section, determine the equation of the indicated tangent line. Write the answer in point-slope form and also in slope-intercept form.

9. $y = x^2$ $(-5, 25)$
10. $y = x^2$ $(0, 0)$
11. $y = x^2$ $(1, 1)$
12. $y = x^2$ $(5, 25)$
13. $y = \dfrac{1}{x}$ $(x = 1)$
14. $y = \dfrac{1}{x}$ $(x = 6)$
15. $y = \sqrt{x}$ $(x = 49)$
16. $y = \sqrt{x}$ $(x = 64)$

Robert Sanderson, of Sanderson Advertising, is preparing an advertising campaign for Fogg Enterprises. He prepares a graph showing expected annual sales S as a function of the amount A spent on advertising. Currently, Fogg does not advertise. On the average, how much will sales increase for each dollar spent on advertising, if Fogg spends the following amounts on advertising?

- 17. $100,000
- 18. $200,000
- 19. $300,000
- 20. $400,000

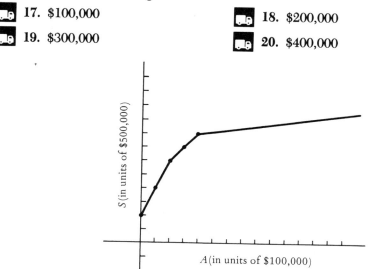

Annual profits for the Richmond Coal Company have been increasing in recent years.

Year	1971	1972	1973	1974	1975	1976	1977	1978	1979	1980
Profit (in millions of dollars)	1.3	1.5	1.8	2.1	2.5	3.0	4.0	4.7	6.2	7.8

What is the average rate of change of profit between

- 21. 1971 and 1976?
- 22. 1976 and 1980?
- 23. 1971 and 1980?
- 24. 1973 and 1977?
- 25. 1974 and 1976?
- 26. 1972 and 1978?

The public's response, in percent, to the question: "Do you approve of the president's performance in office" during the last 12 months is shown in the following table.

Month	Approve	Disapprove	No Opinion
April	42	47	11
May	44	46	10
June	45	44	11
July	48	45	7
August	55	40	5
September	62	35	3
October	60	36	4
November	58	39	3
December	60	37	3
January	60	38	2
February	59	37	4
March	62	34	4

What is the average rate of change, in percent per month, of

27. Approval between June and November?
28. Approval between April and January?
29. Disapproval between May and August?
30. Disapproval between August and February?
31. No opinion between April and January?
32. No opinion between July and December?

Compute the derivative of each of the following functions.

*33. $g(x) = x^3$

*34. $h(v) = v^4$

*35. $f(w) = w^{-2}$

*36. $f(x) = \sqrt{1 + x^2}$

*37. $k(y) = |y|$

*38. $f(y) = \dfrac{y}{|y|}$

Section 10-3 Formulas for Derivatives

Computing derivatives from the definition is slow and hard. From now on, we will compute derivatives using formulas (which are derived in more advanced courses) and using properties of the derivative.

The first formula is called the *power rule* because it gives the derivative of a power of x. In the previous section, we derived the power rule for the functions x^2, x^{-1}, and $x^{1/2}$.

$$\boxed{\dfrac{d}{dx}(x^r) = rx^{r-1}} \quad \text{Power rule}$$

To find the derivative of the function $f(x) = x^3$, use the power rule with $r = 3$ and obtain $f'(x) = 3x^{3-1} = 3x^2$.

Example 1 Find $f'(x)$ if $f(x) = x^{-3/2}$.

Solution Use the power rule with $r = -3/2$.

$$f'(x) = \frac{-3}{2} \cdot x^{(-3/2)-1} = \frac{-3}{2} x^{-5/2}$$

■

The following formulas give the derivatives of the exponential function and the natural logarithm function. Note that the exponential function is its own derivative.

$$\frac{d}{dx}(e^x) = e^x$$
$$\frac{d}{dx}\ln x = 1/x$$

Formulas for derivatives

Example 2 Find the equation of the tangent line to $y = \ln x$ at $x = e$.

Solution At $x = e$, $y = \ln e = 1$. $y'(x) = 1/x$, so that $y'(e) = 1/e$. The equation of the tangent line is $y - 1 = (x - e)/e$ in point-slope form and $y = x/e$ in slope-intercept form.

■

The formula defining a function often has more than one term. To differentiate such a function, a property of the derivative, called *linearity*, is used.

$$(af + bg)' = af' + bg' \quad \text{Linearity of the derivative}$$

Here f and g are functions and a and b are numbers. Linearity allows us to compute the derivative of a complicated expression term by term.

Let's use linearity to compute the derivative of $6x^2 + 3x^{-1}$. Linearity permits us to write this derivative as

$$(6x^2 + 3x^{-1})' = 6(x^2)' + 3(x^{-1})'$$

Now each of the expressions x^2 and x^{-1} may be differentiated by the power rule, so that

$$(6x^2 + 3x^{-1})' = 6(2x) + 3(-1 \cdot x^{-2}) = 12x - 3x^{-2}$$

The last expression was obtained by simplifying.

Example 3 Compute $(24x^{5/3} + 7e^x)'$.

Solution
$$(24x^{5/3} + 7e^x)' = 24(x^{5/3})' + 7(e^x)'$$
$$= 24\left(\frac{5}{3}x^{2/3}\right) + 7(e^x) = 40x^{2/3} + 7e^x$$

Example 4 Compute $(5x^3 - 7x^2 + 3x - 2)'$.

Solution Linearity applies when the formula defining a function has more than two terms, and also when the formula has minus signs, so that

$$(5x^3 - 7x^2 + 3x - 2)' = 5(x^3)' - 7(x^2)' + 3(x)' - (2)'$$
$$= 5(3x^2) - 7(2x) + 3(1) - (0) = 15x^2 - 14x + 3$$

Example 5 Compute $(8\ln x + 3e^x - 12x^5)'$.

Solution
$$(8\ln x + 3e^x - 12x^5)' = 8(\ln x)' + 3(e^x)' - 12(x^5)'$$
$$= 8\left(\frac{1}{x}\right) + 3(e^x) - 12(5x^4) = \frac{8}{x} + 3e^x - 60x^4$$

Example 6 Find the equation of the tangent line to $y = x^3 + 2x^2$ at $x = 4$.

Solution At $x = 4$, $y = 96$. $y' = 3x^2 + 4x$, so that $y'(4) = 64$. The equation of the tangent line is $y - 96 = 64(x - 4)$ in point-slope form and $y = 64x - 160$ in slope-intercept form.

Exercises 10-3

Compute the derivative of each function. Evaluate the derivative at the number shown.

1. $4x + 9$, $x = 1$
2. $-2x + 7$, $x = 5$
3. $x^2 + 3x + 2$, $x = 5$
4. $3x^2 - x - 2$, $x = 1$

5. $4x^{3/2} + 2x$, $x = 9$
6. $2x - x^{-1/2}$, $x = 4$
7. $5e^x + e$, $x = 0$
8. $4e^x + 2ex - e$, $x = 1$
9. $-3\ln x + 6$, $x = 3$
10. $8 - 4\ln x$, $x = 2$
11. $4x^{-1/2} + 8x^{1/2}$, $x = 4$
12. $3x^{1/3} - 10x^{-1/2}$, $x = 1$
13. $12\ln x + 4x$, $x = 3$
14. $12x - 6\ln x$, $x = 2$
15. $e^x + \ln x$, $x = e$
16. $2e^x - 3\ln x$, $x = 1$
17. $x^2 + 4\ln x + 8x^{-1}$, $x = 2$
18. $3x^2 - 2\ln x - 4x^{-1}$, $x = 1$
19. $x^3 - 7x^2 + 4x - 12$, $x = 2$
20. $4x^3 - 17x^2 - 8x + 10$, $x = 0$

Find the equation (in slope-intercept form) of the tangent line at the given point.

21. $y = x^2 + x$ $(2, 6)$
22. $y = x^3 - x^2 + 1$ $(1, 1)$
23. $y = e^x$ $(0, 1)$
24. $y = x^{3/2} + 4$ $(9, 31)$

*25. Use the fact that $\lim_{h \to 0}(e^h - 1)/h = 1$ to prove the formula $(e^x)' = e^x$.
*26. Use the fact that $\lim_{w \to 0}[\ln(1 + w)]/w = 1$ to prove the formula $(\ln x)' = 1/x$.

Section 10-4 Rates of Change

In Section 3, several differentiation formulas were presented. Since these formulas make the computation of derivatives easy, let's begin applying derivatives. Recall the definition of derivative:

$$f'(x) = \lim_{h \to 0} \frac{f(x + h) - f(x)}{h}$$

If f is a function and h is small, $f'(x)$ is approximately

$$\frac{f(x + h) - f(x)}{h}$$

Multiply both quantities in the preceding line by h to obtain:

$$\boxed{f(x + h) - f(x) \quad \text{is approximately } hf'(x) \text{ if } h \text{ is small.}}$$

Therefore, if f is a function and the value of the independent variable changes from x by a small amount h, the value of f changes by about $hf'(x)$.

As an application of this concept, consider a situation in which the specifications for a job are slightly changed. Suppose the Bartle Art Museum is having a circular concrete patio with a radius of 10 ft put in the middle of a garden as part of an improvement project. Ms. Hardison, the director of the project, inspects the garden carefully and decides that she might want the patio to be a little bit larger. She calls Mr. Barber, the contractor, and asks him how much extra the modification would cost.

Mr. Barber usually charges $2 for each extra square foot of patio. Since the area is $A(r) = \pi r^2$, the charge $C(r)$ is given by

$$C(r) = 2\pi r^2$$

Then $C'(r) = 2\pi(2r) = 4\pi r$ and $C'(10) = 4\pi(10) = 40\pi$.

If the radius is increased by h feet, and h is small, then the additional charge is about $40\pi h$ dollars. Mr. Barber decides to state the extra cost in terms of an increase in inches. Since one foot is 12 inches, the extra charge for increasing the radius by i inches is

$$(40\pi) \cdot \left(\frac{i}{12}\right) = \frac{40\pi}{12} i$$

Since $40\pi/12$ is about 10.5, Mr. Barber tells Ms. Hardison that the charge for the modification will be about $10.50 for each inch the radius is increased, and that he would be happy to give her a precise figure when she knows exactly how large she wants the patio.

The use of the derivative allows Mr. Barber to provide quick and convenient estimates. Saying on the phone that each extra inch would cost about $10.50 is better than hanging up and later on mailing out the exact price for one extra inch, two extra inches, and so on, up to 12 extra inches.

Sometimes the rate of change of a function is very important. This is especially true when the function is the position of an object in motion, since the rate of change of position with respect to time is, by definition, speed. For example, one of the tests used in determining car safety consists of driving the car into a brick wall at 30 mph. The 30 mph here is the instantaneous speed of the car at the moment it hits the wall.

Example 1 In the test just discussed, suppose the distance the car travels before hitting the wall is given as a function of time by $s(t) = 1.5t^2$. Here s is measured in feet and t in seconds. How far does the car go before the collision?

Solution The speed is given by $s'(t) = 3t$. The speed is 30 mph (or 44 ft/sec) when

$s'(t) = 44$. Thus, $3t = 44$, and $t = 44/3$ when the car hits the wall. At this time, the distance the car has traveled is $s(44/3) = 1.5(44/3)^2 = 322\frac{2}{3}$ ft.

Increasing and decreasing functions

In the preceding example, if the car goes more than 322 ft 8 in. before hitting the wall, t will exceed $44/3$ sec at the moment of impact. The speed at which the car hits the wall will then exceed 30 mph, because velocity is an increasing function of time here.

It is often important to know whether the rate of change of a function is positive or negative, that is, whether the function is increasing or decreasing.

> **Definition.** The function f is *increasing* at x if $f(x_1) > f(x_2)$ whenever $x_1 > x_2$ and x_1 and x_2 are near enough to x.
> The function f is *decreasing* at x if $f(x_1) < f(x_2)$ whenever $x_1 > x_2$ and x_1 and x_2 are near enough to x.

In Diagram 1, the function is increasing at points, A, E, and F and decreasing at point C. At points B and D, the function is neither increasing nor decreasing. Now we will relate the notions of increasing and decreasing function to the derivative.

Assume x_1 and x_2 are near x, and $x_1 > x_2$. Let $h_1 = x_1 - x$ and $h_2 = x_2 - x$. Then h_1 and h_2 are small, so that

$$f(x_1) - f(x) \quad \text{is approximately} \quad h_1 f'(x)$$
$$f(x_2) - f(x) \quad \text{is approximately} \quad h_2 f'(x)$$

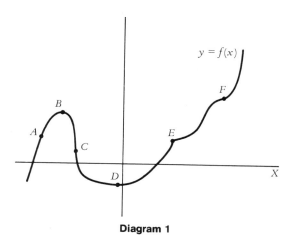

Diagram 1

Subtracting these approximations gives that

$$f(x_1) - f(x_2) \text{ is approximately } (h_1 - h_2)f'(x)$$

Now

$$h_1 - h_2 = (x_1 - x) - (x_2 - x)$$
$$= x_1 - x - x_2 + x$$
$$= x_1 - x_2$$

so that

$$f(x_1) - f(x_2) \text{ is approximately } (x_1 - x_2)f'(x)$$

Because $x_1 - x_2 > 0$, if $f'(x) > 0$, then $f(x_1) - f(x_2) > 0$, so that f is increasing at x. If $f'(x) < 0$, then $f(x_1) - f(x_2) < 0$, so that f is decreasing at x.

> **Theorem.** If $f'(x) > 0$, then f is increasing at x. If $f'(x) < 0$, then f is decreasing at x.

If $f'(x) = 0$ or is undefined, f may be increasing at x (as at point E in Diagram 1), decreasing at x, or neither (as at points B and D).

As an example, consider $y = 3x + 4$. Since $y' = 3$, the derivative is always positive and thus the function is always increasing.

Example 2 Let $y = -2x + 5$. Then $y' = -2$, so that y is always decreasing.

Example 3 Let $f(x) = x^2 - 4x + 3$. Then $f'(x) = 2x - 4$. If $x > 2$, then $f'(x) > 0$, so that f is increasing. If $x < 2$, then $f'(x) < 0$, so that f is decreasing.

Example 4 Let $f(x) = e^x$. Then $f'(x) = e^x > 0$, so that e^x is always increasing.

Example 5 Let $g(x) = \ln x$. Then $g'(x) = 1/x$. Since the domain of g consists of all positive numbers, $g'(x) > 0$ whenever $g(x)$ is defined. Therefore, $\ln x$ is increasing whenever it is defined.

Example 6 Let $h(x) = x^3 - 12x^2 + 36x - 6$. Then $h'(x) = 3x^2 - 24x + 36 = 3(x^2 - 8x + 12) = 3(x - 2)(x - 6)$. By using a sign graph, one obtains

$h'(x) > 0$, and therefore h is increasing if $x < 2$ or $x > 6$. Also, $h'(x) < 0$ and h is therefore decreasing if $2 < x < 6$.

Exercises 10-4

Determine where each function is decreasing, and where it is increasing.

1. $g(t) = 5 - 7t$
2. $h(v) = 5v + 9$
3. $f(x) = 4x^2 - 7x + 2$
4. $z(x) = 6x^2 + 8x - 9$
5. $y = 2x^3 + 15x^2 + 36x + 48$
6. $y = x^3 - 6x^2 - 15x + 12$
7. $y = x^3 + 4x - 5$
8. $y = x^4 - 2x^2$
9. $y = x^3 - 6x^2 - 63x + 27$
10. $y = x^3 + 9x^2 + 24x - 16$
11. $y = \exp(x) - e$
12. $y = \ln(x) - 4$
13. $t = \exp(s) - s$
14. $w = v - \ln v$
15. A car burns $G(s) = (s/2000) + (1.25/s)$ gallons of fuel for each mile traveled at the speed s on an uncrowded highway.
    a. Compute $G'(s)$.
    b. At what speeds is $G(s)$ an increasing function of s?
    c. At what speeds is $G(s)$ a decreasing function of s?
    d. If the driver is going at 40 mph, would increasing his speed a little help or harm fuel efficiency?
16. The height in feet of a ball thrown into the air is given by $h(t) = -16t^2 + 16t + 96$. This formula is valid until the ball lands.
    a. When does the ball land?
    b. At what speed does the ball strike the ground?
    c. When is the height increasing?
    d. When is the height decreasing?
    e. When is the height maximum? What is the value of $h'(t)$ at this time?
    f. How high does the ball get?

Section 10-5 Marginal Analysis

Economists and business executives are often concerned with the cost of producing one additional item. If x items are currently produced, the extra cost involved in producing one more item is

$$C(x + 1) - C(x)$$

The number x is usually large, and 1 can be considered small, so that

$$C(x+1) - C(x) \quad \text{is approximately} \quad 1 \cdot C'(x) = C'(x)$$

Thus this extra cost is approximately $C'(x)$. Economists use the word *marginal* to mean extra, and refer to this extra cost as *marginal cost*.

> Marginal cost (denoted MC) is the derivative of cost.
> Marginal revenue (denoted MR) is the derivative of revenue.
> Marginal profit (denoted MP) is the derivative of profit.

The cost, profit, and revenue functions are related to each other by the equation $P(q) = R(q) - C(q)$. Differentiating this equation gives $P'(q) = R'(q) - C'(q)$, or

$$MP(q) = MR(q) - MC(q)$$

If b items are produced, economists interpret $MC(b)$ as the cost of producing one more item and $hMC(b)$ as the cost of producing h more items —here h must be small. Similarly, $hMR(b)$ is the revenue from the sale of h additional items if currently b items are sold, and $hMP(b)$ is the profit obtained by making and selling h additional items if currently b items are produced and sold.

Example 1 Redstone Novelties makes bouquets of silk roses. The cost and revenue functions are

$$C(q) = 500 + 3q - 0.001q^2 \quad \text{and} \quad R(q) = 5q - 0.002q^2$$

The current production level is 400 bouquets.

PROBLEM A Determine the marginal cost and marginal revenue functions.

Solution $MC(q) = C'(q) = 3 - 0.002q$ and $MR(q) = R'(q) = 5 - 0.004q$.

PROBLEM B How much would costs and revenue change if production were increased by one bouquet?

Solution Costs would increase by $1 \cdot MC(400) = 1 \cdot 2.2 = \$2.20$. Revenue would increase by $1 \cdot MR(400) = 1 \cdot 3.4 = \$3.40$.

10-5 MARGINAL ANALYSIS

PROBLEM C How much would costs and revenue change if three fewer bouquets were produced?

Solution Costs would change by $-3 \cdot MC(400) = -\$6.60$, that is, costs would decrease by \$6.60. Revenue would change by $-3 \cdot MR(400) = -\$10.20$, that is, revenue would decrease by \$10.20.

PROBLEM D Determine the profit and marginal profit functions.

Solution

$$P(q) = R(q) - C(q)$$
$$= (5q - 0.002q^2) - (500 + 3q - 0.001q^2)$$
$$= 5q - 0.002q^2 - 500 - 3q + 0.001q^2$$
$$= -0.001q^2 + 2q - 500$$
$$MP(q) = P'(q) = -0.002q + 2 = 2 - 0.002q$$

The marginal profit function could also be found by using the relationship $MP(q) = MR(q) - MC(q)$.

PROBLEM E At what production levels is profit an increasing function of q? A decreasing function of q?

Solution Profit is an increasing function when marginal profit is positive, that is, when

$$2 - 0.002q > 0$$
$$2 > 0.002q \quad \text{Multiply by 500.}$$
$$1000 > q$$

If $q < 1000$, profits increase whenever an additional bouquet is made. Profit is a decreasing function when marginal profit is negative, that is, when

$$2 - 0.002q < 0$$
$$2 < 0.002q \quad \text{Multiply by 500.}$$
$$1000 < q$$

If $q > 1000$, profits decrease whenever an additional bouquet is made.

PROBLEM F How many bouquets should be made in order to maximize profit?

Solution In problem E, we determined that profit is an increasing function when $q < 1000$, so that $P(q) < P(1000)$ if $q < 1000$.

Profit is a decreasing function when $q > 1000$, so that $P(q) < P(1000)$ if $q > 1000$.

Thus $P(q) < P(1000)$ if $q \neq 1000$. Therefore profit is maximum when $q = 1000$. The maximum possible profit is $P(1000) = \$500.00$.

∎

In Example 1, profit was maximum when marginal profit was neither positive nor negative, that is, when marginal profit was 0. Since $MP(q) = MR(q) - MC(q)$, $MP(q) = 0$ when $MR(q) = MC(q)$. This gives a classical law of economics:

> Profit is maximum when marginal revenue equals marginal cost.

Exercises 10-5

In exercises 1–6, determine the functions $MC(q)$, $MR(q)$, *and* $MP(q)$. *Then determine the maximum possible profit and the value of q at which it is attained.*

1. $C(q) = 100 + 15q$, $\quad R(q) = 25q - \dfrac{q^2}{10}$

2. $C(q) = 20 + 4q$, $\quad R(q) = 8q - \dfrac{q^2}{6}$

3. $C(q) = 750 + 5q + \dfrac{q^2}{1000}$, $\quad R(q) = 8q$

4. $C(q) = 5000 + 10q + \dfrac{q^2}{800}$, $\quad R(q) = 20q$

5. $C(q) = 800 + 8q + \dfrac{q^2}{500}$, $\quad R(q) = 12q$

6. $C(q) = 500 + 2q + \dfrac{q^2}{500}$, $\quad R(q) = 5q$

7. The cost of producing q copies of a record is

$$C(q) = 100{,}000 + 3.24q - 10\sqrt{q}$$

About 40,000 copies of the record are being made. If a few more copies are made, what is the additional cost of producing each of these extra copies?

8. The profit function for Mario's Pizzeria is

$$P(q) = 3q - 5\sqrt{q} - 400$$

where q is the number of pizzas sold each week. How much profit does Mario make on the first 400 pizzas sold this week? On the 401st?

 9. The profit function for Bill's Barber Shop is

$$P(q) = 2q + \left(\frac{q^2}{1000}\right) - 600$$

where q is the number of haircuts done in a week.
  a. What is the profit if 500 haircuts are done in a week?
  b. What is the profit if 510 haircuts are done in a week?
  c. Evaluate $P(q)\big|_{500}^{510}$.
  d. Suppose 500 haircuts are usually done in a week. Use marginal analysis to determine the change in profit if an additional ten haircuts are done. Compare this answer with the answer to part (c) of this exercise.

 10. Ephraim Deans has a small farm. He can sell wheat in any quantity for $12.00 a bushel. He has fixed costs of $1000 and variable costs of $8q + 0.000003q^3$ on q bushels of wheat.
  a. How much wheat should he grow in order to maximize profit? What is his maximum possible profit?
  b. Suppose the price of wheat declines to $10.00 a bushel. How much wheat should he grow now for maximum possible profit? What is his maximum possible profit?

Section 10-6 The Product and Quotient Rules

In Section 10-3 we learned formulas for the derivatives of the functions x^n, e^x, and $\ln x$. We also learned that the derivative has the property of linearity, so that complicated functions can be differentiated one term at a time. Linearity thus enables us to differentiate the sum and difference of two functions. In applications, products and quotients of functions frequently occur. To differentiate the product and quotient of two functions requires additional formulas, called the *product rule* and the *quotient rule*.

The product rule is used to compute the derivative of the function $f(x)g(x)$. Assume f and g are differentiable at x. Then the product function fg is differentiable at x and

$$\boxed{(fg)'(x) = f(x)g'(x) + f'(x)g(x)} \quad \text{Product rule}$$

Let's use the product rule to differentiate the function $x^3 e^x$. This function is the product of the two functions x^3 and e^x. Use the product rule with $f(x) = x^3$ and $g(x) = e^x$ to obtain

$$(x^3 e^x)' = x^3(e^x)' + (x^3)'e^x$$

Since $(e^x)' = e^x$ and $(x^3)' = 3x^2$, we have

$$(x^3 e^x)' = x^3 e^x + 3x^2 e^x$$

Example 1 Compute $(x^2 \ln x)'$.

Solution
$$(x^2 \ln x)' = x^2(\ln x)' + (x^2)' \ln x$$
$$= x^2 \cdot \frac{1}{x} + 2x \ln x = x + 2x \ln x$$

■

To derive the product rule, we need to compute the limit as $h \to 0$ of

$$\frac{f(x+h)g(x+h) - f(x)g(x)}{h}$$

The expression $f(x+h)g(x+h)$ in the numerator is difficult to work with. To handle the numerator more easily, add and subtract the term $f(x+h)g(x)$ in the numerator to obtain

$$\frac{\overbrace{f(x+h)g(x+h) - f(x+h)g(x)} + \overbrace{f(x+h)g(x) - f(x)g(x)}}{h}$$

Now factor $f(x+h)$ out of the first pair of terms in the numerator and $g(x)$ out of the second pair of terms to obtain

$$f(x+h) \cdot \frac{g(x+h) - g(x)}{h} + \frac{f(x+h) - f(x)}{h} \cdot g(x)$$

As $h \to 0$, $f(x+h)$ approaches $f(x)$. By the definition of a derivative $[g(x+h) - g(x)]/h$ approaches $g'(x)$ and $[f(x+h) - f(x)]/h$ approaches

$f'(x)$ as $h \to 0$. Thus

$$(fg)'(x) = \lim_{h \to 0} \frac{f(x+h)g(x+h) - f(x)g(x)}{h} = f(x)g'(x) + f'(x)g(x)$$

The derivation of the product rule was not easy. The derivation of the quotient rule is similar and will be omitted.

The quotient rule is used to compute the derivative of the function $f(x)/g(x)$. Assume f and g are differentiable at x and $g(x) \neq 0$. Then

$$\left(\frac{f}{g}\right)'(x) = \frac{g(x)f'(x) - f(x)g'(x)}{g^2(x)} \quad \text{Quotient rule}$$

Let's use the quotient rule to differentiate

$$h(x) = \frac{x^2 - 1}{x + 5}$$

The function h is the quotient of the two functions $f(x) = x^2 - 1$ and $g(x) = x + 5$. By the quotient rule

$$h'(x) = \frac{(x+5)(x^2-1)' - (x^2-1)(x+5)'}{(x+5)^2}$$

Now $(x^2 - 1)' = 2x$ and $(x + 5)' = 1$. Substitute these derivatives in the formula for h' and simplify to obtain

$$h'(x) = \frac{(x+5)(2x) - (x^2-1)1}{(x+5)^2}$$

$$= \frac{2x^2 + 10x - x^2 + 1}{(x+5)^2} = \frac{x^2 + 10x + 1}{(x+5)^2}$$

Example 2 Compute $\left(\dfrac{x^3}{e^x}\right)'$.

Solution $\left(\dfrac{x^3}{e^x}\right)' = \dfrac{e^x(x^3)' - x^3(e^x)'}{(e^x)^2}$ Use the quotient rule with $f(x) = x^3$ and $g(x) = e^x$.

$$= \frac{e^x(3x^2) - x^3(e^x)}{e^{2x}} \qquad \text{Recall that } (e^x)^2 = e^{2x}.$$

$$= \frac{e^x(3x^2 - x^3)}{e^{2x}} \qquad \text{Factor out } e^x \text{ in the numerator.}$$

$$= \frac{3x^2 - x^3}{e^x} \qquad \text{Cancel out an } e^x. \text{ Note that } \frac{e^x}{e^{2x}} = \frac{1}{e^x}.$$

Example 3 Compute $\left(\dfrac{x^2}{\ln x}\right)'$.

Solution
$$\left(\frac{x^2}{\ln x}\right)' = \frac{\ln x(2x) - x^2\left(\dfrac{1}{x}\right)}{(\ln x)^2} \qquad \text{Use the quotient rule.}$$

$$= \frac{2x \ln x - x}{(\ln x)^2}$$

Exercises 10-6

Find the derivative of each function.

1. $x^2 e^x$
2. $\dfrac{e^x}{x^2}$
3. $x \ln x$
4. $\dfrac{\ln x}{x^2}$
5. $\dfrac{4}{x^2 - 25}$
6. $\dfrac{\ln x}{x^4}$
7. $\sqrt{x} \ln x$
8. $x^{-2/3} \ln x$
9. $\dfrac{x^3}{\ln x}$
10. $6x^4 e^x$
11. $\dfrac{2x - 3}{5x + 6}$
12. $\dfrac{3x + 5}{2x - 1}$
13. $\dfrac{x^2 + 3}{x^2 - 9}$
14. $\dfrac{3x + 5}{x^2 + 2}$
15. $\dfrac{4 + 2x}{5 - 8x}$
16. $\dfrac{2 + 3x - x^2}{2 - x}$
17. $(\ln x)^2 \quad (= \ln x \cdot \ln x)$
18. $e^{2x} \quad (= e^x \cdot e^x)$
19. $(\ln x)^{-1} \quad \left(= \dfrac{1}{\ln x}\right)$
20. $e^{-x} \quad \left(= \dfrac{1}{e^x}\right)$

*21. $x \cdot \ln x \cdot e^x$

*22. $\dfrac{xe^x}{\ln x}$

*23. $(\ln x)^3$

*24. e^{3x}

Write the equation of the tangent line to the graph of each function at the given point.

25. $f(x) = x^5 e^x$ at $(0,0)$
26. $f(x) = x^6 \ln x$ at $(1,0)$
27. $g(x) = \dfrac{x}{e^x}$ at $(2, 2/e^2)$
28. $h(w) = \dfrac{w}{\ln w}$ at (e, e)

29. The supply function for paintings by Marla Ovritch is

$$s(q) = 40 + \frac{q^2 + 10}{q + 5}$$

She is currently producing 25 paintings and selling them at the price $s(25)$. How much would the price of a painting have to increase in order for her to supply 27 paintings?

30. The demand function for ceramic bowls by Oscar Lapointe is

$$d(q) = 100 - \frac{q^2}{9 + q}$$

He currently produces 41 bowls and sells them at the price $d(41)$. How much would he have to decrease the price of each bowl in order to sell 42?

31. The profit function for Harlequin's Hamburgers is

$$P(q) = 0.35q + \frac{q^2}{8q + 1000}$$

Here q is the number of hamburgers sold in a day. How much extra profit is made if sales are increased from 500 hamburgers to 510?

32. The cost function for a production run of couches at Norris Furniture Makers is

$$C(q) = 200q + \frac{1000q}{q^2 + 1} + 5000$$

Usually 300 couches are made in one production run. How much less would it cost if only 295 couches were made?

33. The concentration (in percent) of a drug in a patient's blood is

$$C(t) = \frac{0.01t}{t^2 + 25}$$

t hours after the drug is administered. For which values of t is $C(t)$
a. Increasing?
b. Decreasing?

34. Repeat exercise 33 if

$$C(t) = \frac{0.005t^3}{e^t}$$

*35. Derive the quotient rule.
*36. Verify by using the product rule and the quotient rule that

$$g'(x) = \left[f(x) \cdot \frac{g(x)}{f(x)} \right]'$$

Section 10-7 The Chain Rule

The product and quotient rules greatly extend our ability to differentiate functions. These two rules, together with linearity, make it possible to differentiate any function obtained by combining the functions x^n, $\exp x$, and $\ln x$ with arithmetic operations. However, two functions may also be combined by composition.

For example, composing the exponential function with the linear function $f(x) = kx$ gives the function $\exp(kx)$. Functions of this type are used to compute continuously compounded interest and to study many growth and decay processes. Finding the derivative of a composite function requires the *chain rule*.

$$\boxed{(g \circ f)'(x) = g'[f(x)] f'(x)} \quad \text{Chain rule}$$

For our purposes we need not apply the chain rule itself. Since we are concerned with only a few kinds of function, it will be easier to use three formulas. Each of these formulas is a special case of the chain rule.

> 1. $(f(x)^r)' = rf(x)^{r-1}f'(x)$ Power rule
> 2. $\{\ln[f(x)]\}' = \dfrac{f'(x)}{f(x)}$ Logarithm rule
> 3. $\{\exp[f(x)]\}' = \exp[f(x)]f'(x)$ Exponential rule

To obtain the power rule, let $g(x) = x^r$ in the chain rule. Then

$$(g \circ f)(x) = g[f(x)] = f(x)^r$$

and

$$g'[f(x)]f'(x) = rf(x)^{r-1}f'(x)$$

The chain rule then becomes

$$(f(x)^r)' = rf(x)^{r-1}f'(x)$$

that is, the power rule.

To obtain the logarithm rule, let $g(x) = \ln x$ in the chain rule. Then

$$(g \circ f)(x) = \ln(f(x))$$

and

$$g'[f(x)]f'(x) = \frac{f'(x)}{f(x)}$$

The chain rule then becomes

$$\{\ln[f(x)]\}' = \frac{f'(x)}{f(x)}$$

that is, the logarithm rule. The derivation of the exponential rule is left as an exercise.

The following examples illustrate the power rule, logarithm rule, and exponential rule.

Example 1 Compute $[(2x + 3)^4]'$.

Solution Use the power rule with $f(x) = 2x + 3$ and $r = 4$.

$$[(2x+3)^4]' = 4(2x+3)^3(2x+3)'$$
$$= 4(2x+3)^3 \cdot 2 = 8(2x+3)^3$$

Example 2 Compute $[\ln(x^3 + 4)]'$.

Solution Use the logarithm rule with $f(x) = x^3 + 4$.

$$[\ln(x^3+4)]' = \frac{1}{x^3+4}(x^3+4)' = \frac{3x^2}{x^3+4}$$

Example 3 Compute $(e^{3x})'$.

Solution Use the exponential rule with $f(x) = 3x$.

$$(e^{3x})' = e^{3x}(3x)' = 3e^{3x}$$

Example 4 Compute $\dfrac{d}{dx}(e^x + x)^{1/2}$.

Solution Use the power rule with $f(x) = e^x + x$ and $r = 1/2$.

$$\frac{d}{dx}(e^x + x)^{1/2} = \frac{1}{2}(e^x + x)^{-1/2}(e^x + 1)$$

Example 5 Compute $(x^3 e^{5x})'$.

Solution

$$(x^3 e^{5x})' = x^3(e^{5x})' + (x^3)'e^{5x} \qquad \text{Product rule.}$$
$$= x^3(e^{5x} \cdot 5) + 3x^2 e^{5x} \qquad \text{Exponential rule.}$$
$$= 5x^3 e^{5x} + 3x^2 e^{5x}$$

Example 6 Compute $[x^2 \ln(1 + x^2)]'$.

Solution $[x^2 \ln(1+x^2)]' = x^2[\ln(1+x^2)]' + (x^2)'\ln(1+x^2)$ \qquad Product rule.

$$= x^2\left(\frac{2x}{1+x^2}\right) + 2x\ln(1+x^2) \qquad \text{Logarithm rule.}$$

$$= \frac{2x^3}{1+x^2} + 2x\ln(1+x^2)$$

Example 7 Compute $\dfrac{d}{dx}\left(\dfrac{x^2}{e^{3x}+2}\right)$.

Solution
$$\frac{d}{dx}\left(\frac{x^2}{e^{3x}+2}\right) = \frac{(e^{3x}+2)(x^2)' - x^2(e^{3x}+2)'}{(e^{3x}+2)^2} \quad \text{Quotient rule.}$$

$$= \frac{(e^{3x}+2)2x - x^2(e^{3x}3)}{(e^{3x}+2)^2} \quad \text{Exponential rule.}$$

$$= \frac{2xe^{3x} + 4x - 3x^2 e^{3x}}{(e^{3x}+2)^2}$$

Example 8 The cost function for collecting pearls from an oyster bed is given by

$$C(q) = 100 + 3qe^{0.01q}$$

If 100 pearls have already been found, what is the cost of finding one more?

Solution The cost of finding one more pearl is $MC(100)$.

$$MC(q) = C'(q) = (3q)'e^{0.01q} + 3q(e^{0.01q})'$$
$$= 3e^{0.01q} + 3q(0.01e^{0.01q}) = 3e^{0.01q} + 0.03qe^{0.01q}$$
$$= 3e^{0.01q}(1 + 0.01q)$$
$$MC(100) = 3e^{0.01(100)}[1 + 0.01(100)]$$
$$= 3e(1+1) = 6e$$

The cost of the 101st pearl is about $6e$, or $16.31.

■

One of the reasons given in Chapter 9 for studying the exponential function e^x rather than b^x was that this would make certain formulas in calculus simpler. Let's verify this by finding the derivative of the function

$$f(x) = b^x, \quad b > 0$$

First, write b^x as $(e^{\ln b})^x = \exp(x \ln b)$. Then

$$(b^x)' = [\exp(x \ln b)]' = \ln b \cdot \exp(x \ln b) = \ln b \cdot b^x$$

$$\boxed{(b^x)' = \ln b \cdot b^x, \quad b > 0}$$

This formula is more complicated than the formula $(e^x)' = e^x$, and is one reason why the function e^x is used instead of other exponential functions b^x.

Example 9 Find the derivative of $f(x) = 2^x$.

Solution Use the formula $(b^x)' = \ln b \cdot b^x$ with $b = 2$ to obtain $(2^x)' = \ln 2 \cdot 2^x$.

Exercises 10-7

Find the derivative of each function.

1. $(5x - 2)^{5/2}$
2. $(4x + 3)^7$
3. $(9 - x^2)^{-4}$
4. $(16 + x^2)^{1/2}$
5. e^{10x}
6. e^{-3x}
7. $\exp(x^2)$
8. $\exp(x^2 + 2x - 4)$
9. $\exp(x^3 + 7x^2 - 5x + 3)$
10. $\exp(4 + 3x - 5x^2)$
11. $\ln(4x^2 + 5x + 7)$
12. $\ln(3x^2 - 2x - 9)$
13. $\dfrac{x^2 + x + 1}{2x - 3}$
14. $\dfrac{x^2 - x}{3x + 4}$
15. $(x^2 + 1)^{1/2} x^{50}$
16. $(x^2 + 6x)^{50} x^{3/2}$
17. $\dfrac{9}{\sqrt{x^2 - 25}}$
18. $\dfrac{\sqrt{1 + 3x}}{4 + 7x}$
19. $\dfrac{6e^{2x}}{4x + 7}$
20. $\dfrac{5e^{3x}}{2x - 1}$
21. $6x\sqrt{x - 2}$
22. $5x\sqrt{x + 4}$
23. $\dfrac{2x}{(x^2 - 9)^3}$
24. $\dfrac{x^2 - 25}{\ln(2x + 1)}$
25. $\sqrt{x} \ln(x + 1)$
26. $x^{5/3}(2x + 1)^8$
27. $x^{-2/3} \ln x$
28. $6x^{-2/3} e^{3x}$
29. $\exp(e^x)$
30. $\ln(\ln x)$
31. $(\ln x)^3 + 5(\ln x)^2$
32. $4(\ln x)^2 - 5\ln x + 6$

33. $\sqrt{1 + 4e^x}$

34. $\sqrt{1 + 5\ln x}$

35. $(e + 2\ln x)^{10}$

36. $(5 - 4e^x)^{12}$

37. Trout Lake has become heavily polluted and the Environmental Protection Agency orders all further pollution of the lake immediately stopped. The pollution in the lake will decay according to the equation

$$P(t) = 100 \exp(-0.04t)$$

Here $P(t)$ is the percentage of the current pollution that remains after t months.

a. At what rate is the pollution level declining initially?
b. How quickly is the pollution level declining after one year?
c. How much of the original pollution remains after five years?
d. How quickly is the pollution level declining after five years?

38. Carol Snow has $1500 in a savings account at the Metropolitan Savings and Loan Association. The account pays interest at the rate of 8% compounded continuously. At what instantaneous rate is the amount in the account increasing after

a. Six months?
b. Two years?
c. Five years?
d. Ten years?

A wildlife preserve is big enough to sustain a herd of 1000 deer. A herd is started by moving 50 deer there. The deer population $P(t)$ grows according to the equation

$$P(t) = \frac{1000}{1 + 19e^{-t/4}}$$

where t is the number of years since the herd is started. What is the deer population and at what annual rate is it increasing

39. Initially?
40. After 5 years?
41. After 10 years?
42. After 15 years?
43. After 20 years?
44. After 25 years?

45. After 30 years?

 46. After 35 years?

The demand function for a certain type of music box is $p = d(q) = 10 - 10^{-9} q^3$.

47. Find the inverse function h. Then $q = h(p)$.
48. Find $h'(p)$.
49. The music boxes currently sell for $9 each. How many are sold?
50. If the price is raised 3¢, how many music boxes will be sold? Use the function h' to obtain your answer.
51. If the price is lowered 3¢, how many music boxes will be sold? Use the function h' to obtain your answer.
52. Write a formula for the number n of music boxes that will be sold as a function of the number x of cents by which the price differs from $9. Use the function h' to obtain your answer. The formula will be valid for small values of x.
*53. Derive the exponential rule from the chain rule.

Section 10-8 Higher Order Derivatives

If f is a function, its derivative f' is also a function, and therefore may have a derivative. The derivative of f' is called the *second derivative* of f, and is denoted f'' or $d^2 f / dx^2$. Similarly the derivative of f'' is called the *third derivative* of f and is denoted f''' or $d^3 f / dx^3$. If f is differentiated n times, the resulting function is called the *nth derivative* of f and is denoted $f^{(n)}(x)$ or $d^n f / dx^n$. Note that the nth derivative of f is written $f^{(n)}(x)$ with parentheses around the n to prevent confusion with the nth power $f^n(x)$ of f.

Example 1 Compute all the derivatives of the function

$$f(x) = 10x^3 - 5x^2 + 4x + 3$$

Solution $f'(x) = 30x^2 - 10x + 4$

$f''(x) = [f'(x)]' = (30x^2 - 10x + 4)' = 60x - 10$

$f'''(x) = [f''(x)]' = (60x - 10)' = 60$

10-8 HIGHER ORDER DERIVATIVES

$$f^{(4)}(x) = [f'''(x)]' = 60' = 0$$

■

All derivatives of f beyond the fourth order will also be 0. This example illustrates an interesting fact.

> If $p(x)$ is a polynomial function of degree n, then $p^{(n)}(x)$ is a constant function, and $p^{(m)}(x) = 0$ for $m > n$ for all x.

Thus it is possible to compute all the derivatives of any polynomial.

Example 2 Compute the second derivative of $g(w) = \ln(w^2 + 1)$.

Solution Begin by computing the first derivative.

$$\frac{dg}{dw} = \frac{1}{w^2 + 1} \cdot 2w = \frac{2w}{w^2 + 1}$$

Now compute the second derivative, which is the derivative of the first derivative

$$\frac{d^2g}{dw^2} = \frac{d}{dw}\left(\frac{dg}{dw}\right) = \frac{(w^2 + 1)2 - 2w(2w)}{(w^2 + 1)^2} \quad \text{Quotient rule.}$$

$$= \frac{2w^2 + 2 - 4w^2}{(w^2 + 1)^2} = \frac{2 - 2w^2}{(w^2 + 1)^2}$$

■

In the next chapter, we will use second derivatives in solving optimization problems.

Example 3 Compute the first four derivatives of $f(x) = e^{3x}$.

Solution Begin by computing f'.

$$f'(x) = (e^{3x})' = 3e^{3x}$$

Then

$$f''(x) = \frac{d}{dx}f'(x) = (3e^{3x})' = 9e^{3x}$$

$$f'''(x) = \frac{d}{dx}f''(x) = (9e^{3x})' = 27e^{3x}$$

$$f^{(4)}(x) = \frac{d}{dx}f'''(x) = \frac{d}{dx}(27e^{3x}) = 81e^{3x}$$

Example 4 Compute the first four derivatives of $f(x) = \ln(3x)$.

Solution $f'(x) = \frac{(3x)'}{3x} = \frac{3}{3x} = \frac{1}{x}$

$$f''(x) = \frac{d}{dx}\left(\frac{1}{x}\right) = \frac{d}{dx}(x^{-1}) = -x^{-2}$$

$$f'''(x) = \frac{d}{dx}(-x^{-2}) = -(-2)x^{-3} = 2x^{-3}$$

$$f^{(4)}(x) = \frac{d}{dx}(2x^{-3}) = -6x^{-4}$$

Exercises 10-8

Compute the derivative and second derivative of each function.

1. $y = 4x^2 - 7x + 3$
2. $y = 6x^3 + 8x^2 - 5x + 2$
3. $y = -x^3 + 6x^2 + 12x - 6$
4. $y = -14x^2 + 7x + 3$
5. $y = 6\ln(x + 5)$
6. $y = 3\ln(2x - 1)$
7. $y = 4\exp(-x^2)$
8. $y = 6\exp(9x + 2)$
9. $y = e^x \ln x$
10. $y = (1 + \ln x)^{1/2}$
11. $y = (1 + x^2)^{5/2}$
12. $y = (4 - x^3)^{20}$

Compute all derivatives of each function.

13. $f(x) = x^3 - 3x^2 + 8x + 17$
14. $f(x) = 2x^3 + 4x^2 + 35$
15. $f(x) = 3x^4 + 5x^3 - 7x^2 - 9x - 2$
16. $f(x) = 6x^4 - 2x^3 + 4x^2 + 10 - 7$
*17. $f(x) = \exp(2x)$
*18. $f(x) = \ln(2x)$

Chapter 10 Review

Evaluate the following limits.

1. $\lim_{x \to 2} e^{3x}$
2. $\lim_{x \to 4} \dfrac{x^2 - x}{x^2 + 4x}$

3. $\lim_{x \to 5} \dfrac{x^2 - 25}{x^2 - 2x - 15}$

4. $\lim_{x \to 2} \dfrac{x^2 - 4x + 4}{x^2 + 5x - 14}$

Compute the derivative of each function using only the definition of derivative. Do not use any differentiation formulas.

5. $f(x) = 8x - 12$

6. $g(w) = 3w^2 - 11$

Use differentiation formulas to find the derivative of each function.

7. $12x + 43$

8. $6x^2 - 9x + 2$

9. $5x + 3\ln x$

10. $2x^{3/2} - 5\exp(x)$

11. $5x^6 \ln x$

12. $\dfrac{8e^x}{x^4}$

13. e^{-12x}

14. $\ln(9 + 2x - x^2)$

15. $(5 - 4x)^7$

16. $x^2 \exp(10x)$

17. $\dfrac{x^2 + 3x + 2}{4x - 1}$

18. $\dfrac{\exp(5x)}{5 + x^2}$

Find the equation of the tangent line at the given point.

19. $y = x^2 - 3x + 2$ at $(4, 6)$

20. $y = e^{2x}$ at $(1, e^2)$

21. $y = x \ln x$ at $(1, 0)$

22. $y = \dfrac{x}{x+1}$ at $(1, 1/2)$

Determine where each function is increasing, and where it is decreasing.

23. $y = 7 + 9x$

24. $y = 20 + x - x^2$

25. $f(w) = 2w^3 - 3w^2 - 36w + 24$

26. $f(x) = xe^x$

Compute the derivative and second derivative of each function.

27. $y = 8x^2 - 7x - 3$

28. $f(x) = 4 \ln x$

29. $y = \exp(6 + x^2)$

30. $f(v) = (4 + v^2)^{5/2}$

31. The cost of making a mold and using it to produce q figurines is

$$C(q) = 2000 + 3q - 0.0001q^2$$

What is the cost of the 801st figurine?

32. The profit function for Speer's Gas Station is

$$P(q) = 0.15q - 200 + 0.00001q^2$$

where q is the number of gallons of gasoline sold in a day. How much profit is made if 5000 gallons are sold? How much additional profit is made if instead 5010 gallons are sold?

33. Eggs are selling wholesale for 37¢ a dozen. Robert Alton owns an egg farm and can produce a dozen eggs for $2000 + 0.25q + 0.00000075q^2$ dollars. How many dozen eggs should he produce in order to maximize profit? What is his maximum possible profit?

34. A bacterial culture is being grown for use in industrial research. The mass of the culture in grams after t hours is given by the formula

$$m(t) = 0.01 \exp(0.6t) \qquad 0 \le t \le 5$$

At what rate is the mass of the culture increasing when $t = 2$?

11

Applications of Derivatives

In Chapter 10, we discovered that profit is maximum when marginal revenue equals marginal cost. This is useful, because management usually wants to maximize profit. Other goals of management might be to maximize revenue, maximize market share, minimize average costs, maximize sales or minimize inventory holding costs. In mathematical terms, meeting these goals involves finding the extreme values of a function. In this chapter, we will learn how to use differential calculus to locate these extreme values. Then we will define functions of more than one variable and use calculus to determine the extreme values of functions of two variables.

Section 11-1 Extreme Values

One way to determine extreme values of a function is to evaluate the function at each number in its domain and then select the largest and the smallest values taken by the function. Since a function usually has infinitely many numbers in its domain, this method is impractical. A better method is to show that a function can have extreme values only at certain points, called *critical points*. Thus, *when you are looking for the extreme values of a function, you only need to consider critical points*. Usually a function has only a few critical points.

Definition. The point $(b, f(b))$ is a *critical point* for the function f if b is in the domain of f and either $f'(b) = 0$ or $f'(b)$ is undefined. If $(b, f(b))$ is a critical point, b is called a *critical number*.

Let's find the critical numbers of the function

$$f(x) = 2x^3 - 3x^2 - 12x + 6$$

First, compute and factor the derivative.

$$f'(x) = 6x^2 - 6x - 12$$
$$= 6(x^2 - x - 2) = 6(x - 2)(x + 1)$$

The critical numbers are the numbers where f' is undefined (none in this case) or where $f' = 0$ (in this case, -1 and 2). Thus -1 and 2 are the only critical numbers of f. The graph of f is shown in Diagram 1.

Example 1 Determine the critical numbers of the function $f(x) = x^2 - 4x + 3$.

Solution $f'(x) = 2x - 4$

f' is defined for every x, and $f'(x) = 0$ only when $x = 2$. The only critical number is 2. The graph of f is shown in Diagram 2.

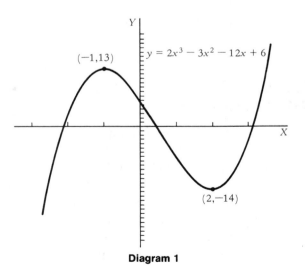

Diagram 1

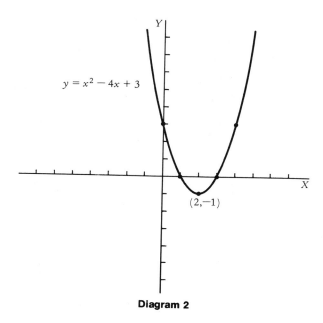

Diagram 2

Example 2 Determine the critical numbers of the function $f(x) = x^4 e^x$.

Solution $f'(x) = x^4 e^x + 4x^3 e^x = x^3 e^x (x + 4)$ Product rule.
f' is defined for every x, and $f'(x) = 0$ only when $x = -4$ or 0. The critical numbers are -4 and 0. The graph of f is shown in Diagram 3.

Example 3 Determine the critical numbers of the function $y = (x - 6)^{1/3}$.

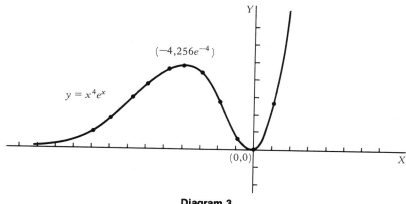

Diagram 3

Solution $y' = \dfrac{1}{3}(x-6)^{-2/3}$

y' is defined except at 6. Since 6 is in the domain of the function $y = (x-6)^{1/3}$, 6 is a critical number. Because y' is never equal to 0, the only critical number is 6. The graph of this function is shown in Diagram 4.

Example 4 Determine the critical numbers of the function $f(x) = x^{1/3}(x-4)$.

Solution
$$f'(x) = \dfrac{1}{3} \cdot x^{-2/3}(x-4) + x^{1/3}$$
$$= \dfrac{x-4}{3x^{2/3}} + x^{1/3}$$
$$= \dfrac{x-4}{3x^{2/3}} + x^{1/3} \cdot \dfrac{3x^{2/3}}{3x^{2/3}}$$
$$= \dfrac{x-4+3x}{3x^{2/3}} = \dfrac{4x-4}{3x^{2/3}}$$
$$= \dfrac{4(x-1)}{3x^{2/3}}$$

The critical numbers of f are 0 (where f' is undefined) and 1 (where $f' = 0$). The graph of f is shown in Diagram 5.

Example 5 Determine the critical numbers of the function $y = \dfrac{1}{x-4}$.

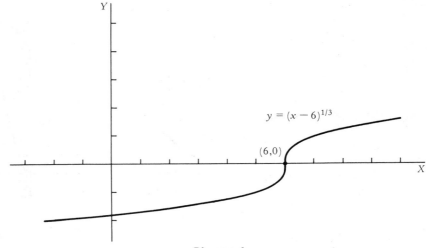

Diagram 4

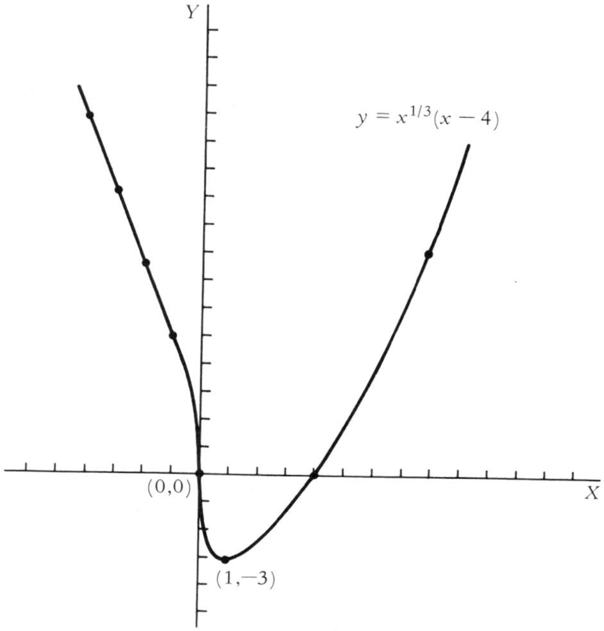

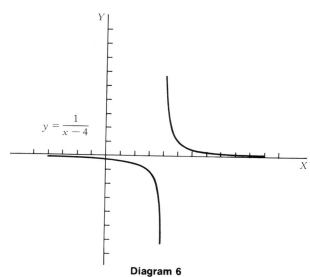

Diagram 5

Solution $y' = \dfrac{-1}{(x-4)^2}$

y' is defined except at 4. However, 4 is not in the domain of the function $y = 1/(x-4)$. y' is never equal to 0. There are no critical numbers. See Diagram 6.

∎

Diagram 6

To continue our discussion, we need some more definitions.

> **Definition.** A *neighborhood* of a number b is an *open interval* $a < x < c$ that contains b.

Usually we will be concerned with small neighborhoods of a number. Examples of neighborhoods of 2 are $1 < x < 3$, $1.9 < x < 2.2$, and $1.999 < x < 2.001$.

> **Definition.** The function f has a *local maximum* at b if $f(b) \geq f(x)$ for all x is some neighborhood of b.
>
> **Definition.** The function f has a *local minimum* at b if $f(b) \leq f(x)$ for all x in some neighborhood of b.
>
> **Definition.** The function f has a *local extremum* at b if f has either a local maximum or a local minimum at b.

A local maximum is shown in Diagram 7, and a local minimum in Diagram 8. The following theorem makes it easy to find the local extrema of a function.

> **Theorem.** If f has a local extremum at b, then b is a critical number for f.

Note that the theorem does not say that the function f has a local extremum at each critical number. The theorem says that if f has a local

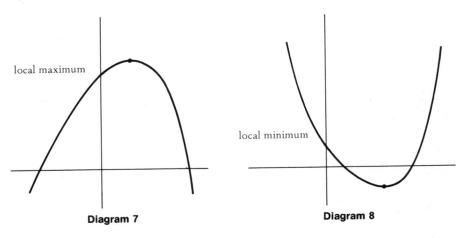

Diagram 7   **Diagram 8**

extremum at b, then b is a critical number. This is equivalent to saying that if b is not a critical number, f cannot have a local extremum at b.

Here is why the theorem is valid. Suppose $f'(b) > 0$. Then f is increasing at b. If x_1 and x_2 are near b, and $x_1 > b > x_2$, then $f(x_1) > f(b) > f(x_2)$. Thus, f cannot have a local extremum at b. Similar reasoning is valid if $f'(b) < 0$. Thus, f cannot have a local extremum at b if $f'(b) > 0$ or $f'(b) < 0$. The only remaining possibilities are that $f'(b) = 0$ or that $f'(b)$ is undefined. Either of these remaining possibilities implies that b is a critical number.

Original equation test

We will now study a method for determining whether a critical number of a function is a local minimum, a local maximum, or not a local extremum. This method is called the *original equation test* because each critical number is tested using the formula for the function. In the next section, we will study two more methods of testing critical numbers.

Original Equation Test. Assume b is a critical number for f. Pick numbers a and c with $a < b < c$ such that b is the only critical number for f between a and c.
1. If $f(a) \leq f(b)$ and $f(c) \leq f(b)$, then f has a local maximum at b.
2. If $f(a) \geq f(b)$ and $f(c) \geq f(b)$, then f has a local minimum at b.
3. If $f(a) > f(b) > f(c)$ or $f(a) < f(b) < f(c)$, then f does not have a local extremum at b.

Consider case 1. The definition of local maximum would require us to compare $f(b)$ with $f(x)$ for each x satisfying $a < x < c$. However, the original equation test makes life much simpler, since $f(b)$ must only be compared with the two numbers $f(a)$ and $f(c)$. See Diagram 9. Note that f is increasing from a to b and decreasing from b to c.

Case 2 is illustrated in Diagram 10. If $f(b) \leq f(a)$ and $f(b) \leq f(c)$, then $f(b) \leq f(x)$ whenever $a < x < c$ and f has a local minimum at b. Note that f is decreasing from a to b and increasing from b to c.

Case 3 is illustrated in Diagram 11. If one of the numbers $f(a)$ and $f(c)$ exceeds $f(b)$ and the other is less than $f(b)$, then f does not have a local extremum at b even though b is a critical number.

Now let's use the original equation test on the function $f(x) = 2x^3 - 3x^2 - 12x + 6$. Earlier in the section, we found that -1 and 2 were the

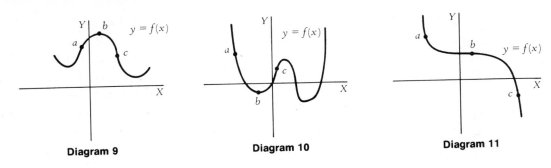

Diagram 9   Diagram 10   Diagram 11

critical numbers. Use -2 and 0 to test -1.

$$f(-2) = 2 \qquad f(-1) = 13 \qquad f(0) = 6$$

Since $f(-2) \leq f(-1)$ and $f(0) \leq f(-1)$, case 1 applies and $(-1, 13)$ is a local maximum point.
Use 0 and 3 to test 2

$$f(0) = 6 \qquad f(2) = -14 \qquad f(3) = -3$$

Since $f(0) \geq f(2)$ and $f(3) \geq f(2)$, case 2 applies and $(2, -14)$ is a local minimum point.

Examples Use the original equation test to find all local extrema of each function. Examples 6–9 use the same functions as Examples 1–4.

Example 6 $f(x) = x^2 - 4x + 3$

Solution The only critical number is 2. Pick a number (for example, 1) less than 2 and a number (for example, 3) greater than 2. Now compute $f(x)$ when $x = 1, 2,$ and 3.

$$f(1) = 0 \qquad f(2) = -1 \qquad f(3) = 0$$

Since $f(1) > f(2)$ and $f(3) > f(2)$, case 2 of the original equation test applies and $(2, -1)$ is a local minimum point.

Example 7 $f(x) = x^4 e^x$

Solution The critical numbers are -4 and 0. Now $-5 < -4 < -1$ and

$$f(-5) = 625e^{-5} = 4.21 \qquad f(-4) = 256e^{-4} = 4.69$$

$$f(-1) = e^{-1} = 0.37$$

Since $f(-5) < f(-4)$ and $f(-1) < f(-4)$, case 1 of the original equation test applies and the function has a local maximum at -4.

To test 0, use -1 and 1.

$$f(-1) = e^{-1} = 0.37 \qquad f(0) = 0 \qquad f(1) = e = 2.72$$

Case 2 applies and the function has a local minimum at 0.

Example 8 $y = (x - 6)^{1/3}$

Solution The only critical number is 6. Now $5 < 6 < 7$ and

$$y(5) = -1 \qquad y(6) = 0 \qquad y(7) = 1$$

Since $y(5) < y(6) < y(7)$, case 3 of the original equation test applies and the function does not have a local extremum at 6.

Example 9 $f(x) = x^{1/3}(x - 4)$

Solution The critical numbers are 0 and 1. Use -1 and $1/2$ to test 0. Now $-1 < 0 < 1/2$ and

$$f(-1) = 5 \qquad f(0) = 0 \qquad f\left(\frac{1}{2}\right) = \left(\frac{1}{2}\right)^{1/3} \cdot (-3.5) = -2.8$$

Since $f(-1) > f(0) > f(1/2)$, case 3 of the original equation test applies and the function does not have a local extremum at 0.

Use $1/2$ and 8 to test 1. Now $1/2 < 1 < 8$ and

$$f\left(\frac{1}{2}\right) = \left(\frac{1}{2}\right)^{1/3} \cdot (-3.5) = -2.8 \qquad f(1) = -3 \qquad f(8) = 8$$

Since $f(1/2) > f(1)$ and $f(8) > f(1)$, case 2 of the original equation test applies and f has a local minimum at 1.

Example 10 The cost function for manufacturing Excello cameras is

$$C(x) = x^2 + 900x + 2500$$

How many cameras should Excello manufacture in order to minimize the average cost of producing a camera?

Solution The average cost function is

$$AC(x) = \frac{C(x)}{x} = x + 900 + \frac{2500}{x}$$

Now let's find the critical numbers of the function AC. First, compute AC'.

$$AC'(x) = 1 - \frac{2500}{x^2}$$

The only critical number is 50, since $AC'(50) = 0$. (Since the domain of AC contains only positive numbers, 0 and -50 cannot be critical numbers.) Let's test 50 with the original equation test using the numbers 10 and 100.

$$AC(10) = 1160 \qquad AC(50) = 1000 \qquad AC(100) = 1025$$

Since $AC(50) < AC(10)$ and $AC(50) < AC(100)$, the average cost is minimum when 50 cameras are made.

■

Let's consider a graphical description of critical points. At a critical point $(b, f(b))$, the tangent line to the graph of f is horizontal if $f'(b) = 0$. If $f'(b)$ is undefined, the tangent line may be vertical, or there may be no tangent line.

In Diagram 12, points A, B, C, and D are critical points. These points are described in Table 1.

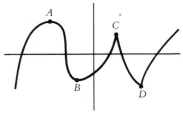

Diagram 12

TABLE 1

Point	Derivative		Tangent Line
A	0	local maximum	horizontal
B	0	local minimum	horizontal
C	undefined	local maximum	vertical
D	undefined	local minimum	none

In Diagram 13, points E and F are critical points. These points are described in Table 2.

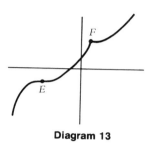

Diagram 13

TABLE 2

Point	Derivative		Tangent Line
E	0	not a local extremum	horizontal
F	undefined	not a local extremum	none

Exercises 11-1

Determine the critical numbers of each function. At each critical number, determine whether the function has a local maximum, local minimum, or does not have a local extremum.

1. $y = x^2$
2. $y = x^3$
3. $y = x^4$
4. $y = x^{2/3}$
5. $y = -x^{4/3}$
6. $y = x^{5/3}$
7. $y = 4x^2 + 8x - 7$
8. $y = -2x^2 + x - 3$
9. $y = x^3 + 9x^2 + 15x + 21$
10. $y = x^3 - 9x^2 - 21x - 15$
11. $y = (x + 4)^3$
12. $y = (2x - 3)^4$
13. $y = xe^x$
14. $y = x^2 e^x$
15. $y = x \ln x$
16. $y = \dfrac{\ln x}{x}$
17. $y = x^3 - 3x + 1$
18. $y = 4x^3 - 3x + 12$

19. $y = (1 - x^2)^5$
20. $y = (1 - x^2)^4$
21. $y = e^{5x}$
22. $y = \ln(4 + x^2)$
23. $y = \exp(-x^2)$
24. $y = \exp(9 + x^2)$
*25. $y = x^{1/3}(x - 7)^2$
*26. $y = x^{2/3}(x - 5)$
*27. $y = (x - 4)^8(x + 6)^{5/3}$
*28. $y = (2x + 3)^4(x - 1)^{1/3}$
*29. $y = x|x|$
*30. $y = \dfrac{x}{|x|}$

31. The concentration (given as a percentage) of an experimental drug in the bloodstream t hours after it is taken is

$$C(t) = \frac{0.05t}{t^2 + 9}$$

What is the maximum concentration of the drug?

32. The revenue function for Ace Rentals is

$$R(q) = 100q^2 \exp(-0.2q)$$

How many items must be rented in order to maximize
a. Revenue?
b. Average revenue per item?

33. The profit function for a sightseeing tour is

$$P(q) = 192q - 0.04q^3$$

where q is the number of customers on the tour. What is the maximum possible profit?

34. The cost function for growing corn on a plot of land is

$$C(q) = q^2 + 100q + 900$$

where q is the amount of corn, measured in thousands of bushels. How much corn should be grown in order to minimize the average cost of a bushel?

35. Sociologists have studied how rumors spread. In one model, the rate R at which a rumor spreads is proportional to the number x of people who have heard the rumor and also to the number who have not heard the rumor. If the total population is N,

$$R = kx(N - x)$$

where k is a constant. Determine the value of x when R is maximum.

36. The reaction $R(D)$ of the body to the dosage D of a drug is given by

$$R(D) = D^2 \left(\frac{C}{2} - \frac{D}{3} \right)$$

where C is the maximum dosage possible. What is the value of D when $R'(D)$ is maximum? (The body is most sensitive to changes in the dosage when $R'(D)$ is maximum.)

37. When you cough, the trachea contracts so that air passes through it faster. The velocity v of air in the trachea is given by

$$v = k(R - r)r^2$$

where k is a constant, R is the normal radius of the trachea, and r is its contracted radius. What value of r maximizes v?

The energy E expended by a fish to swim one mile upstream is given by

$$E(v) = \frac{cv^k}{v - w}$$

where c is a constant, k is a constant depending on the shape of the fish, w is the speed of the water, and v is the speed at which the fish swims.

38. What value of v minimizes E if $k = 2.5$?

39. What value of v minimizes E if $k = 3$?

40. Determine a general formula for the value of v that minimizes E. Assume $k > 2$.

Let $f(x) = c$ be a constant function. Determine

*41. All critical points.

*42. All local maximum points.

*43. All local minimum points.

*44. All local extreme points.

Determine the critical numbers of each function. At each critical number, determine whether the function has a local maximum, local minimum, or does not have a local extremum.

45. $y = 4.72x^2 - 3.25x + 8.74$

46. $y = 2.38x^2 + 4.92x - 7.31$

◆ 47. $y = 3.27x^3 + 5.14x^2 - 9.36x - 11.26$

◆ 48. $y = 1.37x^3 - 4.35x^2 - 1.21x + 4.85$

Section 11-2 The First and Second Derivative Tests

A function can have extreme values only at its critical numbers. We have just learned how to use the original equation test to decide whether a function has a local maximum, a local minimum, or neither at a specific critical number. Two other ways of testing critical numbers, the *first derivative test* and the *second derivative test*, are presented in this section. The first derivative test involves testing a critical point for a function f by just looking at the sign of f' at two nearby points.

> *First Derivative Test.* Assume b is a critical number for f. Pick numbers a and c with $a < b < c$ such that b is the only critical number between a and c.
> 1. If $f'(a) > 0$ and $f'(c) < 0$, then f has a local maximum at b.
> 2. If $f'(a) < 0$ and $f'(c) > 0$, then f has a local minimum at b.
> 3. If $f'(a)$ and $f'(c)$ are either both positive or both negative, then f does not have a local extremum at b.

The three cases in this test correspond to the three cases in the original equation test. To see why this test works, you need only recall that $f'(x)$ is the slope of the tangent line at $(x, f(x))$.

Now let's test the critical numbers of the function $f(x) = 2x^3 - 3x^2 - 12x + 6$. In Section 11-1, we found that -1 and 2 were the critical numbers and that $f'(x) = 6x^2 - 6x - 12$. Use -2 and 0 to test -1.

$$f'(-2) = 24 \qquad f'(0) = -12$$

Since $f'(-2) > 0$ and $f'(0) < 0$, case 1 of the first derivative test applies and $(-1, 13)$ is a local maximum point.

Use 0 and 3 to test 2.

$$f'(0) = -12 \qquad f'(3) = 24$$

Since $f'(0) < 0$ and $f'(3) > 0$, case 2 of the first derivative test applies and $(2, -14)$ is a local minimum point.

Examples Use the first derivative test to determine whether each critical number is a local maximum, a local minimum, or not a local extremum.

Examples 1–4 in this section use the same functions as Examples 1–4 in Section 11-1. The derivatives and critical numbers were determined in Section 11-1.

Example 1 $f(x) = x^2 - 4x + 3$, $f'(x) = 2x - 4$. The only critical number is 2.

Solution Use 1 and 3 to test 2. Since $f'(1) = -2 < 0$ and $f'(3) = 2 > 0$, case 2 of the first derivative test applies and f has a local minimum at 2.

Example 2 $f(x) = x^4 e^x$, $f'(x) = x^3 e^x (x + 4)$. The critical numbers are -4 and 0.

Solution Use -5 and -1 to test -4. Since

$$f'(-5) = 125e^{-5} > 0 \quad \text{and} \quad f'(-1) = -3e^{-1} < 0$$

case 1 of the first derivative test applies and f has a local maximum at -4.
Use -1 and 1 to test 0. Since $f'(-1) < 0$ and $f'(1) = 5e > 0$, case 2 of the first derivative test applies and f has a local minimum at 0.

Example 3 $y = (x - 6)^{1/3}$, $y' = \dfrac{(x - 6)^{-2/3}}{3}$. The only critical number is 6.

Solution Use 5 and 7 to test 6. Since

$$f'(5) = \frac{1}{3} > 0 \quad \text{and} \quad f'(7) = \frac{1}{3} > 0$$

case 3 of the first derivative test applies and $(6, 0)$ is not a local extreme point.

Example 4 $f(x) = x^{1/3}(x - 4)$, $f'(x) = \dfrac{4(x - 1)}{3x^{2/3}}$. The critical numbers are 0 and 1.

Solution Use -1 and $1/2$ to test 0. Since $f'(-1) = -8/3 < 0$ and $f'(1/2) = -2 \cdot 2^{2/3}/3 < 0$, case 3 of the first derivative test applies and $(0, 0)$ is not a local extreme point.
Use $1/2$ and 8 to test 1. Since $f'(1/2) < 0$ and $f'(8) = 28/12 > 0$, case 2 of the first derivative test applies and f has a local minimum at 1.

Second derivative test

The last test is the *second derivative test*. This test is sometimes easier to use than the other tests, especially on polynomials, but it frequently gives no conclusion. This test is applicable only at critical numbers where the derivative is 0. Using this test involves computing the value of f'' at the critical number.

> **Second Derivative Test.** Assume b is a critical number for f, $f'(b) = 0$, and $f''(b)$ is defined.
> 1. If $f''(b) < 0$, then f has a local maximum at b.
> 2. If $f''(b) > 0$, then f has a local minimum at b.
> 3. If $f''(b) = 0$, no conclusion may be reached.

To illustrate case 3, consider the functions x^3, x^4, and $-x^4$. The derivative and second derivative have the value 0 at $x = 0$ for each of these functions. The function x^3 does not have a local extremum at 0, x^4 has a local minimum at 0, and $-x^4$ has a local maximum at 0. Thus no conclusion may be reached when $f''(b) = 0$.

To justify the second derivative test, consider case 1. Since $f''(b) < 0$, f' is decreasing at b. If c is slightly larger than b, then $f'(c) < 0$ since $f'(b) = 0$. Similarly, if a is slightly less than b, then $f'(a) > 0$. Now apply case 1 of the first derivative test. See Diagram 4.

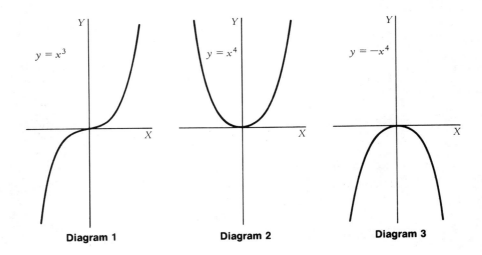

Diagram 1 Diagram 2 Diagram 3

11-2 THE FIRST AND SECOND DERIVATIVE TESTS

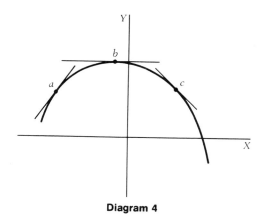

Diagram 4

Now let's use the second derivative test to find the local extreme points of the function $f(x) = 2x^3 - 3x^2 - 12x + 6$. First compute f' and f''.

$$f'(x) = 6x^2 - 6x - 12$$
$$f''(x) = 12x - 6$$

We have already found that -1 and 2 are the critical numbers. Since $f''(-1) = -18 < 0$, case 1 of the second derivative test applies and f has a local maximum at -1. Since $f''(2) = 18 > 0$, case 2 of the second derivative test applies and f has a local minimum at 2.

Examples Use the second derivative test to determine, whenever possible, whether the function has a local maximum or a local minimum at each critical number.

Examples 5–8 use the same functions as the examples using the first derivative test given earlier in this section.

Example 5 $f(x) = x^2 - 4x + 3$, $f'(x) = 2x - 4$.
The only critical number is 2.

Solution $f''(x) = 2$. At the critical number 2, $f' = 0$ and $f'' > 0$. By the second derivative test, the function has a local minimum at 2.

Example 6 $f(x) = x^4 e^x$, $f'(x) = x^3 e^x(x + 4) = (x^4 + 4x^3)e^x$.
The critical numbers are -4 and 0.

Solution Compute $f''(x)$.

$$f''(x) = \left[(x^4 + 4x^3)e^x\right]'$$
$$= (x^4 + 4x^3)(e^x)' + (x^4 + 4x^3)'e^x \quad \text{Product rule.}$$
$$= (x^4 + 4x^3)e^x + (4x^3 + 12x^2)e^x = (x^4 + 8x^3 + 12x^2)e^x$$

At the critical number -4, $f' = 0$ and $f'' = -64e^{-4} < 0$. By the second derivative test, the function has a local maximum at -4.

At the critical number 0, $f' = 0$ and $f'' = 0$. The second derivative test gives no conclusion. This critical number must be tested using either the original equation test or the first derivative test. We have already done so, and found that $(0,0)$ is a local minimum point.

Example 7 $y = (x-6)^{1/3}$, $y' = \dfrac{1}{3}(x-6)^{-2/3}$.

The only critical number is 6.

Solution Since $y'(6)$ is not defined, the second derivative test does not apply at 6. Either the original equation test or the first derivative test must be used. We have already done so, and determined that $(6,0)$ is not a local extreme point.

Example 8 $f(x) = x^{1/3}(x-4)$, $f'(x) = \dfrac{4(x-1)}{3x^{2/3}}$.

The critical numbers are 0 and 1.

Solution Compute $f''(x)$. Write $f'(x)$ as $f'(x) = \dfrac{4}{3}(x-1)x^{-2/3} = \dfrac{4}{3}(x^{1/3} - x^{-2/3})$. Then

$$f''(x) = \dfrac{4}{3}\left(\dfrac{1}{3}x^{-2/3} + \dfrac{2}{3}x^{-5/3}\right)$$
$$= \dfrac{4}{9}(x^{-2/3} + 2x^{-5/3})$$

Since $f'(0)$ is undefined, the second derivative test does not apply at 0. Since $f'(1) = 0$ and $f''(1) = 4/3 > 0$, by the second derivative test the function has a local minimum at 1.

∎

Exercises 11-2
1. Give an argument to justify case 2 of the second derivative test.
2. Do each exercise of the previous section, using the

a. First derivative test. (Omit exercises 41–44.)
b. Second derivative test. (Omit exercises 25–31 and 38–44.)

Section 11-3 Applications to Optimization

In the first two sections of this chapter, powerful tools were developed for finding the maximum and minimum values of a function. Now we will solve several types of applied problems with these tools. The basic procedure is as follows.

Procedure for Applied Optimization Problems
1. Determine the quantity to be maximized or minimized.
2. Assign symbols to all the variables in the problem.
3. Find equations or inequalities relating the variables. (A diagram is often useful here.)
4. Write an equation that expresses the quantity to be optimized as a function of one of the variables.
5. Use the methods of the two previous sections to find the extreme values of this function.

This procedure is a general guideline rather than a precise sequence of steps. There are many different kinds of optimization problems, and different problems must be set up and solved in appropriately different ways.

Example 1 A farmer owns a plot of land that extends 200 feet along a straight river bank and is 100 ft deep. He wishes to enclose as much land as possible in a rectangle. He has enough material to make a fence 240 ft long. No fence is needed along the river bank. How much land can he enclose?

Solution Obviously the river bank will be one side of the rectangle. Let A be the area of the rectangle. Then A is to be maximized. Let x denote the depth of the region. Then the length of the region is $240 - 2x$. See Diagram 1.
Now

$$A(x) = x(240 - 2x) = 240x - 2x^2$$
$$A'(x) = 240 - 4x$$

A' is always defined, and $A'(x) = 0$ only when $x = 60$. Thus 60 is the only critical number. Now test 60 with the first derivative test. (You could

450 CHAPTER 11 APPLICATIONS OF DERIVATIVES

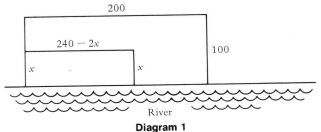

Diagram 1

use the original equation test or the second derivative test instead.) Since $A'(50) = 40 > 0$, and $A'(70) = -40 < 0$, A is maximum when $x = 60$. The maximum area possible is $A(60) = 7200$ sq ft.

Example 2 Change the problem in Example 1 by adding the requirement that the part of the fence parallel to the river bank must be between 40 and 100 ft long. How much land can the farmer enclose now?

Solution The problem is still to maximize the function $A(x) = 240x - 2x^2$, but the values of x are restricted. The part of the fence parallel to the river bank has length $240 - 2x$, so that the inequalities

$$40 \leq 240 - 2x \leq 100$$

must now be satisfied. Solve these inequalities.

$-200 \leq -2x \leq -140$ Subtract 240 from each expression.

$100 \geq x \geq 70$ Multiply by -0.5. Recall that the direction of an inequality sign changes when the inequality is multiplied by a negative number.

or

$70 \leq x \leq 100$

The value $x = 60$ is no longer allowable because 60 does not satisfy the inequalities. Since $A(x)$ has no critical numbers between 70 and 100, the maximum value of A cannot occur at any number between 70 and 100, and must therefore occur at 70 or at 100. Since $A(70) = 7000$ and $A(100) = 4000$, the maximum area now is 7000 sq ft. The graph of the function A is shown in Diagram 2.

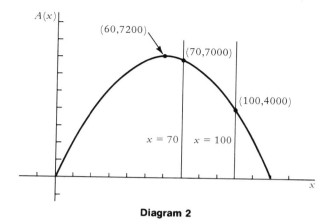

Diagram 2

Example 3 The volume of a closed can is given by $\pi r^2 h$ and the total surface area by $2\pi r^2 + 2\pi rh$, where r is the radius of the top of the can and h is the height of the can. What dimensions should the can have if the surface area is to be a minimum and the volume is to be 54 cu in.?

Solution The volume determines how much the can contains. The surface area determines the amount of material needed to make the can and thus the cost of the can. The problem therefore asks for the dimensions of the cheapest can that can contain 54 cu in. of merchandise. The problem is to minimize the surface area S, which is given by the formula

$$S = 2\pi r^2 + 2\pi rh$$

This formula involves the two variables r and h, but it is possible to express S in terms of one variable. Since the volume is 54,

$$\pi r^2 h = 54 \quad \text{so that} \quad h = \frac{54}{\pi r^2}$$

Now express S as a function of r alone by substituting $54/(\pi r^2)$ for h in the equation for S.

$$S = 2\pi r^2 + 2\pi rh = 2\pi r^2 + 2\pi r \cdot \frac{54}{\pi r^2}$$

$$= 2\pi r^2 + \frac{108}{r}$$

Next compute the derivative of S with respect to r.

$$\frac{dS}{dr} = 4\pi r - \frac{108}{r^2}$$

Since dS/dr is defined for all positive values of r, the only critical numbers will be the solutions of

$$\frac{dS}{dr} = 4\pi r - \frac{108}{r^2} = 0$$

$$4\pi r = \frac{108}{r^2}$$

$$r^3 = \frac{108}{4\pi} = \frac{27}{\pi}$$

$$r = \left(\frac{27}{\pi}\right)^{1/3} = \frac{3}{\pi^{1/3}}$$

The only critical number is $3/\pi^{1/3}$. Now test this number by the second derivative test. (The original equation test or the first derivative test could be used instead.)

$$\frac{d^2S}{dr^2} = 4\pi + \frac{216}{r^3}$$

Since $d^2S/dr^2 > 0$ whenever $r > 0$, S has a local minimum when $r = 3/\pi^{1/3}$. Note that it is not necessary to evaluate d^2S/dr^2 when $r = 3/\pi^{1/3}$. The minimum value of S is $S(3/\pi^{1/3}) = 54\pi^{1/3} = 79.1$ sq in. When S is minimum, $r = 3/\pi^{1/3} = 2.05$ in. and $h = 54/(\pi r^2) = 2r = 4.10$ in. (Verify this.) Thus the cheapest can has height equal to its diameter. (Do you often see such cans in grocery stores?)

Example 4 A book is to have one-inch margins at the top and bottom of each page and 1/2-inch margins on the sides. The area of each page (and thus the amount of paper used) is to be 50 sq in. What dimensions will maximize the printed area on each page (and thereby minimize the number of pages in the book and hence the cost of the book)?

Solution Let h be the height of the page and w be the width. We seek to maximize the printed area P. See Diagram 3.

The printed area has width $w - 1$ and height $h - 2$. Thus,

$$P = (w - 1)(h - 2)$$

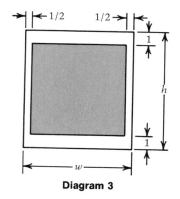

Diagram 3

We have expressed P in terms of the two variables w and h. Now the area of the page is the product wh of its width with its height and is given as 50 sq in. This gives the equation

$$wh = 50$$

Solve this equation for w to obtain $w = 50/h$ and substitute $50/h$ for w in the formula for P to obtain

$$P = \left(\frac{50}{h} - 1\right)(h - 2)$$

$$= 50 - \frac{100}{h} - h + 2 = 52 - h - \frac{100}{h}$$

Then

$$\frac{dP}{dh} = -1 + \frac{100}{h^2}$$

dP/dh is defined whenever $h > 0$. The only positive value for h at which $dP/dh = 0$ is $h = 10$. (Why must h be positive?) Thus 10 is the only critical number for h. Now test this critical number by the original equation test. (The first derivative test or the second derivative test could be used instead.) Now $5 < 10 < 12$ and

$$P(5) = 27 \qquad P(10) = 32 \qquad P(12) = 31\tfrac{2}{3}$$

Since $P(5) < P(10)$ and $P(12) < P(10)$, P is maximum when $h = 10$. When $h = 10$, $w = 50/10 = 5$.

The maximum printed area of 32 sq in. is obtained when the page is 10 in. high and 5 in. wide.

∎

So far in this section, differential calculus has been used to solve specific applied problems. Calculus can also be used to find formulas for solving certain classes of problems. For example, in Section 10-5, the general problem of maximizing profit was studied, and we obtained a classical law of economics—profit is maximum when marginal revenue equals marginal cost.

Now we will study other general problems. The first one is minimizing the average cost function $AC(q)$. When $AC(q)$ is minimum, each item is produced as cheaply as possible, and so the production process is operating at maximum efficiency. Now let's take the derivative of the average cost function.

$$AC(q) = \frac{C(q)}{q}$$

so that

$$AC'(q) = \frac{qC'(q) - C(q)}{q^2} \qquad \text{Quotient rule.}$$

Since $AC(q)$ is defined whenever q is positive, the only critical numbers are the solutions of the equation

$$qC'(q) - C(q) = 0$$

To solve this equation, substitute $qAC(q)$ for $C(q)$.

$$qC'(q) - qAC(q) = 0$$
$$q[C'(q) - AC(q)] = 0$$
$$C'(q) - AC(q) = 0 \qquad \text{Divide by } q \text{ since } q > 0.$$
$$AC(q) = C'(q) = MC(q) \qquad C' \text{ is also called marginal cost.}$$

The critical numbers are the solutions of $AC(q) = C'(q)$. Economists usually assume that there is only one critical number and that average cost is minimum at this critical number.

> Average cost is minimum when
> $$AC(q) = MC(q)$$

Example 5 The cost function for Napo Tools is

$$C(q) = 0.01q^2 + 500q + 8100$$

where q is the number of lathes manufactured. How many lathes should Napo make in order to minimize average cost?

Solution $AC(q) = \dfrac{C(q)}{q} = 0.01q + 500 + \dfrac{8100}{q}$

$MC(q) = 0.02q + 500$

Average costs are minimum when $AC(q) = MC(q)$.

$$0.01q + 500 + \frac{8100}{q} = 0.02q + 500$$

$$\frac{8100}{q} = 0.01q$$

$$810{,}000 = q^2$$

$q = -900$ or $q = 900$ Since q must be positive, -900 is discarded.

Napo should make 900 lathes in order to minimize average cost. ∎

Example 5 could also be done by computing AC' directly and using it to find the critical number of the average cost function.

Another general problem is minimizing the cost of inventories. The maintenance of inventories is very important in business. If inventories are too small, some items run out and sales are lost. If inventories are too large, money and storage space are tied up in an investment (the excess inventory) that pays no return. Here we will consider one aspect of inventory maintenance—how many of an item to order at a time.

Assume a store sells A of a certain item during the year, and that sales are spread out evenly. Suppose the manager orders x of the item each time the store needs more. This means that during the year, the manager will place A/x orders. Each time an order is placed, there is a *reorder cost* of r. This cost pays for paperwork, delivery charges, and so on, and is independent of the size of the order. The total reorder costs during the year are

$$\frac{Ar}{x}$$

Now suppose the orders are placed so that a new supply of items arrives just as the old supply is exhausted. Then the number of items in inventory varies between 0 and x and averages $x/2$. Suppose the *carrying costs* on an item are c. This means that it costs c dollars to keep an item in inventory for one year. Carrying costs include insurance, interest that could otherwise be earned on the wholesale price paid for the item, storage costs, and so on. The total carrying costs during the year are

$$\frac{cx}{2}$$

The total inventory costs K are the sum of the reordering costs and the carrying costs.

$$K(x) = \frac{Ar}{x} + \frac{cx}{2}$$

To minimize K, compute the derivative K'.

$$K'(x) = \frac{-Ar}{x^2} + \frac{c}{2}$$

K' is defined whenever $x > 0$, and the only positive solution of the equation $K'(x) = 0$ is $(2Ar/c)^{1/2}$. Thus there is only one critical number. Since

$$K''(x) = \frac{2Ar}{x^3}$$

and is positive whenever x is positive, K is minimum where $x = (2Ar/c)^{1/2}$. This value for x is called the *economic order quantity* (EOQ).

$$\boxed{x = \sqrt{\frac{2Ar}{c}} \qquad \text{EOQ}}$$

Example 6 Lee-Taylor Video expects to sell 600 color televisions during the year with sales occurring at a fairly constant rate. There is a $210 reorder fee, and the carrying costs on a color television are $70 annually. How many televisions should be ordered each time?

Solution Use the formula $x = \sqrt{\dfrac{2Ar}{c}}$ to obtain

$$x = \sqrt{\frac{2(600)(210)}{70}} = \sqrt{3600} = 60$$

Lee-Taylor Video should order 60 color televisions at a time.

Exercises 11-3

1. Find two numbers whose sum is 12, such that the sum of their squares is as small as possible.
2. Find two numbers whose sum is 20, such that their product is as large as possible.
3. Find the dimensions of the rectangle of largest area that can be enclosed with 100 ft of fencing.
4. Find the dimensions of the rectangle of smallest perimeter with an area of 100 sq ft.
5. A bus tour costs $100 per person if there are exactly 50 people on the tour. For each additional person over this 50, the fare is reduced by $1. What is the maximum possible revenue for the bus company from this tour?
6. A tour can accommodate up to 200 people. If all 200 tickets are sold, the price per ticket is $100. If fewer than 200 tickets are sold, the price of a ticket is increased by $1 for each unsold ticket. What is the maximum possible revenue from the tour?
7. Ray Walters knows that if he plants 100 apple trees per acre in his orchard, he will average 53 bushels of apples per tree each year. For each additional tree planted per acre, the average per tree drops by 1/4 bushel. How many additional trees should he plant per acre in order to maximize apple harvest?
8. E-Z Rental usually rents 40 carpet cleaners a day for $10 each. For each 25¢ decrease in the rental charge, two more carpet cleaners could be rented. For how much should E-Z Rental rent carpet cleaners in order to maximize revenue?
9. A box with a square base and an open top is to have volume 32 cu ft. The material for the box costs 2¢ per square foot. What should be the dimensions of the box, in order to minimize the cost of the material?
10. Repeat exercise 9 if the material for the bottom of the box costs 4¢ per square foot and the material for the sides costs 2¢ per square foot.
11. Repeat exercise 9 if the box is to have a top.
12. Repeat exercise 9 if the box is to have a top, the material for the top and bottom of the box costs 4¢ per square foot, and the material for sides costs 2¢ per square foot.

13. A box without a top is to be made from a 12-in. square sheet of cardboard by cutting out a square from each corner of the cardboard and folding up the sides of the box. What is the maximum possible volume for the box?

14. A box with a top is to be twice as long as it is wide. What dimensions for the box would require the least amount of material? The volume of the box must be 72 cu ft.

15. A farmer wants to enclose two equal rectangular fields as shown. How large an area can the farmer enclose in each rectangle, if 480 yards of fencing are available?

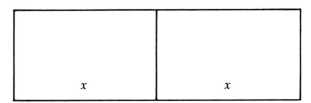

16. The U.S. Postal Service will deliver a package whose height plus girth (distance around) does not exceed 72 inches. What are the dimensions of the largest rectangular box with square bottom that meets this condition?

17. The cost function for a washing machine factory is $C(q) = q^2 + 200q + 6400$, where q is the number of washing machines made daily. What is the minimum average cost of a washing machine?

18. The cost function for a clothes dryer factory is $C(q) = q^2 + 160q + 3600$, where q is the number of dryers made daily. What is the minimum average cost of a dryer?

19. The cost function for a beekeeper is $C(q) = 10q + 50 + 0.1q^{1.5}$, where q is the number of gallons of honey produced. What is the minimum average cost of a gallon of honey?

20. The cost of producing q stereo receivers is $C(q) = 180q + 256 + q^{3/2}$. What is the minimum average cost?

21. Forest Village usually sells 400 vacuum cleaners during the year. Sales occur at an almost constant rate. How many vacuum cleaners should be ordered at a time, if the annual carrying costs on a vacuum cleaner are $10 and there is a $20 reorder cost?

22. Forest Village also sells 750 garbage disposals a year at a fairly constant rate. Annual carrying costs on a garbage disposal are $3, and the reorder cost is $20. How many garbage disposals should be ordered at a time?

23. Tom's Furniture sells 200 model A-78W bedroom sets a year at a

constant rate. Determine the EOQ if the annual carrying costs on a bedroom set are $40 and the reorder cost is $250. How many times a year does Tom's Furniture order bedroom sets?

 24. Tea World sells 5000 pounds of Earl Grey tea annually. Sales occur at a constant rate. Annual carrying costs on a pound of tea are 30¢ and the reorder cost is $30. How much Earl Grey tea should Tea World order at a time?

Section 11-4 Partial Derivatives

So far, we have only considered functions of one variable. If $y = x^2$, the value of the dependent variable y is determined by the value of the independent variable x. In many applications, there is more than one independent variable. For example, consider the simple interest formula $I = Prt$. Here the value of the dependent variable I (interest) is determined by the values of the three independent variables P (principle), r (interest rate), and t (time). Thus I is a *function of more than one variable*.

We will now study functions of more than one variable. Each function will be given by a formula. For example, if $z = f(x, y) = xy + 2y^2$, then $f(3, 4) = 3 \cdot 4 + 2 \cdot 4^2 = 44$. Since $f(4, 3) = 4 \cdot 3 + 2 \cdot 3^2 = 30$, we see that $f(3, 4) \neq f(4, 3)$—the order of the variables is important. If $g(p, q, r) = 3p + 2q - r^2/p$, then $g(4, 1, 8) = 3 \cdot 4 + 2 \cdot 1 - 8^2/4 = -2$ and $g(0, 3, 6)$ is undefined since division by 0 is not permitted.

The instantaneous rate of change of a function of one variable—that is, its derivative—is a very useful concept. For a function of more than one variable, the instantaneous rate of change of the function with respect to one variable, while the other variables are held constant, is very important and is called a *partial derivative*.

Definition. If $z = f(x, y)$, the *partial derivative of z with respect to x* is the function whose value at (x, y) is

$$\lim_{h \to 0} \frac{f(x+h, y) - f(x, y)}{h}$$

This partial derivative is denoted $z_x(x, y)$ or $\partial z/\partial x$ or $f_x(x, y)$ or $\partial f/\partial x$.

Note that the notations for partial derivatives use a "round d" ∂ instead of the usual letter d. This distinguishes partial derivatives from derivatives.

> If $z = f(x, y)$, to compute the partial derivative of z with respect to x, simply treat y as a constant and then use the same differentiation formulas as for functions of one variable. If z is a function of more than two variables, simply treat all variables except x as constants.

A partial derivative can be computed with respect to any variable. If $g = f(x, y, z)$, then g has partial derivatives $g_x(x, y, z)$, $g_y(x, y, z)$, and $g_z(x, y, z)$. These three partial derivatives are called *first-order partial derivatives* because each is obtained by differentiating g once.

Examples Find all first-order partial derivatives of each function.

Example 1 $z = h(x, y) = x^3 + 3x^2y - 5y^3$

Solution $\quad \dfrac{\partial z}{\partial x} = 3x^2 + 6xy \quad$ Treat y as a constant and differentiate with respect to x.

$\dfrac{\partial z}{\partial y} = 3x^2 - 15y^2 \quad$ Treat x as a constant and differentiate with respect to y.

Example 2 $y = p(v, w) = v^2 \ln w$

Solution $\quad p_v = 2v \ln w \quad$ Treat w as a constant.

$p_w = \dfrac{v^2}{w} \quad$ Treat v as a constant.

Example 3 $z = (5x + 1)^3(2y + 1)^4$

Solution $\quad z_x = 3(5x + 1)^2 \cdot 5 \cdot (2y + 1)^4 = 15(5x + 1)^2(2y + 1)^4$

$z_y = (5x + 1)^3 \cdot 4 \cdot (2y + 1)^3 \cdot 2 = 8(5x + 1)^3(2y + 1)^3$

Example 4 $w = f(x, y, z) = x^2y + 3e^{yz} - xz$

Solution $\quad w_x(x, y, z) = 2xy - z \quad$ Treat y and z as constants.

$w_y(x, y, z) = x^2 + 3ze^{yz} \quad$ Treat x and z as constants.

$w_z(x, y, z) = 3ye^{yz} - x \quad$ Treat x and y as constants.

Example 5 The value V of weekly output at Bart's Handicrafts is given by the formula

$$V = 300x + 160y - x^2 - 2y^2 - 4xy$$

where x is the number of skilled workers employed and y is the number of unskilled workers employed. Bart currently employs 40 skilled workers and 25 unskilled workers. How much would V increase if Bart employed one more skilled worker?

Solution Economists call the partial derivative $\partial V/\partial x$ the *marginal productivity of skilled labor* and interpret it as the change in V caused by adding one more skilled worker. Now

$$V_x(x, y) = 300 - 2x - 4y$$

If Bart employed one more skilled worker, V would increase by

$$V_x(40, 25) = 300 - 2(40) - 4(25) = 120$$

that is, by $120 a week.

Example 6 The daily profit function for Ernie's Donut Stop is

$$P(c, d) = 0.15c + 0.08d + 0.0001c^2 + 0.00002d^2 - 40$$

where c is the number of cups of coffee sold in a day and d is the number of donuts sold. Ernie now sells 500 cups of coffee and 300 donuts daily. How much would profit increase if Ernie sold four more donuts each day?

Solution The extra profit from selling one more donut is $P_d(500, 300)$. Now

$$P_d(c, d) = 0.08 + 0.00004d$$
$$P_d(500, 300) = 0.08 + 0.00004(300) = 0.092$$

The extra profit from selling four more donuts in a day is $(0.092) \cdot 4 = 0.368 = 36.8$¢.

■

Second-order partial derivatives

Just as a function of one variable has a second derivative, a function of more than one variable has *second-order partial derivatives*. Functions of more than one variable in fact can have third-order partial derivatives, fourth-order partial derivatives, and so on. However, we will be concerned with only the first-order and second-order partial derivatives. In the next section, second-order partial derivatives will be used to determine the extreme values of functions of two variables.

Suppose $z = f(x, y)$. The *second partial derivative of z with respect to x* is obtained by differentiating z twice with respect to x while treating y as a constant and is denoted z_{xx} or f_{xx} or $\partial^2 z/\partial x^2$ or $\partial^2 f/\partial x^2$. Thus z_{xx} is the partial derivative of z_x with respect to x. Similarly z_{yy} is the partial derivative of z_y with respect to y.

The function z_x can be differentiated with respect to y to yield a function denoted z_{xy} or f_{xy} or $\partial^2 z/\partial y\, \partial x$ or $\partial^2 f/\partial y\, \partial x$. This function is a second-order partial derivative of z and is called a *mixed partial derivative*. There is a fourth second-order partial derivative of z, which is $z_{yx} = \partial(\partial z/\partial y)/\partial x$. If certain very weak hypotheses are placed on z, then z_{xy} and z_{yx} will be equal. These hypotheses are technical in nature and will not be discussed here. However, for all functions that occur in this book, or that you are ever likely to encounter,

$$\boxed{z_{xy} = z_{yx}}$$

Everything just said about second-order partial derivatives applies to functions of more than two variables. If $w = g(x, y, z)$, then w has nine second-order partial derivatives: w_{xx}, w_{xy}, w_{xz}, w_{yx}, w_{yy}, w_{yz}, w_{zx}, w_{zy}, and w_{zz}. However, for the functions to be found in this book, $w_{xy} = w_{yx}$, $w_{xz} = w_{zx}$, and $w_{yz} = w_{zy}$, so the nine partial derivatives reduce to six.

Examples Compute all first-order and second-order partial derivatives of the function.

Example 1 $z = x^3 y^{5/2}$

Solution $z_x = 3x^2 y^{5/2}$ $\qquad z_y = \dfrac{5}{2} x^3 y^{3/2}$

$z_{xx} = \dfrac{\partial}{\partial x}(z_x) = 6xy^{5/2}$ $\qquad z_{yy} = \dfrac{\partial}{\partial y}(z_y) = \dfrac{15}{4} x^3 y^{1/2}$

$z_{yx} = z_{xy} = \dfrac{\partial}{\partial y}(z_x) = \dfrac{\partial}{\partial y}(3x^2 y^{5/2})$

$\qquad = \dfrac{15}{2} x^2 y^{3/2}$

Example 2 $z = \exp(2xy + 3y)$

Solution $z_x = 2y\exp(2xy + 3y)$ $\qquad z_y = (2x + 3)\exp(2xy + 3y)$

$z_{xx} = \dfrac{\partial}{\partial x}(z_x) = 2y(2y)\exp(2xy + 3y)$

$\qquad = 4y^2\exp(2xy + 3y)$

$z_{yy} = \dfrac{\partial}{\partial y}(z_y) = (2x + 3)(2x + 3)\exp(2xy + 3y)$

$\qquad = (2x + 3)^2\exp(2xy + 3y)$

$z_{yx} = z_{xy} = \dfrac{\partial}{\partial y}(z_x) = \dfrac{\partial}{\partial y}[2y\exp(2xy + 3y)]$

$\qquad = 2\exp(2xy + 3y) + 2y(2x + 3)\exp(2xy + 3y) \qquad$ Product rule.

$\qquad = 2(1 + 2xy + 3y)\exp(2xy + 3y)$

Example 3 $z = x^3 y^2 w^4$

Solution $z_x = 3x^2 y^2 w^4 \qquad z_y = 2x^3 y w^4 \qquad z_w = 4x^3 y^2 w^3$

$z_{xx} = \dfrac{\partial}{\partial x}(z_x) = 6xy^2 w^4 \qquad z_{yy} = \dfrac{\partial}{\partial y}(z_y) = 2x^3 w^4$

$z_{ww} = \dfrac{\partial}{\partial w}(z_w) = 12x^3 y^2 w^2 \qquad z_{yx} = z_{xy} = \dfrac{\partial}{\partial y}(z_x) = 6x^2 y w^4$

$z_{wx} = z_{xw} = \dfrac{\partial}{\partial w}(z_x) = 12x^2 y^2 w^3 \qquad z_{wy} = z_{yw} = \dfrac{\partial}{\partial w}(z_y) = 8x^3 y w^3$

∎

Exercises 11-4

Compute all first-order partial derivatives. Evaluate these partial derivatives at the given point.

1. $y = h(v, w) = v^3 e^{5w} \quad (2, 0)$
2. $v = k(x, y) = x^3/y^2 \quad (4, 2)$
3. $z = g(x, y) = 4x^{1/2}\ln(3y + 2) \quad (9, 0)$
4. $z = p(x, y) = \ln(2x + 3y - 7) \quad (4, 5)$
5. $z = c(v, w, x) = v^2 \ln(x)\exp(5w) + 3x \quad (2, 1, 6)$
6. $z = p(b, c, d) = \dfrac{(2b + 3c)^5}{\ln d} \quad (1, 0, e^2)$
7. $y = f(w, x) = w(x + w)^{1/2} \quad (4, 5)$
8. $x = g(v, w) = 3v + 2w - v^2 w \quad (1, 2)$

9. $z = f(w, x, y) = 3w + 4x^2 - 6y + wx^3 \quad (-2, 1, -1)$
10. $z = g(r, s, v) = r^3 - 5s^2 \quad (2, 3, 7)$
11. $w = h(x, y, z) = x^4 \exp(y^2) \quad (-2, 0, 6)$
12. $w = k(x, y, z) = x^3 - y^2 + xz^6 \quad (1, 3, 2)$

Compute all first-order and second-order partial derivatives.

13. $z = f(x, y) = 3x - 2y + 4x^2$
14. $z = f(x, y) = 3x^2y - x^4$
15. $z = f(x, y) = 4x^3y^4 - 6x^2y^2$
16. $z = f(x, y) = 12xy^3 - 5x^2y$
17. $y = g(v, w) = (2v + 3w)^8$
18. $y = h(v, w) = 10 \exp(2v + 3w - 5vw)$
19. $w = g(x, y) = x^3 \exp(xy)$
20. $w = h(x, y) = \ln(3x - 2y)$
21. $w = f(x, y, z) = 3x + 2y - 4z + x^3y^2 \exp(10z)$
22. $w = f(x, y, z) = x^2y^3 \ln z - 7x^5y$
23. $p = h(v, w, x) = v^2w^3 - 5wx^4$
24. $p = h(v, w, x) = (w + 3x)^4 - 7v^3w^5$
25. $w = g(x, y) = x^3y^2 + x^2y^3$
26. $k = f(x, y, z) = xyz^2 + xy^2z + x^2yz$

Scented Candles Inc., makes both large and small candles. The cost function is

$$C(x, y) = 100 + 1.50x + 0.25y + 0.001xy$$

Here x is the number of large candles made in a day and y is the number of small candles. Currently 100 large candles and 200 small candles are made daily. Use partial derivatives to answer the following questions.

27. How much would costs rise if two more large candles were made daily?
28. How much would costs decrease if three fewer small candles were made daily?
29. How much would costs change if each day only 99 large candles were made, but also 205 small candles were made?

Stereo of Halburton finds that its weekly sales are related to advertising expenditures by the formula

$$S = 20{,}000 + 5x + 3y + 4z - 0.001xy$$

The variables represent the weekly amount spent for advertising on television (x), radio (y), and newspapers (z). Currently $x = \$1000$, $y = \$500$, and $z = \$2000$.

30. What are weekly sales now?

31. How much would weekly sales increase for each extra $1 spent on advertising on
    a. Television?   b. Radio?   c. Newspapers?
    (*Note:* Here the extra amount spent must be small.)

32. How much would sales change if $20 less was spent on radio advertising and $10 more was spent on each of television and newspaper advertising?

Ultrashine and Moonbeam dominate part of the market for toothpaste. If Ultrashine charges p_1 for a tube and Moonbeam charges p_2, then Ultrashine will sell q_1 tubes a day and Moonbeam will sell q_2, where

$$q_1(p_1, p_2) = 1000(100 + 30p_2 - 40p_1 + 20p_2^2 - 15p_1^2)$$

and

$$q_2(p_1, p_2) = 1000(80 + 35p_1 - 30p_2 + 20p_1^2 - 20p_2^2).$$

Currently $p_1 = \$1.25$ and $p_2 = \$1.35$.

33. How many tubes does Ultrashine sell daily?

34. How many tubes does Moonbeam sell daily?

35. How many fewer tubes would Ultrashine sell if it raised its price 2¢? How many more would Moonbeam sell than?

36. How many more tubes would Ultrashine sell each day if Moonbeam raised its price 3¢? How many fewer would Moonbeam sell?

Wilson Pharmaceuticals currently spends $3,000,000 annually on research and $7,000,000 on advertising. The revenue function is given by

$$S(A, R) = \$5{,}000{,}000 + 0.3A + 0.4R + 0.000001AR$$

where A is the amount spent on advertising and R is the amount spent on research. Next year, Wilson has $100,000 extra to spend on advertising and research combined. How much would revenue increase if

37. The extra $100,000 was all spent on research?

38. The extra $100,000 was all spent on advertising?

39. The extra $100,000 was evenly divided between research and advertising?

40. $30,000 extra was spent on research and $70,000 extra was spent on advertising?

The pollution level in a city is given by the formula

$$P(N, C) = 20 + 2N + 3C + \frac{C\sqrt{N}}{10}$$

where N is the number of people who live in the city and C is the number of cars in the city during business hours. Here N and C are measured in units of 100,000. Currently $N = 25$ and $C = 14$. Several models predict various values for N and C five years from now. Use partial derivatives to determine how much $P(N, C)$ will change from its current value according to each of these models

41. $N = 27$, $C = 15$

42. $N = 24$, $C = 13$

43. $N = 24$, $C = 16$

44. $N = 26$, $C = 16$

Section 11-5 Extreme Values of Functions of Two Variables

Finding the extreme values of functions of one variable is very useful and has numerous applications. Similarly, finding the maximum and minimum values of functions of two variables is of great importance. We will do this now, starting with some definitions that are similar to the definitions for functions of one variable.

> **Definitions.** Let $z = f(x, y)$
>
> The function f has a *local maximum* at (x_0, y_0) if $f(x_0, y_0) \geq f(x, y)$ for all points (x, y) near (x_0, y_0).
>
> The function f has a *local minimum* at (x_0, y_0) if $f(x_0, y_0) \leq f(x, y)$ for all points (x, y) near (x_0, y_0).
>
> The function f has a *local extremum* at (x_0, y_0) if f has either a local maximum or a local minimum there.
>
> The point (x_0, y_0) is a *critical point* for f if $f_x(x_0, y_0) = 0$ or is undefined *and* $f_y(x_0, y_0) = 0$ or is undefined. (Note that a critical point can arise in four ways.)

> **Theorem.** If f has a local extremum at (x_0, y_0), then (x_0, y_0) is a critical point for f.

Note that the theorem does not say that f has a local extremum at each critical point. The theorem does say that in order to find all the local extrema of f, it is enough just to consider critical points.

To determine whether f has a local maximum or a local minimum or neither at a critical point, the following test may be used. This test is similar to the second derivative test for functions of one variable.

> *Test for Extreme Points*
> **Theorem.** Let $z = f(x, y)$ and (x_0, y_0) be a critical point for f. Let $\Delta(x, y) = f_{xx}(x, y) f_{yy}(x, y) - f_{xy}(x, y)^2$.
> 1. If $\Delta(x_0, y_0) > 0$ and $f_{xx}(x_0, y_0) > 0$, then f has a local minimum at (x_0, y_0).
> 2. If $\Delta(x_0, y_0) > 0$ and $f_{xx}(x_0, y_0) < 0$, then f has a local maximum at (x_0, y_0).
> 2. If $\Delta(x_0, y_0) < 0$, then f does not have a local extremum at (x_0, y_0). The function is said to have a *saddle point* at (x_0, y_0).
> 4. If $\Delta(x_0, y_0) = 0$, no conclusion may be reached by this test.

Cases 1, 2, and 3 of the theorem are illustrated in Diagrams 1, 2, and 3. P is a local minimum point.

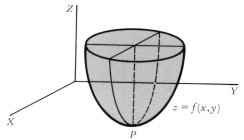

Diagram 1. (From *Modern Mathematics for Business Decision Making*, 2d ed., by Donald R. Williams, copyright 1978 by Wadsworth Publishing Company, Inc., Belmont, California 94002. Reprinted by permission of the publisher.)

P is a local maximum point.

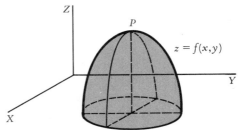

Diagram 2. (From *Modern Mathematics for Business Decision Making*, 2d ed., by Donald R. Williams, copyright 1978 by Wadsworth Publishing Company, Inc., Belmont, California 94002. Reprinted by permission of the publisher.)

P is a saddle point.

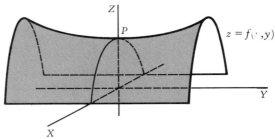

Diagram 3. (From *Modern Mathematics for Business Decision Making*, 2d ed., by Donald R. Williams, copyright 1978 by Wadsworth Publishing Company, Inc., Belmont, California 94002. Reprinted by permission of the publisher.)

You may wonder why, in case 1 and case 2, the sign of $f_{xx}(x_0, y_0)$ is used rather than the sign of $f_{yy}(x_0, y_0)$. The reason is that, in case 1 and case 2, $f_{xx}(x_0, y_0)$ and $f_{yy}(x_0, y_0)$ both have the same sign so that it does not matter which one is used. Now

$$f_{xx}(x_0, y_0)f_{yy}(x_0, y_0) \geq f_{xx}(x_0, y_0)f_{yy}(x_0, y_0) - f_{xy}(x_0, y_0)^2$$
$$= \Delta(x_0, y_0) > 0$$

Thus $f_{xx}(x_0, y_0)f_{yy}(x_0, y_0) > 0$. If the product of two numbers is positive, the two numbers must both have the same sign.

Examples Find all extreme points of each function.

Example 1 $z = f(x, y) = 6x^2 + 4xy + 2y^2 - 28x - 12y + 10$

Solution First find z_x and z_y.

$$z_x(x, y) = 12x + 4y - 28 \qquad z_y(x, y) = 4x + 4y - 12$$

To find the critical points, set $z_x(x, y) = 0$ and $z_y(x, y) = 0$. This gives the system of equations

$$12x + 4y - 28 = 0 \qquad 4x + 4y - 12 = 0$$

The only solution of this system of equations is $x = 2, y = 1$. Thus the only critical point is $(2, 1)$.

Next compute the second-order partial derivatives.

$$z_{xx}(x, y) = 12 \qquad z_{yy}(x, y) = 4 \qquad z_{xy}(x, y) = 4$$

Now test the critical point $(2, 1)$. $\Delta(2, 1) = 12 \cdot 4 - 4^2 = 32 > 0$. Since $z_{xx}(2, 1) = 12 > 0$, f has a local minimum at $(2, 1)$. Case 1 of the test applies here.

Example 2 $z = f(x, y) = x^2 + 3xy + 2y^2 - 5x - 6y + 26$

Solution Begin by finding z_x and z_y

$$z_x = 2x + 3y - 5 \qquad z_y = 3x + 4y - 6$$

To find the critical points, set $z_x = 0$ and $z_y = 0$ and solve to obtain $x = -2, y = 3$. The only critical point is $(-2, 3)$.

Next compute the second-order partial derivatives.

$$z_{xx}(x, y) = 2 \qquad z_{yy}(x, y) = 4 \qquad z_{xy}(x, y) = 3$$

Now test the critical point $(-2, 3)$. $\Delta(-2, 3) = 2 \cdot 4 - 3^2 = -1 < 0$. The function f does not have a local extremum at $(-2, 3)$, and $(-2, 3)$ is a saddle point. Case 3 of the test applies here.

Example 3 $z = \exp(-x^2) - y^3 + 3y^2$

Solution $z_x(x, y) = -2x \exp(-x^2) \qquad z_y(x, y) = -3y^2 + 6y$

Now $z_x(x, y) = 0$ when $x = 0$ and only then. Recall that the range of the exponential function consists of all positive real numbers, so that $\exp(-x^2)$ can never equal 0. Since $z_y(x, y) = 0$ when $y = 0$ or $y = 2$, the critical points are $(0, 0)$ and $(0, 2)$.

Next compute the second-order partial derivatives.

$$z_{xx} = (-2x)(-2x)\exp(-x^2) + (-2)\exp(-x^2) = (4x^2 - 2)\exp(-x^2)$$
$$z_{yy}(x, y) = -6y + 6 \qquad z_{xy}(x, y) = 0$$

Finally, test each critical point. Since

$$\Delta(0, 0) = -2 \cdot 6 - 0^2 = -12 < 0$$

f has a saddle point at $(0, 0)$ and does not have a local extremum there. Since

$$\Delta(0, 2) = (-2) \cdot (-6) - 0^2 = 12 > 0$$

and

$$z_{xx}(0, 2) = -2 < 0$$

f has a local maximum at $(0, 2)$.

Example 4 Harper's Transmissions repairs automobile transmissions. The annual profit function is

$$P(x, y) = 50x + 77y - 0.02x^2 - 0.02y^2 - 0.01xy - 10{,}000$$

where x is the number of manual transmissions repaired and y is the number of automatic transmissions repaired. What is the maximum possible annual profit? How many transmissions of each type must be repaired to maximize profit?

Solution Compute $P_x(x, y)$ and $P_y(x, y)$.

$$P_x(x, y) = 50 - 0.04x - 0.01y \qquad P_y(x, y) = 77 - 0.01x - 0.04y$$

Now set $P_x(x, y) = 0$ and $P_y(x, y) = 0$ and solve the resulting equations. The solution—$x = 820$, $y = 1720$—is the only critical point. Next compute the second-order partial derivatives.

$$P_{xx}(x, y) = -0.04 \qquad P_{yy}(x, y) = -0.04 \qquad P_{xy}(x, y) = -0.01$$

Test the critical point $(820, 1720)$. Since

$$\Delta(820, 1720) = (-0.04)(-0.04) - (0.01)^2 = 0.0015 > 0$$

and $P_{xx}(820, 1720) = -0.04 < 0$, the function has a local maximum. The maximum possible profit is thus $P(820, 1720) = \$76{,}720$ and is obtained by repairing 820 manual transmissions and 1720 automatic transmissions.

∎

Exercises 11-5

Find each critical point of the function and determine all of the local extrema.

1. $z = f(x, y) = 3x^2 - y^2 + 6xy - 24x + 8$
2. $z = f(x, y) = 2x^2 + 5y^2 - 4xy + 24x - 54y + 27$
3. $w = f(x, y) = 3xy - x^2 - 5y^2 + 5x - 2y + 18$
4. $w = f(x, y) = 6xy - 2x^2 - y^2 + 20x - 16y - 8$
5. $r = g(v, w) = 5vw + 3v^2 + 4w^2 - 7v - 2w - 15$
6. $v = h(w, x) = w^2 + 3x^2 - 2wx + 14w - 26x$
7. $w = g(x, y) = 7y + 148 - x - 5xy - 3x^2 - 6y^2$
8. $v = k(p, q) = 2pq - p^2 - 3q^2 - 10p + 30q - 76$
9. $z = f(x, y) = x^3 - 12x - 2y^2 + 6$
10. $z = f(x, y) = 4x^2 - y^3 + 27y - 8$
11. $y = g(w, x) = \exp(wx)$
12. $y = k(v, w) = \exp(v^2 - 6v + w^2 + 4w - 9)$
13. The revenue function for the Ashton Zoo is

    $$R(a, c) = 900a + 1200c - 4a^2 - 10c^2 - 2ac$$

    where a is the cost (in cents) of an adult's ticket and c is the cost (in cents) of a child's ticket. For what prices should tickets sell to maximize revenue?

14. The monthly profit of Lakeview Cinema is related to the amount x it spends on television advertising and to the amount y it spends on newspaper advertising by

    $$P = 2000 + 1.2x + 2.0y - 0.01x^2 - 0.02y^2 - 0.01xy$$

    What is the maximum possible monthly profit? How much should Lakeview Cinema spend on advertising to maximize profit?

15. The labor cost for manufacturing a microscope is

    $$L(x, y) = 60 + x^2 + 2y^2 - 4x - 5y - xy$$

where x is the number of hours of unskilled labor used and y is the number of hours of skilled labor. How much labor of each type should be used to minimize cost?

16. The Cedar Road Nursery sells small cactuses for $2 and large cactuses for $4. The cost function is

$$C(q) = s + 2l + 100 + 0.01s^2 + 0.025l^2$$

where s is the number of small cactuses and l is the number of large cactuses. What is the maximum possible profit? How many cactuses of each size must be sold if profit is to be maximized?

Section 11-6 Lagrange Multipliers

In the last section, we learned how to find the extreme values of a function of two variables. In Chapter 4 (Linear Programming), we learned how to find the extreme values of a function of any number of variables, subject to constraints, provided the function and the constraints are linear. In this section, we will learn a method, call the method of Lagrange multipliers, for finding the extreme values of a function of any number of variables, subject to a constraint; here the function or the constraint or both may be nonlinear. For simplicity, we will only consider the two variable case.

The Method of Lagrange Multipliers

To find the local extrema of a function $f(x, y)$, subject to a constraint $g(x, y) = 0$:

Step 1. Form the function $k(x, y, \lambda) = (x, y) + \lambda g(x, y)$
The symbol λ is the Greek letter *lambda* and is a variable called a Lagrange multiplier.

Step 2. Form the system of equations
$$k_x(x, y, \lambda) = 0$$
$$k_y(x, y, \lambda) = 0$$
$$k_\lambda(x, y, \lambda) = 0$$

Step 3. Solve the preceding system of equations. The desired local extrema will be contained among the solutions.

Now let's find the minimum value of $f(x, y) = 4x^2 + 3y^2$ subject to $2x + y - 10 = 0$.

Step 1 Form the function

$$k(x, y, \lambda) = f(x, y) + \lambda g(x, y)$$
$$= 4x^2 + 3y^2 + 2\lambda x + \lambda y - 10\lambda$$

Step 2 Compute the first-order partial derivatives of k with respect to x, y, and λ, respectively. Set these partial derivatives equal to 0.

$$8x + 2\lambda = 0$$
$$6y + \lambda = 0$$
$$2x + y - 10 = 0$$

Step 3 Solve the above equations. First, solve

$$8x + 2\lambda = 0 \quad \text{for } x \text{ to obtain} \quad x = \frac{-\lambda}{4}$$

Then solve

$$6y + \lambda = 0 \quad \text{for } y \text{ to obtain} \quad y = \frac{-\lambda}{6}.$$

Substitute these values into

$$2x + y - 10 = 0$$

to obtain

$$\frac{-\lambda}{2} - \frac{\lambda}{6} - 10 = 0$$
$$-3\lambda - \lambda - 60 = 0$$
$$-4\lambda = 60$$
$$\lambda = -15$$

Then

$$x = \frac{-\lambda}{4} = \frac{15}{4} = 3.75 \quad \text{and} \quad y = \frac{-\lambda}{6} = \frac{15}{6} = 2.5$$

The minimum value of $f(x, y)$ is

$$f(3.75, 2.5) = 4 \cdot (3.75)^2 + 3(2.5)^2 = 75$$

How do we know that f actually has a minimum at $(3.75, 2.5)$? In more advanced courses, methods are discussed for testing the points given by the method of Lagrange multipliers to determine whether these points are local maximum points, local minimum points, or are not local extreme points. In this book, you may assume that the points you obtain via Lagrange multipliers are the desired extreme points.

Example 1 A pension fund is going to invest x billion dollars in a note paying 1% per month interest and y billion dollars in a riskier note paying 1.5% monthly. The risk factor R is given by the formula

$$R(x, y) = x^2 + 4y^2$$

How much should the pension fund invest in each note in order to maximize income, if a risk factor of 9 is acceptable?

Solution The monthly income I (in units of a million dollars) is given by the formula

$$I = 10x + 15y$$

If a risk factor of 9 is acceptable, the income will be greatest when $R(x, y) = 9$, that is, when $x^2 + 4y^2 = 9$. Since the constraint must be written in the form $g(x, y) = 0$, we write

$$g(x, y) = x^2 + 4y^2 - 9 = 0$$

Now use the method of Lagrange multipliers.

Step 1 Form

$$k(x, y, \lambda) = I(x, y) + \lambda g(x, y)$$
$$= 10x + 15y + \lambda x^2 + 4\lambda y^2 - 9\lambda$$

Step 2 Set $k_x = 0$, $k_y = 0$, and $k_\lambda = 0$.

$$10 + 2\lambda x = 0$$
$$15 + 8\lambda y = 0$$
$$x^2 + 4y^2 - 9 = 0$$

Step 3 Solve the first equation in step 2 for x and the second equation for y.

$$10 + 2\lambda x = 0 \qquad 15 + 8\lambda y = 0$$
$$x = \frac{-5}{\lambda} \qquad y = \frac{-15}{8\lambda}$$

Substitute these values for x and y in the equation

$$x^2 + 4y^2 - 9 = 0$$

to obtain

$$\frac{25}{\lambda^2} + 4 \cdot \frac{225}{64\lambda^2} - 9 = 0$$
$$\frac{25}{\lambda^2} + \frac{225}{16\lambda^2} = 9$$
$$16 \cdot 25 + 225 = 9 \cdot 16 \cdot \lambda^2$$
$$625 = 144\lambda^2$$
$$\lambda^2 = \frac{625}{144}$$
$$\lambda = \pm\sqrt{\frac{625}{144}} = \pm\frac{25}{12}$$

If $\lambda = -25/12$, then $x = -5/\lambda = 12/5 = 2.4$ and $y = -15/(8\lambda) = 9/10 = .9$. If $\lambda = 25/12$, then x and y are negative—so this value of λ must be discarded.

The fund should invest $2.4 billion at 1% per month interest and $.9 billion at 1.5% per month interest. The monthly income from these investments would be $37.5 million.

■

Exercises 11-6

Use Lagrange multipliers to solve the following problems.

1. Minimize $f(x, y) = x^2 + 2y^2$ subject to $2x + 3y - 12 = 0$.
2. Minimize $f(x, y) = 3x^2 + y^2$ subject to $3x + 5y - 8 = 0$.
3. Minimize $f(x, y) = 2x^2 + y^2$ subject to $x - 2y = 4$.
4. Minimize $f(x, y) = x^2 + 4y^2$ subject to $2x - 3y = 6$.
5. Maximize $f(x, y) = xy$ subject to $2x + 5y = 10$.
6. Maximize $f(x, y) = 4xy + 3$ subject to $3x + 4y = 12$.
7. Minimize and maximize $f(x, y) = 3x + 4y + 8$ subject to $x^2 + y^2 = 25$.

8. Minimize and maximize $f(x, y) = 12x + 5y - 9$ subject to $x^2 + y^2 = 169$.

9. A money market fund is going to lend x million dollars to one company at 1.2% interest (for the month) and y million dollars to another company at 1.6% interest. The risk factor R is given by

$$R(x, y) = 2x^2 + 3y^2$$

and a risk factor of 354 is acceptable. How much should the money market fund lend to each company?

10. Repeat exercise 9 if the first loan pays 1.1% for the month, the second loan pays 1.3%, and the risk factor is 4206.

Section 11-7 Implicit Differentiation and Related Rates

Implicit differentiation

So far in calculus, we have been concerned with finding the derivatives of functions. Relations involving x and y can also be differentiated, using a technique called *implicit differentiation*. To differentiate a relation implicitly, we assume that y is a function of x and then use the chain rule in one of the three forms studied earlier. For example, to find y' if

$$x^2 + y^2 = 25$$

differentiate to obtain

$$\frac{d}{dx}(x^2) + \frac{d}{dy}(y^2) = \frac{d}{dx}(25)$$

Since y is assumed to be a function of x, the derivative of y^2 is $2yy'$ by the power rule. Thus

$$2x + 2yy' = 0$$

Now solve this equation for y'.

$$2yy' = -2x$$

$$y' = \frac{-2x}{2y} = -\frac{x}{y}$$

11-7 IMPLICIT DIFFERENTIATION AND RELATED RATES

Sometimes a relation involving x and y is in fact a function. In this case, the derivative may be obtained either by solving the relation for y to obtain a formula of the form $y = f(x)$ and differentiating in the usual way, or by differentiating implicitly.

Examples Use implicit differentiation to find y'.

Example 1 $x^4 = e^{5y} + y^3$

Solution Differentiate implicitly to obtain

$$\frac{d}{dx}(x^4) = \frac{d}{dx}(e^{5y}) + \frac{d}{dx}(y^3)$$

$$4x^3 = 5e^{5y}y' + 3y^2 y'$$

$$= (5e^{5y} + 3y^2)y'$$

$$y' = \frac{4x^3}{5e^{5y} + 3y^2}$$

Example 2 $x^2 y^3 + 4x + 7y = 29$

Solution

$$\frac{d}{dx}(x^2 y^3) + \frac{d}{dx}(4x) + \frac{d}{dx}(7y) = \frac{d}{dx}(29)$$

$$\frac{d}{dx}(x^2) \cdot y^3 + x^2 \frac{d}{dx}(y^3) + 4 + 7y' = 0 \qquad \text{Product rule.}$$

$$2x \cdot y^3 + x^2 \cdot 3y^2 y' + 4 + 7y' = 0$$

$$2xy^3 + 3x^2 y^2 y' + 4 + 7y' = 0$$

$$3x^2 y^2 y' + 7y' = -2xy^3 - 4$$

$$(3x^2 y^2 + 7)y' = -(2xy^3 + 4)$$

$$y' = -\frac{2xy^3 + 4}{3x^2 y^2 + 7}$$

Example 3 Find the equation of the tangent line to $x^2 + 4x - y^2 = 5$ at $(3, 4)$.

Solution Note that the point $(3, 4)$ does satisfy the relation $x^2 + 4x - y^2 = 5$. Now use implicit differentiation to find y'.

$$2x + 4 - 2yy' = 0$$

$$2x + 4 = 2yy'$$

$$y' = \frac{2x+4}{2y} = \frac{x+2}{y}$$

At the point $(3, 4)$, $y' = (3 + 2)/4 = 5/4$. Since the slope of a tangent line to a curve at a point is the value of the derivative at the point, we have

$$y - 4 = \frac{5}{4}(x - 3) \qquad \text{by the point-slope formula}$$

Then

$$y - 4 = \frac{5}{4}x - \frac{15}{4}$$

$$y = \frac{5}{4}x - \frac{15}{4} + 4$$

$$= \frac{5}{4}x + \frac{1}{4}$$

■

Related rates

Implicit differentiation can be used to solve certain problems involving rates of change. For example, suppose a damaged oil tanker is leaking oil at a constant rate and forming a circular oil slick 0.08 ft thick. Determine the rate at which the tanker is losing oil, if the radius of the oil slick is increasing at the rate of 0.75 ft/min when the radius is 300 ft.

First, observe that the volume of the oil lost by the tanker equals the volume V of the oil slick. If R is the radius of the oil slick, V and R are related by the equation

$$V = 0.08\pi R^2$$

since the oil slick is a cylinder of height 0.08 ft.

Since V and R are both functions of time t, we differentiate this equation with respect to t to obtain

$$\frac{dV}{dt} = 0.08\pi \cdot 2R\frac{dR}{dt} = 0.16\pi R\frac{dR}{dt}$$

Now dV/dt is given as constant, and $dR/dt = 0.75$ when $R = 300$, so that

$$\frac{dV}{dt} = (0.16\pi) \cdot (300) \cdot (0.75) = 36\pi = 113$$

The tanker is losing oil at the rate of 113 cu ft/min.

Example 4 The profit function for a burglar alarm company is

$$P(q) = 15q - 0.001q^2 - 20{,}000$$

where q is the number of burglar alarms produced each week. Currently 2000 burglar alarms are produced each week and production is increasing at the rate of 100 alarms per week. How quickly are profits increasing?

Solution Differentiate the profit equation implicitly with respect to t (time in weeks) to obtain

$$\frac{dP}{dt} = 15\frac{dq}{dt} - 0.002q\frac{dq}{dt}$$

Since $q = 2000$ and $dq/dt = 100$, the current value of dP/dt is

$$\frac{dP}{dt} = 15(100) - 0.002(2000)100 = 1100$$

Currently profits are increasing at the rate of $1100 per week.

■

Exercises 11-7

Use implicit differentiation to find y'.

1. $x^3 + y^3 = 64$
2. $x^3 + 3x^2 - y^2 = 9$
3. $x^2 + 4y^2 = 81 - 3y$
4. $x^2 + 2x + y^2 - 4y = 8$
5. $x^2 + e^{3y} = 14$
6. $\exp(x^2) + 14 = y^2$
7. $2x^{3/2} + \ln y = 19$
8. $\ln(3x + 5) + e^{2y} = y^2 + 8$
9. $(4x + 3y)^5 = x^4 + y^4$
10. $\exp(x^2 + y^2) = x^3 + y^3 + 12$

Write the equation of the tangent line at the given point.

11. $x^2 + 2x + 3y^2 = 15$ $(1, 2)$
12. $x^2 - y^2 = 9$ $(5, 4)$
13. $x^2 + y^2 + 2y = 4$ $(2, -2)$
14. $x^3 = y^2 + 5y + 12$ $(2, -1)$

A swimming pool is 30 ft wide and 60 ft long.

15. The pool is being drained so that it can be repaired. Water is leaving at the rate of 3600 cu ft per hour. At what rate is the water level falling?
16. After the repairs are made, the pool is refilled at the rate of 900 cu ft per hour. At what rate is the water level rising?

The cost function for Toriba is $C(q) = 20q + 0.002q^2 + 10,000$, where q is the number of electric drills made weekly. The revenue function is $30q$. Currently $q = 1250$ and q is being increased by 25 each week.

17. At what rate are costs increasing?
18. At what rate is revenue increasing?
19. At what rate are profits increasing?
20. For what value of q is profit maximum?

The cost function for a new factory is $C(q) = 8q + 0.0005q^2 + 3000$ where q is the number of hammers made each week. The revenue function is $R(q) = 11q$. Currently $q = 1800$ and q is being increased by 50 each week.

21. At what rate are costs increasing?
22. At what rate is revenue increasing?
23. At what rate is profit increasing?
24. For what value of q is profit maximum?

Chapter 11 Review

Determine the critical numbers of each function. At each critical number, use the
a. *Original equation test*
b. *First derivative test*
c. *Second derivative test*
to determine whether the function has a local maximum, local minimum, or neither.

1. $y = 8 + 6x - 3x^2$
2. $f(w) = (w^2 - 9)^4$
3. $f(x) = 2x^3 + 3x^2 - 120x + 30$
4. $g(h) = he^{-h}$

5. Part of a yard is to be enclosed in a rectangular fence to form a vegetable patch with area 450 sq ft. One side of the patch faces the street and requires fencing that costs $2 a foot. The fencing for the other sides costs $1 a foot. What is the minimum possible cost of the fencing?

6. The cost function for Moistair Humidifiers is

$$C(q) = 0.2q^2 + 150q + 8000$$

where q is the number of humidifiers made in a week. What is the minimum average cost of a humidifier?

Compute all first-order partial derivatives.

7. $h(x, y) = 4x^3 - 3x^2y + 6xy^2 - 2y^3$
8. $w(x, y) = (2x - 3y)^4$
9. $z = r(v, w, x) = v^2 + 3w^2 - 2vx^3 + wx$
10. $y = g(p, q, r) = p^2 \exp(2q + 4r)$
11. $z = f(x, y) = x^2 \exp(xy^2)$
12. $z = h(w, x, y) = x^3 \ln(w^2 + 4y)$

Find the critical points and determine the local extrema.

13. $f(x, y) = -12x + 4y + 2x^2 + 4y^2 - 4xy + 18$
14. $g(v, w) = v^2 + 3vw + w^2 - 4v - w + 9$
15. $z = x^3 - 24x^2 + 3y^2 - 6y + 8$
16. $z = \exp(-x^2 - y^2)$

Use Lagrange multipliers to

17. Minimize $f(x, y) = 2x^2 + 3y^2$ subject to $x + y = 6$.
18. Maximize $f(x, y) = 4xy - 7$ subject to $5x + 2y = 6$.

Use implicit differentiation to find y'.

19. $2x^2 - 3y^2 = 19$
20. $y^2 - 3x - 2x^2 - 9 = 0$
21. $x^2y^2 + 4x = 6y$
22. $\exp(5x - 4y) = 14x + 8y$

12

Integral Calculus

We have just concluded our study of differential calculus, that is, of the part of calculus involving the derivative. The derivative of a function is its rate of change. The derivative is used to find equations of tangent lines, when a function is increasing and decreasing, and the extreme values of a function.

Now let's begin our study of integral calculus, that is, of the part of calculus involving the integral. The integral is used to find areas, to find functions whose rate of change is known, and to solve differential equations. As with differential calculus, practical applications of integral calculus will be shown in a variety of settings.

Section 12-1 The Antiderivative

Suppose Worley Manufacturing has fixed costs of $10,000 and marginal cost function

$$MC(q) = 200 + 0.02q$$

What is the cost function?

Now the derivative of the cost function is known, so one way to start is to find functions whose derivative is $MC(q)$. After some trial and error, several such functions can be found. For example:

$$200q + 0.01q^2 \qquad 200q + 0.01q^2 + 80 \qquad 200q + 0.01q^2 - 43$$

In fact, if c is any real number, the derivative of

$$200q + 0.01q^2 + c$$

is $MC(q)$. Since fixed costs are \$10,000, the cost function satisfies the relation

$$C(0) = 10{,}000$$

If the cost function is of the form

$$C(q) = 200q + 0.01q^2 + c$$

then substituting 0 for q, we find that $10{,}000 = c$ and thus

$$C(q) = 200q + 0.01q^2 + 10{,}000$$

Obtaining the cost function involved finding a function whose derivative was known. This process is the opposite of differentiation, and is called *antidifferentiation*.

> **Definition.** If f is a function, a function F is an *antiderivative* of f if $F' = f$.

If $F' = f$, and c is any number, $(F + c)' = F' + 0 = F'$, so that $F + c$ is an antiderivative of f. If G is any antiderivative of f, then $(G - F)' = G' - F' = f - f = 0$. Thus $G - F$ is a constant function c and $G = F + c$.

> If $F' = f$, the antiderivatives of f are precisely the functions
> $$F + c \qquad \text{where } c \text{ is any number}$$
> The number c is called the *constant of integration*.

Another notation to describe the relationship $F' = f$ is

$$\int f(x)\,dx = F(x) + c$$

We say that $F(x) + c$ is the *indefinite integral* of f. The symbol $\int$ is called an *integral sign*. The expression dx is regarded as a single symbol and is sometimes called the *differential of* x.

Now let's find the antiderivative of the power function x^r. Since differentiating decreases the exponent by 1, antidifferentiating should increase the exponent by 1. Consider x^{r+1}.

$$(x^{r+1})' = (r+1)x^r, \quad \text{so that} \quad \left(\frac{x^{r+1}}{r+1}\right)' = x^r$$

Thus $x^{r+1}/(r+1)$ is an antiderivative of x^r and so

$$\boxed{\int x^r \, dx = \frac{x^{r+1}}{r+1} + c, \quad r \neq -1 \quad \text{Power rule}}$$

This formula is called the power rule. The requirement $r \neq -1$ is necessary to avoid division by 0.

Examples Use the power rule to evaluate each indefinite integral.

1. $\int x^4 \, dx = \frac{x^{4+1}}{4+1} + c = \frac{x^5}{5} + c \quad r = 4$

2. $\int \sqrt{x} \, dx = \int x^{1/2} \, dx = \frac{x^{3/2}}{3/2} + c = \frac{2}{3} x^{3/2} + c$

3. $\int x^{-3} \, dx = \frac{x^{-2}}{-2} + c = \frac{-x^{-2}}{2} + c = -\frac{1}{2x^2} + c$

4. $\int \frac{1}{x^2} \, dx = \int x^{-2} \, dx = \frac{x^{-1}}{-1} + c = -x^{-1} + c = -\frac{1}{x} + c$

■

(*Note:* Before using the power rule, it is convenient to eliminate radical signs and replace them with exponents, as in Example 2. It is easy to add 1 to an exponent, but how do you add 1 to a radical sign? Similarly, you should use negative exponents instead of division, as in Example 4. Your work will be much more accurate if you make these changes in notation.)

The power rule excludes the case $r = -1$. However,

$$(\ln x)' = \frac{1}{x} = x^{-1}, \quad \text{so that} \quad \int \frac{1}{x} \, dx = \int x^{-1} \, dx = \ln x + c$$

Since ln x is defined only when $x > 0$, this formula is valid only when $x > 0$. A more general formula, valid when $x \neq 0$, is

$$\int \frac{1}{x} dx = \int x^{-1} dx = \ln|x| + c$$

To find the antiderivative of the exponential function, recall that $(e^x)' = e^x$—the exponential function is its own derivative. Therefore, the exponential function is its own antiderivative (to within a constant term).

$$\int e^x \, dx = e^x + c$$

In Chapter 10, we learned that the derivative has the very useful property of linearity:

$$(kf + mg)' = kf' + mg'$$

Here k and m are numbers and f and g are functions. The antiderivative also has this property. Thus the antiderivative of a sum or difference of functions can be computed one term at a time.

$$\int [kf(x) + mg(x)] \, dx = k \int f(x) \, dx + m \int g(x) \, dx \qquad \text{Linearity}$$
(k and m are *numbers*.)

To compute the antiderivative of $5x^3 + 3x^{-3}$, use linearity. Thus,

$$\int (5x^3 + 3x^{-3}) \, dx = 5 \int x^3 \, dx + 3 \int x^{-3} \, dx$$

The antiderivatives of x^3 and x^{-3} can be determined using the power rule.

$$\int x^3 \, dx = \frac{x^4}{4} + c_1 \qquad \int x^{-3} \, dx = \frac{x^{-2}}{-2} + c_2 = -\frac{1}{2x^2} + c_2$$

Then
$$\int (5x^3 + 3x^{-3})\, dx = \frac{5x^4}{4} - \frac{3}{2x^2} + 5c_1 + 3c_2$$
$$= \frac{5x^4}{4} - \frac{3}{2x^2} + c$$

Since c_1 and c_2 are arbitrary constants, their sum $5c_1 + 3c_2$ is an arbitrary constant, which we denote by c.

Examples Use linearity to compute the antiderivative of each function.

5. $\int \left(4 + 7x^2 - \frac{3}{x} + 2e^x\right) dx = 4\int 1\, dx + 7\int x^2\, dx - 3\int \frac{1}{x}\, dx + 2\int e^x\, dx$

$= 4x + \frac{7}{3}x^3 - 3\ln|x| + 2e^x + c$

6. $\int \left(8x^2 + \frac{3}{x^2} + 5x^{-1}\right) dx = \int \left(8x^2 + 3x^{-2} + \frac{5}{x}\right) dx \qquad \frac{1}{x^2} = x^{-2}$

$= \frac{8}{3}x^3 - 3x^{-1} + 5\ln|x| + c$

$= \frac{8}{3}x^3 - \frac{3}{x} + 5\ln|x| + c$

7. $\int (14e^x - 7x + 12x^2 - 10e^x + 4x)\, dx$

$= \int (4e^x - 3x + 12x^2)\, dx \qquad$ Simplify by combining like terms.

$= 4e^x - \frac{3x^2}{2} + 4x^3 + c$

Example 8 The marginal profit function for Wang Inc., is
$$MP(q) = 500 - 2q$$

There are losses of $6000 in the event of no sales. Determine the profit function.

Solution Profit is an antiderivative of marginal profit, so
$$P(q) = \int MP(q)\, dq = \int (500 - 2q)\, dq = 500q - q^2 + c$$

Since there are losses of $6000 in the event of no sales, $P(0) = -6000$. Substitute 0 for q in the equation for profit to obtain $P(0) = -6000 = c$, so that

$$P(q) = 500q - q^2 - 6000$$

■

We conclude this section by remarking that *the derivative of the antiderivative of any function f is f again.*

$$\boxed{\frac{d}{dx}\left(\int f(x)\,dx\right) = f(x)}$$

Thus

$$\frac{d}{dx}\left(\int x^2\,dx\right) = \frac{d}{dx}\left(\frac{x^3}{3} + c\right) = \frac{3x^2}{3} = x^2$$

Similarly

$$\frac{d}{dx}\left[\int \exp(x^2)\,dx\right] = \exp(x^2)$$

Computation of the antiderivative of $\exp(x^2)$ is not needed. Also,

$$\frac{d}{dx}\left[\int xe^x\,dx\right] = xe^x \quad \text{and} \quad \frac{d}{dx}\left[\int x^4\,dx\right] = x^4$$

Exercises 12-1

Compute the following indefinite integrals.

1. $\int (x^2 + 6x - 3)\,dx$
2. $\int (6x^2 - 4x + 5)\,dx$
3. $\int 7x^{5/2}\,dx$
4. $\int (12x^{-3/2})\,dx$
5. $\int \frac{5}{x}\,dx$
6. $\int \frac{-4}{x}\,dx$
7. $\int (8e^x - x^3)\,dx$
8. $\int (12x^2 + 4e^x)\,dx$
9. $\int (x^{5/3} - x^{-5/3})\,dx$
10. $\int (x^{2/5} - x^{-2/5})\,dx$

11. $\int (x^3 + 4x^2 - 3x + 2)\, dx$

12. $\int (x^3 - 6x^2 + 10x - 9)\, dx$

13. $\int \left(\dfrac{4}{x^2} - \dfrac{5}{x^3} \right) dx$

14. $\int (3x^{-1/2} - 2x^{-5/4})\, dx$

15. $\int (5x^3 - 4x + 3x^{-1})\, dx$

16. $\int (-x^2 + 3x + 1 - 2x^{-1})\, dx$

17. $\int (4e^x + 5)\, dx$

18. $\int (3e^x - x^{-1})\, dx$

19. $\int \left(\dfrac{2}{x} + \dfrac{7}{x^3} \right) dx$

20. $\int \left(3x^{-1} + \dfrac{4}{x^{1/2}} \right) dx$

21. $\int \left(4 + \dfrac{3}{x} \right) dx$

22. $\int (-7x^{-1} + 8)\, dx$

23. $\int (6x^{3/2} + 4x^2)\, dx$

24. $\int \left(8x^{-1/2} + \dfrac{3}{x} \right) dx$

25. The marginal cost function for Juan's Tacos is $MC(q) = 0.80 - 0.0004q$. Here q is the number of tacos sold daily. Fixed costs are $150. Determine the cost function.

26. The marginal profit function for Kay's Photography Studio is $MP(q) = 20 + 0.1q$, where q is the number of portraits made in a day. Fixed costs are $80. Determine the profit function.

27. The marginal profit function for a sandwich store is $MP(q) = 0.6 - 0.0002q$, where q is the number of sandwiches sold in a day. Fixed costs are $400. Determine the profit function.

28. The marginal profit function for a shoe store is $MP(q) = 3.5 + 0.001q$, where q is the number of pairs of shoes sold in a day. Fixed costs are $565. Determine the profit function.

29. A ball is 96 ft above the ground at time 0. The height of the ball changes with time according to the formula $h'(t) = -32t + 64$.
    a. Express the height of the ball as a function of time.
    b. How high does the ball get?
    c. When does the ball hit the ground?

30. Repeat exercise 29 if the ball is on the ground at time 0.

Compute the following indefinite integrals.

*31. $\int \dfrac{x^2 + 1}{x}\, dx$

*32. $\int (2x - 1)^2\, dx$

*33. $\int \dfrac{x^2 - 4}{x^2 + 2x}\, dx$

*34. $\int (4x^3 e^x - 3e^x - 4x^3 e^x)\, dx$

Section 12-2 Area

We now leave antiderivatives temporarily, and begin the study of *area*. Although these two concepts appear to be unrelated, in the next section we will learn, via the *Fundamental Theorem of Calculus*, that areas can be computed easily using antiderivatives.

The area of plane regions has several useful properties. Suppose R_1 and R_2 are regions in the plane, with $A(R_1)$ and $A(R_2)$ their respective areas. The following properties apply as common-sense ideas of area.

Properties of Area
1. $A(R_1) \geq 0$.
2. $A(R_1) = A(R_2)$ if R_1 and R_2 are congruent.
3. $A(R_1 \cup R_2) = A(R_1) + A(R_2)$ if R_1 and R_2 do not intersect or only intersect in their boundaries.
4. If $R_1 \subseteq R_2$, then $A(R_1) \leq A(R_2)$.
5. If a rectangle r has height h and length l, then $A(r) = hl$.

From these properties, the area of many types of region can be deduced. In Diagram 1, ABC is a right triangle. If D is chosen so that ABCD is a rectangle, triangles ABC and CDA are congruent and so have the same area by property 2. By property 3, A(triangle ABC) + A(triangle CDA) = A(rectangle ABCD) = lh (by property 5). Thus, A(triangle ABC) = $lh/2$.

By using Diagrams 2 and 3 and the formula for the area of a right triangle, you should derive the formula $lh/2$ for the area of any triangle.

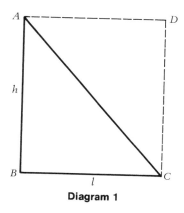

Diagram 1

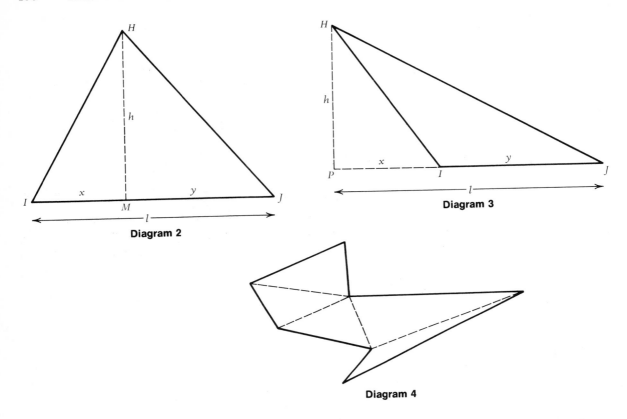

Diagram 2

Diagram 3

Diagram 4

Since any polygon can be divided into triangles (see Diagram 4), we can compute the area of any polygon.

Example 1 Compute the area of the polygon in Diagram 5.

Solution Draw the line from E perpendicular to line AB. See Diagram 6. The line segment EF breaks up the polygon into a rectangle (of area $4 \cdot 1 = 4$) and a right triangle (of area $\frac{1}{2} \cdot 2 \cdot 3 = 3$). Since the rectangle and triangle intersect only in their boundary, the area of the polygon is $4 + 3 = 7$.

∎

Now let us try to compute the area of a circle C of radius r. Let p_n be the regular polygon with 2^n sides inscribed in C, and P_n be the regular polygon with 2^n sides circumscribed about C. See Diagram 7. By considering property 4

$$A(p_i) \leq A(C) \leq A(P_i)$$

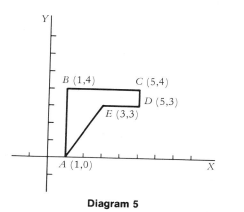

Diagram 5

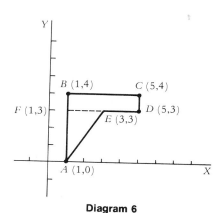

Diagram 6

for any counting numbers i and j. As i and j get larger, $A(p_i)$ increases and $A(P_j)$ decreases, by property 4. It can be shown that only one number, namely πr^2, satisfies the inequalities $A(p_i) \leq x \leq A(P_j)$ for all counting numbers i and j. This number πr^2 is the area of the circle.

We will now describe some more general regions and discuss their areas. Assume that $f(x) \geq 0$ for $a \leq x \leq b$ and that f is continuous for $a \leq x \leq b$. This implies that the graph of f has no jumps or holes and can in fact be drawn without removing your pencil from the paper. Let R be the region bounded by the x-axis and the curve $y = f(x)$ between $x = a$ and $x = b$. See Diagram 8.

A *partition* is a set $\{x_1, x_2, \ldots, x_{n-1}\}$ of points satisfying $a < x_1 < x_2 < \cdots < x_{n-1} < b$. For convenience, put $x_0 = a$ and $x_n = b$. If P is any partition, we can associate with P a region I_P contained in R as follows. For each i from 1 to n, form the rectangle whose base is the line segment from x_{i-1} to x_i on the x-axis, and whose height is the minimum value m_i of $f(x)$ for $x_{i-1} \leq x \leq x_i$. I_P is the union of these rectangles. Since $I_P \subseteq R$, $A(I_P) \leq A(R)$. Note that $A(I_P) = m_1(x_1 - x_0) + m_2(x_2 - x_1) + \cdots + m_n(x_n - x_{n-1})$. See Diagram 9.

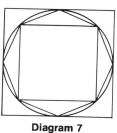

Diagram 7

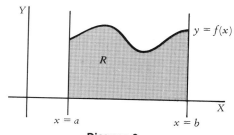

Diagram 8

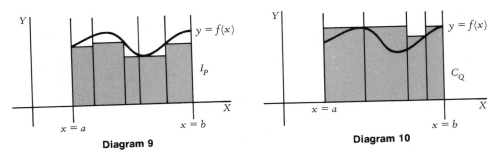

Diagram 9   Diagram 10

Similarly, associate with each partition Q a region C_Q containing R as follows. For each i from 1 to n, form the rectangle whose base is the line segment from x_{i-1} to x_i on the x-axis and whose height is the maximum value M_i of $f(x)$ for $x_{i-1} \leq x \leq x_i$. C_Q is the union of these rectangles. Since $R \subseteq C_Q$, $A(R) \leq A(C_Q)$. Note that $A(C_Q) = M_1(x_1 - x_0) + M_2(x_2 - x_1) + \cdots + M_n(x_n - x_{n-1})$. See Diagram 10.

Note that, by property 4, $A(I_P) \leq A(R) \leq A(C_Q)$ for any partitions P and Q. Now the sum $m_1(x_1 - x_0) + m_2(x_2 - x_1) + \cdots + m_n(x_n - x_{n-1})$ can be defined even if the condition $f(x) \geq 0$ for $a \leq x \leq b$ is omitted—call this sum $s(P)$. Similarly, $M_1(x_1 - x_0) + M_2(x_2 - x_1) + \cdots + M_n(x_n - x_{n-1})$ is called $S(Q)$. Now

$$\boxed{s(P) \leq S(Q) \quad \text{for any partitions } P \text{ and } Q.}$$

The following theorem is proved in more rigorous calculus courses.

> **Theorem.** Let f be continuous on $a \leq t \leq b$. Then there is a unique number (called $\int_a^b f(t)\,dt$) such that
> $$s(P) \leq \int_a^b f(t)\,dt \leq S(Q)$$
> for all partitions P and Q of $a \leq t \leq b$.

If $f(x) \geq 0$ for $a \leq x \leq b$, the number mentioned in the theorem is $A(R)$. Otherwise the number cannot be interpreted yet. The expression

$$\int_a^b f(t)\,dt$$

is called the *definite integral of f from a to b*, and is spoken of as the definite integral. In the next section, we find a simple way to compute the definite integral using antiderivatives.

Example 2 Consider the function $y = x^2$ on the interval $1 \le x \le 4$ and the partition $\{1.5, 2, 2.5, 3, 3.5\}$.

PROBLEM A Compute I_P.

Solution The graph of f and I_P are shown in Diagram 11.

$$A(I_P) = 1^2(1.5 - 1) + (1.5)^2(2 - 1.5) + 2^2(2.5 - 2) + 2.5^2(3 - 2.5)$$
$$+ 3^2(3.5 - 3) + 3.5^2(4 - 3.5)$$
$$= 17.375$$

PROBLEM B Compute C_P.

Solution The graph of f and C_P are shown in Diagram 12.

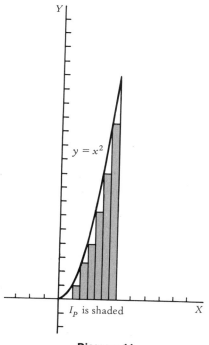

I_P is shaded

Diagram 11

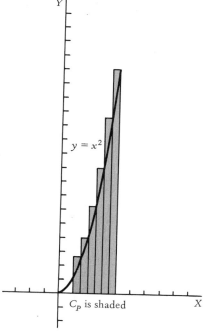

C_P is shaded

Diagram 12

$$A(C_P) = 1.5^2(1.5 - 1) + 2^2(2 - 1.5) + 2.5^2(2.5 - 2) + 3^2(3 - 2.5)$$
$$+ 3.5^2(3.5 - 3) + 4^2(4 - 3.5)$$
$$= 24.875$$

∎

Exercises 12-2

Use the properties of area to derive the formula $A = lh/2$ *for*
1. An acute triangle. (See Diagram 2.)
2. An obtuse triangle. (See Diagram 3.)

Determine the area of each polygon.

3.
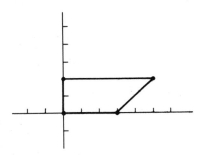

4.
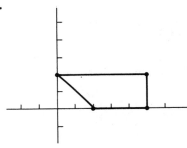

5.
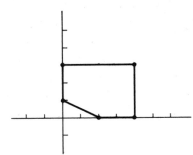

6.
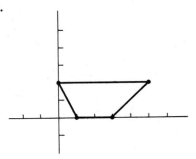

For each function f and partition P, compute I_P *and* C_P.

7. $f(x) = x^3$, $0 \le x \le 4$ $P = \{1, 2, 3\}$
8. $f(x) = x^3$, $0 \le x \le 4$ $P = \{0.5, 1, 1.5, 2, 2.5, 3, 3.5\}$
9. $f(x) = x^2$, $1 \le x \le 4$
    $P = \{1.25, 1.5, 1.75, 2, 2.25, 2.5, 2.75, 3, 3.25, 3.5, 3.75\}$

Section 12-3 The Fundamental Theorem of Calculus

We will now prove the Fundamental Theorem of Calculus and learn how to compute definite integrals and areas using antiderivatives.

> **Fundamental Theorem of Calculus.** Let f be continuous on $a \leq t \leq b$. Let F be any antiderivative of f. Then
> $$\int_a^b f(t)\, dt = F(t)\Big|_a^b = F(b) - F(a)$$

This theorem is extremely important because it connects the two branches of calculus: differential calculus and integral calculus. To make the proof simpler, assume $f(t) \geq 0$ for $a \leq t \leq b$.

1. Let $A(x) = \int_a^x f(t)\, dt$, so that $A(x)$ is the area between the x-axis and the graph of f from the vertical line at a to the vertical line at x. See Diagram 1. The first part of the proof is to show that $A' = f$.
2. Assume $h > 0$. (This is not really necessary.) If m is the minimum value taken by f between x and $x + h$ and M is the maximum value, then
$$mh \leq A(x+h) - A(x) \leq Mh$$
by property 4 of area. See Diagram 2. Divide this last inequality by h to obtain
$$m \leq \frac{A(x+h) - A(x)}{h} \leq M$$
3. As h approaches 0, m and M must both approach $f(x)$ since f is continuous, and $[A(x+h) - A(x)]/h$ must approach $A'(x)$, by the defi-

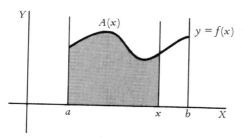

Diagram 1

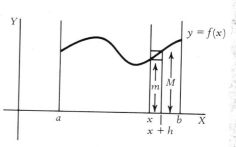

Diagram 2

nition of derivative. Therefore, $f(x) \leq A'(x) \leq f(x)$, so that $A'(x) = f(x)$. Thus A is an antiderivative of f. This ends the first part of the proof.

4. Since $A(x) = \int_a^x f(t)\,dt$, substituting b for x gives $A(b) = \int_a^b f(t)\,dt$.

5. $A(a)$ is the area of the line segment joining the points $(a, 0)$ and $(a, f(a))$. The area of a line segment is 0, so that $A(a) = 0$. Then

$$\int_a^b f(t)\,dt = A(b) - A(a)$$

6. The final part of the proof is to show that the last equation is valid if A is replaced by any antiderivative of f.

7. Let F be any antiderivative of f. Then there is a number c such that $A(x) = F(x) + c$. Thus,

$$\int_a^b f(t)\,dt = A(b) - A(a) = [F(b) + c] - [F(a) + c]$$
$$= F(b) + c - F(a) - c = F(b) - F(a)$$

This completes the proof.

When $f(x) \geq 0$ for $a \leq x \leq b$, the Fundamental Theorem of Calculus can be used to compute areas.

If $f(x) \geq 0$ and is continuous for $a \leq x \leq b$, the area of the region R bounded by the x-axis and the graph of f between $x = a$ and $x = b$ is given by

$$A(R) = \int_a^b f(x)\,dx = F(b) - F(a)$$

where F is any antiderivative of f.

Let's find the area of the region R bounded by the x-axis and the graph of $y = x^2$ from $x = -3$ to $x = 4$. (See Diagram 3.) Since $x^2 \geq 0$ for $-3 \leq x \leq 4$, the definite integral represents area. Then

$$A(R) = \int_{-3}^{4} x^2\,dx$$

$$= \frac{x^3}{3}\Big|_{-3}^{4} \qquad \text{Fundamental Theorem of Calculus}$$

$$= \frac{4^3}{3} - \frac{(-3)^3}{3} = 30\tfrac{1}{3}$$

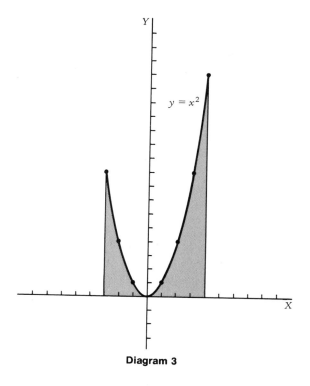

Diagram 3

Example 1 Find the area between the x-axis and the graph of $y = 1/x$ from $x = 1$ to $x = e$. See Diagram 4.

Solution Note that $1/x > 0$ for $x > 0$, so that the definite integral represents area.

$$A = \int_1^e \frac{1}{x} dx$$

$$= \ln|x|\Big|_1^e \qquad \text{Fundamental Theorem of Calculus}$$

$$= \ln e - \ln 1 = 1 - 0 = 1$$

Example 2 Find the area between the x-axis and the graph of $y = 5e^x$ from $x = -1$ to $x = 1$. See Diagram 5.

Solution Note that $5e^x > 0$ for all x.

$$A = \int_{-1}^1 5e^x \, dx$$

$$= 5e^x\Big|_{-1}^1 = 5e - \frac{5}{e} = 11.8$$

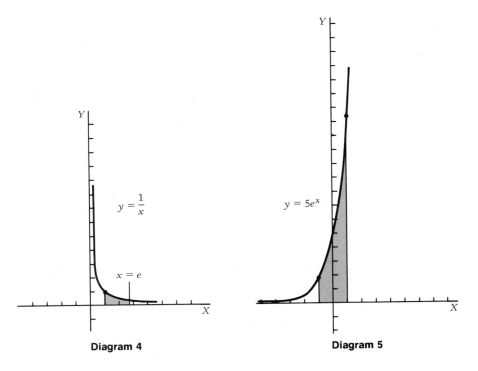

Diagram 4

Diagram 5

Example 3 Find the area between the x-axis and the graph of $y = x^3 - 3x^2$ from $x = 3$ to $x = 4$. See Diagram 6.

Solution Note that $x^3 - 3x^2 = x^2(x - 3) \geq 0$ for $3 \leq x \leq 4$.

$$A = \int_3^4 (x^3 - 3x^2)\,dx$$

$$= \left(\frac{x^4}{4} - x^3\right)\bigg|_3^4 = \left(\frac{4^4}{4} - 4^3\right) - \left(\frac{3^4}{4} - 3^3\right) = 6.75$$

■

The Fundamental Theorem of Calculus can be used to evaluate definite integrals. For example, to evaluate $\int_1^2 (x^2 - 4x + 3)\,dx$, use the Fundamental Theorem to obtain

$$\int_1^2 (x^2 - 4x + 3)\,dx = \left(\frac{x^3}{3} - 2x^2 + 3x\right)\bigg|_1^2$$

$$= \left(\frac{8}{3} - 8 + 6\right) - \left(\frac{1}{3} - 2 + 3\right) = -\frac{2}{3}$$

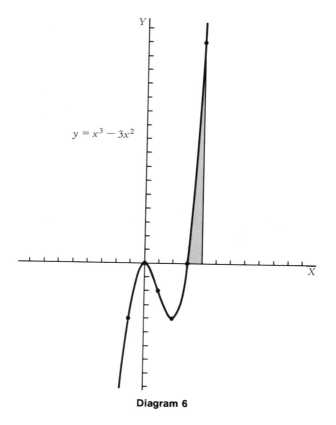

Diagram 6

Example 4 Evaluate $\int_2^5 \left(\dfrac{3}{x} - e^x\right) dx$.

Solution Note that this integral does not represent area since the function $3/x - e^x$ takes negative values for $2 \le x \le 5$. (For example, substituting $x = 2$ gives $3/2 - e = -1.2$.) This integral can nevertheless be evaluated using the Fundamental Theorem of Calculus.

$$\begin{aligned}
\int_2^5 \left(\dfrac{3}{x} - e^x\right) dx &= (3\ln|x| - e^x)\Big|_2^5 \\
&= (3\ln 5 - e^5) - (3\ln 2 - e^2) \\
&= 3\ln 5 - e^5 - 3\ln 2 + e^2 \\
&= 3(\ln 5 - \ln 2) + e^2 - e^5 \\
&= 3\ln(2.5) + e^2 - e^5 = -138.28
\end{aligned}$$

■

Another application of the Fundamental Theorem of Calculus is to determine the change in a function whose derivative is known. In the theorem, substitute g for F and g' for f to obtain

$$g(b) - g(a) = \int_a^b g'(t)\,dt$$

The following examples use this formula.

Example 5 A new worker on Moresco's assembly line works more quickly as he gets more experienced. In fact, to process his yth radiator takes him $4 + y^{-1/2}$ minutes. Suppose he has already processed 400 radiators. How long will he take to process the next 1200?

Solution Let $T(y)$ be the time required to process y radiators. $T'(y)$ can be interpreted as the time required to process one more radiator, if y radiators are already processed. Thus $T'(y) = 4 + y^{-1/2}$. The problem asks for the change in T between $y = 400$ and $y = 1600$ ($400 + 1200 = 1600$), that is, for

$$T(1600) - T(400) = \int_{400}^{1600} T'(y)\,dy$$

$$= \int_{400}^{1600} (4 + y^{-1/2})\,dy = \left(4y + \frac{y^{1/2}}{1/2}\right)\Big|_{400}^{1600}$$

$$= (4y + 2y^{1/2})\Big|_{400}^{1600}$$

$$= \left[4(1600) + 2(1600)^{1/2}\right] - \left[4(400) + 2(400)^{1/2}\right]$$

$$= 4840$$

The worker requires 4840 minutes, or $80\frac{2}{3}$ hours, to process the next 1200 radiators.

Example 6 The revenue on the tth business day at a new electronics store is $500 + 3t$. How much is the total revenue for the first 100 business days? For the second 100 business days?

Solution Suppose $R(t)$ is the total revenue for the first t days of business. Then $R'(t)$ is the revenue on the tth day of business, so that $R'(t) = 500 + 3t$.

The total revenue for the first 100 business days is

$$\int_0^{100} R'(t)\,dt = \int_0^{100}(500 + 3t)\,dt$$

$$= 500t + \frac{3t^2}{2}\bigg|_0^{100} = \$65,000$$

The revenue for the second 100 business days is

$$\int_{100}^{200} R'(t)\,dt = \int_{100}^{200}(500 + 3t)\,dt$$

$$= 500t + \frac{3t^2}{2}\bigg|_{100}^{200}$$

$$= 100,000 + 60,000 - 50,000 - 15,000 = \$95,000$$

∎

The definite integral has several useful properties. These properties are immediate consequences of the Fundamental Theorem of Calculus.

Properties of the Definite Integral

1. $\int_a^c f(t)\,dt = \int_a^b f(t)\,dt + \int_b^c f(t)\,dt \qquad a < b < c$

2. $\int_a^a f(t)\,dt = 0$

3. $\int_a^b [cf(t) + kg(t)]\,dt = c\int_a^b f(t)\,dt + k\int_a^b g(t)\,dt$ if c and k are numbers.

Exercises 12-3

Compute the area between the x-axis, the graph of the function, and the given vertical lines. In each case, verify that the function is nonnegative between the vertical lines.

1. $y = 3x, \quad x = 1, \quad x = 5$
2. $y = 6 - x, \quad x = 0, \quad x = 6$
3. $y = x^2, \quad x = 0, \quad x = 4$
4. $y = x^3, \quad x = 1, \quad x = 3$
5. $y = 9 - x^2, \quad x = 1, \quad x = 3$
6. $y = 16 - 2x, \quad x = 4, \quad x = 6$
7. $y = \dfrac{4}{x}, \quad x = 1, \quad x = 6$
8. $y = \dfrac{6}{x^2}, \quad x = 1, \quad x = 3$
9. $y = \dfrac{10}{x^3}, \quad x = 1, \quad x = 2$
10. $y = 4 - \dfrac{2}{x}, \quad x = 2, \quad x = 4$

11. $y = 2x + \dfrac{2}{x}$, $x = 3$, $x = 6$
12. $y = 8 - x^3$, $x = -2$, $x = 2$
13. $y = e^x$, $x = 1$, $x = 2$
14. $y = e^x$, $x = 0$, $x = 1$
15. $y = e^x - e$, $x = 1$, $x = 2$
16. $y = e - e^x$, $x = 0$, $x = 1$

Compute the definite integrals.

17. $\displaystyle\int_{-1}^{1} x^3\, dx$
18. $\displaystyle\int_{-2}^{2} x^2\, dx$
19. $\displaystyle\int_{0}^{1} (6x^2 - 2x - 3)\, dx$
20. $\displaystyle\int_{0}^{2} (12x^2 + 6x - 9)\, dx$
21. $\displaystyle\int_{-1}^{1} (t^2 - 4t - 5)\, dt$
22. $\displaystyle\int_{2}^{3} (4x^3 - 6x^2)\, dx$
23. $\displaystyle\int_{0}^{4} (3e^x - 10)\, dx$
24. $\displaystyle\int_{-2}^{3} (4e^x - x)\, dx$

25. A new worker for Magiclawn works faster as she gets more experienced. To process (mow and spray chemicals on) her nth lawn takes $50 + (40/n)$ minutes. This formula is valid for $n \geq 10$. How long does she take to process her 21st through 50th lawn?

26. A new worker at a shirt factory takes $3 + (6/\sqrt{n})$ minutes to sew the buttons on his nth shirt. Suppose he has already worked on nine shirts. How long will he need for the next 16 shirts?

27. A new ice cream store sells $100 + 2t$ cones on its tth day of business. How many cones does it sell in
    a. The first 30 days?
    b. The second 30 days?

28. The revenue on the nth day of a 30-day going out of business sale is $1000 - 20n$. How much is the revenue for
    a. The first 15 days of the sale?
    b. The whole sale?

29. The cost of maintenance per mile for a car that has been driven x miles is

$$0.01 + 0.00005\sqrt{x}$$

What is the maintenance cost for
    a. The first 10,000 mi?
    b. The next 20,000 mi, if the car has been driven 50,000 mi already?

30. The cost of maintenance per month for a factory that is t months old is $2000 + 10t$. What is the maintenance cost of the factory for its
    a. First year?
    b. Tenth year?

31. The population density alongside an east-west interstate highway is given by

$$D(t) = \frac{500{,}000}{x}, \quad 10 \le x \le 50$$

where x is the distance from the center of a major city. How many people live alongside the highway between
   a. 10 and 20 miles east of the city?
   b. 25 and 50 miles east of the city?

32. Repeat exercise 31 if

$$D(t) = \frac{100{,}000}{\sqrt{x}}, \quad 10 \le x \le 50$$

Compute each definite integral. Use trial and error to find the antiderivatives.

*33. $\int_0^1 e^{5x}\,dx$

*34. $\int_0^2 \frac{x}{4+x^2}\,dx$

*35. $\int_0^5 (3x+1)^{1/2}\,dx$

*36. $\int_0^2 x\sqrt{4-x^2}\,dx$

*37. $\int_0^1 \frac{x^2}{(1+x^3)^2}\,dx$

*38. $\int_{-1}^1 x\exp(x^2)\,dx$

Section 12-4 The Area Between Curves

With the aid of the Fundamental Theorem of Calculus, we can now compute the areas of certain regions quite easily. Such a region must be bounded by two vertical lines $x=a$ and $x=b$, the x-axis, and the graph of a function $y=f(x)$. Besides this, it is necessary that $f(x) \ge 0$ for $a \le x \le b$.

We will now learn how to find the areas of more general regions. Begin by assuming that $f(x) \le 0$ (instead of $f(x) \ge 0$) for $a \le x \le b$.

Let R and R_1 be the regions indicated on Diagram 1. Then R and R_1 are congruent, so that $A(R) = A(R_1)$. Since $(-f)(x) \ge 0$ for $a \le x \le b$, $A(R_1) = \int_a^b (-f)(x)\,dx$ by the Fundamental Theorem of Calculus. Then $A(R) = \int_a^b (-f)(x)\,dx = -\int_a^b f(x)\,dx$.

Example 1 Find the area between the x-axis and the graph of $y = x^3$ from $x = -2$ to $x = -1$. See Diagram 2.

Solution Since $x^3 \le 0$ whenever $-2 \le x \le -1$,

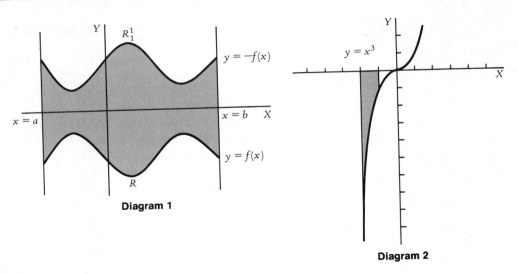

Diagram 1

Diagram 2

$$A(R) = -\int_{-2}^{-1} x^3 \, dx$$

$$= \left. \frac{-x^4}{4} \right|_{-2}^{-1}$$

$$= -\left[\frac{(-1)^4}{4} - \frac{(-2)^4}{4} \right]$$

$$= -\left(\frac{1}{4} - \frac{16}{4} \right) = -\left(\frac{-15}{4} \right) = \frac{15}{4}$$

∎

Now consider the situation in Diagram 3. The area between the x-axis and the graph of f from $x = a$ to $x = c$ equals $A(R_1) + A(R_2)$ since R_1 and R_2 meet only in the point $(b, 0)$. Thus, this area is equal to

$$A(R_1) + A(R_2) = \int_a^b f(x) \, dx + \int_b^c -f(x) \, dx$$

$$= \int_a^b |f(x)| \, dx + \int_b^c |f(x)| \, dx = \int_a^c |f(x)| \, dx$$

since

$$|f(x)| = \begin{cases} f(x) & a \le x \le b \\ -f(x) & b \le x \le c \end{cases}$$

12-4 THE AREA BETWEEN CURVES

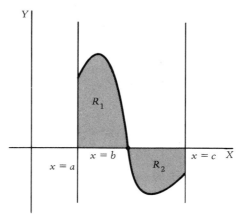

Diagram 3

Theorem. Let f be continuous on $a \leq x \leq b$. The area between the x-axis and the graph of f from $x = a$ to $x = b$ is

$$\int_a^b |f(x)|\, dx \qquad \text{Area between a curve and the x-axis.}$$

The formula in this theorem is not used directly to compute area, since it is difficult to compute the antiderivative of $|f(x)|$. To compute area, start by breaking the interval from a to b up into subintervals on which f is always ≥ 0 or always ≤ 0.

Example 2 Find the area between the x-axis and the graph of $y = x^2 - x$ from $x = 0$ to $x = 2$. See Diagram 4.

Solution First determine where $x^2 - x$ is ≥ 0 and ≤ 0. This can be done by drawing

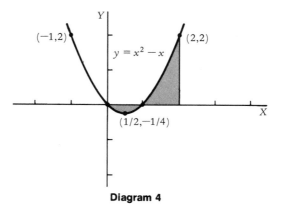

Diagram 4

the graph of $y = x^2 - x$, or by using a sign graph.

For $0 \le x \le 1$, $x^2 - x \le 0$. Therefore, for x between 0 and 1, $|x^2 - x| = -(x^2 - x)$. For $1 \le x \le 2$, $x^2 - x \ge 0$, so that $|x^2 - x| = x^2 - x$. Thus the area equals

$$\int_0^1 -(x^2 - x)\,dx + \int_1^2 (x^2 - x)\,dx = -\left(\frac{x^3}{3} - \frac{x^2}{2}\right)\bigg|_0^1 + \left(\frac{x^3}{3} - \frac{x^2}{2}\right)\bigg|_1^2 = 1$$

Example 3 Find the area between the x-axis and the graph of $y = (1/x) - (1/2)$ from $x = 1/4$ to $x = 8$. See Diagram 5.

Solution When $x = 2$, $y = 0$. If $1/4 \le x \le 2$, then $y \ge 0$. If $2 \le x \le 8$, then $y \le 0$. The area equals

$$\int_{1/4}^2 \left(\frac{1}{x} - \frac{1}{2}\right) dx - \int_2^8 \left(\frac{1}{x} - \frac{1}{2}\right) dx = \left(\ln x - \frac{x}{2}\right)\bigg|_{1/4}^2 - \left(\ln x - \frac{x}{2}\right)\bigg|_2^8$$

$$= \left\{(\ln 2 - 1) - \left[\ln\left(\frac{1}{4}\right) - \frac{1}{8}\right]\right\} - \left[(\ln 8 - 4) - (\ln 2 - 1)\right]$$

$$= \left[\ln 2 - \ln\left(\frac{1}{4}\right) - 1 + \frac{1}{8}\right] - (\ln 8 - \ln 2 - 4 + 1)$$

$$= \left(\ln 8 - \frac{7}{8}\right) - (\ln 4 - 3) = \ln 8 - \frac{7}{8} - \ln 4 + 3 = \ln 2 + 2\tfrac{1}{8} = 2.82$$

Example 4 Find the area between the x-axis and $y = x^3 - 3x^2 + 2x$ from $x = -1$ to $x = 3$. See Diagram 6.

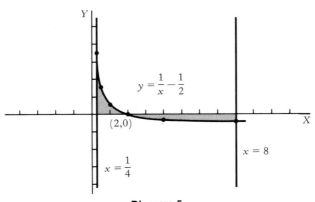

Diagram 5

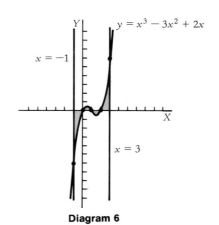

Diagram 6

Solution We will express this area in terms of definite integrals but will not compute it. $x^3 - 3x^2 + 2x = x(x^2 - 3x + 2) = x(x-1)(x-2)$. Now draw a sign graph.

```
x - 2            —    |    —    |    —    |    +
x - 1            —    |    —    |    +    |    +
x                —    |    +    |    +    |    +
x(x - 1)(x - 2)  —    0    +    0    —    0    +
                      0         1         2
```

From the sign graph, observe that $y \leq 0$ if $-1 \leq x \leq 0$, $y \geq 0$ if $0 \leq x \leq 1$, $y \leq 0$ if $1 \leq x \leq 2$, and $y \geq 0$ if $2 \leq x \leq 3$. Then

$$A = -\int_{-1}^{0} (x^3 - 3x^2 + 2x)\,dx + \int_{0}^{1} (x^3 - 3x^2 + 2x)\,dx$$
$$- \int_{1}^{2} (x^3 - 3x^2 + 2x)\,dx + \int_{2}^{3} (x^3 - 3x^2 + 2x)\,dx$$

■

So far, we have studied the area between a curve and the x-axis. Now let's study the area between two curves. In Diagram 7, R_1 is the region between the graph of f and the graph of g from $x = a$ to $x = b$. Since R_1 and R_2 intersect only in the graph of g,

$$A(R_1 \cup R_2) = A(R_1) + A(R_2)$$
$$\int_{a}^{b} f(x)\,dx = \int_{a}^{b} g(x)\,dx + A(R_2)$$
$$A(R_2) = \int_{a}^{b} f(x)\,dx - \int_{a}^{b} g(x)\,dx = \int_{a}^{b} [f(x) - g(x)]\,dx$$

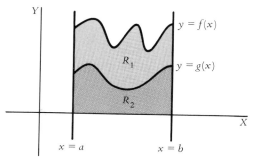

Diagram 7

This formula is valid even if f and g take negative values. However, it is required that $f(x) - g(x) \geq 0$ for $a \leq x \leq b$. In general,

> **Theorem.** The area between the graphs of $y = f(x)$ and $y = g(x)$ from $x = a$ to $x = b$ equals
> $$\int_a^b |f(x) - g(x)| \, dx \qquad \text{Area between curves}$$

To actually compute the area, first determine where $f(x) - g(x) \geq 0$ and $f(x) - g(x) \leq 0$.

Example 5 Find the area between $y = x^2 - 4x$ and $y = 2x^2$ from $x = 0$ to $x = 3$.

Solution As we see from Diagram 8, $2x^2 \geq x^2 - 4x$ if $0 \leq x \leq 3$, so the area is

$$\int_0^3 [2x^2 - (x^2 - 4x)] \, dx = \int_0^3 (x^2 + 4x) \, dx$$
$$= \left(\frac{x^3}{3} + 2x^2 \right) \Big|_0^3 = 27$$

Example 6 Using the same functions, find the area from $x = -6$ to $x = -2$.

Solution For $-6 \leq x \leq -4$, $2x^2 \geq x^2 - 4x$. For $-4 \leq x \leq -2$, $x^2 - 4x \geq 2x^2$. Thus

$$A = \int_{-6}^{-4} [2x^2 - (x^2 - 4x)] \, dx + \int_{-4}^{-2} [(x^2 - 4x) - 2x^2] \, dx$$
$$= \int_{-6}^{-4} (x^2 + 4x) \, dx + \int_{-4}^{-2} (-x^2 - 4x) \, dx$$
$$= \left(\frac{x^3}{3} + 2x^2 \right) \Big|_{-6}^{-4} + \left(\frac{-x^3}{3} - 2x^2 \right) \Big|_{-4}^{-2}$$
$$= \left(\frac{-64}{3} + 32 \right) - (-72 + 72) + \left(\frac{8}{3} - 8 \right) - \left(\frac{64}{3} - 32 \right)$$
$$= -21\tfrac{1}{3} + 32 + 2\tfrac{2}{3} - 8 - 21\tfrac{1}{3} + 32 = 16$$

Example 7 Find the area enclosed by $y = x^2$ and $y = 5x - 6$.

Solution First find where the curves meet. Let $x^2 = 5x - 6$ and solve to obtain $x = 2$ or $x = 3$. The integral goes from $x = 2$ to $x = 3$. For $2 < x < 3$, $x^2 < 5x - 6$.

12-4 THE AREA BETWEEN CURVES

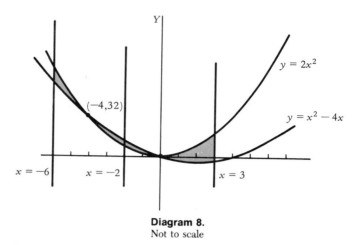

Diagram 8.
Not to scale

To see this, compute both functions at any number, say 2.5, between 2 and 3. Since the functions are equal only at 2 and 3, whichever function has the larger value at 2.5 has the larger value for $2 < x < 3$. The inequality $x^2 < 5x - 6$ for $2 < x < 3$ can also be obtained by drawing a graph. See Diagram 9. The area equals

$$\int_2^3 (5x - 6 - x^2)\,dx = \left(\frac{5x^2}{2} - 6x - \frac{x^3}{3} \right)\Bigg|_2^3 = \frac{1}{6}$$

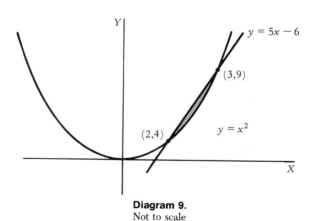

Diagram 9.
Not to scale

Example 8 Find the area enclosed by $y = 16 - x^2$ and $y = x^2 - 9$. See Diagram 10.

Solution To find where the curves meet, set $x^2 - 9 = 16 - x^2$ and solve for x.

$$2x^2 = 25$$

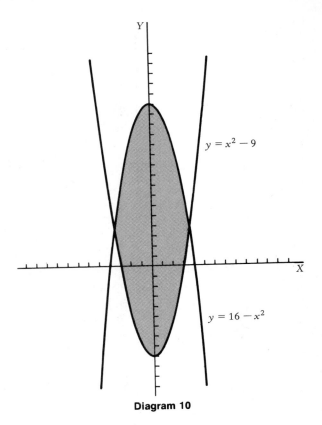

Diagram 10

$$x^2 = \frac{25}{2}$$

$$x = \frac{-5}{\sqrt{2}} \quad \text{or} \quad x = \frac{5}{\sqrt{2}}$$

Since $16 - x^2 > x^2 - 9$ when $x = 0$, $16 - x^2 > x^2 - 9$ when $\frac{-5}{\sqrt{2}} < x < \frac{5}{\sqrt{2}}$, so that the area equals

$$\int_{-5/\sqrt{2}}^{5/\sqrt{2}} [(16 - x^2) - (x^2 - 9)]\,dx = \int_{-5/\sqrt{2}}^{5/\sqrt{2}} (25 - 2x^2)\,dx$$

$$= \left(25x - \frac{2x^3}{3}\right)\Bigg|_{-5/\sqrt{2}}^{5/\sqrt{2}} = \frac{250\sqrt{2}}{3} = 118$$

Exercises 12-4

Find the area between the given curve and the x-axis between the given values of x.

1. $y = x^2 + x$, $x = -2$ to $x = 0$
2. $y = x^2 - 3x$, $x = 1$ to $x = 4$
3. $y = 3x + 3$, $x = -2$ to $x = 2$
4. $y = 6 - 2x$, $x = 0$ to $x = 4$
5. $y = x^2 - 5x + 4$, $x = 0$ to $x = 4$
6. $y = x^2 + 3x + 2$, $x = -4$ to $x = 0$
7. $y = x^2 + 2x + 3$, $x = 0$ to $x = 1$
8. $y = x^2 - 2x + 5$, $x = -2$ to $x = 0$
9. $y = \dfrac{4}{x} - 2$, $x = 1$ to $x = 8$
10. $y = \dfrac{3}{x} - 3$, $x = \dfrac{1}{2}$ to $x = 5$
11. $y = e^x - 1$, $x = -1$ to $x = 1$
12. $y = e^x - e$, $x = 1$ to $x = 3$

Find the area between the given curves between the given values of x.

13. $y = e^x$, $y = 1$, $x = 1$ to $x = 2$
14. $y = e^x$, $y = x$, $x = 0$ to $x = 1$
15. $y = x^2$, $y = x$, $x = -4$ to $x = 1$
16. $y = 4e^x$, $y = e^x$, $x = -1$ to $x = 1$
17. $y = \sqrt{x}$, $y = x + 6$, $x = 0$ to $x = 4$
18. $y = \sqrt{x}$, $y = 1 + x^2$, $x = 0$ to $x = 9$

Find the area enclosed by the given curves.

19. $y = x^2$, $y = x$
20. $y = x^3$, $y = x$
21. $y = x^2$, $y = x^3$
22. $y = x^2 + 1$, $y = x + 3$
23. $y = x^2 + 2x$, $y = 8x - 8$
24. $y = x^2 - 2x - 3$, $y = 4x - 8$

Section 12-5 Computation of More Antiderivatives

So far, we have not found antiderivatives for composite functions, such as $(2x + 1)^{1/2}$ or $\exp(6x)$. Finding these antiderivatives involves using the chain rule in reverse. Now

$$\left[f(x)^{r+1}\right]' = (r+1)f(x)^r f'(x) \quad \text{Chain rule.}$$

so that

$$\left[\frac{f(x)^{r+1}}{r+1}\right]' = f(x)^r f'(x) \quad \text{Divide by } r+1.$$

CHAPTER 12 INTEGRAL CALCULUS

In terms of antiderivatives,

$$\int f(x)^r f'(x)\,dx = \frac{f(x)^{r+1}}{r+1} + c, \quad r \neq -1 \quad \text{Power rule}$$

This formula is called the power rule. The condition $r \neq -1$ is needed to prevent division by 0.

Examples Use the power rule to find the antiderivative of each function.

Example 1 $\int (3x^4 + 5)^{10} 12x^3\,dx = \int (3x^4 + 5)^{10}(3x^4 + 5)'\,dx$ The derivative of $3x^4 + 5$ is $12x^3$.

$$= \frac{(3x^4 + 5)^{11}}{11} + c$$

Power rule: $f(x) = 3x^4 + 5$ and $r = 10$.

Example 2 $\int (5x + 3)^{-1/2} 5\,dx = \int (5x + 3)^{-1/2}(5x + 3)'\,dx$ The derivative of $5x + 3$ is 5.

$$= \frac{(5x + 3)^{1/2}}{1/2} + c$$

Power rule: $f(x) = 5x + 3$ and $r = -1/2$.

$$= 2(5x + 3)^{1/2} + c$$

Simplify.

■

The power rule excluded the case $r = -1$. This case is covered by the following formula.

$$\int \frac{1}{f(x)} f'(x)\,dx = \ln|f(x)| + c \quad \text{Logarithm rule}$$

To verify the logarithm rule, all you need to do is observe that

$$\frac{d}{dx}\ln|f(x)| = \frac{f'(x)}{f(x)} \quad \text{Chain rule.}$$

Examples Use the logarithm rule to find the antiderivative of each function.

Example 3 $\int \dfrac{2x}{x^2+1} \, dx = \int \dfrac{1}{x^2+1}(x^2+1)' \, dx$ The derivative of $x^2 + 1$ is $2x$.

$\phantom{\int \dfrac{2x}{x^2+1} \, dx} = \ln(x^2+1) + c$ Logarithm rule. (Why are there no absolute value signs here?)

Example 4 $\int \dfrac{3x^2}{x^3+8} \, dx = \int \dfrac{1}{x^3+8}(x^3+8)' \, dx$

$\phantom{\int \dfrac{3x^2}{x^3+8} \, dx} = \ln|x^3+8| + c$ Logarithm rule.

■

Composite functions involving the exponential function may be antidifferentiated with the exponential rule.

$$\boxed{\int \exp[f(x)] f'(x) \, dx = \exp[f(x)] + c \quad \text{Exponential rule}}$$

To verify the exponential rule, you need only note that

$$\dfrac{d}{dx} \exp[f(x)] = f'(x) \exp[f(x)] \quad \text{Chain rule.}$$

Examples Use the exponential rule to find the antiderivative of each function.

Example 5 $\int e^{3x} 3 \, dx = \int e^{3x}(3x)' \, dx$ The derivative of $3x$ is 3.

$\phantom{\int e^{3x} 3 \, dx} = e^{3x} + c$ Exponential rule.

Example 6 $\int 2x \exp(x^2) \, dx = \int \exp(x^2)(x^2)' \, dx$ The derivative of x^2 is $2x$.

$ = \exp(x^2) + c$ Exponential rule.

■

In most problems, it is necessary to multiply and divide by a constant in order to use the formulas of this section. Let us try to evaluate

$$\int (5x^2+1)^{-1/2} x \, dx.$$

Since $(5x^2+1)' = 10x$, the power rule cannot be used directly. However,

$$\int (5x^2 + 1)^{-1/2} x\, dx$$

$$= \int \frac{1}{10}(5x^2 + 1)^{-1/2} 10x\, dx \qquad \text{The function to be integrated has been multiplied and divided by 10.}$$

$$= \frac{1}{10}\int (5x^2 + 1)^{-1/2} 10x\, dx \qquad \text{Use linearity in the form } \int kf(x)\, dx = k\int f(x)\, dx \text{ when } k \text{ is a number.}$$

$$= \frac{1}{10}\int (5x^2 + 1)^{-1/2}(5x^2 + 1)'\, dx \quad (5x^2 + 1)' = 10x.$$

$$= \frac{1}{10}\frac{(5x^2 + 1)^{1/2}}{\frac{1}{2}} + c \qquad \text{Power rule.}$$

$$= \frac{1}{5}(5x^2 + 1)^{1/2} + c \qquad \text{Simplify.}$$

The key step in solving this problem was the use of linearity to move the factor of $1/10$ from the right of the integral sign to its left. Only constant factors may be moved this way. *A variable cannot be moved from one side of an integral sign to the other.*

Examples Evaluate each integral.

Example 7 $\int_4^{12}(2x + 1)^{1/2}\, dx$

$$= \frac{1}{2}\int_4^{12}(2x + 1)^{1/2} 2\, dx \qquad \text{Multiply and divide by 2 since } (2x + 1)' = 2$$

$$= \frac{1}{2}\int_4^{12}(2x + 1)^{1/2}(2x + 1)'\, dx$$

$$= \frac{1}{2}\frac{(2x + 1)^{3/2}}{3/2}\bigg|_4^{12} \qquad \text{Power rule.}$$

$$= \frac{(2x + 1)^{3/2}}{3}\bigg|_4^{12} = \frac{25^{3/2}}{3} - \frac{9^{3/2}}{3} = \frac{5^3}{3} - \frac{3^3}{3} = \frac{98}{3} = 32\frac{2}{3}$$

Example 8 $\int_0^8 6x\sqrt{100 - x^2}\, dx$

$$= \frac{6}{-2} \int_0^8 (100 - x^2)^{1/2}(-2x)\, dx \qquad \text{Move 6 to left of integral sign; multiply and divide by } -2 \text{ since } (100 - x^2)' = -2x.$$

$$= -3 \int_0^8 (100 - x^2)^{1/2}(100 - x^2)'\, dx$$

$$= -3 \frac{(100 - x^2)^{3/2}}{3/2} \Big|_0^8 = -2(100 - x^2)^{3/2} \Big|_0^8 \qquad \text{Use the power rule and simplify.}$$

$$= -2(36)^{3/2} + 2(100)^{3/2}$$

$$= -2(6)^3 + 2(10)^3 = 1568$$

Example 9 $\int \frac{1}{4x + 3}\, dx = \frac{1}{4} \int \frac{1}{4x + 3} 4\, dx$ Multiply and divide by 4 since $(4x + 3)' = 4$.

$$= \frac{1}{4} \int \frac{1}{4x + 3}(4x + 3)'\, dx$$

$$= \frac{1}{4} \ln|4x + 3| + c \qquad \text{Logarithm rule.}$$

Example 10 $\int_1^3 \frac{3x + 4}{3x^2 + 8x - 2}\, dx$

$$= \frac{1}{2} \int_1^3 \frac{1}{3x^2 + 8x - 2}(6x + 8)\, dx \qquad \text{Multiply and divide by 2 since } (3x^2 + 8x - 2)' = 6x + 8.$$

$$= \frac{1}{2} \int_1^3 \frac{1}{3x^2 + 8x - 2}(3x^2 + 8x - 2)'\, dx$$

$$= \frac{1}{2} \ln|3x^2 + 8x - 2| \Big|_1^3 \qquad \text{Logarithm rule.}$$

$$= \frac{1}{2} \ln 49 - \frac{1}{2} \ln 9$$

$$= \ln 7 - \ln 3 = \ln\left(\frac{7}{3}\right) = .847$$

Example 11 $\int \exp(6x)\, dx = \frac{1}{6} \int \exp(6x) 6\, dx$ Multiply and divide by 6 since $(6x)' = 6$.

$$= \frac{1}{6} \int \exp(6x)(6x)'\, dx$$

$$= \frac{1}{6} \exp(6x) + c \qquad \text{Exponential rule.}$$

Example 12 $\displaystyle\int_0^2 5x^2 \exp(x^3)\, dx = \frac{5}{3}\int_0^2 \exp(x^3)(3x^2)\, dx \qquad$ Multiply and divide by 3 since $(x^3)' = 3x^2$.

$$= \frac{5}{3}\int_0^2 \exp(x^3)(x^3)'\, dx$$

$$= \frac{5}{3} \exp(x^3)\Big|_0^2 \qquad \text{Exponential rule.}$$

$$= \frac{5}{3}\exp 8 - \frac{5}{3} = 4.97 \times 10^3 \qquad \exp(0) = 1.$$

Example 13 Woofco has marginal profit function $MP(q) = 10q/\sqrt{q^2 + 90{,}000}$, where q is the number of sanders produced. How much does profit change if q increases from 400 to 600?

Solution The change in profit is

$$P(600) - P(400) = \int_{400}^{600} MP(q)\, dq$$

$$= \int_{400}^{600} \frac{10q}{\sqrt{q^2 + 90{,}000}}\, dq$$

$$= \int_{400}^{600} 10q(q^2 + 90{,}000)^{-1/2}\, dq$$

$$= \frac{10}{2}\int_{400}^{600}(q^2 + 90{,}000)^{-1/2} 2q\, dq$$

$$= 5\frac{(q^2 + 90{,}000)^{1/2}}{1/2}\bigg|_{400}^{600} \qquad \text{Power rule.}$$

$$= 10(600^2 + 90{,}000)^{1/2} - 10(400^2 + 90{,}000)^{1/2}$$

$$= 10\sqrt{450{,}000} - 10\sqrt{250{,}000} = \$1702.20$$

Profit increases by \$1702.20 if q increases from 400 to 600.

Example 14 A gold mine produces gold at the rate of $t\exp(-t^2/100)$ tons per year after it has been operating for t years. How much gold will the mine produce during its first 20 years of service?

Solution The amount of gold produced during the first 20 years of service is

$$\int_0^{20} t\exp\left(\frac{-t^2}{100}\right) dt = -50 \int_0^{20} \exp\left(\frac{-t^2}{100}\right)\left(\frac{-t}{50}\right) dt \qquad \left(\frac{-t^2}{100}\right)' = \frac{-t}{50}$$

$$= -50 \exp\left(\frac{-t^2}{100}\right)\Big|_0^{20} \qquad \text{Exponential rule.}$$

$$= -50 \exp(-4) - [-50 \exp(0)]$$

$$= 49.08$$

The mine will produce 49.08 tons of gold during its first 20 years of service.

Exercises 12-5

Evaluate the following.

1. $\int_0^4 \sqrt{9 + 4x}\, dx$

2. $\int_0^1 e^{5x}\, dx$

3. $\int_0^3 \frac{x}{x^2 + 16}\, dx$

4. $\int_1^3 \frac{1}{(2x + 3)^2}\, dx$

5. $\int (2x - 1)^8\, dx$

6. $\int_1^6 \frac{1}{2x + 3}\, dx$

7. $\int (3x - 2)^{2/3}\, dx$

8. $\int e^{x/2}\, dx$

9. $\int \frac{10}{(2x + 3)^{1/2}}\, dx$

10. $\int_0^9 (100 - 4x)^{-1}\, dx$

11. $\int_0^9 (100 - 4x)^{-1/2}\, dx$

12. $\int \frac{1}{5x + 10}\, dx$

13. $\int_0^2 \frac{1}{(5x + 10)^2}\, dx$

14. $\int \exp(4x + 8)\, dx$

15. $\int \frac{1}{(3x + 4)^{1.5}}\, dx$

16. $\int_3^6 \frac{6x}{x^2 + 3}\, dx$

17. $\int \frac{8x}{(x^2 + 4)^2}\, dx$

18. $\int_0^1 6x\exp(x^2)\, dx$

19. $\int_1^5 x^2(x^3 + 1)^{-1}\, dx$

20. $\int (25 - x)^{1/2}\, dx$

21. $\int (x - 1)\exp(x^2 - 2x + 3)\, dx$

22. $\int \exp(4t + 2)\, dt$

23. $\int \exp(3-v)\,dv$

24. $\int \exp(12v-4)\,dv$

25. $\int_0^3 \dfrac{1}{8-2x}\,dx$

26. $\int_0^5 (7-x)^{-1}\,dx$

27. $\int_0^1 x\sqrt{x^2+1}\,dx$

28. $\int x^2(4-x^3)^5\,dx$

29. $\int_0^1 \exp(4-5t)\,dt$

30. $\int_{-2}^0 \exp(3t+6)\,dt$

31. $\int \dfrac{w}{4w^2+9}\,dw$

32. $\int \dfrac{w^2}{(4w^3+10)^2}\,dw$

33. $\int t(7-t^2)^{-1}\,dt$

34. $\int_2^6 (v-4)(v^2-8v+12)^3\,dv$

35. $\int_0^2 t(7-t^2)^{-2}\,dt$

36. $\int (8t-8)^{1/3}\,dt$

37. The debt of Hillside County is now ten million dollars and is expected to keep increasing at the rate of $e^{0.3t} \times 10^6$ dollars each year. Here t is the number of years from the present. How much is the debt expected to be 5 years from now?

38. A bacterial sample originally weighs 5 mg. The weight of the sample increases at the rate of $\exp(0.1t)$ mg/hr. Here t is measured in hours. How much will the sample weigh after 2 days?

39. Nevertear Paper Towels is a new product. Its initial market share of 5.0% increases at the rate of $6\exp(-0.2t)$ percent a month. Here t is measured in months. What is Nevertear's market share after
    a. Three months?
    b. Six months?
    c. One year?
    d. Two years?
    e. Three years?
    f. More than three years?

40. A new factory undergoes a break-in period during which it is not very efficient. The factory starts operating at 50% efficiency and during the tth month of operation, the efficiency increases by $5\exp(-0.1t)$ percent. How efficiently is the factory operating after a year?

41. Ace Industries has marginal cost function $MC(q) = 10q/\sqrt{q^2+400}$. Currently 30 items are produced. How much would costs increase if 20 additional items were produced?

42. An oil tanker hits a rock and immediately starts to leak. A circular oil slick results. Its radius is 0 when the leak begins and increases at the rate of $4/(t+5)$ ft/min. What is the radius of the oil slick after

a. 30 minutes?
b. 12 hours?

Chapter 12 Review

Evaluate each of the following.

1. $\int (x^2 - 4x + 7)\, dx$
2. $\int (3e^x - 4)\, dx$
3. $\int (x^{-2} - x^{-1})\, dx$
4. $\int (x^3 - 4x^2 - 3x + 8)\, dx$
5. $\int \left(\frac{6}{x^3} - \frac{4}{x}\right) dx$
6. $\int \left(\frac{1}{\sqrt{x}} + 3\right) dx$
7. $\int_1^3 v^2\, dv$
8. $\int_4^{12} \frac{2}{x}\, dx$
9. $\int_{-2}^3 (6x^2 - 4x + 1)\, dx$
10. $\int_0^1 (2e^w - 1)\, dw$
11. $\int \sqrt{7 - 3x}\, dx$
12. $\int \exp(8t - 3)\, dt$
13. $\int_6^8 x(x^2 - 25)^{-1}\, dx$
14. $\int 12x^2(x^3 - 4)^5\, dx$
15. $\int_{-1}^1 \exp(-5v)\, dv$
16. $\int_0^1 \frac{x}{(1 + x^2)^2}\, dx$

Compute the area between the x-axis, the graph of the function, and the given vertical lines.

17. $y = x^2$, $x = -2$, $x = 4$
18. $y = e^{2x}$, $x = 0$, $x = 4$
19. $y = \dfrac{3}{x^2}$, $x = 1$, $x = 3$
20. $y = x^2 + 2x$, $x = 1$, $x = 5$
21. $f(x) = e^x - e$, $x = 0$, $x = 2$
22. $g(x) = x^2 - x - 2$, $x = 1$, $x = 4$
23. Find the area between $y = 1/x$ and $y = x$ from $x = 1$ to $x = 4$.
24. Find the area between $y = x$ and $y = x^3$ from $x = -1$ to $x = 1$.
25. Find the area enclosed by $y = x^2$ and $y = 4$.
26. Find the area enclosed by $y = x^2 + x$ and $y = 10 - 2x$.
27. The marginal cost function for Allweather Storm Windows is $MC(q) = 15 - 0.04q$ and the fixed costs are $120. Determine the cost function.
28. The revenue for an expanding new business in its nth month of operation is $R(n) = 2000 + 50n + n^2$. What is the total revenue during its first two years?
29. The debt on the Cranesbill Turnpike is now $10,000,000 and will decrease by $50,000 + 30,000t$ in the tth coming year. How much will

the debt decrease by in the next ten years? How much debt will remain after ten years?

30. An oil well produces oil at the rate of $100{,}000 - 5000t$ barrels per year after it has been in operation for t years. How much oil will the well produce before it runs dry? (It runs dry when the production rate becomes 0.)

13

Applications of Integration

In Chapter 12, we developed the antiderivative and the definite integral and used them to find a function whose rate of change is known. Here we will apply integral calculus to economics, differential equations, and probability.

Section 13-1 Applications to Economics

We have already used the integral to find functions whose derivative is known. Finding cost, revenue, and profit functions from the corresponding marginal functions is useful enough to be worth special study.

Since marginal cost is the derivative of cost, the two functions $C(q)$ and $\int_0^q MC(t)\,dt$ differ by a constant because they are both antiderivatives of marginal cost. Thus,

$$C(q) = \int_0^q MC(t)\,dt + K \qquad K \text{ is a constant.}$$

Substitute 0 for q to obtain $C(0) = 0 + K = K$. This gives the formula

$$\boxed{C(q) = C(0) + \int_0^q MC(t)\,dt}$$

Recall that $C(0)$ is called *fixed costs*. The expression

$$\int_0^q MC(t)\,dt \quad \text{is the } \textit{variable cost}$$

of producing q items. Now let's find the cost function for Ralph's Restaurant.

Example 1 Ralph's Restaurant serves complete dinners at a fixed price. His fixed costs are $1000 a day and the marginal cost function is $MC(q) = 6 - 0.01q$, where q is the number of dinners sold. Determine the cost function.

Solution Substitute directly into the formula to obtain

$$C(q) = 1000 + \int_0^q (6 - 0.01t)\,dt$$
$$= 1000 + (6t - 0.005t^2)\Big|_0^q = 1000 + 6q - 0.005q^2$$

■

Now we will consider the revenue function. The functions

$$R(q) \quad \text{and} \quad \int_0^q MR(t)\,dt$$

differ by a constant because they are both antiderivatives of revenue. Since these functions both have the value 0 when $q = 0$, this constant is 0, so that

$$\boxed{R(q) = \int_0^q MR(t)\,dt}$$

Example 2 The marginal revenue function for Ralph's is $MR(q) = 12 - 0.025q$. Determine the revenue function.

Solution By the formula just discussed,

$$R(q) = \int_0^q (12 - 0.025t)\,dt = (12t - 0.0125t^2)\Big|_0^q = 12q - 0.0125q^2$$

Ralph's profit function can now be obtained by using the formula $P(q) = R(q) - C(q)$. This gives

$$P(q) = 6q - 0.0075q^2 - 1000$$

Example 3 What is the maximum profit Ralph can make? To make maximum profit, how many dinners does Ralph sell and what is their price?

Solution Recall that profit is maximum when marginal cost equals marginal revenue. This gives the equation

$$6 - 0.01q = 12 - 0.025q$$
$$0.015q = 6$$
$$q = \frac{6}{0.015} = 400$$

The maximum possible profit is obtained by selling 400 dinners and is $P(400) = \$200$. His total revenue on these dinners is $R(400) = \$2800$. Since the dinners are all sold at the same fixed price, this price is $R(400)/400 = \$7$.

■

The profit function can also be obtained from marginal profit by using the formula

$$\boxed{P(q) = \int_0^q MP(t)\,dt - C(0)}$$

Example 4 Determine Ralph's profit using the marginal profit function.

Solution Recall that $MP(q) = MR(q) - MC(q)$, so that

$$MP(q) = (12 - 0.025q) - (6 - 0.01q)$$
$$= 12 - 0.025q - 6 + 0.01q = 6 - 0.015q$$

Since $C(0)$ was given as \$1000 (see Example 1),

$$P(q) = \int_0^q (6 - 0.015t)\,dt - 1000$$
$$= (6t - 0.0075t^2)\Big|_0^q - 1000$$
$$= 6q - 0.0075q^2 - 1000$$

■

Often one is interested in the change in profit caused by an increase in production. In Section 12-3, we derived the formula

$$g(b) - g(a) = \int_a^b g'(t)\,dt$$

This formula, as applied to the special case of profit, is

$$\boxed{P(b) - P(a) = \int_a^b MP(q)\,dq \qquad \text{Change in profit}}$$

Thus, as production changes from a items to b items, profit changes by $\int_a^b MP(q)\,dq$. For example, if Ralph's Restaurant currently sells 200 dinners daily and increased this to 300, the profit would change by

$$\int_{200}^{300}(6 - 0.015q)\,dq = \left.(6q - 0.0075q^2)\right|_{200}^{300}$$
$$= (1800 - 675) - (1200 - 300) = \$225$$

Consumers' surplus and producers' surplus

We now discuss an entirely different application of integral calculus, one that exploits the concept of area.

In Diagram 1, $p = S(q)$ is the supply function for some item, $p = D(q)$ is the demand function, and (q^*, p^*) is the equilibrium point. (Supply and demand functions and equilibrium point were discussed in Section 9-1.) In a free economy, q^* items would be traded, all at the price p^*. Thus, each

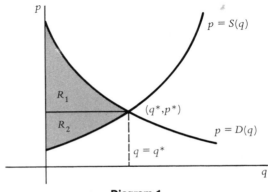

Diagram 1

consumer would pay p^* for an item. Some consumers, however, would be willing to pay a higher price than p^* and thus receive more satisfaction from this item than it costs. The total value of this extra satisfaction is called *consumers' surplus*.

The area between the demand curve and the q-axis from $q = 0$ to $q = q^*$ represents the value to consumers of the q^* items traded. However, consumers pay only p^*q^* for these items. Thus consumers' surplus is the area of region R_1 and is given by the formula

$$\text{C.S.} = \int_0^{q^*} D(q)\,dq - p^*q^* \qquad \text{Consumers' surplus}$$

Similarly some producers would be willing to sell the item at a lower price than p^*. In fact, the area between the supply curve and the q-axis from $q = 0$ to $q = q^*$ represents the value to producers of the q^* items traded. However, producers receive p^*q^* for these items. The difference between the amount received and the value to producers is called *producers' surplus* and represents extra profit for the producers. Producers' surplus is the area of region R_2 and is given by the formula

$$\text{P.S.} = p^*q^* - \int_0^{q^*} S(q)\,dq \qquad \text{Producers' surplus}$$

Note that consumers' surplus is determined by the demand function and the equilibrium point. The supply function is not needed to compute consumers' surplus. Similarly the demand function is not needed to compute producers' surplus.

Let's compute consumers' surplus and producers' surplus when $S(q) = 5 + 3q$ and $D(q) = 15 - 2q$. To find the equilibrium point, set $S(q) = D(q)$ and solve to obtain $q = 2$. The equilibrium point is $(2, 11)$. Now use the formulas for consumers' surplus and producers' surplus.

$$\text{C.S.} = \int_0^2 (15 - 2q)\,dq - 11 \cdot 2 = \left. (15q - q^2) \right|_0^2 - 22 = 26 - 22 = 4$$

$$\text{P.S.} = 11 \cdot 2 - \int_0^2 (5 + 3q)\,dq = 22 - \left[\frac{(5q + 3q^2)}{2} \right]_0^2$$

$$= 22 - [(10 + 6) - (0 + 0)] = 6$$

Example 4 Determine consumers' surplus and producers' surplus if

$$S(q) = 10 - \frac{5}{q+1} \quad \text{and} \quad D(q) = 8 + \frac{15}{q+1}$$

Solution First find the equilibrium quantity by solving the equation $S(q) = D(q)$.

$$10 - \frac{5}{q+1} = 8 + \frac{15}{q+1}$$

$$2 = \frac{20}{q+1}$$

$$2(q+1) = 20$$

$$q+1 = 10$$

$$q = 9$$

The equilibrium point is $(9, 9.5)$ since $S(9) = D(9) = 9.5$.
Now calculate consumers' surplus and producers' surplus.

$$\text{C.S.} = \int_0^9 \left(8 + \frac{15}{q+1}\right) dq - (9.5)(9)$$

$$= \left[8q + 15\ln(q+1)\right]\Big|_0^9 - 85.5$$

$$= (72 + 15\ln 10) - (0 + 15\ln 1) - 85.5 = 15\ln 10 - 13.5 = 21.0$$

$$\text{P.S.} = (9.5)(9) - \int_0^9 \left(10 - \frac{5}{q+1}\right) dq$$

$$= 85.5 - \left[10q - 5\ln(q+1)\right]\Big|_0^9$$

$$= 85.5 - \left[(90 - 5\ln 10) - (0 - 5\ln 1)\right] = 85.5 - 90 + 5\ln 10$$

$$= 5\ln 10 - 4.5 = 7.01$$

Exercises 13-1

Determine the functions $C(q)$, $R(q)$, and $P(q)$. Determine the maximum possible profit.

1. $C(0) = 40$, $MC(q) = 10 + 3q$, $MR(q) = 40 - 2q$
2. $C(0) = 50$, $MC(q) = 10 + 2q$, $MR(q) = 60 - 3q$
3. $C(0) = 30$, $MC(q) = 100 - 2q$, $MP(q) = 10 - q$
4. $C(0) = 80$, $MC(q) = 70 - 3q$, $MP(q) = 30 - 2q$

5. $C(0) = 50$, $MR(q) = 60$, $MP(q) = 50 - q$
6. $C(0) = 80$, $MR(q) = 50 - 3q$, $MP(q) = 20 - q$

Determine fixed costs.

7. $MC(q) = 20 - q$, $C(10) = 180$
8. $MC(q) = 100 - q - 0.1q^2$, $C(20) = 2100$
9. $MC(q) = 80 - 2q - 0.1q^2$, $C(30) = 2500$
10. $MC(q) = 40 - 2q$, $C(25) = 500$

Determine the change in profit if production changes from q_1 to q_2.

11. $MP(q) = 7 - 0.1q$, $q_1 = 5$, $q_2 = 80$
12. $MP(q) = 10 - 0.2q$, $q_1 = 10$, $q_2 = 50$
13. $MP(q) = 10 - 0.4q$, $q_1 = 25$, $q_2 = 30$
14. $MP(q) = 20 - 0.5q$, $q_1 = 10$, $q_2 = 30$

Determine the equilibrium point, consumers' surplus, and producers' surplus.

15. $S(q) = 3 + 4q$, $D(q) = 27 - 2q$
16. $S(q) = 5 + 5q$, $D(q) = 47 - 2q$
17. $S(q) = 10 + 3q$, $D(q) = 50 - 2q$
18. $S(q) = 8 + 2q$, $D(q) = 33 - 3q$
19. $S(q) = 9 + 4q$, $D(q) = 30 - 3q$
20. $S(q) = 5 + 6q$, $D(q) = 37 - 2q$
21. The marginal profit function for ILCO is

$$MP(q) = 20 + \frac{4}{q+5}$$

Fixed costs are $500. Determine
a. The profit if 50 items are traded.
b. The change in profit if the number of items traded goes from 30 to 45.

22. The marginal cost function for Arland Enterprises is

$$MC(q) = 15 - 6\exp(-0.1q)$$

Determine the change in costs if production is changed
a. From 10 units to 20 units.
b. From 20 units to 10 units.

23. The marginal revenue function for Marlowe's Mints is

$$MR(q) = 150 + \frac{500}{q+3}$$

and the marginal cost function is

$$MC(q) = 155$$

Fixed costs are $1000. Here q is the number of 120-pound boxes of mints made in a day. Determine

a. $C(q)$
b. $R(q)$
c. $P(q)$
d. The maximum possible profit.
e. The profit per box and number of boxes sold when profit is maximum.

24. National Switches can sell any number of toggle switches for 95¢ each. The marginal cost function for toggle switches is

$$MC(q) = 0.70 + 0.0001q$$

Fixed costs are $50. Determine

a. $C(q)$
b. $P(q)$
c. The maximum possible profit.
d. How many toggle switches must be made to maximize profit.

Section 13-2 Separable Differential Equations

A *differential equation* is an equation involving derivatives. Differential equations have numerous applications, especially to growth and decay processes. Here we will study *separable differential equations* and learn how to solve them using the method of *separation of variables*. Let's begin by solving the differential equation

$$\frac{dy}{dx} + \frac{3x}{y} = 0$$

To solve this equation, consider dy and dx as algebraic symbols. Now manipulate the equation into the form

$$M(y)\,dy + N(x)\,dx = 0$$

Here M and N are functions. A differential equation that can be put in this form is said to be *separable*, since the x's can be separated from the y's. For the equation we are solving, multiplying both sides by $y\,dx$ gives

$$\frac{dy}{dx}(y\,dx) + \frac{3x}{y}(y\,dx) = 0(y\,dx)$$

or

$$y\,dy + 3x\,dx = 0$$

The variables are now *separated*. The term $y\,dy$ involves y and dy but not x. The term $3x\,dx$ involves x and dx but not y.

Now integrate the equation to obtain

$$\int y\,dy + \int 3x\,dx = c$$

$$\frac{y^2}{2} + \frac{3x^2}{2} = c \quad \text{or} \quad y^2 + 3x^2 = 2c$$

The solution of the equation is thus $y^2 + 3x^2 = 2c$. Note that the solution involves an arbitrary constant c.

Example 1 Solve $\dfrac{dy}{dx} + \exp(2x - y) = 0$.

Solution

$\dfrac{dy}{dx} + e^{2x} \cdot e^{-y} = 0$ Simplify.

$\dfrac{dy}{dx}(e^y\,dx) + e^{2x}e^{-y}(e^y\,dx) = 0(e^y\,dx)$ Separate the variables.

$e^y\,dy + e^{2x}\,dx = 0$

$\int e^y\,dy + \int e^{2x}\,dx = c$ Integrate.

$e^y + \tfrac{1}{2}e^{2x} = c$

$e^y = c - \tfrac{1}{2}e^{2x}$ Start solving for y.

$\ln(e^y) = \ln(c - \tfrac{1}{2}e^{2x})$ Apply ln to both sides of the equation.

$y = \ln(c - \tfrac{1}{2}e^{2x})$

■

Note that the equation $e^y + (e^{2x}/2) = c$, which was obtained in Example 1 by integrating, determines y as a function of x. Whenever this happens, you should solve for y. Earlier we solved the differential equation $(dy/dx) + (3x/y) = 0$ but did not solve for y. Why not?

The solution of each of the differential equations we have solved involved an arbitrary constant. If one point on the graph of the solution of the differential equation is given, the value of this constant is no longer arbitrary and can be determined.

Now let's solve $(dy/dx) + (5e^x/y^2) = 0$ if $y = 1$ when $x = 0$. First separate variables and integrate as before.

$$\frac{dy}{dx} + \frac{5e^x}{y^2} = 0$$

$$y^2\,dy + 5e^x\,dx = 0 \qquad \text{Multiply both sides by } y^2\,dx \text{ and simplify.}$$

$$\int y^2\,dy + \int 5e^x\,dx = c \qquad \text{Integrate.}$$

$$\frac{y^3}{3} + 5e^x = c$$

Now substitute the value $y = 1$ and $x = 0$ to obtain c.

$$\tfrac{1}{3} + 5e^0 = c \qquad c = 5\tfrac{1}{3} = 16/3$$

Replace c with its value of $16/3$ and proceed to solve for y.

$$\frac{y^3}{3} + 5e^x = 16/3$$

$$y^3 + 15e^x = 16$$

$$y^3 = 16 - 15e^x$$

$$y = (16 - 15e^x)^{1/3}$$

The solution is $y = (16 - 15e^x)^{1/3}$. This solution does not involve an arbitrary constant.

Example 2 Solve $x^2\dfrac{dy}{dx} + 3y + 5 = 0$, given $y = 2e - \dfrac{5}{3}$ when $x = 3$.

Solution

$$x^2\frac{dy}{dx} + 3y + 5 = 0$$

$$\frac{1}{3y+5}dy + \frac{1}{x^2}dx = 0 \qquad \text{Multiply both sides by } \frac{1}{x^2(3y+5)}dx \text{ and simplify.}$$

13-2 SEPARABLE DIFFERENTIAL EQUATIONS

$$\int \frac{1}{3y+5}\,dy + \int \frac{1}{x^2}\,dx = c \qquad \text{Integrate.}$$

$$\frac{1}{3}\int \frac{1}{3y+5}\,3\,dy + \int x^{-2}\,dx = c$$

$$\frac{1}{3}\ln(3y+5) - x^{-1} = c$$

Now determine c.

$$\frac{1}{3}\ln\left[3\left(2e-\frac{5}{3}\right)+5\right] - 3^{-1} = c \qquad \text{Substitute } y = 2e - \frac{5}{3} \text{ and } x = 3.$$

$$\frac{1}{3}\ln(6e) - \frac{1}{3} = c \qquad \text{Simplify.}$$

$$\frac{1}{3}\ln 6 + \frac{1}{3}\ln e - \frac{1}{3} = c \qquad \ln(pq) = \ln p + \ln q.$$

$$\frac{1}{3}\ln 6 + \frac{1}{3}(1) - \frac{1}{3} = \frac{1}{3}\ln 6 = c \qquad \ln e = 1.$$

$$\frac{1}{3}\ln(3y+5) - \frac{1}{x} = \frac{1}{3}\ln 6 \qquad \text{Substitute } \frac{1}{3}\ln 6 \text{ for } c.$$

$$\frac{1}{3}\ln(3y+5) = \frac{1}{x} + \frac{1}{3}\ln 6 \qquad \text{Start solving for } y.$$

$$\ln(3y+5) = \frac{3}{x} + \ln 6$$

$$\exp[\ln(3y+5)] = \exp\left(\frac{3}{x} + \ln 6\right)$$

$$3y+5 = \exp\left(\frac{3}{x}\right) \cdot \exp(\ln 6) \qquad \text{exp and ln are inverses.}$$

$$= 6\exp\left(\frac{3}{x}\right) \qquad \exp(p+q) = \exp(p)\cdot\exp(q)$$

$$3y = 6\exp\left(\frac{3}{x}\right) - 5$$

$$y = 2\exp\left(\frac{3}{x}\right) - \frac{5}{3}$$

■

We will now study some applications of differential equations. Suppose the mass of a colony of bacteria is increasing at a rate proportional to itself. Initially (at time $t = 0$), the mass is 10 mg. The mass doubles in three hours. What is the mass at time t?

Let $m(t)$ be the mass at time t. Since $m(t)$ is increasing at a rate proportional to itself,

$$\frac{dm}{dt} = km$$

Here k is a constant. This constant must be positive because the mass is increasing so that $dm/dt > 0$. We are given $m(0) = 10$ and $m(3) = 20$. Now solve the differential equation

$$\frac{dm}{dt} = km$$

$$\frac{1}{m} dm = k\, dt$$

$$\int \frac{1}{m} dm = \int k\, dt$$

$$\ln m = kt + c \qquad \text{(Why is } \ln m \text{ correct as well as } \ln|m|\text{?)}$$

Now use $m(0) = 10$ to obtain

$$\ln 10 = k(0) + c \quad \text{so that } c = \ln 10$$

Now use $m(3) = 20$ to obtain

$$\ln 20 = 3k + \ln 10$$

$$3k = \ln 20 - \ln 10 = \ln\left(\frac{20}{10}\right) = \ln 2$$

$$k = \frac{\ln 2}{3}$$

Substitute the values for c and k and then solve for m.

$$\ln m = \frac{\ln 2}{3} \cdot t + \ln 10$$

$$= \ln(2^{t/3}) + \ln 10$$

$$= \ln(10 \cdot 2^{t/3})$$

$$m = 10 \cdot 2^{t/3} \qquad \text{The function ln is one-to-one.}$$

Thus the solution is $m = 10 \cdot 2^{t/3}$.

Example 3 A country has ten billion coins in circulation, and an average of 20 million coins pass through the banks each day. The old coins are to be replaced by new coins. Starting today, each old coin that passes through a bank is destroyed and is replaced by a new coin. How many new coins will be in circulation after t days?

Solution Let $A(t)$ be the number of new coins in circulation at time t. Then $10^{10} - A(t)$ is the number of old coins in circulation at time t. Each day

$$20{,}000{,}000 \cdot \frac{10^{10} - A(t)}{10^{10}} = \frac{10^{10} - A(t)}{500}$$

old coins reach the bank and are replaced. Thus,

$$\frac{dA}{dt} = \frac{10^{10} - A(t)}{500}$$

$$\frac{1}{10^{10} - A} dA = \frac{1}{500} dt \qquad \text{Separate the variables.}$$

$$\int \frac{1}{10^{10} - A} dA = \int \frac{1}{500} dt$$

$$-\ln(10^{10} - A) = \frac{1}{500} t + c$$

At time 0, no new coins are in circulation. Thus, $A = 0$ when $t = 0$, so that $-\ln(10^{10}) = c$.

$$-\ln(10^{10} - A) = \frac{1}{500} t - \ln(10^{10})$$

$$\ln(10^{10} - A) = \ln(10^{10}) - \frac{1}{500} t$$

$$10^{10} - A = \exp\left[\ln(10^{10}) - \frac{1}{500} t\right]$$

$$= \exp[\ln(10^{10})] \cdot \exp\left[\frac{-1}{500} t\right]$$

$$= 10^{10} \exp\left(\frac{-t}{500}\right)$$

$$A = 10^{10} - 10^{10} \exp\left(\frac{-t}{500}\right) = 10^{10}\left[1 - \exp\left(\frac{-t}{500}\right)\right]$$

After two years ($t = 730$), $A = 7.68$ billion, so that slightly more than 3/4 of the coins are new. After five years ($t = 1825$), $A = 9.74$ billion, so that only about 260 million old coins remain in circulation.

∎

Exercises 13-2

Solve the differential equations.

1. $\dfrac{dy}{dx} + \dfrac{x^2}{y^2} = 0$
2. $y\dfrac{dy}{dx} + x^3 = 0$
3. $5\dfrac{dy}{dx} + \dfrac{e^x}{y^2} = 0$
4. $3\dfrac{dy}{dx} + xy = 0$
5. $4\dfrac{dy}{dx} - \dfrac{\sqrt{x}}{y+5} = 0$
6. $x\dfrac{dy}{dx} - 8\sqrt{y} = 0$
7. $\dfrac{dy}{dx} + 4y - 12 = 0$
8. $\dfrac{dy}{dx} + (x+3)y = 0$
9. $xy\dfrac{dy}{dx} + 6 = 0$
10. $\exp x \cdot \dfrac{dy}{dx} + 5 = 0$

Find the solution of the differential equation that goes through the given point.

11. $x^2\dfrac{dy}{dx} + 4 = 0$ $(2,3)$
12. $xy^2\dfrac{dy}{dx} + 8 = 0$ $(1,3)$
13. $\dfrac{dy}{dx} + \dfrac{y+5}{x} = 0$ $(2,-1)$
14. $\dfrac{dy}{dx} - \dfrac{x^2}{y} = 0$ $(3,4)$
15. $x\dfrac{dy}{dx} + y^2 = 0$ $(e,1)$
16. $x^2 y^2\dfrac{dy}{dx} + 1 = 0$ $(1,-2)$
17. $y^2\dfrac{dy}{dx} - x^2 = 0$ $(1,1)$
18. $y\dfrac{dy}{dx} - 4x = 0$ $(-1,2)$
19. $\exp(x+y)\dfrac{dy}{dx} + 2 = 0$ $(0,1)$
20. $y\exp(x)\dfrac{dy}{dx} + y^2 = 0$ $(0,4)$

21. A bacterial colony weighs 50 mg. An antibiotic is introduced into the colony at $t = 0$, and the weight m of the colony decreases according to the equation

$$\frac{dm}{dt} = -mt$$

Express m as a function of t.

22. Do exercise 21 if

$$\frac{dm}{dt} = -m\sqrt{t}$$

23. The debt D of the State Highway Authority is currently $250 million and is changing according to the differential equation

$$\frac{dD}{dt} = -0.09Dt$$

Here t is the time measured in years, starting now. Solve for D.

24. The amount A of income tax collected annually by Ashtown satisfies the differential equation

$$\frac{dA}{dt} = At^{-1/2}$$

and is currently $500,000. Express A as a function of the time t, measured in years.

25. The rate of decay of a radioactive substance is proportional to the mass $m(t)$. Initially there are 50 kg of the substance. Half of it decays in eight hours. Find a formula for $m(t)$.

26. Solve exercise 25 if the initial mass is 80 lb and half of it decays in 12 hours.

27. A hot object brought into a room where the temperature is 60° cools according to the equation $du/dt = -k(u - 60)$, where t is the time in minutes and u is the temperature. The object cools from 100° to 90° in three min. What is the temperature after six min?

28. In exercise 27, what is the temperature after 15 min, if the object cools from 180° to 150° in five min?

Section 13-3 Probability Density Functions

When we studied probability in Chapters 6, 7, and 8, each experiment discussed had only finitely many different possible outcomes, that is, had a finite sample space. Frequently an experiment has infinitely many possible outcomes, and in fact the result can be any number in some interval $a \leq x \leq b$. For example, the weight of a "one-pound" loaf of bread might be *any* number between 15 and 18 ounces. Waiting time at a dentist's office might be anywhere from 0 to 4 hours. Even when the sample space is a large but finite set of numbers, it may be convenient to pretend the sample space is an interval. Examples would be the number of paid admissions to a football game and the number of cars sold by General Motors in one year.

When the sample space is an interval, we do not ask the probability of a specific number, but instead ask for the probability that $c \leq x \leq d$. This probability is given by an integral involving a *probability density function*.

Definition. The function p is a *probability density function* (abbreviated p.d.f.) on the interval $a \leq x \leq b$ if $p(x)$ is continuous for $a \leq x \leq b$,
$$p(x) \geq 0 \text{ whenever } a \leq x \leq b, \text{ and } \int_a^b p(x)\,dx = 1.$$

In terms of area, the area under the graph of p is 1. See Diagram 1. $P(c \leq x \leq d)$ is the area under the graph of p between the vertical lines $x = c$ and $x = d$.

If $a \leq c \leq d \leq b$ and p is a p.d.f. on the interval $a \leq x \leq b$, then
$$P(c \leq x \leq d) = \int_c^d p(x)\,dx$$

In particular, $P(a \leq x \leq b) = 1$ because p is a p.d.f. Note that the

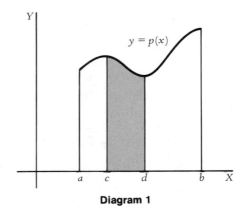

Diagram 1

13-3 PROBABILITY DENSITY FUNCTIONS

probability of any number c, that is, $P(x = c)$ is $\int_c^c p(x)\,dx = 0$, even though $p(c) \neq 0$. Consequently, $P(c \leq x \leq d) = P(c < x < d)$.

We can think of $p(x)$ as the relative likelihood that the outcome of the experiment is near x. If $p(2) = 5$ and $p(3) = 20$, the outcome of the experiment is $20/5 = 4$ times as likely to be near 3 as to be near 2.

As an example, consider the function $p(x) = 3x^2/7$, $1 \leq x \leq 2$. This function is a p.d.f. because

1. $\dfrac{3x^2}{7} \geq 0$ for all x and in particular for $1 \leq x \leq 2$

2. $\displaystyle\int_1^2 \dfrac{3x^2}{7}\,dx = \dfrac{3}{7}\dfrac{x^3}{3}\Big|_1^2 = \dfrac{x^3}{7}\Big|_1^2 = \dfrac{8}{7} - \dfrac{1}{7} = \dfrac{7}{7} = 1$

Thus, to find the probability that $1.3 \leq x \leq 1.6$, integrate $p(x)$ from 1.3 to 1.6.

$$P(1.3 \leq x \leq 1.6) = \int_{1.3}^{1.6} p(x)\,dx$$

$$= \int_{1.3}^{1.6} \dfrac{3x^2}{7}\,dx$$

$$= \dfrac{x^3}{7}\Big|_{1.3}^{1.6} = \dfrac{1.6^3}{7} - \dfrac{1.3^3}{7} = 0.27$$

Now let's find the probability that $1 \leq x \leq 1.5$.

$$P(1 \leq x \leq 1.5) = \int_1^{1.5} \dfrac{3x^2}{7}\,dx$$

$$= \dfrac{x^3}{7}\Big|_1^{1.5} = \dfrac{1.5^3}{7} - \dfrac{1^3}{7} = 0.34$$

Example 1 Let $p(x) = 3x\sqrt{1 - x^2}$, $0 \leq x \leq 1$.

PROBLEM A Verify that p is a p.d.f.

Solution Clearly $p(x) \geq 0$, for $0 \leq x \leq 1$. Now we must verify that $\int_0^1 p(x)\,dx = 1$.

$$\int_0^1 p(x)\,dx = \int_0^1 3x\sqrt{1-x^2}\,dx$$

$$= \frac{-3}{2}\int_0^1 (1-x^2)^{1/2}(-2x)\,dx \qquad \text{Multiply and divide by } -2.$$

$$= \frac{-3}{2} \left.\frac{(1-x^2)^{3/2}}{\frac{3}{2}}\right|_0^1 \qquad \text{Power rule.}$$

$$= -(1-x^2)^{3/2}\Big|_0^1$$

$$= -(1-1^2)^{3/2} + (1-0^2)^{3/2}$$

$$= -0 + 1 = 1$$

PROBLEM B Determine $P(0.6 \leq x \leq 0.8)$.

Solution
$$P(0.6 \leq x \leq 0.8) = \int_{0.6}^{0.8} 3x\sqrt{1-x^2}\,dx$$

$$= -(1-x^2)^{3/2}\Big|_{0.6}^{0.8} \qquad \text{In problem A, the antiderivative of } p \text{ was determined.}$$

$$= -0.36^{3/2} + 0.64^{3/2}$$

$$= -0.216 + 0.512 = 0.296$$

Example 2 The function $k(x^2 - x)$ is a p.d.f. on $2 \leq x \leq 5$. What is the value of k?

Solution First note that $x^2 - x = x(x-1)$ is nonnegative for $2 \leq x \leq 5$. This function is a p.d.f. if $\int_2^5 k(x^2 - x)\,dx = 1$. Thus

$$1 = \int_2^5 k(x^2 - x)\,dx = k\left(\frac{x^3}{3} - \frac{x^2}{2}\right)\Big|_2^5 = 28.5k$$

Since $28.5k = 1$, $k = 1/28.5 = 0.035$.

Example 3 The waiting time (in hours) at a dentist's office has the p.d.f. $3(x-4)^2/64$, $0 \leq x \leq 4$. What is the probability of having to wait more than one hour?

Solution The question asks for $P(1 \leq x)$. Since waiting time is at most four hours, this is the same as $P(1 \leq x \leq 4)$.

$$P(1 \leq x \leq 4) = \int_1^4 \frac{3}{64}(x-4)^2\,dx$$

$$= \frac{3}{64} \frac{(x-4)^3}{3} \Big|_1^4 = \frac{0^3}{64} - \frac{(-3)^3}{64} = \frac{27}{64} = 0.422$$

The probability of having to wait more than one hour is $27/64 = 0.422$. ∎

Expected value

You are likely to be concerned with how long you will have to wait at the dentist's office. On some occasions, the dentist will see you almost immediately; on others you will have to wait a long time. Finding the average wait involves the concept of expected value. We studied expected value in Section 8-1 for experiments with finite sample spaces. When the sample space is the interval $a \leq x \leq b$,

> The *expected value* of x is $\int_a^b xp(x)\,dx$ and is written $\mathrm{EV}(x)$.

At the dentist's office, the expected waiting time is

$$\mathrm{EV}(x) = \int_0^4 xp(x)\,dx = \int_0^4 x\left[\frac{3}{64}(x-4)^2\right]dx$$

$$= \frac{3}{64}\int_0^4 x(x-4)^2\,dx = \frac{3}{64}\int_0^4 x(x^2 - 8x + 16)\,dx$$

$$= \frac{3}{64}\int_0^4 (x^3 - 8x^2 + 16x)\,dx = \frac{3}{64}\left(\frac{x^4}{4} - \frac{8x^3}{3} + \frac{16x^2}{2}\right)\Big|_0^4$$

$$= \frac{3}{64} \cdot \frac{64}{3} = 1$$

The expected waiting time at the dentist's office is one hour.

Example 4 The weight of a "one-pound" loaf of bread has the p.d.f.

$$p(x) = \frac{4 - (x-16)^2}{9} \quad 15 \leq x \leq 18$$

PROBLEM A What is the probability that a loaf selected at random weighs at least one pound (16 ounces)?

Solution
$$P(x \geq 16) = \int_{16}^{18} \left[\frac{4}{9} - \frac{(x-16)^2}{9} \right] dx$$

$$= \frac{4x}{9} - \frac{(x-16)^3}{27} \bigg|_{16}^{18} = \left(\frac{72}{9} - \frac{2^3}{27} \right) - \left(\frac{64}{9} - \frac{0^3}{27} \right)$$

$$= \frac{72}{9} - \frac{8}{27} - \frac{64}{9} = \frac{8}{9} - \frac{8}{27} = \frac{24}{27} - \frac{8}{27} = \frac{16}{27}$$

The probability that a loaf selected at random weighs at least one pound is $16/27 = 0.59$.

PROBLEM B What is the expected weight of a loaf of bread?

Solution
$$\text{EV}(x) = \int_{15}^{18} x p(x) dx$$

$$= \int_{15}^{18} \frac{4x - x(x-16)^2}{9} dx = \int_{15}^{18} \frac{4x - x^3 + 32x^2 - 256x}{9} dx$$

$$= \int_{15}^{18} \frac{-x^3 + 32x^2 - 252x}{9} dx = \frac{1}{9} \left(\frac{-x^4}{4} + \frac{32x^3}{3} - \frac{252x^2}{2} \right) \bigg|_{15}^{18}$$

$$= 16.25$$

The expected weight of a loaf of bread is 16.25 ounces.
∎

Exercises 13-3

1. Check that the function $p(x) = x^2/9$ is a p.d.f. on $0 \leq x \leq 3$. What is $P(1 \leq x \leq 3)$? What is $\text{EV}(x)$?

2. Check that the function $p(x) = x/12$ is a p.d.f. on $1 \leq x \leq 5$. What is $P(1 \leq x \leq 2)$? What is $\text{EV}(x)$?

3. Check that the function $p(x) = x^{-1/2}/6$ is a p.d.f. on $1 \leq x \leq 16$. What is $P(4 < x < 9)$? What is $\text{EV}(x)$?

4. Check that the function $p(x) = 1/x$ is a p.d.f. on $e \leq x \leq e^2$. What is $P(3 \leq x \leq 6)$? What is $\text{EV}(x)$?

5. Check that the function $(x^2 + 1)/78$ is a p.d.f. on $0 \leq x \leq 6$. What is $P(1 \leq x \leq 2)$? What is $\text{EV}(x)$?

6. Check that the function $3(x^2 - 2x)/20$ is a p.d.f. on $2 \leq x \leq 4$. What is

$P(3 \le x \le 4)$? What is $EV(x)$?

7. Check that the function $p(x) = 1/(x \ln 4)$ is a p.d.f. on $1 \le x \le 4$. What is $P(1 \le x \le 2)$? What is $EV(x)$?

8. Check that the function $p(x) = 2/5$ is a p.d.f. on $4 \le x \le 6.5$. What is $P(5 \le x \le 6)$? What is $EV(x)$?

9. For what value of k is $k(2x + 3)$ a p.d.f. on $1 \le x \le 3$?

10. For what value of k is ke^x a p.d.f. on $0 \le x \le 2$?

11. For what value of k is ke^{3x} a p.d.f. on $1 \le x \le 4$?

12. For what value of k is $kx/(1 + x^2)$ a p.d.f. on $0 \le x \le 1$?

13. For what value of k is $kx \exp(-x^2)$ a p.d.f. on $0 \le x \le 1$?

14. For what value of k is $kx\sqrt{9 + x^2}$ a p.d.f. on $0 \le x \le 4$?

15. For what value of k is $kx^2/(5 + x^3)$ a p.d.f. on $1 \le x \le 3$?

16. For what value of k is ke^{-2x} a p.d.f. on $0 \le x \le 4$?

17. The 32W bus is always late. The p.d.f. for the number of minutes m that it is late is

$$p(m) = \frac{4(40 - m)^3}{40^4} \quad 0 \le m \le 40$$

   a. What is the probability that a bus is less than 20 minutes late?
   b. What is the expected number of minutes that a bus is late?

18. The amount of gasoline x (in thousands of gallons) that Al's Service Station can sell in a day has the p.d.f.

$$p(x) = 0.02 + 0.002x \quad 5 \le x \le 25$$

   a. What is the probability that Al can sell more than 20,000 gallons tomorrow?
   b. How many gallons should Al expect to sell tomorrow?

19. Westland Steel employs 5000 workers at its Willinocket plant. The number x of workers who call in sick on Monday has the p.d.f.

$$p(x) = \frac{200 - x}{11{,}250} \quad 50 \le x \le 200$$

   a. What is the probability that fewer than 100 workers call in sick next Monday?

b. On the average, how many workers call in sick on a Monday?

20. The number of phone calls x for the time-of-day in Boston has p.d.f.

$$p(x) = \frac{16 - x}{128} \quad 0 \leq x \leq 16$$

Here x is the number (in thousands) of calls made in a day.
a. What is the probability that more than 8000 calls for time-of-day will be made next Wednesday?
b. On the average, how many calls are made in a day?

21. The percentage x of impurities in a batch of peanut butter at Nut's Butter has the p.d.f.

$$p(x) = \frac{e^{1-x}}{e - 1} \quad 0 \leq x \leq 1$$

a. What is the probability that a batch of peanut butter has less than 0.5% impurities?
b. What is the expected percentage of impurities in a batch of peanut butter?
(*Hint:* $-(1 + x)e^{1-x}$ is an antiderivative of xe^{1-x}.)

22. The waiting time t at the check-out lines at Sally's Savings Store has p.d.f.

$$p(t) = \frac{3}{256} \cdot (12t - t^2 - 20) \quad 2 \leq t \leq 10$$

where t is measured in minutes.
a. What is the probability of waiting less than five minutes?
b. How long can a customer expect, on the average, to wait in the check-out lines?

23. The number of minutes a laboratory rat needs to solve a certain maze has the p.d.f.

$$p(t) = \frac{14t - t^2 - 40}{36} \quad 4 \leq t \leq 10$$

a. What is the probability that a rat will solve the maze in less than six minutes?
b. What is the expected time for a rat to solve the maze?

Section 13-4 Integration by Parts

In practical applications, probability density functions often involve exponential functions. In such cases, determining the expected value involves finding the antiderivative of a function of the form $x\exp(kx)$, where k is a constant. This antiderivative can be found using a technique called *integration by parts*.

Integration by parts is the reverse of the product rule and is often useful *when the product of two functions must be integrated*. To derive the formula for integration by parts, first recall the product rule for derivatives

$$[f(x)g(x)]' = f(x)g'(x) + f'(x)g(x)$$

Rewrite this formula as

$$f(x)g'(x) = [f(x)g(x)]' - f'(x)g(x)$$

If two functions are equal, they have the same antiderivatives, so that

$$\int f(x)g'(x)\,dx = \int [f(x)g(x)]'\,dx - \int f'(x)g(x)\,dx$$

However, any function is an antiderivative of its derivative. In particular, $f(x)g(x)$ is an antiderivative of $[f(x)g(x)]'$. Thus,

$$\boxed{\int f(x)g'(x)\,dx = f(x)g(x) - \int f'(x)g(x)\,dx} \qquad \text{Integration by parts}$$

Example 1 Evaluate $\int x \ln x \, dx$.

Solution Use integration by parts with $f(x) = \ln x$ and $g'(x) = x$. Then, $f'(x) = 1/x$ and $g(x) = x^2/2$, so that

$$\int x \ln x \, dx = \ln x \cdot \frac{x^2}{2} - \int \frac{1}{x} \cdot \frac{x^2}{2} \, dx$$

$$= \frac{1}{2}x^2 \ln x - \frac{1}{2} \int x \, dx$$

$$= \frac{1}{2}x^2 \ln x - \frac{1}{2} \cdot \frac{x^2}{2} + c$$
$$= \frac{1}{2}x^2 \ln x - \frac{1}{4}x^2 + c$$

Example 2 Evaluate $\int_1^e \ln x \, dx$

Solution First, find the antiderivative of $\ln x$. Use integration by parts with $f(x) = \ln x$ and $g'(x) = 1$. Then $f'(x) = 1/x$ and $g(x) = x$.

$$\int \ln x \, dx = \int \ln x \cdot 1 \cdot dx = \ln x \cdot x - \int \frac{1}{x} x \, dx$$
$$= x \ln x - \int 1 \cdot dx$$
$$= x \ln x - x + c$$

By the Fundamental Theorem of Calculus,

$$\int_1^e \ln x \, dx = x \ln x - x \Big|_1^e$$
$$= (e \ln e - e) - (1 \ln 1 - 1)$$
$$= (e \cdot 1 - e) - (1 \cdot 0 - 1) = 1$$

Example 3 Evaluate $\int x e^x dx$.

Solution Use integration by parts with $f(x) = x$ and $g'(x) = e^x$. Then $f'(x) = 1$ and $g(x) = e^x$, so that

$$\int x e^x \, dx = x e^x - \int e^x \cdot 1 \, dx$$
$$= x e^x - \int e^x \, dx$$
$$= x e^x - e^x + c$$

Example 4 A manufacturer provides a 90-day warranty on radios. For those radios that fail within 90 days, the probability of failing on day t is given by

$$p(t) = \frac{e^2}{45(e^2-1)} \exp\left(\frac{-t}{45}\right) \qquad 0 \le t \le 90$$

If a radio fails within 90 days, how many days can it be expected to last?

Solution This problem asks for the expected value of t, that is, for

$$\mathrm{EV}(t) = \int_0^{90} \frac{e^2}{45(e^2-1)} t \exp\left(\frac{-t}{45}\right) dt$$

To evaluate this integral, first use integration by parts to find the antiderivative of $t\exp(-t/45)$. Let $f(t) = t$ and $g'(t) = \exp(-t/45)$. Then $f'(t) = 1$ and (verify this!) $g(t) = -45\exp(-t/45)$, so that

$$\int t \exp\left(\frac{-t}{45}\right) dt = t \cdot \left[-45\exp\left(\frac{-t}{45}\right)\right] - \int -45\exp\left(\frac{-t}{45}\right) \cdot 1 \, dt$$

$$= -45t\exp\left(\frac{-t}{45}\right) + 45\int \exp\left(\frac{-t}{45}\right) dt$$

$$= -45t\exp\left(\frac{-t}{45}\right) + 45 \cdot \left[-45\exp\left(\frac{-t}{45}\right)\right] + c$$

$$= -45t\exp\left(\frac{-t}{45}\right) - 2025\exp\left(\frac{-t}{45}\right) + c$$

Thus

$$\mathrm{EV}(t) = \int_0^{90} \frac{e^2}{45(e^2-1)} t\exp\left(\frac{-t}{45}\right) dt$$

$$= \frac{e^2}{45(e^2-1)} \int_0^{90} t\exp\left(\frac{-t}{45}\right) dt$$

$$= \frac{e^2}{45(e^2-1)} \cdot \left[-45t\exp\left(\frac{-t}{45}\right) - 2025\exp\left(\frac{-t}{45}\right)\right]\Big|_0^{90}$$

$$= \frac{e^2}{(e^2-1)} \left[-t\exp\left(\frac{-t}{45}\right) - 45\exp\left(\frac{-t}{45}\right)\right]\Big|_0^{90}$$

$$= \frac{e^2}{e^2-1} \left[-90\exp(-2) - 45\exp(-2) + 0 + 45\exp(0)\right]$$

$$= \frac{e^2}{e^2 - 1} \cdot (45 - 135e^{-2}) = 30.9$$

If a radio fails within 90 days, it can be expected to last 30.9 days.

■

Exercises 13-4

Use integration by parts to evaluate each integral.

1. $\int x^2 \ln x \, dx$
2. $\int \sqrt{x} \ln x \, dx$
3. $\int_1^e x^3 \ln x \, dx$
4. $\int_1^e x^4 \ln x \, dx$
5. $\int xe^{-x} \, dx$
6. $\int xe^{2x} \, dx$
7. $\int_0^1 xe^{-2x} \, dx$
8. $\int_1^2 xe^{3x} \, dx$

Evaluate each of the following integrals. Some of these problems require integration by parts and some do not.

9. $\int x(4 + x) \, dx$
10. $\int_0^2 xe^{x^2} \, dx$
11. $\int xe^{5x} \, dx$
12. $\int (x + 1) \ln x \, dx$
13. $\int_0^4 x\sqrt{9 + x^2} \, dx$
14. $\int_0^1 xe^{-4x} \, dx$
15. $\int_0^3 \frac{x}{x^2 + 16} \, dx$
16. $\int \frac{1}{x \ln x} \, dx$

17. A drill bit's lifetime L (in years) has the p.d.f.

$$p(L) = e^{-L} \quad 0 \leq L \leq 10$$

What is the expected lifetime of the drill bit?

18. The marginal revenue function for a company is $R(q) = q \exp(0.001q)$. Determine the revenue from the first 100 items sold.

A new worker processes his nth component in $10 + 10ne^{-n}$ minutes.

19. How long does it take him to process his first eight components?
20. How long does it take him to process his next eight components?

Evaluate the following integrals using integration by parts twice.

*21. $\int x^2 e^x \, dx$
*22. $\int (\ln x)^2 \, dx$

*23. $\int x^2 e^{3x}\, dx$ *24. $\int x(\ln x)^2\, dx$

Section 13-5 The Trapezoidal Rule

So far, we have evaluated definite integrals using the Fundamental Theorem of Calculus. This involves finding an antiderivative of the function being integrated. Because finding antiderivatives is often difficult, mathematicians have developed several numerical procedures that can be used to compute an approximation to the value of a definite integral. Here we will study one such procedure, called the *trapezoidal rule*. The basis of this rule is to approximate a region by a union of trapezoids. A *trapezoid* is a four-sided figure with two parallel sides.

In the derivation of the trapezoidal rule and in Diagram 1, assume (to make things simpler) that $f(x) > 0$ for $a < x < b$. Select a counting number n, and divide the segment of the x-axis from a to b into n parts, each of width w. Then

$$w = \frac{b-a}{n}$$

The dividing points are on the x-axis at a, $a + w$, $a + 2w, \ldots,$ $a + (n-1)w, a + nw\, (= b)$.

In Diagram 1, the region R below the graph of f from $x = a$ to $x = b$ is very well approximated by the union of the n trapezoids shown, and so the

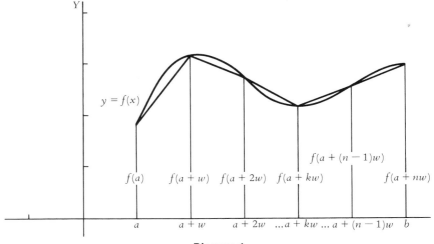

Diagram 1

area $A(R)$ can be approximated by the sum of the areas of these n trapezoids. The area of a trapezoid is half the product of its width with the sum of the parallel sides. (See exercise 9 at the end of this section.)

Since $A(R) = \int_a^b f(x)\,dx$, $\int_a^b f(x)\,dx$ can be approximated by

$$\frac{w}{2}[f(a) + f(a+w)]$$
$$+ \frac{w}{2}[f(a+w) + f(a+2w)]$$
$$+ \cdots$$
$$+ \frac{w}{2}[f(a+(n-1)w) + f(b)]$$
$$= \frac{w}{2}[f(a) + 2f(a+w) + 2f(a+2w) + \cdots + 2f(a+(n-1)w) + f(b)]$$

Trapezoidal Rule

$\int_a^b f(x)\,dx$ can be approximated by

$$\frac{w}{2}[f(a) + 2f(a+w) + \cdots + 2f(a+(n-1)w) + f(b)]$$

where $w = (b-a)/n$.

As an example, let's approximate $\int_1^2 x^2\,dx$ using the trapezoidal rule with $n = 10$. Note that $w = (2-1)/10 = 0.1$. It is convenient to construct a table, such as Table 1. In Table 1, the numbers in the first column are the

TABLE 1

x	$f(x) = x^2$	Factor	Product
1	1	1	1
1.1	1.21	2	2.42
1.2	1.44	2	2.88
1.3	1.69	2	3.38
1.4	1.96	2	3.92
1.5	2.25	2	4.5
1.6	2.56	2	5.12
1.7	2.89	2	5.78
1.8	3.24	2	6.48
1.9	3.61	2	7.22
2	4	1	4

$$\text{sum} = 46.7$$
$$\frac{w}{2}(= 0.05) \times \text{sum} = 2.335$$

values used for x, in this case, the numbers $1, 1.1, 1.2, \ldots, 2$. The second column contains the values of the function $f(x) = x^2$ at the numbers in the first column. The third column contains the factor by which each function value is multiplied in the trapezoidal rule. The fourth column contains the product of the second and third columns. The sum of the numbers in the fourth column is obtained and is multiplied by $w/2$ to obtain the result of the trapezoidal rule.

The trapezoidal rule with $n = 10$ gives a result of 2.335. A more accurate approximation could be obtained by using $n = 20$ to obtain 2.33375. The exact value of $\int_1^2 x^2 \, dx$ is $2\frac{1}{3} = 2.333333\ldots$, and is obtained by using the Fundamental Theorem of Calculus.

Example 1 Use the trapezoidal rule with $n = 8$ to evaluate $\int_1^2 e^{-x^2/2} \, dx$.

Solution Definite integrals involving the function $\exp(-x^2/2)$ are extremely important in statistics. Such integrals cannot be conveniently evaluated via the Fundamental Theorem of Calculus. The solution is contained in Table 2.

TABLE 2

x	$f(x) = e^{-x^2}$	Factor	Product
1	0.3678794	1	0.3678794
1.125	0.282063	2	0.5641259
1.25	0.2096114	2	0.4192228
1.375	0.1509774	2	0.3019548
1.5	0.1053992	2	0.2107984
1.625	0.0713167	2	0.1426334
1.75	0.0467706	2	0.0935412
1.875	0.0297292	2	0.0594584
2	0.0183156	1	0.0183156
		sum =	2.1779301
	$\frac{w}{2}(= 0.0625) \times$ sum =		0.1361206

Since the trapezoidal rule is only an approximation, we will give the answer as 0.136.

Exercises 13-5

Use the trapezoidal rule with the given value of n to approximate each definite integral, and then evaluate each definite integral using the Fundamental Theorem of Calculus.

1. $\int_0^3 x^2\, dx, \quad n = 6$
2. $\int_0^3 x^2\, dx, \quad n = 12$
3. $\int_0^{10} \sqrt{x}\, dx, \quad n = 5$
4. $\int_0^{10} \sqrt{x}\, dx, \quad n = 10$
5. $\int_1^2 e^x\, dx, \quad n = 5$
6. $\int_1^2 e^x\, dx, \quad n = 10$
7. $\int_1^6 \ln x\, dx, \quad n = 5$
8. $\int_1^6 \ln x\, dx, \quad n = 10$

9. Derive the formula for the area of a trapezoid

$$\text{area} = \frac{1}{2} \cdot \text{width} \cdot (\text{sum of parallel sides})$$

Chapter 13 Review

The marginal cost function for a given manufacturer's drill presses is $MC(q) = 300 + q$, and the marginal revenue function is $MR(q) = 500$. Fixed costs are $3000. Determine

1. $C(q)$
2. $R(q)$
3. $P(q)$
4. The maximum possible profit.
5. The change in profit if production is changed from 170 drill presses to 180.
6. Determine the equilibrium point, consumers' surplus, and producers' surplus if

$$S(q) = 13 + 5q \quad \text{and} \quad D(q) = 49 - q^2$$

7. Solve the differential equation $\dfrac{dy}{dx} + (x - 2)y^2 = 0$.
8. Find the solution of the differential equation

$$x\sqrt{2y + 1}\,\frac{dy}{dx} + 4 = 0$$

that contains the point $(1, 4)$.

The weekly revenue R of Jacques, Inc., is increasing and satisfies the differential equation

$$\frac{dR}{dt} = 10{,}000 - R$$

Here t is measured in weeks and the revenue the first week ($t = 1$) is $2000.

9. Solve for R.
10. What is the revenue the fifth week?
11. What is the weekly revenue, after a year?
12. The function $p(x) = (x^2 - 2x)/36$ is a p.d.f. for $3 \leq x \leq 6$. What is $P(x > 4)$? Determine EV(x).
13. For what value of k is $kx/(4 + x^2)$ a p.d.f. for $0 \leq x \leq 4$?
14. Your favorite submarine shop always has 250 meatball subs prepared before lunch hour. The number of meatball subs ordered by customers has the p.d.f. $p(x) = (2x - 275)/15{,}000$, $200 \leq x \leq 275$. On what proportion of days does the store sell all the meatball subs that have been prepared? What is the expected demand for meatball subs each lunch hour?

Use integration by parts to evaluate the following integrals.

15. $\int xe^{8x}\, dx$
16. $\int x^{3/2} \ln x\, dx$

Use the trapezoidal rule with the given value of n to approximate each integral.

17. $\int_0^1 x^3\, dx, \quad n = 5$
18. $\int_0^4 e^{-x}\, dx, \quad n = 8$

APPENDIX:
Tables for Mathematics of Finance*

*Reprinted by permission from A. Mizrahi and M. Sullivan, *Mathematics for Business and Social Sciences: An Applied Approach*, John Wiley and Sons, Inc., 1976.

TABLE 1 Amount at compound interest:
i = interest rate per payment period
n = number of payment periods

$$(1 + i)^n$$

n	$\frac{1}{4}\%$	$\frac{1}{2}\%$	$\frac{3}{4}\%$	1%	$1\frac{1}{4}\%$	$1\frac{1}{2}\%$	$1\frac{3}{4}\%$	2%	$2\frac{1}{2}\%$	3%
1	1.002500	1.005000	1.007500	1.010000	1.012500	1.015000	1.017500	1.020000	1.025000	1.030000
2	1.005006	1.010025	1.015056	1.020100	1.025156	1.030225	1.035306	1.040400	1.050625	1.060900
3	1.007519	1.015075	1.022669	1.030301	1.037971	1.045678	1.053424	1.061208	1.076891	1.092727
4	1.010038	1.020151	1.030339	1.040604	1.050945	1.061364	1.071859	1.082432	1.103813	1.125509
5	1.012563	1.025251	1.038067	1.051010	1.064082	1.077284	1.090617	1.104081	1.131408	1.159274
6	1.015094	1.030378	1.045852	1.061520	1.077383	1.093443	1.109702	1.126162	1.159693	1.194052
7	1.017632	1.035529	1.053696	1.072135	1.090850	1.109845	1.129122	1.148686	1.188686	1.229874
8	1.020176	1.040707	1.061599	1.082857	1.104486	1.126493	1.148882	1.171659	1.218403	1.266770
9	1.022726	1.045911	1.069561	1.093685	1.118292	1.143390	1.168987	1.195093	1.248863	1.304773
10	1.025283	1.051140	1.077583	1.104622	1.132271	1.160541	1.189444	1.218994	1.280085	1.343916
11	1.027846	1.056396	1.085664	1.115668	1.146424	1.177949	1.210260	1.243374	1.312087	1.384234
12	1.030416	1.061678	1.093807	1.126825	1.160755	1.195618	1.231439	1.268242	1.344889	1.425761
13	1.032922	1.066986	1.102010	1.138093	1.175264	1.213552	1.252989	1.293607	1.378511	1.468534
14	1.035574	1.072321	1.110276	1.149474	1.189955	1.231756	1.274917	1.319479	1.412974	1.512590
15	1.038163	1.077683	1.118603	1.160969	1.204829	1.250232	1.297228	1.345868	1.448298	1.557967
16	1.040759	1.083071	1.126992	1.172579	1.219890	1.268986	1.319929	1.372786	1.484506	1.604706
17	1.043361	1.088487	1.135445	1.184304	1.235138	1.288020	1.343028	1.400241	1.521618	1.652848
18	1.045969	1.093929	1.143960	1.196147	1.250577	1.307341	1.366531	1.428246	1.559659	1.702433
19	1.048584	1.099399	1.152540	1.208109	1.266210	1.326951	1.390445	1.456811	1.598650	1.753506
20	1.051205	1.104896	1.161134	1.220190	1.282037	1.346855	1.414778	1.485947	1.638616	1.806111
21	1.053833	1.110420	1.169893	1.232392	1.298063	1.367058	1.439537	1.515666	1.679582	1.860295
22	1.056468	1.115972	1.178667	1.244716	1.314288	1.387564	1.464729	1.545980	1.721571	1.916103
23	1.059109	1.121552	1.187507	1.257163	1.330717	1.408377	1.490361	1.576899	1.764611	1.973586
24	1.061757	1.127160	1.196414	1.269735	1.347351	1.429503	1.516443	1.608437	1.808726	2.032794
25	1.064411	1.132796	1.205387	1.282432	1.364193	1.450945	1.542981	1.640606	1.853944	2.093778
26	1.067072	1.138460	1.214427	1.295256	1.381245	1.472710	1.569983	1.673418	1.900293	2.156591
27	1.069740	1.144152	1.223535	1.308209	1.398511	1.494800	1.597457	1.706886	1.947800	2.221289
28	1.072414	1.149873	1.232712	1.321291	1.415992	1.517222	1.625413	1.741024	1.996495	2.287928
29	1.075096	1.155622	1.241957	1.334504	1.433692	1.539981	1.653858	1.775845	2.046407	2.356565
30	1.077783	1.161400	1.251272	1.347849	1.451613	1.563080	1.682800	1.811362	2.097568	2.427262
31	1.080478	1.167207	1.260656	1.361327	1.469758	1.586526	1.712249	1.847589	2.150007	2.500080
32	1.083179	1.173043	1.270111	1.374941	1.488130	1.610324	1.742213	1.884541	2.203757	2.575083
33	1.085887	1.178908	1.279637	1.388690	1.506732	1.634479	1.772702	1.922231	2.258851	2.652335
34	1.088602	1.184803	1.289234	1.402577	1.525566	1.658996	1.803724	1.960676	2.315322	2.731905
35	1.091323	1.190727	1.298904	1.416603	1.544636	1.683881	1.835290	1.999889	2.373205	2.813862
36	1.094051	1.196681	1.308645	1.430769	1.563944	1.709140	1.867407	2.039887	2.432535	2.898278
37	1.096786	1.202664	1.318460	1.445076	1.583493	1.734777	1.900087	2.080685	2.493349	2.985227
38	1.099528	1.208677	1.328349	1.459527	1.603287	1.760798	1.933338	2.122299	2.555682	3.074783
39	1.102277	1.214721	1.338311	1.474122	1.623328	1.787210	1.967172	2.164745	2.619574	3.167027
40	1.105033	1.220794	1.348349	1.488864	1.643619	1.814018	2.001597	2.208040	2.685064	3.262038
41	1.107796	1.226898	1.358461	1.503752	1.664165	1.841229	2.036625	2.252200	2.752190	3.359899
42	1.110565	1.233033	1.368650	1.518790	1.684967	1.868847	2.072266	2.297244	2.820995	3.460696
43	1.113341	1.239198	1.378915	1.533978	1.706029	1.896880	2.108531	2.343189	2.891520	3.564517
44	1.116125	1.245394	1.389256	1.549318	1.727354	1.925333	2.145430	2.390053	2.963808	3.671452
45	1.118915	1.251621	1.399676	1.564811	1.748946	1.954213	2.182975	2.437854	3.037902	3.781596
46	1.121712	1.257879	1.410173	1.580459	1.770808	1.983526	2.221177	2.486611	3.113851	3.895044
47	1.124517	1.264168	1.420750	1.596263	1.792943	2.013279	2.260048	2.536343	3.191697	4.011895
48	1.127328	1.270489	1.431405	1.612226	1.815355	2.043478	2.299599	2.587070	3.271489	4.132252
49	1.130146	1.276842	1.442141	1.628348	1.838047	2.074130	2.339842	2.638812	3.353277	4.256219
50	1.132972	1.283226	1.452957	1.644632	1.861022	2.105242	2.380789	2.691588	3.437109	4.381906

TABLE 1 Amount at compound interest: (Continued)

i = interest rate per payment period
n = number of payment periods

$$(1+i)^n$$

n	$3\frac{1}{2}\%$	4%	$4\frac{1}{2}\%$	5%	$5\frac{1}{2}\%$	6%	$6\frac{1}{2}\%$	7%	$7\frac{1}{2}\%$	8%
1	1.035000	1.040000	1.045000	1.050000	1.055000	1.060000	1.065000	1.070000	1.075000	1.080000
2	1.071225	1.081600	1.092025	1.102500	1.113025	1.123600	1.134225	1.144900	1.155625	1.166400
3	1.108718	1.124864	1.141166	1.157625	1.174241	1.191016	1.207950	1.225043	1.242297	1.259712
4	1.147523	1.169859	1.192519	1.215506	1.238825	1.262477	1.286466	1.310796	1.335469	1.360489
5	1.187686	1.216653	1.246182	1.276282	1.306960	1.338226	1.370087	1.402552	1.435629	1.469328
6	1.229255	1.265319	1.302260	1.340096	1.378843	1.418519	1.459142	1.500730	1.543302	1.586874
7	1.272279	1.315932	1.360862	1.407100	1.454679	1.503630	1.553987	1.605781	1.659045	1.713824
8	1.316809	1.368569	1.422101	1.477455	1.534687	1.593848	1.654996	1.718186	1.783478	1.850930
9	1.362897	1.423321	1.486095	1.551328	1.619094	1.689479	1.762570	1.838459	1.917239	1.999005
10	1.410599	1.480244	1.552969	1.628895	1.708144	1.790848	1.877137	1.967151	2.061032	2.158925
11	1.459970	1.539454	1.622853	1.710339	1.802092	1.898299	1.999151	2.104852	2.215609	2.331639
12	1.511069	1.601032	1.695881	1.795856	1.901207	2.012196	2.129096	2.252192	2.381780	2.518170
13	1.563956	1.665073	1.772196	1.885649	2.005774	2.132928	2.267487	2.409845	2.560413	2.719624
14	1.618695	1.731676	1.851945	1.979932	2.116091	2.260904	2.414874	2.578534	2.752444	2.937194
15	1.675349	1.800943	1.935282	2.078928	2.232476	2.396558	2.571841	2.759032	2.958877	3.172169
16	1.733986	1.872981	2.022370	2.182875	2.355263	2.540352	2.739011	2.952164	3.180793	3.425943
17	1.794676	1.947900	2.113377	2.292018	2.484802	2.692773	2.917046	3.158815	3.419353	3.700018
18	1.857489	2.025816	2.208479	2.406619	2.621466	2.854339	3.106654	3.379932	3.675804	3.996019
19	1.922501	2.106849	2.307860	2.526950	2.765647	3.025599	3.308587	3.616527	3.951489	4.315701
20	1.989789	2.191123	2.411714	2.653298	2.917757	3.207135	3.523645	3.869684	4.247851	4.660957
21	2.059431	2.278768	2.520241	2.785963	3.078234	3.399564	3.752682	4.140562	4.566440	5.033834
22	2.131512	2.369919	2.633652	2.925261	3.247537	3.603537	3.996606	4.430402	4.908923	5.436540
23	2.206114	2.464715	2.752166	3.071524	3.426152	3.819750	4.256386	4.740530	5.277092	5.871464
24	2.283328	2.563304	2.876014	3.225100	3.614590	4.048935	4.533051	5.072367	5.672874	6.341181
25	2.363245	2.665836	3.005434	3.386355	3.813392	4.291871	4.827699	5.427433	6.098340	6.848475
26	2.445959	2.772470	3.140679	3.555673	4.023125	4.549383	5.141500	5.807353	6.555715	7.396353
27	2.531567	2.883368	3.282010	3.733456	4.244401	4.822346	5.475697	6.213867	7.047394	7.988061
28	2.620172	2.998703	3.429700	3.920129	4.477843	5.111687	5.831617	6.648838	7.575948	8.627106
29	2.711878	3.118651	3.584037	4.116136	4.724124	5.418388	6.210672	7.114257	8.144144	9.317275
30	2.806794	3.243397	3.745318	4.321942	4.983951	5.743491	6.614366	7.612255	8.754955	10.062657
31	2.905031	3.373133	3.913857	4.538039	5.258068	6.088101	7.044300	8.145113	9.411577	10.867669
32	3.006708	3.508059	4.089981	4.764941	5.547262	6.453387	7.502179	8.715271	10.117445	11.737083
33	3.111942	3.648381	4.274030	5.003188	5.852362	6.840590	7.989821	9.325340	10.876253	12.676049
34	3.220860	3.794316	4.466362	5.253348	6.174242	7.251025	8.509160	9.978113	11.691972	13.690133
35	2.333590	3.946089	4.667348	5.516015	6.513825	7.686087	9.062255	10.676581	12.568870	14.785344
36	3.450266	4.103932	4.877376	5.791816	6.872085	8.147252	9.651301	11.423942	13.511535	15.968171
37	3.571025	4.268090	5.096861	6.081407	7.250050	8.636087	10.278636	12.223618	14.524901	17.245625
38	3.696011	4.438813	5.326219	6.385477	7.648803	9.154252	10.946747	13.079271	15.614268	18.625275
39	3.825372	4.616366	5.565890	6.704751	8.069487	9.703507	11.658286	13.994820	16.785338	20.115297
40	3.959260	4.801020	5.816365	7.039989	8.513309	10.285718	12.416075	14.974457	18.044239	21.724521
41	4.097834	4.993061	6.078101	7.391988	8.981541	10.902861	13.223119	16.022669	19.397557	23.462483
42	4.241258	5.192784	6.351616	7.761587	9.475525	11.557032	14.082622	17.144256	20.852373	25.339481
43	4.389702	5.400495	6.637449	8.149667	9.996679	12.250454	14.997993	18.344354	22.416301	27.366640
44	4.543341	5.616515	6.936123	8.557150	10.546496	12.985482	15.972862	19.628459	24.097524	29.555971
45	4.702358	5.841175	7.248248	8.985008	11.126554	13.764610	17.011098	21.002451	25.904838	31.920449
46	4.866941	6.074822	7.574420	9.434258	11.738514	14.590487	18.116820	22.472622	27.847701	34.474084
47	5.037284	6.317815	7.915269	9.905971	12.384132	15.465916	19.294413	24.045706	29.936278	37.232011
48	5.213589	6.570528	8.271456	10.401269	13.065260	16.393871	20.548550	25.728905	32.181498	40.210572
49	5.396064	6.833349	8.643671	10.921333	13.783849	17.377504	21.884205	27.529929	34.595112	43.427418
50	5.584927	7.106683	9.032636	11.467400	14.541961	18.420154	23.306679	29.457024	37.189745	46.901611

TABLE 2 Present value:
i = interest rate per payment period
n = number of payment periods

$$(1+i)^{-n}$$

n	$\frac{1}{4}\%$	$\frac{1}{2}\%$	$\frac{3}{4}\%$	1%	$1\frac{1}{4}\%$	$1\frac{1}{2}\%$	$1\frac{3}{4}\%$	2%	$2\frac{1}{2}\%$	3%
1	0.997506	0.995025	0.992556	0.990099	0.987654	0.985222	0.982801	0.980392	0.975610	0.970874
2	0.995019	0.990075	0.985167	0.980296	0.975461	0.970662	0.965898	0.961169	0.951814	0.942596
3	0.992537	0.985149	0.977833	0.970590	0.963418	0.956317	0.949285	0.942322	0.928599	0.915142
4	0.990062	0.980248	0.970554	0.960980	0.951524	0.942184	0.932959	0.923845	0.905951	0.888487
5	0.987593	0.975371	0.963329	0.951466	0.939777	0.928260	0.916913	0.905731	0.883854	0.862609
6	0.985130	0.970518	0.956158	0.942045	0.928175	0.914542	0.901143	0.887971	0.862297	0.837484
7	0.982674	0.965690	0.949040	0.932718	0.916716	0.901027	0.885644	0.870560	0.841265	0.813092
8	0.980223	0.960885	0.941975	0.923483	0.905398	0.887711	0.870412	0.853490	0.820747	0.789409
9	0.977779	0.956105	0.934963	0.914340	0.894221	0.874592	0.855441	0.836755	0.800728	0.766417
10	0.975340	0.951348	0.928003	0.905287	0.883181	0.861667	0.840729	0.820348	0.781198	0.744094
11	0.972908	0.946615	0.921095	0.896324	0.872277	0.848933	0.826269	0.804263	0.762145	0.722421
12	0.970482	0.941905	0.914238	0.887449	0.861509	0.836387	0.812058	0.788493	0.743556	0.701380
13	0.968062	0.937219	0.907432	0.878663	0.850873	0.824027	0.798091	0.773033	0.725420	0.680951
14	0.965648	0.932556	0.900677	0.869963	0.840368	0.811849	0.784365	0.757875	0.707727	0.661118
15	0.963240	0.927917	0.893973	0.861349	0.829993	0.799852	0.770875	0.743015	0.690466	0.641862
16	0.960837	0.923300	0.887318	0.852821	0.819746	0.788031	0.757616	0.728446	0.673625	0.623167
17	0.958441	0.918707	0.880712	0.844377	0.809626	0.776385	0.744586	0.714163	0.657195	0.605016
18	0.956051	0.914136	0.874156	0.836017	0.799631	0.764912	0.731780	0.700159	0.641166	0.587395
19	0.953667	0.909588	0.867649	0.827740	0.789759	0.753607	0.719194	0.686431	0.625528	0.570286
20	0.951289	0.905063	0.861190	0.819544	0.780009	0.742470	0.706825	0.672971	0.610271	0.553676
21	0.948917	0.900560	0.854779	0.811430	0.770379	0.731498	0.694668	0.659776	0.595386	0.537549
22	0.946550	0.896080	0.848416	0.803396	0.760868	0.720688	0.682720	0.646839	0.580865	0.521893
23	0.944190	0.891622	0.842100	0.795442	0.751475	0.710037	0.670978	0.634156	0.566697	0.506692
24	0.941835	0.887186	0.835831	0.787566	0.742197	0.699544	0.659438	0.621722	0.552875	0.491934
25	0.939486	0.882772	0.829609	0.779768	0.733034	0.689206	0.648096	0.609531	0.539391	0.477606
26	0.937144	0.878380	0.823434	0.772048	0.723984	0.679021	0.636950	0.597579	0.526235	0.463695
27	0.934806	0.874010	0.817304	0.764404	0.715046	0.668986	0.625995	0.585862	0.513400	0.450189
28	0.932475	0.869662	0.811220	0.756836	0.706219	0.659099	0.615228	0.574375	0.500878	0.437077
29	0.930150	0.865335	0.805181	0.749342	0.697500	0.649359	0.604647	0.563112	0.488661	0.424346
30	0.927830	0.861030	0.799187	0.741923	0.688889	0.639762	0.594248	0.552071	0.476743	0.411987
31	0.925517	0.856746	0.793238	0.734577	0.680384	0.630308	0.584027	0.541246	0.465115	0.399987
32	0.923209	0.852484	0.787333	0.727304	0.671984	0.620993	0.573982	0.530633	0.453771	0.388337
33	0.920906	0.848242	0.781472	0.720103	0.663688	0.611816	0.564111	0.520229	0.442703	0.377026
34	0.918610	0.844022	0.775654	0.712973	0.655494	0.602774	0.554408	0.510028	0.431905	0.366045
35	0.916319	0.839823	0.769880	0.705914	0.647402	0.593866	0.544873	0.500028	0.421371	0.355383
36	0.914034	0.835645	0.764149	0.698925	0.639409	0.585090	0.535502	0.490223	0.411094	0.345032
37	0.911754	0.831487	0.758461	0.692005	0.631515	0.576443	0.526292	0.480611	0.401067	0.334983
38	0.909481	0.827351	0.752814	0.685153	0.623719	0.567924	0.517240	0.471187	0.391285	0.325226
39	0.907213	0.823235	0.747210	0.678370	0.616019	0.559531	0.508344	0.461948	0.381741	0.315754
40	0.904950	0.819139	0.741648	0.671653	0.608413	0.551262	0.499601	0.452890	0.372431	0.306557
41	0.902694	0.815064	0.736127	0.665003	0.600902	0.543116	0.491008	0.444010	0.363347	0.297628
42	0.900443	0.811008	0.730647	0.658419	0.593484	0.535089	0.482563	0.435304	0.354485	0.288959
43	0.898197	0.806974	0.725208	0.651900	0.586157	0.527182	0.474264	0.426769	0.345839	0.280543
44	0.895957	0.802959	0.719810	0.645445	0.578920	0.519391	0.466107	0.418401	0.337404	0.272372
45	0.893723	0.798964	0.714451	0.639055	0.571773	0.511715	0.458090	0.410197	0.329174	0.264439
46	0.891494	0.794989	0.709133	0.632728	0.564714	0.504153	0.450212	0.402154	0.321146	0.256737
47	0.889271	0.791034	0.703854	0.626463	0.557742	0.496702	0.442469	0.394268	0.313313	0.249259
48	0.887053	0.787098	0.698614	0.620260	0.550857	0.489362	0.434858	0.386538	0.305671	0.241999
49	0.884841	0.783182	0.693414	0.614119	0.544056	0.482130	0.427379	0.378958	0.298216	0.234950
50	0.882635	0.779286	0.688252	0.608039	0.537339	0.475005	0.420029	0.371528	0.290942	0.228107

TABLE 2 Present value: (Continued)

i = interest rate per payment period
n = number of payment periods

$$(1+i)^{-n}$$

n	3½%	4%	4½%	5%	5½%	6%	6½%	7%	7½%	8%
1	0.966184	0.961538	0.956938	0.952381	0.947867	0.943396	0.938967	0.934579	0.930233	0.925926
2	0.933511	0.924556	0.915730	0.907029	0.898452	0.889996	0.881659	0.873439	0.865333	0.857339
3	0.901943	0.888996	0.876297	0.863838	0.851614	0.839619	0.827849	0.816298	0.804961	0.793832
4	0.871442	0.854804	0.838561	0.822702	0.807217	0.792094	0.777323	0.762895	0.748801	0.735030
5	0.841973	0.821927	0.802451	0.783526	0.765134	0.747258	0.729881	0.712986	0.696559	0.680583
6	0.813501	0.790315	0.767896	0.746215	0.725246	0.704961	0.685334	0.666342	0.647962	0.630170
7	0.785991	0.759918	0.734828	0.710681	0.687437	0.665057	0.643506	0.622750	0.602755	0.583490
8	0.759412	0.730690	0.703185	0.676839	0.651599	0.627412	0.604231	0.582009	0.560702	0.540269
9	0.733731	0.702587	0.672904	0.644609	0.617629	0.591898	0.567353	0.543934	0.521583	0.500249
10	0.708919	0.675564	0.643928	0.613913	0.585431	0.558395	0.532726	0.508349	0.485194	0.463193
11	0.684946	0.649581	0.616199	0.584679	0.554911	0.526788	0.500212	0.475093	0.451343	0.428883
12	0.661783	0.624597	0.589664	0.556837	0.525982	0.496969	0.469683	0.444012	0.419854	0.397114
13	0.639404	0.600574	0.564272	0.530321	0.498561	0.468839	0.441017	0.414964	0.390562	0.367698
14	0.617782	0.577475	0.539973	0.505068	0.472569	0.442301	0.414100	0.387817	0.363313	0.340461
15	0.596891	0.555265	0.516720	0.481017	0.447933	0.417265	0.388827	0.362446	0.337966	0.315242
16	0.576706	0.533908	0.494469	0.458112	0.424581	0.393646	0.365095	0.338735	0.314387	0.291890
17	0.557204	0.513373	0.473176	0.436297	0.402447	0.371364	0.342813	0.316574	0.292453	0.270269
18	0.538361	0.493628	0.452800	0.415521	0.381466	0.350344	0.321890	0.295864	0.272049	0.250249
19	0.520156	0.474642	0.433302	0.395734	0.361579	0.330513	0.302244	0.276508	0.253069	0.231712
20	0.502566	0.456387	0.414643	0.376889	0.342729	0.311805	0.283797	0.258419	0.235413	0.214548
21	0.485571	0.438834	0.396787	0.358942	0.324862	0.294155	0.266476	0.241513	0.218989	0.198656
22	0.469151	0.421955	0.379701	0.341850	0.307926	0.277505	0.250212	0.225713	0.203711	0.183941
23	0.453286	0.405726	0.363350	0.325571	0.291873	0.261797	0.234941	0.210947	0.189498	0.170315
24	0.437957	0.390121	0.347703	0.310068	0.276657	0.246979	0.220602	0.197147	0.176277	0.157699
25	0.423147	0.375117	0.332731	0.295303	0.262234	0.232999	0.207138	0.184249	0.163979	0.146018
26	0.408838	0.360689	0.318402	0.281241	0.248563	0.219810	0.194496	0.172195	0.152539	0.135202
27	0.395012	0.346817	0.304691	0.267848	0.235605	0.207368	0.182625	0.160930	0.141896	0.125187
28	0.381654	0.333477	0.291571	0.255094	0.223322	0.195630	0.171479	0.150402	0.131997	0.115914
29	0.368748	0.320651	0.279015	0.242946	0.211679	0.184557	0.161013	0.140563	0.122788	0.107328
30	0.356278	0.308319	0.267000	0.231377	0.200644	0.174110	0.151186	0.131367	0.114221	0.099377
31	0.344230	0.296460	0.255502	0.220359	0.190184	0.164255	0.141959	0.122773	0.106252	0.092016
32	0.332590	0.285058	0.244500	0.209866	0.180269	0.154957	0.133295	0.114741	0.098839	0.085200
33	0.321343	0.274094	0.233971	0.199873	0.170871	0.146186	0.125159	0.107235	0.091943	0.078889
34	0.310476	0.263552	0.223896	0.190355	0.161963	0.137912	0.117520	0.100219	0.085529	0.073045
35	0.299977	0.253415	0.214254	0.181290	0.153520	0.130105	0.110348	0.093663	0.079562	0.067635
36	0.289833	0.243669	0.205028	0.172657	0.145516	0.122741	0.103613	0.087535	0.074011	0.062625
37	0.280032	0.234297	0.196199	0.164436	0.137930	0.115793	0.097289	0.081809	0.068847	0.057986
38	0.270562	0.225285	0.187750	0.156605	0.130739	0.109239	0.091351	0.076457	0.064044	0.053690
39	0.261413	0.216621	0.179665	0.149148	0.123924	0.103056	0.085776	0.071455	0.059576	0.049713
40	0.252572	0.208289	0.171929	0.142046	0.117463	0.097222	0.080541	0.066780	0.055419	0.046031
41	0.244031	0.200278	0.164525	0.135282	0.111319	0.091719	0.075625	0.062412	0.051553	0.042621
42	0.235779	0.192575	0.157440	0.128840	0.105535	0.086527	0.071010	0.058329	0.047956	0.039464
43	0.227806	0.185168	0.150661	0.122704	0.100033	0.081630	0.066676	0.054513	0.044610	0.036541
44	0.220102	0.178046	0.144173	0.116861	0.094818	0.077009	0.062606	0.050946	0.041498	0.033834
45	0.212659	0.171198	0.137964	0.111297	0.089875	0.072650	0.058785	0.047613	0.038603	0.031328
46	0.205468	0.164614	0.132023	0.105997	0.085190	0.068538	0.055197	0.044499	0.035910	0.029007
47	0.198520	0.158283	0.126338	0.100949	0.080748	0.064658	0.051828	0.041587	0.033404	0.026859
48	0.191806	0.152195	0.120898	0.096142	0.076539	0.060998	0.048665	0.038867	0.031074	0.024869
49	0.185320	0.146341	0.115692	0.091564	0.072549	0.057546	0.045695	0.036324	0.028906	0.023027
50	0.179053	0.140713	0.110710	0.087204	0.068767	0.054288	0.042906	0.033948	0.026889	0.021321

TABLE 3 Amount of an Annuity

i = interest rate per payment period
n = number of payment periods

$$s_{\overline{n}|i} = \frac{(1+i)^n - 1}{i}$$

n	1/4%	1/2%	3/4%	1%	1 1/4%	1 1/2%	1 3/4%	2%	2 1/2%	3%
1	1.000000	1.000000	1.000000	1.000000	1.000000	1.000000	1.000000	1.000000	1.000000	1.000000
2	2.002499	2.005000	2.007500	2.010000	2.012500	2.015000	2.017500	2.020000	2.025000	2.030000
3	3.007505	3.015025	3.022556	3.030100	3.037656	3.045225	3.052806	3.060400	3.075625	3.090900
4	4.015023	4.030100	4.045225	4.060401	4.075627	4.090903	4.106230	4.121608	4.152515	4.183627
5	5.025060	5.050251	5.075564	5.101005	5.126572	5.152267	5.178089	5.204040	5.256328	5.309136
6	6.037623	6.075502	6.113631	6.152015	6.190654	6.229551	6.268706	6.308121	6.387737	6.468410
7	7.052717	7.105880	7.159483	7.213535	7.268037	7.322994	7.378408	7.434283	7.547430	7.662462
8	8.070347	8.141409	8.213179	8.285670	8.358888	8.432839	8.507530	8.582969	8.736116	8.892336
9	9.090523	9.182116	9.274778	9.368527	9.463374	9.559332	9.656412	9.754628	9.954518	10.159106
10	10.113249	10.228027	10.344339	10.462212	10.581666	10.702722	10.825399	10.949720	11.203381	11.463879
11	11.138532	11.279167	11.421921	11.566834	11.713936	11.863262	12.014844	12.168715	12.483466	12.807795
12	12.166377	12.335563	12.507586	12.682502	12.860361	13.041211	13.225103	13.412089	13.795552	14.192029
13	13.196793	13.397241	13.601393	13.809327	14.021115	14.236830	14.456542	14.680331	15.140441	15.617790
14	14.229784	14.464227	14.703403	14.947421	15.196379	15.450382	15.709532	15.973937	16.518952	17.086324
15	15.265359	15.536549	15.813679	16.096895	16.386334	16.682138	16.984449	17.293416	17.931926	18.598913
16	16.303521	16.614231	16.932281	17.257864	17.591163	17.932370	18.281676	18.639284	19.380224	20.156881
17	17.344280	17.697302	18.059273	18.430442	18.811052	19.201355	19.601606	20.012070	20.864730	21.761587
18	18.387640	18.785789	19.194717	19.614747	20.046190	20.489376	20.944634	21.412311	22.386348	23.414435
19	19.433609	19.879718	20.338678	20.810894	21.296767	21.796716	22.311165	22.840558	23.946006	25.116868
20	20.482192	20.979116	21.491218	22.019003	22.562977	23.123667	23.701610	24.297369	25.544656	26.870374
21	21.533398	22.084012	22.652402	23.239193	23.845014	24.470522	25.116388	25.783316	27.183273	28.676485
22	22.587230	23.194432	23.822295	24.471585	25.143077	25.837580	26.555925	27.298982	28.862855	30.536780
23	23.643699	24.310404	25.000962	25.716301	26.457365	27.225143	28.020654	28.844962	30.584426	32.452883
24	24.702807	25.431957	26.188469	26.973463	27.788082	28.633521	29.511015	30.421861	32.349036	34.426469
25	25.764564	26.559116	27.384883	28.243198	29.135433	30.063024	31.027458	32.030298	34.157762	36.459263
26	26.828975	27.691912	28.590269	29.525630	30.499626	31.513969	32.570438	33.670904	36.011706	38.553041
27	27.896046	28.830372	29.804696	30.820886	31.880871	32.986678	34.140421	35.344322	37.911999	40.709632
28	28.965785	29.974524	31.028231	32.129095	33.279382	34.481478	35.737878	37.051208	39.859799	42.930921
29	30.038200	31.124396	32.260943	33.450386	34.695374	35.998701	37.363291	38.792232	41.856294	45.218849
30	31.113295	32.280018	33.502900	34.784890	36.129066	37.538681	39.017148	40.568077	43.902701	47.575414
31	32.191078	33.441419	34.754172	36.132739	37.580679	39.101761	40.699948	42.379438	46.000268	50.002677
32	33.271555	34.608626	36.014828	37.494066	39.050438	40.688288	42.412197	44.227027	48.150275	52.502757
33	34.354734	35.781669	37.284939	38.869006	40.538568	42.298612	44.154411	46.111568	50.354032	55.077840
34	35.440620	36.960577	38.564576	40.257696	42.045300	43.933091	45.927113	48.033799	52.612883	57.730175
35	36.529221	38.145380	39.853810	41.660276	43.570866	45.592088	47.730837	49.994475	54.928205	60.462080
36	37.620543	39.336107	41.152714	43.076876	45.115502	47.275969	49.566127	51.994364	57.301409	63.275942
37	38.714594	40.532788	42.461359	44.507645	46.679446	48.985109	51.433534	54.034251	59.733945	66.174221
38	39.811380	41.735452	43.779819	45.952721	48.262939	50.719885	53.333621	56.114936	62.227293	69.159447
39	40.910909	42.944129	45.108168	47.412248	49.866225	52.480683	55.266959	58.237235	64.782976	72.234231
40	42.013185	44.158850	46.446479	48.886371	51.489553	54.267894	57.234131	60.401979	67.402550	75.401258
41	43.118218	45.379644	47.794828	50.375234	53.133172	56.081912	59.235728	62.610019	70.087614	78.663295
42	44.226012	46.606543	49.153289	51.878987	54.797337	57.923141	61.272353	64.862219	72.839804	82.023194
43	45.336577	47.839575	50.521938	53.397776	56.482304	59.791988	63.344619	67.159464	75.660799	85.483890
44	46.449918	49.078873	51.900853	54.931754	58.188332	61.688868	65.453150	69.502653	78.552318	89.048406
45	47.566043	50.324167	53.290109	56.481072	59.915686	63.614201	67.598580	71.892706	81.516126	92.719858
46	48.684957	51.575788	54.689785	58.045882	61.664632	65.568414	69.781555	74.330560	84.554030	96.501454
47	49.806669	52.833667	56.099958	59.626341	63.435440	67.551940	72.002732	76.817171	87.667880	100.396498
48	50.931185	54.097835	57.520707	61.222604	65.228383	69.565219	74.262780	79.353514	90.859577	104.408392
49	52.058510	55.368324	58.952113	62.834830	67.043738	71.608698	76.562378	81.940584	94.131066	108.540644
50	53.188659	56.645166	60.384253	64.463178	68.881786	73.682828	78.902220	84.579396	97.484343	112.796863

TABLE 3 Amount of an annuity: (Continued)

i = interest rate per payment period
n = number of payment periods

$$\frac{(1+i)^n - 1}{i}$$

n	3½%	4%	4½%	5%	5½%	6%	6½%	7%	7½%	8%
1	1.000000	1.000000	1.000000	1.000000	1.000000	1.000000	1.000000	1.000000	1.000000	1.000000
2	2.035000	2.040000	2.045000	2.050000	2.055000	2.060000	2.065000	2.070000	2.075000	2.080000
3	3.106225	3.121600	3.137025	3.152500	3.168025	3.183600	3.199225	3.214900	3.230625	3.246400
4	4.214943	4.246464	4.278191	4.310125	4.342266	4.374616	4.407175	4.439943	4.472922	4.506112
5	5.362466	5.416322	5.470710	5.525631	5.581091	5.637093	5.693641	5.750739	5.808391	5.866601
6	6.550152	6.632975	6.716892	6.801913	6.888051	6.975318	7.063728	7.153291	7.244020	7.335929
7	7.779407	7.898294	8.019152	8.142008	8.266894	8.393838	8.522870	8.654021	8.787322	8.922803
8	9.051687	9.214226	9.380014	9.549109	9.721573	9.897468	10.076857	10.259802	10.446371	10.636628
9	10.368496	10.582795	10.802114	11.026564	11.256259	11.491316	11.731852	11.977989	12.229849	12.487558
10	11.731393	12.006107	12.288209	12.577892	12.875354	13.180795	13.494423	13.816448	14.147087	14.486562
11	13.141992	13.486351	13.841179	14.206787	14.583498	14.971642	15.371560	15.783599	16.208119	16.645487
12	14.601961	15.025805	15.464032	15.917126	16.385590	16.869941	17.370711	17.888451	18.423728	18.977126
13	16.113030	16.626837	17.159913	17.712983	18.286798	18.882137	19.499808	20.140643	20.805507	21.495296
14	17.676986	18.291911	18.932110	19.598632	20.292572	21.015066	21.767295	22.550487	23.365920	24.214920
15	19.295680	20.023587	20.784054	21.578563	22.408663	23.275970	24.182169	25.129022	26.118364	27.152114
16	20.971029	21.824530	22.719337	23.657492	24.641139	25.672528	26.754010	27.888053	29.077242	30.324283
17	22.705015	23.697511	24.741707	25.840366	26.996402	28.212879	29.493021	30.840217	32.258035	33.750225
18	24.499691	25.645412	26.855084	28.132384	29.481204	30.905652	32.410067	33.999032	35.677387	37.450243
19	26.357180	27.671228	29.063563	30.539004	32.102670	33.759991	35.516722	37.378964	39.353191	41.446263
20	28.279681	29.778077	31.371423	33.065954	34.868317	36.785591	38.825309	40.995491	43.304681	45.761964
21	30.269470	31.969200	33.783137	35.719251	37.786075	39.992726	42.348954	44.865176	47.552532	50.422921
22	32.328901	34.247968	36.303378	38.505214	40.864309	43.392289	46.101636	49.005738	52.118972	55.456754
23	34.460413	36.617887	38.937030	41.430475	44.111846	46.995827	50.098242	53.436140	57.027894	60.893295
24	36.666527	39.082602	41.689197	44.501998	47.537997	50.815576	54.354628	58.176669	62.304987	66.764758
25	38.949855	41.645906	44.565210	47.727098	51.152587	54.864511	58.887679	63.249036	67.977861	73.105939
26	41.313100	44.311742	47.570645	51.113453	54.965979	59.156381	63.715378	68.676469	74.076200	79.954414
27	43.759059	47.084212	50.711324	54.669126	58.989108	63.705764	68.856877	74.483821	80.631915	87.350767
28	46.290626	49.967580	53.993334	58.402582	63.233509	68.528110	74.332575	80.697689	87.679309	95.338828
29	48.910798	52.966284	57.423034	62.322711	67.711352	73.639797	80.164192	87.346527	95.255257	103.965934
30	51.622675	56.084935	61.007070	66.438846	72.435476	79.058184	86.374864	94.460783	103.399401	113.283209
31	54.429469	59.328332	64.752388	70.760789	77.419427	84.801676	92.989230	102.073039	112.154356	123.345866
32	57.334500	62.701465	68.666246	75.298828	82.677496	90.889776	100.033530	110.218151	121.565933	134.213535
33	60.341208	66.209524	72.756227	80.063770	88.224758	97.343163	107.535710	118.933422	131.683377	145.950617
34	63.453150	69.857905	77.030257	85.066958	94.077119	104.183752	115.525531	128.258761	142.559631	158.626667
35	66.674010	73.652221	81.496619	90.320306	100.251361	111.434777	124.034691	138.236874	154.251603	172.316800
36	70.007600	77.598309	86.163966	95.836321	106.765186	119.120864	133.096946	148.913455	166.820473	187.102144
37	73.457866	81.702242	91.041345	101.628137	113.637271	127.268116	142.748247	160.337397	180.332008	203.070315
38	77.028891	85.970331	96.138206	107.709544	120.887320	135.904202	153.026883	172.561014	194.856909	220.315940
39	80.724903	90.409144	101.464425	114.095021	128.536123	145.058455	163.973630	185.640285	210.471177	238.941216
40	84.550274	95.025510	107.030324	120.799772	136.605610	154.761961	175.631916	199.635105	227.256515	259.056512
41	88.509534	99.826530	112.846689	127.839761	145.118918	165.047679	188.047991	214.609562	245.300754	280.781033
42	92.607367	104.819591	118.924790	135.231749	154.100458	175.950540	201.271110	230.632231	264.698310	304.243515
43	96.848625	110.012375	125.276405	142.993336	163.575984	187.507572	215.353732	247.776488	285.550683	329.582997
44	101.238326	115.412870	131.913843	151.143003	173.572663	199.758026	230.351725	266.120841	307.966984	356.949636
45	105.781668	121.029384	138.849966	159.700153	184.119159	212.743508	246.324508	285.749300	332.064508	386.505607
46	110.484026	126.870560	146.098215	168.685160	195.245712	226.508118	263.335686	306.751749	357.969346	418.426055
47	115.350967	132.945382	153.672635	178.119419	206.984207	241.098605	281.452505	329.224373	385.817046	452.900140
48	120.388251	139.263197	161.587903	188.025389	219.368358	256.564521	300.746919	353.270078	415.753325	490.132150
49	125.601839	145.833724	169.859359	198.426659	232.433618	272.958392	321.295468	378.998984	447.934824	530.342722
50	130.997904	152.667073	178.503030	209.347992	246.217467	290.335895	343.179673	406.528912	482.529936	573.770138

TABLE 4 Sinking fund payments:

i = interest rate per payment period
n = number of payment periods

$$\frac{i}{(1+i)^n - 1}$$

n	1/4%	1/2%	3/4%	1%	1 1/4%	1 1/2%	1 3/4%	2%	2 1/2%	3%
1	1.000000	1.000000	1.000000	1.000000	1.000000	1.000000	1.000000	1.000000	1.000000	1.000000
2	0.499376	0.498753	0.498132	0.497512	0.496894	0.496278	0.495663	0.495050	0.493827	0.492611
3	0.332502	0.331672	0.330846	0.330024	0.329201	0.328383	0.327567	0.326755	0.325137	0.323530
4	0.249065	0.248133	0.247205	0.246281	0.245361	0.244445	0.243532	0.242624	0.240818	0.239027
5	0.199003	0.198010	0.197022	0.196040	0.195062	0.194089	0.193121	0.192158	0.190247	0.188355
6	0.165628	0.164595	0.163569	0.162548	0.161534	0.160525	0.159523	0.158526	0.156550	0.154598
7	0.141789	0.140729	0.139675	0.138628	0.137589	0.136556	0.135531	0.134512	0.132495	0.130506
8	0.123910	0.122829	0.121756	0.120690	0.119633	0.118584	0.117543	0.116510	0.114467	0.112456
9	0.110005	0.108907	0.107819	0.106740	0.105671	0.104610	0.103558	0.102515	0.100457	0.098434
10	0.098880	0.097771	0.096671	0.095582	0.094503	0.093434	0.092375	0.091327	0.089259	0.087231
11	0.089778	0.088659	0.087551	0.086454	0.085368	0.084294	0.083230	0.082178	0.080106	0.078077
12	0.082194	0.081066	0.079951	0.078849	0.077758	0.076680	0.075614	0.074560	0.072487	0.070462
13	0.075776	0.074642	0.073522	0.072415	0.071321	0.070240	0.069173	0.068118	0.066048	0.064030
14	0.070275	0.069136	0.068011	0.066901	0.065805	0.064723	0.063656	0.062602	0.060537	0.058526
15	0.065508	0.064364	0.063236	0.062124	0.061026	0.059944	0.058877	0.057825	0.055766	0.053767
16	0.061336	0.060189	0.059059	0.057945	0.056847	0.055765	0.054700	0.053650	0.051599	0.049611
17	0.057656	0.056506	0.055373	0.054258	0.053160	0.052080	0.051016	0.049970	0.047928	0.045953
18	0.054384	0.053232	0.052098	0.050982	0.049885	0.048806	0.047745	0.046702	0.044670	0.042709
19	0.051457	0.050303	0.049167	0.048052	0.046955	0.045878	0.044821	0.043782	0.041761	0.039814
20	0.048823	0.047666	0.046531	0.045415	0.044320	0.043246	0.042191	0.041157	0.039147	0.037216
21	0.046439	0.045282	0.044145	0.043031	0.041937	0.040865	0.039815	0.038785	0.036787	0.034872
22	0.044273	0.043114	0.041977	0.040864	0.039772	0.038703	0.037656	0.036631	0.034647	0.032747
23	0.042295	0.041135	0.039998	0.038886	0.037797	0.036731	0.035688	0.034668	0.032696	0.030814
24	0.040481	0.039321	0.038185	0.037073	0.035987	0.034924	0.033886	0.032871	0.030913	0.029047
25	0.038813	0.037652	0.036516	0.035407	0.034322	0.033263	0.032230	0.031220	0.029276	0.027428
26	0.037273	0.036112	0.034977	0.033869	0.032787	0.031732	0.030703	0.029699	0.027769	0.025938
27	0.035847	0.034686	0.033552	0.032446	0.031367	0.030315	0.029291	0.028293	0.026377	0.024564
28	0.034523	0.033362	0.032229	0.031124	0.030049	0.029001	0.027982	0.026990	0.025088	0.023293
29	0.033291	0.032129	0.030997	0.029895	0.028822	0.027779	0.026764	0.025778	0.023891	0.022115
30	0.032141	0.030979	0.029848	0.028748	0.027679	0.026639	0.025630	0.024650	0.022788	0.021019
31	0.031065	0.029903	0.028774	0.027676	0.026609	0.025574	0.024570	0.023596	0.021739	0.019999
32	0.030056	0.028895	0.027766	0.026671	0.025608	0.024577	0.023578	0.022611	0.020768	0.019047
33	0.029108	0.027947	0.026820	0.025727	0.024668	0.023641	0.022648	0.021687	0.019859	0.018156
34	0.028216	0.027056	0.025931	0.024840	0.023784	0.022762	0.021774	0.020819	0.019007	0.017322
35	0.027375	0.026215	0.025092	0.024004	0.022951	0.021934	0.020951	0.020002	0.018206	0.016539
36	0.026581	0.025422	0.024300	0.023214	0.022165	0.021152	0.020175	0.019233	0.017452	0.015804
37	0.025830	0.024671	0.023551	0.022468	0.021423	0.020414	0.019443	0.018507	0.016741	0.015112
38	0.025118	0.023960	0.022842	0.021761	0.020720	0.019716	0.018750	0.017821	0.016070	0.014459
39	0.024443	0.023286	0.022169	0.021092	0.020054	0.019055	0.018094	0.017171	0.015436	0.013844
40	0.023802	0.022646	0.021530	0.020456	0.019421	0.018427	0.017472	0.016556	0.014836	0.013262
41	0.023192	0.022036	0.020923	0.019851	0.018821	0.017831	0.016882	0.015972	0.014268	0.012712
42	0.022611	0.021456	0.020345	0.019276	0.018249	0.017264	0.016321	0.015417	0.013729	0.012192
43	0.022057	0.020903	0.019793	0.018727	0.017705	0.016725	0.015787	0.014890	0.013217	0.011698
44	0.021529	0.020375	0.019268	0.018204	0.017186	0.016210	0.015278	0.014388	0.012730	0.011230
45	0.021023	0.019871	0.018765	0.017705	0.016690	0.015720	0.014793	0.013910	0.012268	0.010785
46	0.020540	0.019389	0.018285	0.017228	0.016217	0.015251	0.014330	0.013453	0.011827	0.010363
47	0.020078	0.018927	0.017825	0.016771	0.015764	0.014803	0.013888	0.013018	0.011407	0.009961
48	0.019634	0.018485	0.017385	0.016334	0.015331	0.014375	0.013466	0.012602	0.011006	0.009578
49	0.019209	0.018061	0.016963	0.015915	0.014916	0.013965	0.013061	0.012204	0.010623	0.009213
50	0.018801	0.017654	0.016558	0.015513	0.014518	0.013572	0.012674	0.011823	0.010258	0.008865

TABLE 4 Sinking fund payments: (Continued)

i = interest rate per payment period
n = number of payment periods

$$\frac{i}{(1+i)^n - 1}$$

n	3½%	4%	4½%	5%	5½%	6%	6½%	7%	7½%	8%
1	1.000000	1.000000	1.000000	1.000000	1.000000	1.000000	1.000000	1.000000	1.000000	1.000000
2	0.491401	0.490196	0.488998	0.487805	0.486618	0.485437	0.484262	0.483092	0.481928	0.480769
3	0.321934	0.320349	0.318773	0.317209	0.315654	0.314110	0.312576	0.311052	0.309538	0.308034
4	0.237251	0.235490	0.233744	0.232012	0.230294	0.228591	0.226903	0.225228	0.223568	0.221921
5	0.186481	0.184627	0.182792	0.180975	0.179176	0.177396	0.175635	0.173891	0.172165	0.170456
6	0.152668	0.150762	0.148878	0.147017	0.145179	0.143363	0.141568	0.139796	0.138045	0.136315
7	0.128544	0.126610	0.124701	0.122820	0.120964	0.119135	0.117331	0.115553	0.113800	0.112072
8	0.110477	0.108528	0.106610	0.104722	0.102864	0.101036	0.099237	0.097468	0.095727	0.094015
9	0.096446	0.094493	0.092574	0.090690	0.088839	0.087022	0.085238	0.083486	0.081767	0.080080
10	0.085241	0.083291	0.081379	0.079505	0.077668	0.075868	0.074105	0.072378	0.070686	0.069029
11	0.076092	0.074149	0.072248	0.070389	0.068571	0.066793	0.065055	0.063357	0.061697	0.060076
12	0.068484	0.066552	0.064666	0.062825	0.061029	0.059277	0.057568	0.055902	0.054278	0.052695
13	0.062062	0.060144	0.058275	0.056456	0.054684	0.052960	0.051283	0.049651	0.048064	0.046522
14	0.056571	0.054669	0.052820	0.051024	0.049279	0.047585	0.045940	0.044345	0.042797	0.041297
15	0.051825	0.049941	0.048114	0.046342	0.044626	0.042963	0.041353	0.039795	0.038287	0.036830
16	0.047685	0.045820	0.044015	0.042270	0.040583	0.038952	0.037378	0.035858	0.034391	0.032977
17	0.044043	0.042199	0.040418	0.038699	0.037042	0.035445	0.033906	0.032425	0.031000	0.029629
18	0.040817	0.038993	0.037237	0.035546	0.033920	0.032357	0.030855	0.029413	0.028029	0.026702
19	0.037940	0.036139	0.034407	0.032745	0.031150	0.029621	0.028156	0.026753	0.025411	0.024128
20	0.035361	0.033582	0.031876	0.030243	0.028679	0.027185	0.025756	0.024393	0.023092	0.021852
21	0.033037	0.031280	0.029601	0.027996	0.026465	0.025005	0.023613	0.022289	0.021029	0.019832
22	0.030932	0.029199	0.027546	0.025971	0.024471	0.023046	0.021691	0.020406	0.019187	0.018032
23	0.029019	0.027309	0.025682	0.024137	0.022670	0.021278	0.019961	0.018714	0.017535	0.016422
24	0.027273	0.025587	0.023987	0.022471	0.021036	0.019679	0.018398	0.017189	0.016050	0.014978
25	0.025674	0.024012	0.022439	0.020952	0.019549	0.018227	0.016981	0.015811	0.014711	0.013679
26	0.024205	0.022567	0.021021	0.019564	0.018193	0.016904	0.015695	0.014561	0.013500	0.012507
27	0.022852	0.021239	0.019719	0.018292	0.016952	0.015697	0.014523	0.013426	0.012402	0.011448
28	0.021603	0.020013	0.018521	0.017123	0.015814	0.014593	0.013453	0.012392	0.011405	0.010489
29	0.020445	0.018880	0.017415	0.016046	0.014769	0.013580	0.012474	0.011449	0.010498	0.009619
30	0.019371	0.017830	0.016392	0.015051	0.013805	0.012649	0.011577	0.010586	0.009671	0.008827
31	0.018372	0.016855	0.015443	0.014132	0.012917	0.011792	0.010754	0.009797	0.008916	0.008107
32	0.017442	0.015949	0.014563	0.013280	0.012095	0.011002	0.009997	0.009073	0.008226	0.007451
33	0.016572	0.015104	0.103745	0.012490	0.011335	0.010273	0.009299	0.008408	0.007594	0.006852
34	0.015760	0.014315	0.012982	0.011755	0.010630	0.009598	0.008656	0.007797	0.007015	0.006304
35	0.014998	0.013577	0.012270	0.011072	0.009975	0.008974	0.008062	0.007234	0.006483	0.005803
36	0.014284	-0.012887	0.011606	0.010434	0.009366	0.008395	0.007513	0.006715	0.005994	0.005345
37	0.013613	0.012240	0.010984	0.009840	0.008800	0.007857	0.007005	0.006237	0.005545	0.004924
38	0.012982	0.011632	0.010402	0.009284	0.008272	0.007358	0.006535	0.005795	0.005132	0.004539
39	0.012388	0.011061	0.009856	0.008765	0.007780	0.006894	0.006099	0.005387	0.004751	0.004185
40	0.011827	0.010523	0.009343	0.008278	0.007320	0.006462	0.005694	0.005009	0.004400	0.003860
41	0.011298	0.010017	0.008862	0.007822	0.006891	0.006059	0.005318	0.004660	0.004077	0.003561
42	0.010798	0.009540	0.008409	0.007395	0.006489	0.005683	0.004968	0.004336	0.003778	0.003287
43	0.010325	0.009090	0.007982	0.006993	0.006113	0.005333	0.004644	0.004036	0.003502	0.003034
44	0.009878	0.008665	0.007581	0.006616	0.005761	0.005006	0.004341	0.003758	0.003247	0.002802
45	0.009453	0.008262	0.007202	0.006262	0.005431	0.004700	0.004060	0.003500	0.003011	0.002587
46	0.009051	0.007882	0.006845	0.005928	0.005122	0.004415	0.003797	0.003260	0.002794	0.002390
47	0.008669	0.007522	0.006507	0.005614	0.004831	0.004148	0.003553	0.003037	0.002592	0.002208
48	0.008306	0.007181	0.006189	0.005318	0.004559	0.003898	0.003325	0.002831	0.002405	0.002040
49	0.007962	0.006857	0.005887	0.005040	0.004302	0.003664	0.003112	0.002639	0.002232	0.001886
50	0.007634	0.006550	0.005602	0.004777	0.004061	0.003444	0.002914	0.002460	0.002072	0.001743

TABLE 5 Present Value of an Annuity:

i = interest rate per payment period
n = number of payment periods

$$\frac{1 - (1+i)^{-n}}{i}$$

n	¼%	½%	¾%	1%	1¼%	1½%	1¾%	2%	2½%	3%
1	0.997506	0.995025	0.992556	0.990099	0.987654	0.985222	0.982801	0.980392	0.975610	0.970874
2	1.992524	1.985099	1.977723	1.970395	1.963115	1.955883	1.948699	1.941561	1.927424	1.913470
3	2.985061	2.970248	2.955556	2.940985	2.926534	2.912200	2.897984	2.883883	2.856024	2.828611
4	3.975123	3.950496	3.926110	3.901965	3.878058	3.854385	3.830942	3.807729	3.761974	3.717098
5	4.962716	4.925867	4.889439	4.853431	4.817835	4.782645	4.747855	4.713459	4.645828	4.579707
6	5.947846	5.896385	5.845597	5.795476	5.746010	5.697187	5.648997	5.601431	5.508125	5.417191
7	6.930519	6.862074	6.794637	6.728194	6.662725	6.598214	6.534641	6.471991	6.349390	6.230283
8	7.910741	7.822960	7.736613	7.651677	7.568124	7.485925	7.405053	7.325481	7.170137	7.019692
9	8.888520	8.779064	8.671576	8.566017	8.462344	8.360517	8.260494	8.162236	7.970865	7.786109
10	9.863860	9.730412	9.599579	9.471304	9.345525	9.222184	9.101223	8.982585	8.752064	8.530203
11	10.836767	10.677027	10.520674	10.367628	10.217803	10.071118	9.927492	9.786848	9.514208	9.252624
12	11.807249	11.618933	11.434912	11.255077	11.079311	10.907505	10.739549	10.575341	10.257764	9.954004
13	12.775310	12.556152	12.342345	12.133740	11.930184	11.731532	11.537641	11.348373	10.983185	10.634955
14	13.740957	13.488708	13.243022	13.003702	12.770552	12.543381	12.322005	12.106248	11.690912	11.296073
15	14.704197	14.416626	14.136994	13.865052	13.600545	13.343233	13.092880	12.849263	12.381377	11.937935
16	15.665033	15.339926	15.024312	14.717873	14.420291	14.131264	13.850496	13.577709	13.055002	12.561102
17	16.623475	16.258863	15.905024	15.562251	15.229918	14.907649	14.595082	14.291871	13.712197	13.166118
18	17.579525	17.172769	16.779180	16.398268	16.029548	15.672561	15.326862	14.992031	14.353363	13.753513
19	18.533192	18.082357	17.646829	17.226008	16.819307	16.426168	16.046056	15.678462	14.978891	14.323799
20	19.484480	18.987420	18.508019	18.045552	17.599315	17.168639	16.752881	16.351433	15.589162	14.877475
21	20.433396	19.887980	19.362798	18.856982	18.369694	17.900137	17.447549	17.011209	16.184548	15.415024
22	21.379946	20.784060	20.211214	19.660379	19.130562	18.620824	18.130269	17.658048	16.765413	15.936916
23	22.324136	21.675682	21.053314	20.455820	19.882036	19.330861	18.801247	18.292204	17.332110	16.443608
24	23.265970	22.562867	21.889145	21.243386	20.624233	20.030405	19.460685	18.913925	17.884985	16.935542
25	24.205456	23.445639	22.718754	22.023155	21.357268	20.719611	20.108781	19.523456	18.424376	17.413147
26	25.142599	24.324019	23.542188	22.795203	22.081252	21.398632	20.745731	20.121035	18.950611	17.876842
27	26.077405	25.198029	24.359492	23.559607	22.796298	22.067717	21.371726	20.706897	19.464010	18.327031
28	27.009879	26.067691	25.170711	24.316442	23.502517	22.726717	21.986954	21.281272	19.964888	18.764108
29	27.940030	26.933025	25.975892	25.065784	24.200016	23.376076	22.591601	21.844384	20.453549	19.188454
30	28.867859	27.794055	26.775079	25.807707	24.888905	24.015838	23.185849	22.396455	20.930292	19.600441
31	29.793376	28.650802	27.568317	26.542284	25.569289	24.646146	23.769876	22.937701	21.395407	20.000428
32	30.716584	29.503285	28.355649	27.269588	26.241273	25.267139	24.343858	23.468334	21.849177	20.388765
33	31.637490	30.351527	29.137121	27.989691	26.904961	25.878954	24.907969	23.988563	22.291880	20.765792
34	32.556099	31.195550	29.912775	28.702665	27.560455	26.481728	25.462377	24.498591	22.723786	21.131836
35	33.472417	32.035373	30.682655	29.408579	28.207857	27.075595	26.007250	24.998619	23.145157	21.487220
36	34.386451	32.871018	31.446804	30.107504	28.847266	27.660684	26.542752	25.488842	23.556251	21.832252
37	35.298205	33.702505	32.205264	30.799509	29.478781	28.237127	27.069044	25.969453	23.957318	22.167235
38	36.207685	34.529856	32.958079	31.484662	30.102500	28.805052	27.586284	26.440640	24.348603	22.492461
39	37.114898	35.353091	33.705289	32.163032	30.718518	29.364583	28.094628	26.902588	24.730344	22.808215
40	38.019848	36.172230	34.446937	32.834685	31.326932	29.915845	28.594229	27.355478	25.102775	23.114772
41	38.922541	36.987293	35.183064	33.499688	31.927834	30.458961	29.085237	27.799489	25.466121	23.412400
42	39.822983	37.798302	35.913711	34.158107	32.521317	30.994050	29.567801	28.234793	25.820606	23.701359
43	40.721180	38.605275	36.638919	34.810007	33.107474	31.521232	30.042064	28.661562	26.166445	23.981902
44	41.617136	39.408234	37.358729	35.455452	33.686394	32.040622	30.508171	29.079962	26.503849	24.254274
45	42.510859	40.207198	38.073180	36.094507	34.258167	32.552337	30.966262	29.490159	26.833023	24.518712
46	43.402353	41.002187	38.782312	36.727235	34.822881	33.056490	31.416473	29.892313	27.154169	24.775449
47	44.291624	41.793221	39.486166	37.353698	35.380623	33.553192	31.858942	30.286581	27.467482	25.024708
48	45.178676	42.580320	40.184780	37.973958	35.931479	34.042554	32.293800	30.673119	27.773153	25.266706
49	46.063518	43.363502	40.878194	38.588077	36.475535	34.524683	32.721180	31.052077	28.071369	25.501657
50	46.946152	44.142788	41.566445	39.196116	37.012874	34.999688	33.141209	31.423605	28.362311	25.729764

TABLE 5 Present value of an annuity: (Continued)

i = interest rate per payment period
n = number of payment periods

$$\frac{1-(1+i)^{-n}}{i}$$

n	3½%	4%	4½%	5%	5½%	6%	6½%	7%	7½%	8%
1	0.966184	0.961538	0.956938	0.952381	0.947867	0.943396	0.938967	0.934579	0.930233	0.925926
2	1.899694	1.886095	1.872668	1.859410	1.846320	1.833393	1.820626	1.808018	1.795565	1.783265
3	2.801637	2.775091	2.748964	2.723248	2.697933	2.673012	2.648475	2.624316	2.600526	2.577097
4	3.673079	3.629895	3.587526	3.545950	3.505150	3.465106	3.425799	3.387211	3.349326	3.312127
5	4.515052	4.451822	4.389977	4.329477	4.270284	4.212364	4.155679	4.100197	4.045885	3.992710
6	5.328553	5.242137	5.157873	5.075692	4.995530	4.917324	4.841014	4.766540	4.693846	4.622880
7	6.114544	6.002055	5.892701	5.786373	5.682967	5.582381	5.484520	5.389289	5.296601	5.206370
8	6.873955	6.732745	6.595886	6.463213	6.334566	6.209794	6.088751	5.971298	5.857304	5.746639
9	7.607686	7.435331	7.268791	7.107822	6.952195	6.801692	6.656104	6.515232	6.378887	6.246888
10	8.316605	8.110896	7.912718	7.721735	7.537626	7.360087	7.188830	7.023581	6.864081	6.710081
11	9.001551	8.760477	8.528917	8.306414	8.092536	7.886875	7.689042	7.498674	7.315424	7.138964
12	9.663334	9.385074	9.118581	8.863252	8.618518	8.383844	8.158725	7.942686	7.735278	7.536078
13	10.302738	9.985648	9.682852	9.393573	9.117078	8.852683	8.599742	8.357651	8.125840	7.903776
14	10.920520	10.563123	10.222825	9.898641	9.589648	9.294984	9.013842	8.745468	8.489154	8.244237
15	11.517411	11.118387	10.739546	10.379658	10.037581	9.712249	9.402669	9.107914	8.827120	8.559479
16	12.094117	11.652296	11.234015	10.837770	10.462162	10.105895	9.767764	9.446649	9.141507	8.851369
17	12.651320	12.165669	11.707191	11.274066	10.864608	10.477260	10.110577	9.763223	9.433960	9.121638
18	13.189682	12.659297	12.159992	11.689587	11.246074	10.827603	10.432466	10.059087	9.706009	9.371887
19	13.709837	13.133939	12.593294	12.085321	11.607653	11.158116	10.734710	10.335595	9.959078	9.603599
20	14.212403	13.590326	13.007937	12.462210	11.950382	11.469921	11.018507	10.594014	10.194491	9.818147
21	14.697974	14.029160	13.404724	12.821153	12.275244	11.764077	11.284983	10.835527	10.413480	10.016803
22	15.167125	14.451115	13.784425	13.163003	12.583170	12.041582	11.535196	11.061240	10.617191	10.200744
23	15.620410	14.856841	14.147775	13.488574	12.875042	12.303379	11.770137	11.272187	10.806689	10.371059
24	16.058367	15.246963	14.495478	13.798642	13.151699	12.550357	11.990739	11.469334	10.982967	10.528758
25	16.481514	15.622080	14.828209	14.093945	13.413933	12.783356	12.197877	11.653583	11.146946	10.674776
26	16.890352	15.982769	15.146611	14.375185	13.662495	13.003166	12.392373	11.825779	11.299485	10.809978
27	17.285364	16.329585	15.451303	14.643034	13.898100	13.210534	12.574998	11.986709	11.441381	10.935165
28	17.667019	16.663063	15.742874	14.898127	14.121422	13.406164	12.746477	12.137111	11.573378	11.051078
29	18.035767	16.983714	16.021889	15.141074	14.333101	13.590721	12.907490	12.277674	11.696165	11.158406
30	18.392045	17.292033	16.288889	15.372451	14.533745	13.764831	13.058676	12.409041	11.810386	11.257783
31	18.736276	17.588493	16.544391	15.592810	14.723929	13.929086	13.200635	12.531814	11.916638	11.349799
32	19.068865	17.873551	16.788891	15.802677	14.904198	14.084043	13.333929	12.646555	12.015478	11.434999
33	19.390208	18.147645	17.022862	16.002549	15.075069	14.230230	13.459089	12.753790	12.107421	11.513888
34	19.700684	18.411197	17.246758	16.192904	15.237033	14.368141	13.576609	12.854009	12.192950	11.586934
35	20.000661	18.664613	17.461012	16.374194	15.390552	14.498246	13.686957	12.947672	12.272511	11.654568
36	20.290494	18.908282	17.666041	16.546852	15.536068	14.620987	13.790570	13.035208	12.346522	11.717193
37	20.570525	19.142579	17.862240	16.711287	15.673998	14.736780	13.887859	13.117017	12.415370	11.775179
38	20.841087	19.367864	18.049990	16.867893	15.804738	14.846019	13.979210	13.193473	12.479414	11.828869
39	21.102500	19.584485	18.229656	17.017041	15.928662	14.949075	14.064986	13.264928	12.538989	11.878582
40	21.355072	19.792774	18.401584	17.159086	16.046125	15.046297	14.145527	13.331709	12.594409	11.924613
41	21.599103	19.993052	18.566109	17.294368	16.157464	15.138016	14.221152	13.394120	12.645962	11.967235
42	21.834883	20.185627	18.723550	17.423208	16.262999	15.224543	14.292162	13.452449	12.693918	12.006699
43	22.062688	20.370795	18.874210	17.545912	16.363032	15.306173	14.358837	13.506962	12.738528	12.043240
44	22.282791	20.548841	19.018383	17.662773	16.457851	15.383182	14.421443	13.557908	12.780026	12.077074
45	22.495450	20.720040	19.156347	17.774070	16.547726	15.455832	14.480228	13.605522	12.818629	12.108401
46	22.700918	20.884654	19.288371	17.880066	16.632915	15.524370	14.535426	13.650020	12.854539	12.137409
47	22.899438	21.042936	19.414700	17.981016	16.713664	15.589028	14.587254	13.691608	12.887943	12.164267
48	23.091244	21.195131	19.535607	18.077158	16.790203	15.650027	14.635919	13.730474	12.919017	12.189137
49	23.276564	21.341472	19.651298	18.168722	16.862751	15.707572	14.681615	13.766799	12.947922	12.212163
50	23.455618	21.482184	19.762008	18.255925	16.931518	15.761861	14.724521	13.800746	12.974812	12.233485

TABLE 6 Amortization:

i = interest rate per payment period
n = number of payment periods

$$\frac{i}{1-(1+i)^{-n}}$$

n	$\frac{1}{4}$%	$\frac{1}{2}$%	$\frac{3}{4}$%	1%	$1\frac{1}{4}$%	$1\frac{1}{2}$%	$1\frac{3}{4}$%	2%	$2\frac{1}{2}$%	3%
1	1.002500	1.005000	1.007500	1.010000	1.012500	1.015000	1.017500	1.020000	1.025000	1.030000
2	0.501876	0.503753	0.505632	0.507512	0.509394	0.511278	0.513163	0.515050	0.518827	0.522611
3	0.335002	0.336672	0.338346	0.340022	0.341701	0.343383	0.345067	0.346755	0.350137	0.353530
4	0.251565	0.253133	0.254705	0.256281	0.257861	0.259445	0.261032	0.262624	0.265818	0.269027
5	0.201503	0.203010	0.204522	0.206040	0.207562	0.209089	0.210621	0.212158	0.215247	0.218355
6	0.168128	0.169595	0.171069	0.172548	0.174034	0.175525	0.177023	0.178526	0.181550	0.184598
7	0.144289	0.145729	0.147175	0.148628	0.150089	0.151556	0.153031	0.154512	0.157495	0.160506
8	0.126410	0.127829	0.129256	0.130690	0.132133	0.133584	0.135043	0.136510	0.139467	0.142456
9	0.112505	0.113907	0.115319	0.116740	0.118171	0.119610	0.121058	0.122515	0.125457	0.128434
10	0.101380	0.102771	0.104171	0.105582	0.107003	0.108434	0.109875	0.111327	0.114259	0.117231
11	0.092278	0.093659	0.095051	0.096454	0.097868	0.099294	0.100730	0.102178	0.105106	0.108077
12	0.084694	0.086066	0.087451	0.088849	0.090258	0.091680	0.093114	0.094560	0.097487	0.100462
13	0.078276	0.079642	0.081022	0.082415	0.083821	0.085240	0.086673	0.088118	0.091048	0.094030
14	0.072775	0.074136	0.075511	0.076901	0.078305	0.079723	0.081156	0.082602	0.085537	0.088526
15	0.068008	0.069364	0.070736	0.072124	0.073526	0.074944	0.076377	0.077825	0.080766	0.083767
16	0.063836	0.065189	0.066559	0.067945	0.069347	0.070765	0.072200	0.073650	0.076599	0.079611
17	0.060156	0.061506	0.062873	0.064258	0.065660	0.067080	0.068516	0.069970	0.072928	0.075953
18	0.056884	0.058232	0.059598	0.060982	0.062385	0.063806	0.065245	0.066702	0.069670	0.072709
19	0.053957	0.055303	0.056667	0.058052	0.059455	0.060878	0.062321	0.063782	0.066761	0.069814
20	0.051323	0.052666	0.054031	0.055415	0.056820	0.058246	0.059691	9.061157	0.064147	0.067216
21	0.048939	0.050282	0.051645	0.053031	0.054437	0.055865	0.057315	0.058785	0.061787	0.064872
22	0.046773	0.048114	0.049477	0.050864	0.052272	0.053703	0.055156	0.056631	0.059647	0.062747
23	0.044795	0.046135	0.047498	0.048886	0.050297	0.051731	0.053188	0.054668	0.057696	0.060814
24	0.042981	0.044321	0.045685	0.047073	0.048487	0.049924	0.051386	0.052871	0.055913	0.059047
25	0.041313	0.042652	0.044016	0.045407	0.046822	0.048263	0.049730	0.051220	0.054276	0.057428
26	0.039773	0.041112	0.042477	0.043869	0.045287	0.046732	0.048203	0.049699	0.052769	0.055938
27	0.038347	0.039686	0.041052	0.042446	0.043867	0.045315	0.046791	0.048293	0.051377	0.054564
28	0.037023	0.038362	0.039729	0.041124	0.042549	0.044001	0.045482	0.046990	0.050088	0.053293
29	0.035791	0.037129	0.038497	0.039895	0.041322	0.042779	0.044264	0.045778	0.048891	0.052115
30	0.034641	0.035979	0.037348	0.038748	0.040179	0.041639	0.043130	0.044650	0.047778	0.051019
31	0.033565	0.034903	0.036274	0.037676	0.039109	0.040574	0.042070	0.043596	0.046739	0.049999
32	0.032556	0.033895	0.035266	0.036671	0.038108	0.039577	0.041078	0.042611	0.045768	0.049047
33	0.031608	0.032947	0.034320	0.035727	0.037168	0.038641	0.040148	0.041687	0.044859	0.048156
34	0.030716	0.032056	0.033431	0.034840	0.036284	0.037762	0.039274	0.040819	0.044007	0.047322
35	0.029875	0.031215	0.032592	0.034004	0.035451	0.036934	0.038451	0.040002	0.043206	0.046539
36	0.029081	0.030422	0.031800	0.033214	0.034665	0.036152	0.037675	0.039233	0.042452	0.045804
37	0.028330	0.029671	0.031051	0.032468	0.033923	0.035414	0.036943	0.038507	0.041741	0.045112
38	0.027618	0.028960	0.030342	0.031761	0.033220	0.034716	0.036250	0.037821	0.041070	0.044459
39	0.026943	0.028286	0.029669	0.031092	0.032554	0.034055	0.035594	0.037171	0.040436	0.043844
40	0.026302	0.027646	0.029030	0.030456	0.031921	0.033427	0.034972	0.036556	0.039836	0.043262
41	0.025692	0.027036	0.028423	0.029851	0.031321	0.032831	0.034382	0.035972	0.039268	0.042712
42	0.025111	0.026456	0.027845	0.029276	0.030749	0.032264	0.033821	0.035417	0.038729	0.042192
43	0.024557	0.025903	0.027293	0.028727	0.030205	0.031725	0.033287	0.034890	0.038217	0.041698
44	0.024029	0.025375	0.026768	0.028204	0.029686	0.031210	0.032778	0.034388	0.037730	0.041230
45	0.023523	0.024871	0.026265	0.027705	0.029190	0.030720	0.032293	0.033910	0.037268	0.040785
46	0.023040	0.024389	0.025785	0.027228	0.028717	0.030251	0.031830	0.033453	0.036827	0.040363
47	0.022578	0.023927	0.025325	0.026771	0.028264	0.029803	0.031388	0.033018	0.036407	0.039961
48	0.022134	0.023485	0.024885	0.026334	0.027831	0.029375	0.030966	0.032602	0.036006	0.039578
49	0.021709	0.023061	0.024463	0.025915	0.027416	0.028965	0.030561	0.032204	0.035623	0.039213
50	0.021301	0.022654	0.024058	0.025513	0.027018	0.028572	0.030174	0.031823	0.035258	0.038865

TABLE 6 Amortization: (Continued)

i = interest rate per payment period
n = number of payment periods

$$\frac{i}{1-(1+i)^{-n}}$$

n	3½%	4%	4½%	5%	5½%	6%	6½%	7%	7½%	8%
1	1.035000	1.040000	1.045000	1.050000	1.055000	1.060000	1.065000	1.070000	1.075000	1.080000
2	0.526401	0.530196	0.533998	0.537805	0.541618	0.545437	0.549262	0.553092	0.556928	0.560769
3	0.356934	0.360349	0.363773	0.367209	0.370654	0.374110	0.377576	0.381052	0.384538	0.388034
4	0.272251	0.275490	0.278744	0.282012	0.285294	0.288591	0.291903	0.295228	0.298568	0.301921
5	0.221481	0.224627	0.227792	0.230975	0.234176	0.237396	0.240635	0.243891	0.247165	0.250456
6	0.187668	0.190762	0.193878	0.197017	0.200179	0.203363	0.206568	0.209796	0.213045	0.216315
7	0.163544	0.166610	0.169701	0.172820	0.175964	0.179135	0.182331	0.185553	0.188800	0.192072
8	0.145477	0.148528	0.151610	0.154722	0.157864	0.161036	0.164237	0.167468	0.170727	0.174015
9	0.131446	0.134493	0.137574	0.140690	0.143839	0.147022	0.150238	0.153486	0.156767	0.160080
10	0.120241	0.123291	0.126379	0.129505	0.132668	0.135868	0.139105	0.142378	0.145686	0.149029
11	0.111092	0.114149	0.117248	0.120389	0.123571	0.126793	0.130055	0.133357	0.136697	0.140076
12	0.103484	0.106552	0.109666	0.112825	0.116029	0.119277	0.122568	0.125902	0.129278	0.132695
13	0.097062	0.100144	0.103275	0.106456	0.109684	0.112960	0.116283	0.119651	0.123064	0.126522
14	0.091571	0.094669	0.097820	0.101024	0.104279	0.107585	0.110940	0.114345	0.117797	0.121297
15	0.086825	0.089941	0.903114	0.096342	0.099626	0.102963	0.106353	0.109795	0.113287	0.116830
16	0.082685	0.085820	0.089015	0.092270	0.095583	0.098952	0.102378	0.105858	0.109391	0.112977
17	0.079043	0.082199	0.085418	0.088699	0.092042	0.095445	0.098906	0.102425	0.106000	0.109629
18	0.075817	0.078993	0.082237	0.085546	0.088920	0.092357	0.095855	0.099413	0.103029	0.106702
19	0.072940	0.076139	0.079407	0.082745	0.086150	0.089621	0.093156	0.096753	0.100411	0.104128
20	0.070361	0.073582	0.076876	0.080243	0.083679	0.087185	0.090756	0.094393	0.098092	0.101852
21	0.068037	0.071280	0.074601	0.077996	0.081465	0.085005	0.088613	0.092289	0.096029	0.099832
22	0.065932	0.069199	0.072546	0.075971	0.079471	0.083046	0.086691	0.090406	0.094187	0.098032
23	0.064019	0.067309	0.070682	0.074137	0.077670	0.081278	0.084961	0.088714	0.092535	0.096422
24	0.062273	0.065587	0.068987	0.072471	0.076036	0.079679	0.083398	0.087189	0.091050	0.094978
25	0.060674	0.064012	0.067439	0.070952	0.074549	0.078227	0.081981	0.085811	0.089711	0.093679
26	0.059205	0.062567	0.066021	0.069564	0.073193	0.076904	0.080695	0.084561	0.088500	0.092507
27	0.057852	0.061239	0.064719	0.068292	0.071952	0.075697	0.079523	0.083426	0.087402	0.091448
28	0.056603	0.060013	0.063521	0.067123	0.070814	0.074593	0.078453	0.082392	0.086405	0.090489
29	0.055445	0.058880	0.062415	0.066046	0.069769	0.073580	0.077474	0.081449	0.085498	0.089619
30	0.054371	0.057830	0.061392	0.065051	0.068805	0.072649	0.076577	0.080586	0.084671	0.088827
31	0.053372	0.056855	0.060443	0.064132	0.067917	0.071792	0.075754	0.079797	0.083916	0.088107
32	0.052442	0.055949	0.059563	0.063280	0.067095	0.071002	0.074997	0.079073	0.083226	0.087451
33	0.051572	0.055104	0.058745	0.062490	0.066335	0.070273	0.074299	0.078408	0.082594	0.086852
34	0.050760	0.054315	0.057982	0.061755	0.065633	0.069598	0.073656	0.077797	0.082015	0.086304
35	0.049998	0.053577	0.057270	0.061072	0.064975	0.068974	0.073062	0.077234	0.081483	0.085803
36	0.049284	0.052887	0.056606	0.060434	0.064366	0.068395	0.072513	0.076715	0.080994	0.085345
37	0.048613	0.052240	0.055984	0.059840	0.063800	0.067857	0.072005	0.076237	0.080545	0.084924
38	0.047982	0.051632	0.055402	0.059284	0.063272	0.067358	0.071535	0.075795	0.080132	0.084539
39	0.047388	0.051061	0.054856	0.058765	0.062780	0.066894	0.071099	0.075387	0.079751	0.084185
40	0.046827	0.050523	0.054343	0.058278	0.062320	0.066462	0.070694	0.075009	0.079400	0.083860
41	0.046298	0.050017	0.053862	0.057822	0.061891	0.066059	0.070318	0.074660	0.079077	0.083561
42	0.045798	0.049540	0.053409	0.057395	0.061489	0.065683	0.069968	0.074336	0.078778	0.083287
43	0.045325	0.049090	0.052982	0.056993	0.061113	0.065333	0.069644	0.074036	0.078502	0.083034
44	0.044878	0.048665	0.052581	0.056616	0.060761	0.065006	0.069341	0.073758	0.078247	0.082802
45	0.044453	0.048262	0.052202	0.056262	0.060431	0.064700	0.069060	0.073500	0.078011	0.082587
46	0.044051	0.047882	0.051845	0.055928	0.060122	0.064415	0.068797	0.073260	0.077794	0.082390
47	0.043669	0.047522	0.051507	0.055614	0.059831	0.064148	0.068553	0.073037	0.077592	0.082208
48	0.043306	0.047181	0.051189	0.055318	0.059559	0.063898	0.068325	0.072831	0.077405	0.082040
49	0.042962	0.046857	0.050887	0.055040	0.059302	0.063664	0.068112	0.072639	0.077232	0.081886
50	0.042634	0.046550	0.050602	0.054777	0.059061	0.063444	0.067914	0.072460	0.077072	0.081743

Answers to Odd-Numbered Exercises and Review Exercises

Chapter 1

Section 1-1, page 5

1. $\in$ 3. $\notin$ 5. $\in$ 7. $=$ 9. $\neq$
11. $\neq$ 13. $\not\subseteq$ 15. $\subseteq$ 17. $\subseteq$ 19. $\{c\}$
21. $\{a, c, e, f, g, h\}$ 23. $\varnothing$ 25. $\{c, e\}$ 27. $\{b, d\}$
29. $\{a, b, d, f, g, h\}$ 31. U 33. $\{b\}$
35. M' is the set of female employees of Fifth Federal.
37. $E \cap M$ is the set of exempt male employees of Fifth Federal.
39. $M \cap E \cap S$ is the set of exempt male employees of Fifth Federal with at least 20 years seniority.
41. $M' \cap E' \cap S$ is the set of nonexempt female employees of Fifth Federal with at least 20 years seniority.
43. $E \cap S$ is the set of exempt employees of Fifth Federal with at least 20 years seniority.
45. $\varnothing, \{a\}, \{b\}, \{c\}, \{a, b\}, \{a, c\}, \{b, c\}, \{a, b, c\}$

Section 1-2, page 10

1. $\dfrac{7}{6}$ 3. $-\dfrac{3}{5}$ 5. $\dfrac{28}{15}$ 7. $\dfrac{34}{35}$
9. -1 11. $\dfrac{27}{10}$ 13. $\dfrac{8}{21}$ 15. $-\dfrac{45}{44}$
17. $\dfrac{15}{56}$ 19. $\dfrac{3}{2}$ 21. $-\dfrac{15}{22}$ 23. $-\dfrac{4}{15}$
25. 25   27. -6 29. 0   31. $x > 43{,}000{,}000$
33. $7500 \leq x \leq \$18{,}462$ 35. $0.4999 \leq x \leq 0.5001$
37. Rational   39. Irrational
41. Either rational or irrational

Section 1-3, page 20

1. 125   3. $\dfrac{1}{49}$ 5. 1
7. 64   9. $\dfrac{16}{81}$ 11. $\dfrac{8}{125}$
13. $-\dfrac{1}{243}$ 15. $-\dfrac{125}{8}$ 17. $625x^4$
19. $-\dfrac{x^3}{64}$ 21. $x^7 y^7$ 23. $x^6 y^3$
25. $27c^3$ 27. $-125b^9$ 29. $-\dfrac{8}{y^3}$
31. $-64b^3$ 33. 1.23×10^1 35. -5×10^{-5}
37. 4.73×10^8 39. 1.414×10^0 41. 23,600
43. -0.00137 45. $2{,}100{,}000{,}000{,}000$ 47. -0.00000346
49. 4   51. 11   53. Undefined
55. $\dfrac{4}{5}$ 57. $\dfrac{125}{8}$ 59. $\dfrac{4}{9}$
61. 216   63. $\dfrac{9}{25}$ 65. $x^{14/3}$
67. x^2 69. x^2 71. $x^2 y^2$
73. $\dfrac{x}{y}$ 75. $-xy$ 77. $\dfrac{1}{x^3}$
79. x^6 81. Undefined   83. $\dfrac{y^4}{x^2}$

Section 1-4, page 30

1. $-5x^2 + 3x + 2$, quadratic
3. $x^2 + 2x + 6$, quadratic, monic
5. $x^5 - 3x^2 + 5$, monic
7. $2x - 3$, linear, binomial
9. $4x^3 + 7$, cubic, binomial
11. $-6x^8 + x^6 + 1$
13. $5x^2 + 3x + 1$
15. $2x - 7$
17. $4x^2 + 3x - 10$
19. $x^3 + 1$
21. $x^2 - 25$
23. $8x^2 + 24x - 16$
25. $-2x^2 - 3x - 20$
27. $x^2 - 2x + 7$
29. 13
31. $2x^2 + 3$
33. $4x^2 + 12x + 9$
35. q^2
37. $6y - 5$
39. 5
41. w
43. $-2w + 3$
45. 256
47. $2x + 2$
49. $4x + 8y$
51. $2x^2 + 3xy + y - y^2$
53. $x^2 + 3xy + 5x$
55. $4x^2 + 4x - x^2y - xy^2$
57. $6x^2 + 5xy - y^2$
59. $2x^2 + 4xy$
61. $3x - 3y$
63. $b^2 + b - c^2 + c$
65. $6x^2 + 7xy - 20y^2$
67. $4x^2 - xy + 4y^2$
69. $b^2 - 4bc + 10c^2$
71. $2x^2 + 3xy - 6y^2 - 12y$
73. $-12x^3y^6$
75. $24a^2b^3c$

Section 1-5, page 36

1. $3(5bc + 4y)$
3. $4c(b - 2d)$
5. $(b - 5)^2$
7. $(y - 4k)(y + 4k)$
9. $x^2(x - 4)$
11. $(x - 2)(x - 7)$
13. $(3x - 1)(2x + 1)$
15. $(x - 2)(x + 1)$
17. $(c - 2)(c - 5)$
19. $(7c - 8d)(7c + 8d)$
21. $c(b + 6)^2$
23. $y(y - 12)(y + 3)$
25. $5(x - 3)^2$
27. $b(x - 9)(x + 2)$
29. $(2x - 3)(3x + 4)$
31. $4(3x^2 - 2)$
33. $(5x - 1)(3x + 2)$
35. $2(2b - y)(2b + y)$
37. $(x + 1)(x^2 - x + 1)$
39. $(x - 2)(x + 2)(x^2 + 4)$
41. $(3x + 2y)(9x^2 - 6xy + 4y^2)$
43. $(b - c)^2(b + c)^2$

Section 1-6, page 43

1. $\dfrac{x^2}{x-1}$
3. $\dfrac{1}{3}$
5. $\dfrac{x+1}{x-2}$
7. $\dfrac{ac^2}{2b^3}$
9. $\dfrac{x-4}{6}$
11. $\dfrac{3b+1}{5}$
13. $\dfrac{2x+1}{2(5x+3)}$
15. $\dfrac{4xz+3y^2}{yz}$
17. $\dfrac{(b-3)(b+1)}{b+4}$
19. $\dfrac{2(2x+3)}{8x+9}$
21. $\dfrac{-y+5}{y-1}$
23. $\dfrac{(y+4)(y-1)}{(y-3)(y+2)}$
25. $\dfrac{x(x^2-2x-5)}{(x+1)(x+2)}$
27. $\dfrac{2(x^2+5x-5)}{(x+4)(x-5)}$
29. $\dfrac{-9x-2}{x^2-4}$
31. $\dfrac{56xz}{25y^5}$
33. $\dfrac{2(x^2-x+2)}{x-3}$
35. $\dfrac{6x^3}{vw^3}$

Section 1-7, page 49

1. $12x^{3/2} - 3x^{1/2}$
3. $10x^2 - 7x\sqrt{y} - 12y$
5. $x^{2/3} - 4x + 3$
7. x
9. $25x^3 - x^{2/3}$
11. $-2c^{1/2} + c^{3/2}$
13. $7 + 2\sqrt{x} - 6x$
15. $\dfrac{1}{2b^{1/2}c}$
17. $\dfrac{23}{4(5 + \sqrt{2})}$
19. $\dfrac{61}{35 + 12\sqrt{3}}$
21. $\dfrac{x^2 - 4y}{x^2 + 3x\sqrt{y} + 2y}$
23. $\dfrac{5 - x}{5 - 3\sqrt{5x} + 2x}$
25. $2(\sqrt{5} + \sqrt{3})$
27. $\dfrac{c + 2\sqrt{5cd} + 5d}{c - 5d}$
29. $\dfrac{4\sqrt{15}}{15}$
31. $\dfrac{5c(4 + \sqrt{5c})}{16 - 5c}$

Chapter 1 Review, page 50

1. True
2. True
3. False
4. True
5. $\{5,6,7,8,9\}$
6. $\{8\}$
7. $\{4,6,9\}$
8. $\{5,6,7,9\}$
9. $\dfrac{22}{35}$
10. $\dfrac{1}{12}$
11. $\dfrac{8}{5}$
12. 12
13. $\dfrac{81}{16}$
14. $\dfrac{125}{8}$
15. $\dfrac{27}{8}$
16. $\dfrac{9}{4}$
17. 0.0475
18. 1.89×10^5
19. $81x^8$
20. $\dfrac{y^6}{x^3}$
21. $\dfrac{y^4}{x^2}$
22. $(x + 2y)^2$
23. $4x^2 + 5x - 17$
24. $6x^2 + 5x - 4$
25. $3x^2 + 5y^2$
26. $b^2 + 9bc - 7c^2$
27. 24

28. 276   29. $3x + 20$ 30. 71
31. $3x(x - 6)$ 32. $2(b - 3c)(b + 3c)$
33. $(v + 3w)^2$ 34. $(r - 7)(r + 2)$
35. $y(y - 9)(y + 2)$ 36. $(bc - 2)^2$
37. $\dfrac{2x^2 + y^2}{xy}$ 38. $\dfrac{(x + 4)(x + 2)}{(x + 3)(x - 2)}$
39. $\dfrac{3w}{w + 6}$ 40. $\dfrac{x^2 + 33x - 47}{(x - 5)(x + 2)(x + 6)}$
41. $x^4 + 2x^3 - 8x^2$ 42. $2bc^3/3$
43. $11 + 6\sqrt{w} + w$ 44. $7x^{1/3}$
45. $\dfrac{7\sqrt{7} - 17}{3}$ 46. $\dfrac{k + 3\sqrt{3jk} + 6j}{k - 3j}$
47. $\dfrac{1}{12 + 4\sqrt{7}}$ 48. $\dfrac{1}{2(\sqrt{13} + \sqrt{11})}$

27. $-3, 3$ 29. $\dfrac{-3 - \sqrt{17}}{4}, \dfrac{-3 + \sqrt{17}}{4}$
31. $-4 - \sqrt{11}, -4 + \sqrt{11}$ 33. $-6 - \sqrt{43}, -6 + \sqrt{43}$
35. $-\dfrac{17}{4}$ 37. No solutions, inconsistent
39. All real numbers, indentity   41. $-3, 5$
43. $\dfrac{3 - \sqrt{5}}{2}, \dfrac{3 + \sqrt{5}}{2}$ 45. $4 - \sqrt{26}, 4 + \sqrt{26}$
47. $-5, 2$ 49. After 2 sec
51. 16 ft   53. $\left(\dfrac{8}{4.75\pi}\right)^{1/2} = 0.73$ m
55. \$4 million   57. $-\sqrt{3}, \sqrt{3}$
59. $-2, -1, 1, 2$ 61. $-5, 5$
63. $-0.33, -1.23$ 65. $-0.70, -14.78$

Chapter 2

Section 2-1, page 56

1. $2\tfrac{1}{2}$ 3. 7   5. -1
7. No solutions, inconsistent   9. $\dfrac{12}{7}$
11. $\dfrac{26}{7}$ 13. All real numbers, identity
15. $\dfrac{2}{3}$ 17. No solutions, inconsistent
19. -4 21. -3
23. No solutions, inconsistent   25. 2
27. All real numbers, identity   29. $-\dfrac{7}{2}$
31. 409   33. 740   35. 10.68
37. $V = \dfrac{(P_0 - P)b}{P + a}$, $P = \dfrac{P_0 b - aV}{V + b}$
39. 85%   41. 200   43. 42 mph
45. 4.56   47. -8.565

Section 2-2, page 64

1. $-4, 4$ 3. $-3, 2$ 5. $-9, 9$
7. $-1, 10$ 9. $1, 4$ 11. -7
13. $-\dfrac{1}{2}, -\dfrac{1}{3}$ 15. $-10, -5$ 17. $\dfrac{1}{2}$
19. $\dfrac{-5 - \sqrt{13}}{2}, \dfrac{-5 + \sqrt{13}}{2}$ 21. $3 - \sqrt{19}, 3 + \sqrt{19}$
23. No solutions, inconsistent   25. -5

Section 2-3, page 71

1. $-3, 0, 3$ 3. $-5, -2, 0, 1$ 5. $\dfrac{1}{2}$
7. 1   9. No solutions, inconsistent
11. $1, 5$ 13. No solutions, inconsistent
15. $-2, 0, 7$ 17. No solutions, inconsistent
19. $4, -4 + \sqrt{5}, -4 - \sqrt{5}$ 21. 2   23. $-1, 3$
25. $a = S(1 - r)$, $r = 1 - \dfrac{a}{S}$
27. $r = \dfrac{ab}{a + b}$, $a = \dfrac{br}{b - r}$, $b = \dfrac{ar}{a - r}$
29. $a = p - (n - 1)d$, $d = \dfrac{p - a}{n - 1}$, $n = \dfrac{p - a}{d} + 1$
31. $x = -6 \pm \sqrt{30 - y}$, $y = -x^2 - 12x - 6$
33. $v = c\sqrt{1 - \dfrac{m_0^2}{m^2}}$, 2.98×10^8 m/sec
35. $p = \dfrac{RT}{v - b} - \dfrac{a}{v^2}$ 37. $p_2 = p_1 - \dfrac{4vLs}{a^2 - r^2}$
39. $a = \left[\dfrac{8vLQ}{\pi(p_1 - p_2)}\right]^{1/4}$ 41. $-1, -2, 4$
43. $-1, 1$ 45. 7

Section 2-4, page 78

1. $x > 2$ 3. $x \geq 1$ 5. $x < 1$
7. $x \leq \dfrac{3}{13}$ 9. $x < 2$ 11. $x > 2$
13. $2 < x < 3$ 15. $x < 1$ or $2 < x < 3$
17. $-3 \leq x \leq 0$ or $x > 1$ 19. $x < -2$ or $x > 5$
21. $x \leq -2$ or $x \geq 7$

23. $x < -1$ or $2 < x < 3$
25. $-2 < x < 0$ or $x > 7$
27. $-5 < x < 0$ or $0 < x < 5$
29. $x \leq -2$ or $x \geq 2$
31. $200 + 0.05x < 100 + 0.07x$, $x > 5000$
33. $18 + 0.25x > 30 + 0.15x$, $x > 120$
35. $100 < q < 400$ 37. $2 < t < 5$
39. $-4 \leq x \leq -2$ or $3 \leq x$ 41. $-3 \leq x \leq 3$
43. $x < -2.5$ or $x > 1.5$ 45. $-3 < x < \dfrac{17}{3}$

5.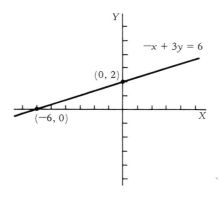

Section 2-5, page 86

1. $20,000,000   3. $7000   5. 300 mi
7. 440 mi   9. $3\frac{1}{3}$ qt   11. 70
13. $10,500   15. 45 hr   17. 5 in.
19. 37°   21. 68°

Section 2-6, page 92

1.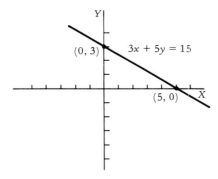

7.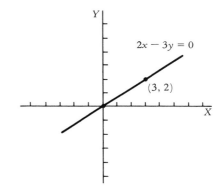

3.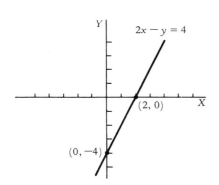

9.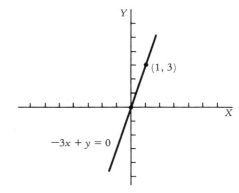

11.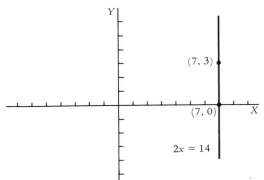

17.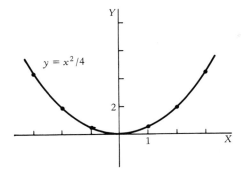

13.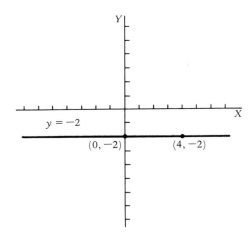

19.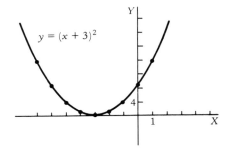

21.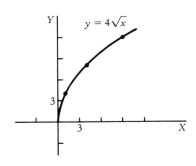

15.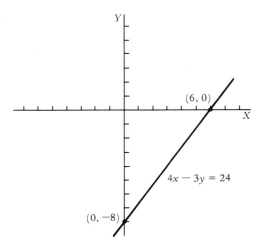

23.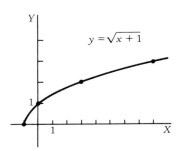

ANSWERS TO ODD-NUMBERED EXERCISES AND REVIEW EXERCISES   571

25.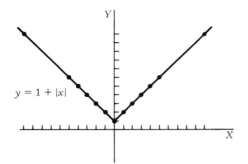

27.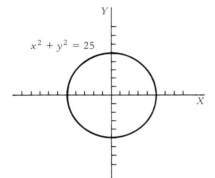

29.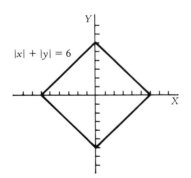

31.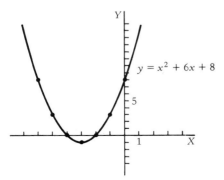

33. a. ∅   b. the whole plane

Chapter 2 Review, page 93

1. 5
2. -2
3. 10
4. All real numbers, identity
5. -1
6. $-8, 8$
7. $\frac{-7}{3}, 1$
8. $-7, 3$
9. -11
10. $-2, 8$
11. $\frac{-7 - \sqrt{5}}{2}, \frac{-7 + \sqrt{5}}{2}$
12. $2 - \sqrt{10}, 2 + \sqrt{10}$
13. No solutions, inconsistent
14. $0, 6$
15. No solutions, inconsistent
16. $-4, 14$
17. $-15, 0, 5$
18. $\frac{1}{3}$
19. 5
20. $-1.8, 2$
21. 9
22. 1
23. $x > -\frac{1}{3}$
24. $x > \frac{17}{2}$
25. $x \le -4$ or $x \ge 8$
26. $x < -4$ or $2 < x < 5$
27. $x \le -1$ or $x > 8$
28. $x \le -4$ or $-2 \le x < 1$
29. 35,000 cu ft
30. \$390,000
31. $> \$80.00$
32. $0 < x \le 500$

33.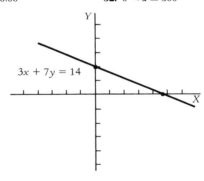

34.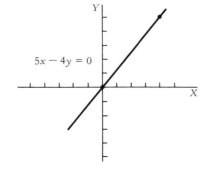

35.

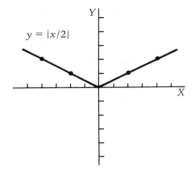
$y = |x/2|$

36.

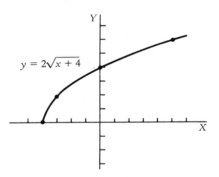
$y = 2\sqrt{x+4}$

Chapter 3

Section 3-1, page 100

1. 2×3 3. 4×3 5. 2×2 7. -4
9. 3 11. 1 13. 6
15. $\begin{bmatrix} 3 & -4 & 10 \\ 11 & 2 & -2 \end{bmatrix}$ 17. $\begin{bmatrix} 8 & 0 \\ 1 & 6 \end{bmatrix}$
19. $\begin{bmatrix} -21 & -12 \\ 3 & 18 \\ -6 & -9 \end{bmatrix}$ 21. $\begin{bmatrix} 29 & 6 \\ 22 & 33 \end{bmatrix}$
23. Undefined 25. $\begin{bmatrix} -2 & -6 & -8 \\ 32 & -18 & 20 \end{bmatrix}$
27. $\begin{bmatrix} -46 & 7 & 5 \\ -18 & -41 & 38 \\ -44 & -12 & 19 \end{bmatrix}$ 29. $\begin{bmatrix} 4 & 2 & 3 & 1 \\ 2 & 3 & 1 & 2 \end{bmatrix}$
31. $\begin{bmatrix} 4 & 2 & 3 & 1 \\ 2 & 3 & 1 & 2 \end{bmatrix} + \begin{bmatrix} 0 & 3 & 0 & 0 \\ 1 & 0 & 0 & 0 \end{bmatrix} = \begin{bmatrix} 4 & 5 & 3 & 1 \\ 3 & 3 & 1 & 2 \end{bmatrix}$
33. $\begin{bmatrix} 1.33 & 1.71 \\ 2.34 & 2.25 \end{bmatrix}$

35. $\begin{matrix} \text{Plant E} & \text{Plant C} & \text{Plant W} \end{matrix}$
$\begin{bmatrix} 1.47 & 1.94 \\ 2.74 & 2.63 \end{bmatrix} \begin{bmatrix} 1.52 & 1.90 \\ 2.64 & 2.52 \end{bmatrix} \begin{bmatrix} 1.48 & 1.90 \\ 2.48 & 2.41 \end{bmatrix}$

Section 3-2, page 114

1. -72 3. Undefined 5. Undefined 7. 4×3
9. 3×3 11. 3×2 13. $\begin{bmatrix} 26 & 5 \\ 22 & 31 \end{bmatrix}$
15. Undefined 17. $\begin{bmatrix} 0 & -5 & 0 \\ 28 & 26 & 7 \\ 12 & -68 & 31 \end{bmatrix}$
19. $\begin{bmatrix} 34 & 75 & 29 \\ 16 & 30 & 26 \\ 18 & 48 & 12 \end{bmatrix}$ 21. $[124 \quad 122 \quad 66]$
23. $\begin{bmatrix} 20 & 0 & 0 & 0 \\ 0 & -18 & 0 & 0 \\ 0 & 0 & -54 & 0 \\ 0 & 0 & 0 & 96 \end{bmatrix}$ 25. $\begin{bmatrix} 17 & -27 \\ 66 & -26 \end{bmatrix}$
27. $\begin{bmatrix} -1 & 3 \\ -7 & 1 \end{bmatrix}$ 29. $\begin{bmatrix} -6 & 7 & -37 \\ -1 & 12 & -26 \\ 19 & 53 & -33 \end{bmatrix}$
31. $PS = [1452.00 \quad 2178.00 \quad 1835.50]$
33. $SP = \begin{bmatrix} 33 & 29 & 10 & 26 \\ 7 & 11 & 6 & 6 \\ 18 & 14 & 4 & 14 \end{bmatrix}$
35. $[D(SP)]C = \$4952.40$
37. $S(PC) = \begin{bmatrix} 245.8 \\ 88.6 \\ 120.4 \end{bmatrix}$
39. Person 1 41. Persons 3, 4
43. Persons 2, 3 45. 4
47. $\begin{bmatrix} 53.1236 & 12.0222 \\ 94.0260 & -41.3892 \end{bmatrix}$
49. $\begin{bmatrix} 70.0097 & 11.4539 & 5.3747 \\ 56.9263 & -37.7081 & -49.0894 \\ -3.4272 & 30.9027 & 13.6707 \end{bmatrix}$

Section 3-3, page 126

1. $x = 6, \quad y = 2$ 3. No solution
5. $a = \dfrac{1}{2}, \quad b = \dfrac{3}{2}$ 7. $x = 2, \quad y = -1$
9. $x = -3, \quad y = 3$ 11. Infinitely many solutions
13. $x = \dfrac{1}{2}, \quad y = 1$ 15. No solution
17. $p = 2.5, \quad q = -2.5$ 19. $x = -1, \quad y = 2$
21. 13 23. \$83,000

25. 26 metal, 11 wood
27. 60,000
29. 100 woolen, 240 cotton
31. 8 Vitaful, 5 Orangerust
33. 35
35. $(-\sqrt{7}, -3), (-\sqrt{7}, 3), (\sqrt{7}, -3), (\sqrt{7}, 3)$
37. $(3, 4), (-4, -3)$
39. $(-\sqrt{2}, 1), (\sqrt{2}, 1)$
41. $x = 2.12, y = 3.17$

Section 3-4, page 134

1. $\begin{bmatrix} 2 & -1 & | & 5 \\ 3 & 2 & | & 6 \end{bmatrix}$

3. $\begin{bmatrix} 2 & 3 & 6 & | & 9 \\ 4 & 3 & 6 & | & 11 \\ 4 & 3 & 6 & | & 11 \end{bmatrix}$

5. $\begin{bmatrix} 1 & -1 & 3 & | & 7 \\ -2 & 0 & 5 & | & 12 \\ 0 & 3 & 2 & | & 9 \\ 4 & 2 & 0 & | & 15 \end{bmatrix}$

7. $\begin{bmatrix} -2 & -4 & 9 & | & 5 \\ 3 & 7 & 2 & | & 11 \\ 4 & -5 & 0 & | & 15 \end{bmatrix}$

9. $4x + y + 2z = -3$
   $3x - 5y + 7z = 5$
   $2y + 6z = 9$

11. $x + 3y + 7z = 15$
    $-9x - 4y + 2z = 11$
    $-6x + 3y + 8z = 2$

13. $\begin{bmatrix} 1 & 3 & 6 \\ 4 & 8 & 12 \\ 6 & 3 & 9 \end{bmatrix}$

15. $\begin{bmatrix} 4 & 8 & 12 \\ 1 & 3 & 6 \\ 2 & 1 & 3 \end{bmatrix}$

17. $\begin{bmatrix} 1 & 3 & 6 \\ 4 & 8 & 12 \\ 5 & 10 & 21 \end{bmatrix}$

19. $\begin{bmatrix} 1 & 3 & 6 \\ 4 & 8 & 12 \\ 0 & -5 & -9 \end{bmatrix}$

Section 3-5, page 142

1. $x = 4, y = 1, z = -2$
3. $x = 0, y = 5, z = 2$
5. Dependent
7. $x = -4, y = 3, z = 6$
9. $x = 4, y = 0, z = -3$
11. $x = -1, y = 2, z = 12$
13. Dependent
15. $x = 3, y = 1, z = 7$
17. 200 bookcases, 100 tables, and 400 armchairs
19. $12,000 in grade AA, $6,000 in grade A, and $9,000 in grade B
21. $x_1 = 2, x_2 = 4, x_3 = -1, x_4 = 7$

Section 3-6, page 152

1. $\begin{bmatrix} \frac{4}{34} & -\frac{3}{34} \\ \frac{2}{34} & \frac{7}{34} \end{bmatrix}$

3. $\begin{bmatrix} -\frac{10}{80} & \frac{8}{80} \\ \frac{5}{80} & \frac{4}{80} \end{bmatrix}$

5. No inverse

7. $\frac{1}{6}\begin{bmatrix} 6 & -8 \\ -3 & 5 \end{bmatrix}$

9. $\begin{bmatrix} -9 & 14 & -6 \\ -2 & 1 & 0 \\ 3 & -3 & 1 \end{bmatrix}$

11. $\begin{bmatrix} -28 & 64 & -31 \\ -16 & 37 & -18 \\ 1 & -2 & 1 \end{bmatrix}$

13. No inverse

15. $\frac{1}{10}\begin{bmatrix} -6 & 8 & 1 \\ 3 & -4 & 2 \\ 1 & 2 & -1 \end{bmatrix}$

17. $D^{-1}C^{-1}B^{-1}A^{-1}$
19. $0.25 D^{-1}B^{-1}$
21. $x = 3, y = 5$
23. $x = 2, y = -2, z = 4$
25. 200 bookcases, 100 tables, 400 armchairs
27. 300 bookcases, no tables, 400 armchairs
29. 190 bookcases, 90 tables, 420 armchairs
31. $A^{-1} = (-3A - 5I)/2$
33. $\begin{bmatrix} 0.54 & -0.26 \\ -0.10 & 0.15 \end{bmatrix}$
35. $\begin{bmatrix} -1.03 & 1.80 \\ 2.32 & 0.59 \end{bmatrix}$

Section 3-7, page 156

1. $4000 of agricultural output and $4000 of manufacturing output.

3. $\begin{bmatrix} \frac{40}{22} & \frac{35}{22} \\ \frac{20}{22} & \frac{45}{22} \end{bmatrix} \begin{bmatrix} 600 \\ 200 \end{bmatrix} = \begin{bmatrix} 1409 \\ 955 \end{bmatrix}$

   A must produce $1409 of output.
   B must produce $ 955 of output.

5. Technological matrix $\begin{bmatrix} 0.3 & 0.4 \\ 0.2 & 0.1 \end{bmatrix}$

   C must produce $1182 of output.
   D must produce $ 818 of output.

Chapter 3 Review, page 157

1. 2×4
2. 4
3. $\begin{bmatrix} 7 & 6 & 6 \\ 0 & 11 & 1 \\ -1 & -2 & 1 \end{bmatrix}$
4. Undefined
5. $\begin{bmatrix} 12 & 6 & 22 \\ 20 & 9 & 37 \end{bmatrix}$
6. $\begin{bmatrix} 30 & 11 & 53 \\ 23 & 10 & 38 \\ 27 & 10 & 49 \end{bmatrix}$
7. Undefined
8. $\begin{bmatrix} 2 & -1 \\ -\frac{3}{4} & \frac{1}{2} \end{bmatrix}$
9. $\begin{bmatrix} 3 & 4 & -2 \\ -3 & -1 & 1 \\ -1 & -2 & 1 \end{bmatrix}$

10. Undefined
11. $x = 6$, $y = 1$
12. $x = 1$, $y = 5$
13. $x = 2$, $y = 7$, $z = 6$
14. $x = 5$, $y = -4$, $z = 2$
15. $x = 72$, $y = -31$, $z = -33$
16. P must produce $43,600 of output, Q must produce $23,200 of output.

17. $\begin{bmatrix} 20 & 30 & 25 \\ 15 & 25 & 35 \\ 10 & 20 & 15 \end{bmatrix} - \begin{bmatrix} 6 & 8 & 12 \\ 0 & 5 & 15 \\ 2 & 6 & 0 \end{bmatrix} = \begin{bmatrix} 14 & 22 & 13 \\ 15 & 20 & 20 \\ 8 & 14 & 15 \end{bmatrix}$

18. $\begin{bmatrix} 100 & 50 & 30 \end{bmatrix} \begin{bmatrix} 20 & 30 & 25 \\ 15 & 25 & 35 \\ 10 & 20 & 15 \end{bmatrix} =$

Downtown East West
$\begin{bmatrix} 3050 & 4850 & 4700 \end{bmatrix}$

Chapter 4

Section 4-1, page 166

1.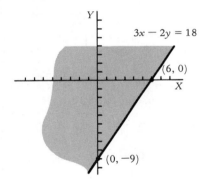

3.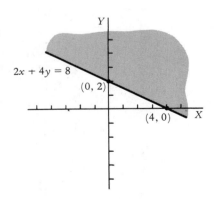

5.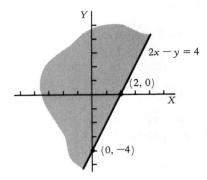

7.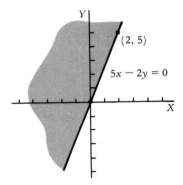

9.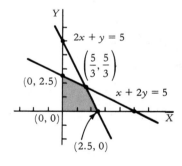

11.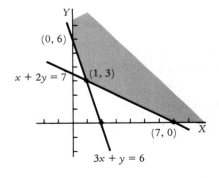

ANSWERS TO ODD-NUMBERED EXERCISES AND REVIEW EXERCISES   575

13.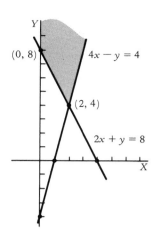

19.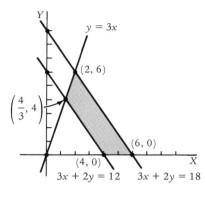

15.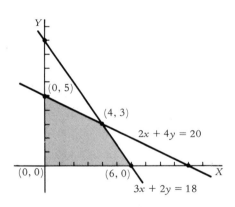

21.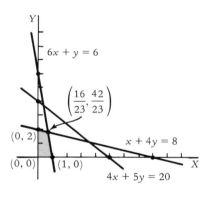

17.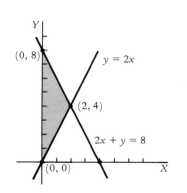

23.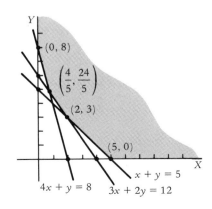

25. Let s = number of standard trucks made in a day.
Let d = number of deluxe trucks made in a day.
$15s + 20d \leq 480$ assembly
$10s + 30d \leq 570$ painting
$s \geq 0, d \geq 0$ nonnegativity

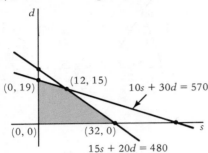

27. Let f = number of Ferritol pills taken daily.
Let v = number of Vitamint pills taken daily.

$3f + v \geq 20$ thiamine
$8f + 6v \geq 60$ iron
$f \geq 0, v \geq 0$ nonnegativity

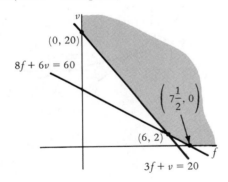

29.

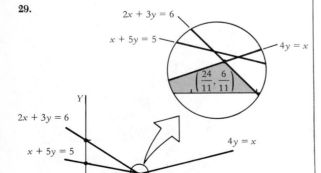

Section 4-2, page 174

1. Yes
3. No. The sides are not parts of lines.
5. Yes
7. No. Not convex.
9. 60 at $(0, 2.5)$ 11. 6 at $(0, 6)$
13. 16 at $(0, 8)$
15. 20 at $(0, 5)$, $(4, 3)$ and all points of the line segment joining them.
17. 18 at $(2, 4)$ 19. $\frac{8}{3}$ at $\left(\frac{4}{3}, 4\right)$
21. 100 at $\left(\frac{16}{23}, \frac{42}{23}\right)$
23. 32 at $\left(\frac{4}{5}, \frac{24}{5}\right)$
25. Make 12 standard toy trucks and 15 deluxe toy trucks each day, for a maximum possible profit of $165.
27. Take 6 Feritol and 2 Vitamint each day, at a cost of 58¢.
29. 18 at $(3, 0)$

Section 4-3, page 180

1. a. 3 magazine racks, 2 bookcases
   b. No
   c. $31
   d. 6 magazine racks, no bookcases
   One plywood board is left over; no time is left over; profit is $36.
3. a. $40
   b. 25 lb of Grow, 10 lb of Bloom
   c. $2.50
5. 18 boxes of Deluxe Mix, 14 boxes of Elegant Mix; revenue is $319; no leftovers.
7. a. $175,000
   b. 350 two-person boats
   c. 450 work-hours of cutting
9. $2304—use 360 pounds of bluegrass and 240 pounds of rye.
11. a. 400   b. $15,600   c. 500, $17,000

Section 4-4, page 194

1.
$$\begin{aligned} 2x + y + u &= 8 \\ x + 4y + v &= 12 \\ F - 5x - 2y &= 0 \end{aligned}$$

$$\begin{array}{cccc} x & y & u & v \\ \left[\begin{array}{cccc|c} \boxed{2} & 1 & 1 & 0 & 8 \\ 1 & 4 & 0 & 1 & 12 \\ -5 & -2 & 0 & 0 & 0 \end{array}\right] \end{array}$$

$x \geq 0, y \geq 0, u \geq 0, v \geq 0$

3.
$$\begin{aligned} x + y + u &= 12 \\ 3x + 2y + v &= 18 \\ 2x + y + w &= 14 \\ F - 2x - 8y &= 0 \end{aligned}$$

$$\begin{array}{ccccc} x & y & u & v & w \\ \left[\begin{array}{ccccc|c} 1 & 1 & 1 & 0 & 0 & 12 \\ 3 & \boxed{2} & 0 & 1 & 0 & 18 \\ 2 & 1 & 0 & 0 & 1 & 14 \\ -2 & -8 & 0 & 0 & 0 & 0 \end{array}\right] \end{array}$$

$x \geq 0, y \geq 0, u \geq 0, v \geq 0, w \geq 0$

5.
$$\begin{aligned} 3x + 2y + 4z + u &= 34 \\ x + 5y + 2z + v &= 42 \\ F - x - 4y - 5z &= 0 \end{aligned}$$

$$\begin{array}{ccccc} x & y & z & u & v \\ \left[\begin{array}{ccccc|c} 3 & 2 & \boxed{4} & 1 & 0 & 34 \\ 1 & 5 & 2 & 0 & 1 & 42 \\ -1 & -4 & -5 & 0 & 0 & 0 \end{array}\right] \end{array}$$

$x \geq 0, y \geq 0, z \geq 0, u \geq 0, v \geq 0$

7.
$$\begin{aligned} x + y + 2z + u &= 12 \\ 3x + 5y + v &= 16 \\ 2x + 3y + 6z + w &= 18 \\ F - 4x - 2y - 7z &= 0 \end{aligned}$$

$x \geq 0, y \geq 0, z \geq 0, u \geq 0, v \geq 0, w \geq 0$

$$\begin{array}{ccccc} x & y & z & u & v & w \\ \left[\begin{array}{cccccc|c} 1 & 1 & 2 & 1 & 0 & 0 & 12 \\ 3 & 5 & 0 & 0 & 1 & 0 & 16 \\ 2 & 3 & \boxed{6} & 0 & 0 & 1 & 18 \\ -4 & -2 & -7 & 0 & 0 & 0 & 0 \end{array}\right] \end{array}$$

9. 60 at $(0, 2.5)$

11. 5 at $(1, 2)$

13. 72 at $(12, 0, 0)$

15. 40 at $(0, 0, 8)$

17. 8.4 at $\left(\dfrac{26}{5}, \dfrac{8}{5}, 0\right)$

19. $\dfrac{168}{11}$ at $\left(0, \dfrac{32}{11}, \dfrac{36}{11}\right)$

21. a. $170

b. $\dfrac{5}{2}$ stools, $\dfrac{20}{3}$ chairs, $\dfrac{5}{3}$ coffee tables. A fraction means that the item is started one day and finished the next day.

c. No

23. 20 acres of corn, 40 acres of lettuce, no tomatoes, 100 hours at planting time, 100 hours at harvest time.

25. 185 bass, 25 trout, revenue is $1420

27. 6 cars, 12 vans, and 18 minibuses; capacity is 240 passengers

Section 4-5, page 207

1. Maximize $8p + 4q$

subject to $3p + q \leq 2$

$p + 2q \leq 5$

$p \geq 0, \quad q \geq 0$

3. Maximize $12p + 16q$

subject to $2p + q \leq 1$

$p + 3q \leq 3$

$4p + 2q \leq 2$

$p \geq 0, \quad q \geq 0$

5. Maximize $12p + 14q + 8r$

subject to $2p + q + r \leq 3$

$p + 3q + r \leq 2$

$p \geq 0, \quad q \geq 0, \quad r \geq 0$

7. Maximize $11p + 9q + 6r$

subject to $p + 2q + r \leq 1$

$5p + 3q + r \leq 2$

$q + r \leq 5$

$p \geq 0, \quad q \geq 0, \quad r \geq 0$

9. 6 at $(0, 6)$

11. 32 at $\left(\dfrac{4}{5}, \dfrac{24}{5}\right)$

13. 20 at $(0, 10)$

15. $\dfrac{32}{3}$ at $\left(\dfrac{4}{3}, 0, \dfrac{10}{3}\right)$

17. 6 at $(0, 6, 0)$

19. $\dfrac{145}{17}$ at $\left(\dfrac{20}{17}, 0, \dfrac{25}{17}\right)$

21. 8 at $(0, 8, 0)$

23. a. No seeds or nuts, $6\tfrac{2}{3}$ pounds of grain

b. $66\frac{2}{3}$¢

c. No more vitamins, $4\frac{1}{3}$ more ounces of protein, $1666\frac{2}{3}$ more calories

25. 220 acres of corn, 440 acres of soybeans, 340 acres of tomatoes, cost is $17,300.

Chapter 4 Review, page 210

1. 44 attained at (2, 10)
2. 72 attained at any point of the line segment joining (4, 3) with (0, 6).
3. 20 pairs of shoes and 50 pairs of boots; revenue is $5990; all available labor is used.
4. 95 pairs of shoes and no boots; revenue is $6650; 25 hours of unskilled labor are not used.
5. 35 attained at (5, 0)
6. $\frac{64}{3}$ attained at $\left(\frac{16}{3}, 0, \frac{4}{3}\right)$
7. 16 attained at (0, 8, 0)
8. 300 at $\left(\frac{4}{3}, 1, \frac{4}{3}\right)$
9. A: 10 tons, B: 40 tons, C: 15 tons; revenue is $5100. All facilities are used to capacity.

Chapter 5

Section 5-1, page 219

1. $787.50
3. $54.00
5. $12.50
7. $844.87
9. $1772.95
11. $1516.44
13. 5.095%
15. 10.250%
17. 9 years
19. $9\frac{1}{2}$ years
21. 16%
23. $200, $800
25. 4.4×10^7, 1.4×10^8
27. Northern State Bank, $187.01

Section 5-2, page 226

1. $3450.44
3. $24,597.29
5. $1747.22
7. $13,480.51
9. $123.36
11. $4620.41
13. $14,486.56
15. $259,056.52
17. $77,380.98
19. $129,528.26
21. $123.36
23. $126.83
25. $18,098
27. $17,046
29. $4855
31. $3316
33. $1363.15
35. $328.71
37. $252.03
39. $1159.04
41. $414.61
43. $456.79
45. $4117.47
47. $5163.80
49. $2057.64
51. $1,670,925
53. $340,589
55. $639.61
57. $305.48
59. $243,930
61. $22,509.23
63. a. $136,183.02
    b. $110,446.90, $25,736.15
    c. $1223.81

Section 5-3, page 233

1. $16,098.66
3. $15,327.85
5. $10,257.65
7. $8820.36
9. $11,337.86
11. $7673.91
13. $932.96
15. $706.83
17. $1840.02
19. $1536.14
21. $98,963.87
23. $66,658.55
25. $218.89
27. $206.24
29. $68,636.34
31. $53,326.84
33. $248.85
35. $278.31
37. $270.16
39. $284.10
41. $63.74
43. $70.58
45. $143.10
47. $155.99
49. $120,342
51. $234,337
53. a. $626.53
    b. $559.28
    c. $504.51
55. Sam Miller Sports Cars—payments are $242.66 versus $245.83 at Square Deal Cars
57. $7026.81

Section 5-4, page 240

1. $205.63
3. $1539.01
5. $1264.44
7. $1214.43
9. $858.20
11. $362.61
13. $678.25
15. $1487.07
17.

Period	Outstanding Principal at Start of Period	Payment	Interest	Paid	Outstanding Principal at End of Period
1	2500.00	728.21	350.00	378.21	2121.79
2	2121.79	728.21	297.05	431.16	1690.63
3	1690.63	728.21	236.69	491.52	1199.11
4	1199.11	728.21	167.88	560.33	638.78
5	638.78	728.21	89.43	638.78	0.00

19.

Period	Outstanding Principal at Start of Period	Payment	Interest	Principal Paid	Outstanding Principal at End of Period
1	1000.00	90.54	13.00	77.54	922.46
2	922.46	90.54	11.99	78.55	843.91
3	843.91	90.54	10.97	79.57	764.34
4	764.34	90.54	9.94	80.60	683.74
5	683.74	90.54	8.89	81.65	602.09
6	602.09	90.54	7.83	82.71	519.38
7	519.38	90.54	6.75	83.79	435.59
8	435.59	90.54	5.66	84.88	350.71
9	350.71	90.54	4.56	85.98	264.73
10	264.73	90.54	3.44	87.10	177.63
11	177.63	90.54	2.31	88.23	89.40
12	89.40	90.56	1.16	89.40	0.00

Chapter 5 Review, page 242

1. $27.50 **2.** $120.00 **3.** $1482.05
4. $31.62 **5.** 12.683% **6.** $11\frac{1}{2}$ years
7. $1675.15 **8.** $21,945.80 **9.** $10,289
10. $91,640 **11.** $25,726.05 **12.** $393.78
13. $12,037.04 **14.** $7186.30 **15.** $249.62
16. $549.26 **17.** $280.00 **18.** $253.65
19. $4423.80, after 47 payments **20.** $345,145
21. $2500 now; the other option has a present value of $2369
22. $74.86, $633.52, $147.48
23. $822.89, $57,355.94, $573.56, $249.33

Chapter 6

Section 6-1, page 248

1. .35 **3.** 1 **5.** .30
7. 0 **9.** $\frac{13}{52}$ **11.** $\frac{4}{52}$
13. $\frac{8}{52}$ **15.** $\frac{19}{52}$ **17.** $\frac{18}{40}$
19. $\frac{15}{40}$ **21.** $\frac{33}{40}$ **23.** 0

25. 35 **27.** 140 **29.** $\frac{1}{100}, 230$
31. $\frac{900}{1200} = .75$ **33.** $\frac{600}{1200} = .5$ **35.** $\frac{1200}{2200}$
37. $\frac{2}{2200}$ **39.** $\frac{5000}{125,000} = .04$ **41.** $\frac{107}{5000} = .021$
43. $\frac{3412}{3442} = .991$ **45.** $\frac{721}{1066} = .676$

Section 6-2, page 253

1. $\{(x,a),(x,b),(x,x),(z,a),(z,b),(z,x)\}$
3. $\{(1,a,a),(1,a,1),(1,a,c),(1,c,a),(1,c,1),(1,c,c),$
$(2,a,a),(2,a,1),(2,a,c),(2,c,a),(2,c,1),(2,c,c)\}$
5. $10 \cdot 12 = 120$ **7.** $\frac{1 \cdot 26 \cdot 26}{26^3} = \frac{1}{26}$
9. $\frac{5^4}{26^4} = .00137$ **11.** $\frac{5 \cdot 5 \cdot 5}{10 \cdot 10 \cdot 10} = \frac{1}{8}$
13. $\frac{3}{1000}$ (The numerals satisfying the condition are 998, 989, 899.)
15. $\frac{9^3}{10^3} = .729$ **17.** $\frac{3^4}{10^4} = \frac{81}{10,000} = .0081$
19. $5 \cdot 3 \cdot 4 = 60$ **21.** $3 \cdot 5 \cdot 3 = 45$
23. $\frac{2}{2 \cdot 26 \cdot 26 \cdot 26} = .000057$ **25.** $2 \cdot 3 \cdot 5 \cdot 3 = 90$
27. 2^n **29.** 3,037,606

Section 6-3, page 259

1. II, III, IV **3.** II, III **5.** III, IV **7.** IV
9. 0 **11.** 8 **13.** 20
15.

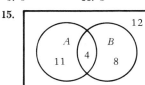

a. 20 b. 23 c. 31
d. 23 e. 12 f. 12

17.

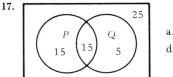

a. 15 b. 45 c. 25
d. 45 e. 40 f. 35

580 ANSWERS TO ODD-NUMBERED EXERCISES AND REVIEW EXERCISES

19.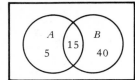
a. 5   b. 20
c. 40   d. 45

21.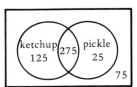
a. 275   b. 125
c. 25   d. 425

23. 22   25. 18   27. 94

Section 6-4, page 266

1. $1 - \dfrac{25^3}{26^3} = .111$ 3. $1 - \dfrac{5^5}{10^5} = \dfrac{31}{32}$
5. $P(A \cup B \cup C \cup D) = .88$,
   $P[(A \cup C \cup D)'] = 1 - .81 = .19$
7.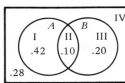
$P(A \cap B') = .42$
$P(A \cup B') = .42 + .10 + .28 = .80$
$P(A' \cup B) = .10 + .20 + .28 = .58$

9.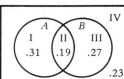
$P(A \cap B) = .19$
$P(A' \cap B) = .27$
$P(A' \cup B) = .19 + .27 + .23 = .69$

11.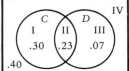
$P(C) = .30 + .23 = .53$
$P(D) = .23 + .07 = .30$
$P(C' \cup D) = .23 + .07 + .40 = .70$

13. a. .49   b. .87   c. .74   d. .87
15. .1

17. .05
19. .09
21. .12

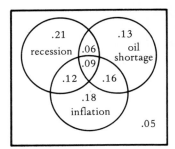

Section 6-5, page 271

1. No   3. Yes   5. No   7. .10, .60, .10
9. $\left(\dfrac{1}{2}\right)^5 = \dfrac{1}{32}$ 11. $\left(\dfrac{5}{26}\right)^3 = .0071$, $\left(\dfrac{21}{26}\right)^3 = .53$
13. $\left(\dfrac{13}{52}\right) \cdot \left(\dfrac{13}{52}\right) = \dfrac{1}{16}$ 15. $\left(\dfrac{4}{52}\right) \cdot \left(\dfrac{4}{52}\right) = \dfrac{1}{169}$
17. $\left(\dfrac{4}{52}\right) \cdot \left(\dfrac{4}{52}\right) = \dfrac{1}{169}$ 19. $\left(\dfrac{8}{20}\right)^2 = \dfrac{4}{25}$
21. $\left(\dfrac{15}{20}\right)^2 = \dfrac{9}{16}$ 23. $(.7) \cdot (.2) = .14$
25. $1 - (.7)(.25) = .825$ 27. $(.7)^2 = .49$
29. $[(.7) \cdot (.25)]^2 = .031$
31. a. $(.8)^5 = .328$ b. $1 - (.2)^5 = .99968$

Section 6-6, page 279

1. $\dfrac{13}{39} = \dfrac{1}{3}$ 3. $\dfrac{1}{13}$ 5. $\dfrac{4}{12} = \dfrac{1}{3}$
7. $\dfrac{.2}{.6} = \dfrac{1}{3}$
9.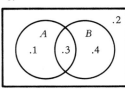
$P(A|B) = \dfrac{.3}{.7} = \dfrac{3}{7}$
$P(B|A) = \dfrac{.3}{.4} = \dfrac{3}{4}$
$P(A'|A' \cup B) = \dfrac{.6}{.9} = \dfrac{2}{3}$
$P(A|A \cup B') = \dfrac{.4}{.6} = \dfrac{2}{3}$

11.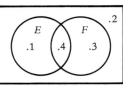
$P(E|F) = \dfrac{.4}{.7} = \dfrac{4}{7}$
$P(F|E) = \dfrac{.4}{.5} = \dfrac{4}{5}$
$P(F'|E \cup F) = \dfrac{.1}{.8} = \dfrac{1}{8}$
$P(E \cup F|F') = \dfrac{.1}{.3} = \dfrac{1}{3}$

13. $\left(\dfrac{9}{20}\right) \cdot \left(\dfrac{8}{19}\right) \cdot \left(\dfrac{7}{18}\right) \cdot \left(\dfrac{6}{17}\right) = .026$

15. $\left(\dfrac{3}{20}\right) \cdot \left(\dfrac{3}{20}\right) \cdot \left(\dfrac{3}{20}\right) = .0034$

17. $\left(\dfrac{8}{20}\right) \cdot \left(\dfrac{12}{19}\right) \cdot \left(\dfrac{11}{18}\right) \cdot \left(\dfrac{10}{17}\right) = .091$

19. $\dfrac{3}{11}$

21. $\dfrac{500}{4180} = .120$ 23. $\dfrac{.57}{.70} = \dfrac{57}{70}$

25. $\dfrac{.49}{.70} = \dfrac{49}{70}$ 27. $(.23) \cdot (.008) = .00184$

29. $\dfrac{(.23)(1 - .008) + (.06)(1 - .014)}{.23 + .06} = .991$

Section 6-7, page 284

1. $\dfrac{30}{64}, \dfrac{32}{64}, \dfrac{2}{64}$ 3. $\dfrac{7}{49}, \dfrac{3}{49}, \dfrac{24}{49}, \dfrac{15}{49}$

5. $\dfrac{\left(\dfrac{1}{16}\right) \cdot 1}{\left(\dfrac{1}{16}\right) \cdot 1 + \left(\dfrac{15}{16}\right) \cdot \left(\dfrac{1}{32}\right)} = \dfrac{32}{47}$

7. $\dfrac{5}{19}$ 9. $\dfrac{10}{19}$ 11. $\dfrac{4}{19}$ 13. $\dfrac{2}{5}$ 15. 0

17. a. $\dfrac{(.70) \cdot (.85)}{(.70) \cdot (.85) + (.30) \cdot (.75)} = .726$

   b. $\dfrac{(.70) \cdot (.15)}{(.70) \cdot (.15) + (.30) \cdot (.25)} = .583$

19. a. $\dfrac{(.35) \cdot (.98)}{(.50) \cdot (.99) + (.35) \cdot (.98) + (.15) \cdot (.90)} = .353$

   b. $\dfrac{(.15) \cdot (.10)}{(.50) \cdot (.01) + (.35) \cdot (.02) + (.15) \cdot (.10)} = .556$

   c. $(.50) \cdot (.01) + (.35) \cdot (.02) + (.15) \cdot (.10) = .027$
   d. $1 - .556 = .444$

21. a. $\dfrac{(.4) \cdot (.9)}{(.4) \cdot (.9) + (.6) \cdot (.2)} = .75$

   b. $\dfrac{(.4) \cdot (.1)}{(.4) \cdot (.1) + (.6) \cdot (.8)} = \dfrac{1}{13}$

23. a. .206 b. .272 c. .522

Chapter 6 Review, page 287

1. $\dfrac{1}{200}, \dfrac{199}{200}$ 2. 700 3. .6

4. .65 5. .45 6. $\dfrac{.2}{.25} = \dfrac{4}{5}$

7. .55 8. $\dfrac{64}{125}$ 9. $\dfrac{61}{125}$

10. $\dfrac{1}{125}$ 11. Dependent 12. Independent

13. $\dfrac{2}{3}$ 14. $\dfrac{3}{8}$ 15. $\dfrac{1}{13}$

16. $\dfrac{16}{52}$ 17. $\dfrac{4}{44}$ 18. $\dfrac{4}{19}$

19. a. $\dfrac{1}{16}$ b. $\dfrac{1}{17}$ 20. a. $\dfrac{9}{16}$ b. $\dfrac{19}{34}$

21. a. $\dfrac{7}{16}$ b. $\dfrac{15}{34}$ 22. a. $\left(\dfrac{1}{52}\right)^2$ b. 0

23. a. $\dfrac{1}{4}$ b. $\dfrac{13}{51}$ 24. a. $\dfrac{1}{4}$ b. $\dfrac{12}{51}$

25. a. $\dfrac{3}{4}$ b. $\dfrac{39}{51}$ 26. $\dfrac{10}{39}$

27. $\dfrac{140}{235} = .60$, $\dfrac{5}{65} = .077$

Chapter 7

Section 7-1, page 293

1. 120 3. 90 5. $\dfrac{1}{7}$ 7. 210

9. $2n(2n - 1) = 4n^2 - 2n$ 11. $2n + 3$

13. $8! = 40320$ 15. $10 \cdot 9 \cdot 8 = 720$

17. $7! = 5040$ 19. $\dfrac{1}{4!} = \dfrac{1}{24}$

21. $\dfrac{13 \cdot 12 \cdot 11}{39 \cdot 38 \cdot 37} = .031$ 23. $25 \cdot 24 \cdot 23 = 13{,}800$

25. $\dfrac{15 \cdot 14 \cdot 13}{25 \cdot 24 \cdot 23} = .198$ 27. $\dfrac{5 \cdot 4 \cdot 3}{15 \cdot 14 \cdot 13} = .022$

29. $\dfrac{5 \cdot 4 \cdot 3 \cdot 2}{10 \cdot 9 \cdot 8 \cdot 7} = \dfrac{1}{42}$ 31. $2 \cdot 5! = 240$

33. $4 \cdot 3 \cdot 4! = 288$

Section 7-2, page 299

1. $C(31, 3) = 4495$ 3. $C(150, 4) = 20{,}260{,}275$

5. $\dfrac{C(6,2) \cdot C(14,3)}{C(20,5)} = .352$

7. $\dfrac{C(4,2) \cdot C(48,3)}{C(52,5)} = 0.40$

9. $\dfrac{C(4,3)\cdot C(44,2)}{C(48,5)} = .0022$

11. $\dfrac{3}{52}$

13. $\dfrac{C(13,4)\cdot C(13,3)^3}{C(52,13)} = .026$

15. $\dfrac{C(9,3)\cdot C(6,2) + C(9,4)\cdot C(6,1) + C(9,5)}{C(15,5)}$
$= \dfrac{102}{143} = .713$

17. a. $\dfrac{C(7,3)\cdot C(8,6)}{C(25,9)} = .00048$

b. $\dfrac{C(7,3)\cdot C(10,3)\cdot C(8,3)}{C(25,9)} = .115$

c. $\dfrac{C(10,4)\cdot C(7,5)}{C(17,9)} = .18$

d. $\dfrac{10}{C(25,9)} = 4.89 \times 10^{-6}$

19. a. 1
b. 1

c. $\dfrac{C(18,4) + C(18,3)\cdot C(2,1)}{C(20,4)} = .968$

d. $\dfrac{C(17,3)\cdot C(3,1) + C(17,4)}{C(20,4)} = .912$

e. $\dfrac{C(16,4) + C(16,3)\cdot C(4,1)}{C(20,4)} = .838$

21. $\dfrac{4!}{2!2!} = 6$ 23. $\dfrac{4!}{2!} = 12$

25. $\dfrac{9!}{4!3!2!} = 1260$ 27. $\dfrac{11!}{4!4!2!} = 34650$

Section 7-3, page 304

1. a. $C(10,6)(.2)^6(.8)^4 = .0055$
b. $C(10,9)(.2)^9(.8) + .2^{10} = 4.20 \times 10^{-6}$
c. 0
d. $C(10,7)(.2)^7(.8)^3 + C(10,8)(.2)^8(.8)^2 = 8.6 \times 10^{-4}$

3. $1 - (.8)^5 = .672$

5. a. $(.6)^{20} = 3.66 \times 10^{-5}$
b. $(.4)^{20} = 1.10 \times 10^{-8}$
c. $C(20,10)(.6)^{10}(.4)^{10} = .12$

d. $C(20,14)(.6)^{14}(.4)^6 = .124$
e. $1 - (.4)^{20} - C(20,1)(.6)(.4)^{19} = .9999997$
f. $C(20,17)(.6)^{17}(.4)^3 + C(20,18)(.6)^{18}(.4)^2$
$+ C(20,19)(.6)^{19}(.4) + (.6)^{20} = .016$
g. $C(20,3)(.6)^3(.4)^{17} + C(20,2)(.6)^2(.4)^{18} +$
$C(20,1)(.6)(.4)^{19} + (.4)^{20} = 4.7 \times 10^{-5}$
h. $(.4)^{20} + C(20,1)(.6)(.4)^{19} = 3.4 \times 10^{-7}$

7. a. $C(10,9)(.6)^9(.4) + (.6)^{10} = .046$
b. $1 - (1 - .046)^{20} = .61$

Section 7-4, page 309

1. $\dfrac{13C(38,3)}{C(52,5)} = .042$

3. $\dfrac{C(13,3)\cdot C(39,2) + 39C(13,4) + C(13,5)}{C(52,5)} = .093$

5. $1 - \dfrac{C(26,5)}{C(39,5)} - \dfrac{13C(26,4)}{C(39,5)} = .548$

7. $\dfrac{3}{20}$ 9. $3\left(\dfrac{1}{20}\right)\cdot\left(\dfrac{1}{19}\right) = .0079$

11. $\dfrac{1}{20}$ 13. $1 - .9^{12} = .72$

15. $\dfrac{C(7,2)\cdot C(18,2)\cdot 5}{C(30,5)} = .113$

17. $1 - \left(\dfrac{12}{30}\right)^6 = .996$

19. $\left(\dfrac{5}{28}\right)\cdot\left(\dfrac{4}{27}\right)\cdot\left(\dfrac{3}{26}\right) = .00305$

21. $\dfrac{C(18,2)\cdot C(7,2)}{C(25,4)} = .254$

Chapter 7 Review, page 310

1. 210 2. 380
3. $n(n-1)(n-2)$ 4. 560
5. $(2n+4)(2n+3)$ 6. $6! = 720$
7. $\dfrac{1}{24}$ 8. $P(15,5) = 360,360$
9. $C(20,3) = 1140$
10. $\dfrac{C(10,2)\cdot C(15,3)}{C(25,5)} = \dfrac{195}{506} = .39$

ANSWERS TO ODD-NUMBERED EXERCISES AND REVIEW EXERCISES 583

11. $.8^{10} = .11$
12. $.2^{10} = .0000001$
13. $1 - .2^{10}$
14. $C(10,4)(.8)^4(.2)^6 = .0055$
15. $\dfrac{C(40,10)}{C(50,10)} = .083$
16. $\dfrac{C(10,10)}{C(50,10)} = 9.7 \times 10^{-11}$
17. $1 - 9.7 \times 10^{-11}$
18. $\dfrac{C(40,4) \cdot C(10,6)}{C(50,10)} = .0019$

17. No, $\begin{bmatrix} 4 & -3 & -10 & -2 \\ 1 & 0 & 3 & -1 \end{bmatrix}$ is an example.

Chapter 8

Section 8-1, page 317

1. 4.5
3. -0.55
5. $15\frac{1}{6}$
7. $\$0.50$
9. $\$5, \$250$
11. $4:1, \$5$
13. $1:1$
15. $\dfrac{5}{7} = \dfrac{P(E)}{1 - P(E)}$, $P(E) = 5/12$
17. 11.58 minutes
19. $0.005 \cdot (5{,}000{,}000 - 30{,}000) + 0.995 \cdot (-30{,}000) = -\$5{,}000, \$5{,}000$
21. $(\$1{,}530{,}000{,}000) \cdot (.025) = \$38{,}250{,}000$
23. 700

Section 8-2, page 321

1. Buy the land.
5. Either 2 or 3. The expected revenue is $3800 in either case.
7. Expand.
9. Small, established companies
11. Yes, his expected value otherwise is $190,000.

Section 8-3, page 326

1. 4
3. 6
5. Not strictly determined
7. $(1,2), 2$
9. $(2,2), 0$
11. $(1,2), (1,4), (3,2),$ and $(3,4), -1$
13. General Alexander attacks both north and south. General Khan places his reserves in the north.
15. a. None
   b. None on optimum strategies, value is doubled.

Section 8-4, page 333

1. row 1: .9, row 2: .1, column 1: .4, column 2: .6; value is 1.4
3. row 1: $\dfrac{1}{13}$, row 2: $\dfrac{12}{13}$, column 1: $\dfrac{6}{13}$, column 2: $\dfrac{7}{13}$; value is $\dfrac{19}{13}$
5. Strictly determined, saddle point is $(3,2)$; value is -1
7. row 1: $\dfrac{1}{12}$, row 2: $\dfrac{11}{12}$, row 3: 0, column 1: $\dfrac{5}{6}$, column 2: $\dfrac{1}{6}$; value is $6\frac{1}{3}$.
9. Strictly determined, saddle point is $(3,3)$; value is 2.
11. row 1: 0, row 2: .6, row 3: 0, row 4: .4, column 1: .4, column 2: 0, column 3: .6, column 4: 0; value is 5.2.
13. Each player should show heads 50% of the time and tails 50% of the time. Value is 0.
15. Gnome: radio $\dfrac{7}{12}$, television $\dfrac{5}{12}$. Oriental: radio $\dfrac{5}{12}$, television $\dfrac{7}{12}$. Value is $-\dfrac{11}{12}$% to Gnome.
17. No effect on strategies; value is doubled.

Chapter 8 Review, page 335

1. 4.7
2. $\dfrac{76}{6}$
3. $3:2$
4. $\dfrac{2}{7}$
5. $-\$30, \$30$
6. The second one
7. row 1: $\dfrac{7}{8}$, row 2: $\dfrac{1}{8}$, column 1: $\dfrac{3}{4}$, column 2: $\dfrac{1}{4}$, value is $\dfrac{13}{4}$
8. row 1: $\dfrac{2}{7}$, row 2: $\dfrac{5}{7}$, column 1: $\dfrac{4}{7}$, column 2: $\dfrac{3}{7}$, value is $\dfrac{1}{7}$
9. Strictly determined, saddle point is $(3,1)$; value is 1.
10. row 1: $\dfrac{2}{7}$, row 2: 0, row 3: $\dfrac{5}{7}$, column 1: 0, column 2: $\dfrac{3}{7}$, column 3: $\dfrac{4}{7}$, value is $-\dfrac{6}{7}$.
11. row 1: 0, row 2: $\dfrac{1}{2}$, row 3: $\dfrac{1}{2}$, column 1: $\dfrac{1}{6}$, column 2: 0, column 3: 0, column 4: $\dfrac{5}{6}$; value is $-\dfrac{3}{2}$.
12. Strictly determined, saddle point is $(4,1)$; value is 0.
13. row 1: $\dfrac{6}{13}$, row 2: $\dfrac{7}{13}$, column 1: $\dfrac{5}{13}$, column 2: $\dfrac{8}{13}$; Democrat gains $\dfrac{9}{13}$%.

Chapter 9

Section 9-1, page 343

1. Function; domain $\{-1,4,3\}$; range $\{4,-1,0\}$
3. Not a function; domain $\{4,6,7\}$; range $\{2,3,6,8\}$
5. Function; domain $\{0,1,2,3\}$; range $\{0,1,2\}$
7. $\{s\,|\,s>0\}$ 9. $\{v\,|\,v\neq 2\}$
11. $\{y\,|\,-2\leq y\leq 2\}$
13. All real numbers   15. $\{t\,|\,t<-5 \text{ or } t>1\}$
17. $\frac{1}{v^2}+\frac{2}{v}+3$ 19. $\frac{4}{v}-5$ 21. $4v^2+8v+7$
23. y^2+2y+3 25. $9x^2+6x+3$ 27. $16t-25$
29. $8v-1$ 31. $\frac{1}{2x-7}$ 33. 51
35. 18 37. $y=8-\frac{2x}{3}$ 39. No
41. $y=\frac{9}{x^2}$ 43. No   45. $y=(x^3-9)^{1/3}$
47. $y=x^{2/3}-x$ 49. No
51. a. 1000   b. 1600   c. 5800   d. 14.5
    e. 1280   f. 16   g. 16   h. -600
    i. -6 j. 3000   k. 3   l. 250
53. a. 120,000   b. 130,000   c. 200,000   d. 500
    e. 30,400   f. 380   g. 400   h. $-100,000$
    i. -1000 j. 980,000   k. 980   l. 300
55. a. 500   b. 220   c. (30, 290)
57. a. 20,000   b. 7600   c. (50, 7500)
59. a. 11,000   b. 200   c. (20, 600)
61. a. $c=.03n+15.00$ b. $30.00   c. 1500 sheets
    d. $.18   e. $.045   f. $.03

Section 9-2, page 352

1. $\frac{6}{4}$ 3. 0   5. $-\frac{3}{2}$
7. 0   9. Undefined   11. $-\frac{1}{7}$
13. a. $y-8=-(x-2)$ b. $y=-x+10$
    c. $x+y=10$
15. a. $y-5=3(x+1)$ b. $y=3x+8$
    c. $3x-y=-8$
17. a. $y-4=0$ b. $y=4$
    c. $y=4$
19. a. $y+6=2(x+3)$ b. $y=2x$
    c. $2x-y=0$
21. a. $y-7=-2(x-3)$ b. $y=-2x+13$
    c. $2x+y=13$
23. a. $y-4=3(x-1)$ b. $y=3x+1$
    c. $3x-y=-1$
25. a. $y-1=-(x+1)$ b. $y=-x$
    c. $x+y=0$
27. a. $y=3x$ b. $y=3x$
    c. $3x-y=0$
29. a. $y=2(x-4)$ b. $y=2x-8$
    c. $-2x+y=-8$
31. a. $y-3=x+3$ b. $y=x+6$
    c. $x-y=-6$
33. $y=\left(-\frac{2}{3}\right)x+2, \quad m=-\frac{2}{3}, \quad b=2$
35. $y=\left(\frac{1}{5}\right)x-1, \quad m=\frac{1}{5}, \quad b=-1$
37. No slope; no y-intercept
39. $y=5x-7, \quad m=5, \quad b=-7$
41. $y=-4x-1, \quad m=-4, \quad b=-1$
43. $b=4.00+2.50x$ 45. $C=500+150q$
47. $14\frac{1}{3}$ million   49. 100
51. $p=214-9t, \quad 0\leq t\leq 14$ 53. 6:40 P.M.
55. $450,000   57. $6000
59. $B=6000-1500t$ 61. $1800
63. $600

Section 9-3, page 359

1. $y=x^2$

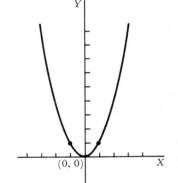

3. $y = x^2 - 2x + 1$

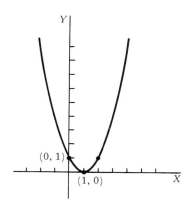

5. $y = -x^2$

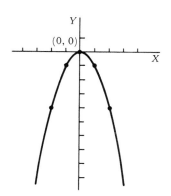

7. $y = -x^2 + 2x$

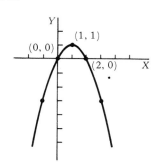

9. $y = x^2 - x - 2$

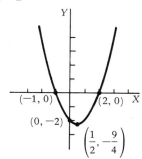

11. $y = 2x^2 - 3x - 2$

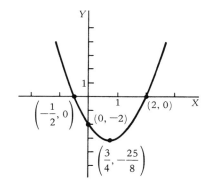

13. $3400

15. a. $AR(q) = 150 - 3q$ b. 25
c. $C(q) = 30q + 500$ d. $120q - 3q^2 - 500$
e. $700 f. 20

17. $(30, 160)$ **19.** $450,000 **21.** 8 at $w = \pm\sqrt{2}$

23. None; functions $f(x) = kx$ satisfy the condition but are not quadratic.

25. 2.83 **27.** -119.8

Section 9-4, page 364

1. $7 + 3w$ **3.** 66 **5.** $w^{5/2} + 2w^{3/2}$
7. $5w^{-1/2} - w^{1/2}$ **9.** $22 - 4w$
11. $4w^2 + 8w + 2$ **13.** $4w^{1/2} + 2$
15. $5 - w^{1/2}$ **17.** 20
19. 7 **21.** 30 **23.** 2
25. a. $C(q) = -.025q^2 + 12.5q + 300$
b. $-.025q + 12.5 + \dfrac{300}{q}$

c. $P(q) = -.275q^2 + 27.5q - 300$
d. $387.50   e. $7.75   f. +$82.50
g. +$27.50   h. -$110.00

27. a. +$45.00   b. $80.00   c. +$35.00

29. $f(x) = x$, $f(x) = b - x$, b is any real number

Section 9-5, page 368

1. No
3. $\{(5,1),(6,-3),(7,-4),(0,0)\}$
5. $\{(0,0),(1,1),(5,5),(9,7)\}$
7. $y = 3 - \dfrac{3x}{2}$
9. $y = \dfrac{x-3}{2}$
11. $y = 80 + \dfrac{4x}{5}$
13. No; $f(0) = f(1) = 0$
15. $y = x^{1/3}$
17. $y = \dfrac{1}{x}$
19. No; $y = x^{2/3}$; $x = \pm y^{3/2}$
21. $y = (x-4)^{-3}$
23. $y = x^{2/3}$; domain $\{x \mid x \geq 0\}$
25. Yes   27. No   29. Yes
31. $S = 1{,}250{,}000 + 25P$
33. $q = 5000 - 5000p$, where p is the price, 1000
35. $W = f(H) = 320 + 12H$, $H \geq 0$; $H = g(W) = (W - 320)/12$, $W \geq 320$; f and g are inverses of each other.
37. 70°F   39. $.2f^{-1}$ 41. $f^{-1} \circ g^{-1}$

Section 9-6, page 374

1. 2.72   3. 2.06×10^{-9} 5. 9.97
7. 1.59×10^{15} 9. $\exp(17)$ 11. $\exp(2x)$
13. $\exp(5w)$ 15. $\exp(5x + x^3)$

17.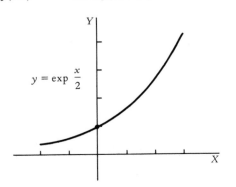

19.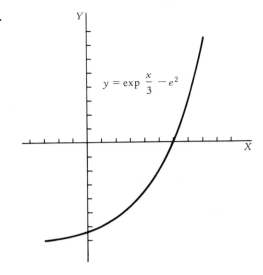

21. 0.00016 g, 0.040 g   23. 8.329%, 12.750%
25. a. 632   b. 909
   c. Yes; he can process (to the nearest integer) 1000 components a day.
27. 0.016   29. 0.078

Section 9-7, page 381

1. $\ln 4$ 3. $\ln\left(\dfrac{10}{3}\right)$ 5. $\ln 16$
7. $\ln\left(\dfrac{9}{125}\right)$ 9. $\ln[(x+1)/(x-2)]$ 11. $\ln\left(\dfrac{x^2}{y^3}\right)$
13. $\ln \dfrac{x^2}{yz^4}$ 15. $\ln(5e^2)$ 17. $\ln 7$
19. $\ln 22$ 21. 0.693   23. -11.5
25. 57.6   27. 5.20   29. 2.50×10^{33}
31. 2.11   33. $1 + e^x$ 35. $\dfrac{5 - \exp(x/4)}{2}$
37. $\ln(-x) - 3$ 39. $\dfrac{\ln(x/4) - 1}{2}$

41.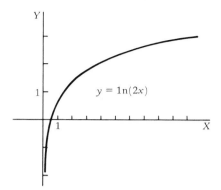

43.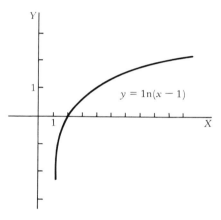

45. 1729 years 47. After about 46 hours
49. 13.4 years 51. $0.47

Section 9-8, page 386

1. $\log 115$
3. $\log\left(\dfrac{1}{8}\right)$
5. $\log 64$
7. $\log\left(\dfrac{x^2}{y}\right)$
9. $10^{4.5} = 3.2 \times 10^4$
11. $10^{11} = 100{,}000{,}000{,}000$
13. $10^{6.5} = 3.2 \times 10^6$
15. 7
17. 4.7
19. 3.5

Chapter 9 Review, page 387

1. Yes; domain $\{3, 4, 5, 6, 7\}$; range $\{2, 3, 5, 6\}$
2. $x \le \dfrac{5}{3}$
3. All numbers except -3 and 3
4. a. $16w^2 + 8w - 2$ b. $4w^2 - 11$
   c. $x^4 + 2x^2 - 2$ d. $-4v + 13$
5. Yes, $y = \dfrac{3x}{x^2 - 5}$ 6. No
7. Yes, $y = (27 - x^3)^{1/3}$
8. a. $y + 3 = 7(x - 2)$ b. $y = 7x - 17$
9. a. $y - 2 = -2(x - 9)$ b. $y = -2x + 20$
10. $m = \dfrac{5}{2},\ b = -20$

11.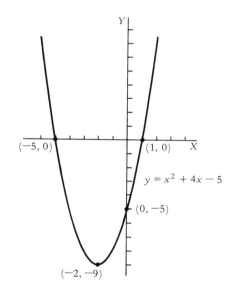

12.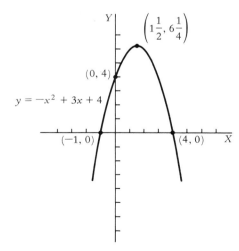

13. $v^2 + 11v + 2$
14. $-4v^2 - 26v + 4$
15. $3v^3 + 26v^2 + 16v$
16. $3v^2 + 24v + 2$
17. $9v^2 + 36v + 20$
18. $\dfrac{3v + 2}{v^2 + 8v}$
19. 13
20. $\dfrac{v-2}{3}$
21. $y = \dfrac{5}{3}x + 10$
22. No inverse
23. $y = \exp\left(\dfrac{x}{2}\right) - 3$
24. $y = \dfrac{\ln x - 19}{2}$
25. $\exp(5x - 6)$
26. $\exp(x^2 - 2x)$
27. $\ln 12$
28. $\ln 4$
29. $\ln\left(\dfrac{x+3}{x-3}\right)$
30. $\ln 2.5$
31. $\log 9$
32. $\log(x^{1/2} y^3)$
33. $5000 + 15q$
34. $\dfrac{5000}{q} + 15$
35. $26q - \dfrac{q^2}{500}$
36. $\dfrac{-q^2}{500} + 11q - 5000$
37. $\dfrac{-q}{500} + 11 - \dfrac{5000}{q}$
38. 500 and 5000
39. $10,125
40. 2750
41. + $5000
42. + $22,560
43. $B = 0.06A + 4.50,\ A \geq 0$
44. $A = \dfrac{100B}{6} - 75,\ B \geq 4.50$
The functions are inverses of each other.
45. $3547.67, about 6.9 years, about 11 years

Chapter 10

Section 10-1, page 395

1. -8
3. 4
5. e^4
7. 0
9. $\dfrac{2}{30}$
11. $\dfrac{1}{4}$
13. Undefined
15. Undefined
17. $\dfrac{1}{4}$
19. Undefined
21. $\dfrac{1}{8}$
23. 0
25. 0
27. 1
29. 4
31. 3

Section 10-2, page 402

1. $f'(x) = 2,\ f'(16) = 2$
3. $f'(t) = 4,\ f'(16) = 4$
5. $h'(s) = 8s,\ h'(16) = 128$
7. $j'(v) = \dfrac{3}{\sqrt{v}},\ j'(16) = \dfrac{3}{4}$
9. $y - 25 = -10(x + 5),\ y = -10x - 25$
11. $y - 1 = 2(x - 1),\ y = 2x - 1$
13. $y - 1 = -(x - 1),\ y = -x + 2$
15. $y - 7 = \dfrac{x - 49}{14},\ y = \dfrac{x}{14} + \dfrac{7}{2}$
17. $10
19. $8.33
21. $.34 million
23. $0.72 million
25. $0.45 million
27. 2.6
29. -2
31. -1
33. $g'(x) = 3x^2$
35. $f'(w) = -\dfrac{2}{w^3}$
37. $k'(y) = \begin{cases} -1 & y < 0 \\ 1 & y > 0 \end{cases}$

Section 10-3, page 406

1. 4, 4
3. $2x + 3, 13$
5. $6x^{1/2} + 2, 20$
7. $5e^x, 5$
9. $-\dfrac{3}{x}, -1$
11. $-2x^{-3/2} + 4x^{-1/2}, \dfrac{7}{4}$
13. $\dfrac{12}{x} + 4, 8$
15. $e^x + \dfrac{1}{x},\ e^e + \dfrac{1}{e}$
17. $2x + 4x^{-1} - 8x^{-2}, 4$
19. $3x^2 - 14x + 4, -12$
21. $y = 5x - 4$
23. $y = x + 1$

Section 10-4, page 411

1. Decreasing: all real numbers
3. Decreasing: $x < \dfrac{7}{8}$; increasing: $x > \dfrac{7}{8}$
5. Decreasing: $-3 < x < -2$; increasing: $x < -3$ or $x > -2$
7. Increasing: all real numbers
9. Decreasing: $-3 < x < 7$; increasing: $x < -3$ or $x > 7$
11. Increasing: all real numbers
13. Decreasing: $s < 0$; increasing: $s > 0$
15. a. $G'(s) = \dfrac{1}{2000} - \dfrac{1.25}{s^2}$ b. $s > 50$ mph
c. $0 \leq s < 50$ mph d. Help

Section 10-5, page 414

1. $MC(q) = 15,\ MR(q) = 25 - \dfrac{q}{5},\ MP(q) = 10 - \dfrac{q}{5}$. The maximum possible profit is $150 when $q = 50$.
3. $MC(q) = 5 + \dfrac{q}{500},\ MR(q) = 8,\ MP(q) = 3 - \dfrac{q}{500}$. The

maximum possible profit is $1500 when $q = 1500$.
5. $MC(q) = 8 + \frac{q}{250}$, $MR(q) = 12$, $MP(q) = 4 - \frac{q}{250}$. The maximum possible profit is $1200 when $q = 1000$.
7. $3.215
9. a. $650.00   9. b. $680.10
9. c. $30.10   9. d. $30.00

Section 10-6, page 418

1. $e^x(x^2 + 2x)$ 3. $1 + \ln x$ 5. $-\dfrac{8x}{(x^2 - 25)^2}$

7. $\dfrac{2 + \ln x}{2\sqrt{x}}$ 9. $\dfrac{3x^2 \ln x - x^2}{(\ln x)^2}$ 11. $\dfrac{27}{(5x + 6)^2}$

13. $\dfrac{-24x}{(x^2 - 9)^2}$ 15. $\dfrac{42}{(5 - 8x)^2}$ 17. $\dfrac{2\ln x}{x}$

19. $\dfrac{-1}{x(\ln x)^2}$ 21. $e^x(1 + \ln x + x\ln x)$

23. $\dfrac{3(\ln x)^2}{x}$ 25. $y = 0$

27. $y = \dfrac{-x}{e^2} + \dfrac{4}{e^2}$ 29. $1.92   31. $4.70

33. a. $0 \leq t < 5$ b. $t > 5$

Section 10-7, page 424

1. $12.5(5x - 2)^{3/2}$ 3. $8x(9 - x^2)^{-5}$
5. $10e^{10x}$ 7. $2x\exp(x^2)$
9. $(3x^2 + 14x - 5)\exp(x^3 + 7x^2 - 5x + 3)$
11. $\dfrac{8x + 5}{4x^2 + 5x + 7}$ 13. $\dfrac{2x^2 - 6x - 5}{(2x - 3)^2}$

15. $50(x^2 + 1)^{1/2}x^{49} + (x^2 + 1)^{-1/2}x^{51}$

17. $\dfrac{-9x}{(x^2 - 25)^{3/2}}$ 19. $\dfrac{12e^{2x}(4x + 5)}{(4x + 7)^2}$

21. $\dfrac{9x - 12}{\sqrt{x - 2}}$ 23. $-\dfrac{10x^2 + 18}{(x^2 - 9)^4}$

25. $\dfrac{\ln(x + 1)}{2\sqrt{x}} + \dfrac{\sqrt{x}}{x + 1}$ 27. $(3 - 2\ln x)/(3x^{5/3})$

29. $e^x \exp(e^x)$ 31. $\dfrac{3(\ln x)^2}{x} + \dfrac{10\ln x}{x}$

33. $\dfrac{2e^x}{\sqrt{1 + 4e^x}}$ 35. $\dfrac{20(e + 2\ln x)^9}{x}$

37. a. 4% a month   b. 2.48% a month
    c. 9.07%   d. 0.36% a month
39. 50, 12   41. 391, 60   43. 887, 25   45. 990, 3

47. $h(p) = 1000(10 - p)^{1/3}$ 49. 1000   51. 1010

Section 10-8, page 428

1. $y' = 8x - 7$, $y'' = 8$
3. $y' = -3x^2 + 12x + 12$, $y'' = -6x + 12$
5. $y' = \dfrac{6}{x + 5}$, $y'' = -\dfrac{6}{(x + 5)^2}$
7. $y' = -8x\exp(-x^2)$, $y'' = (16x^2 - 8)\exp(-x^2)$
9. $y' = e^x\left(\ln x + \dfrac{1}{x}\right)$, $y'' = e^x\left(\ln x + \dfrac{2}{x} - \dfrac{1}{x^2}\right)$
11. $y' = 5x(1 + x^2)^{3/2}$, $y'' = (20x^2 + 5)(1 + x^2)^{1/2}$
13. $f'(x) = 3x^2 - 6x + 8$, $f''(x) = 6x - 6$, $f'''(x) = 6$, $f^{(n)}(x) = 0$ for $n \geq 4$
15. $f'(x) = 12x^3 + 15x^2 - 14x - 9$, $f''(x) = 36x^2 + 30x - 14$,
    $f'''(x) = 72x + 30$, $f^{(4)}(x) = 72$, $f^{(n)}(x) = 0$ for $n \geq 5$
17. $f^{(n)}(x) = 2^n \exp(2x)$, $n \geq 1$

Chapter 10 Review, page 428

1. e^6 2. $\dfrac{12}{32}$ 3. $\dfrac{10}{8}$
4. 0   5. 8   6. $6w$
7. 12   8. $12x - 9$
9. $5 + \dfrac{3}{x}$ 10. $3x^{1/2} - 5\exp(x)$
11. $5x^5(6\ln x + 1)$ 12. $\dfrac{8e^x(x - 4)}{x^5}$
13. $-12e^{-12x}$ 14. $\dfrac{2 - 2x}{9 + 2x - x^2}$
15. $-28(5 - 4x)^6$ 16. $(2x + 10x^2)\exp(10x)$
17. $\dfrac{4x^2 - 2x - 11}{(4x - 1)^2}$ 18. $\dfrac{\exp(5x)(5x^2 - 2x + 25)}{(5 + x^2)^2}$
19. $y = 5x - 14$ 20. $y = 2e^2x - e^2$
21. $y = x - 1$ 22. $y = 0.25x + 0.25$
23. Increasing: all real numbers
24. Increasing: $x < \dfrac{1}{2}$; decreasing: $x > \dfrac{1}{2}$
25. Increasing: $x < -2$ or $x > 3$; decreasing: $-2 < x < 3$
26. Increasing: $x > -1$; decreasing: $x < -1$
27. $y' = 16x - 7$, $y'' = 16$
28. $f'(x) = \dfrac{4}{x}$, $f''(x) = -\dfrac{4}{x^2}$
29. $y' = 2x\exp(6 + x^2)$, $y'' = (2 + 4x^2)\exp(6 + x^2)$

30. $f'(v) = 5v(4+v^2)^{3/2}$, $f''(v) = 20(1+v^2)(4+v^2)^{1/2}$
31. $2.84   32. $800.00, $2.50   33. 80,000, $2800
34. 0.020 grams per hour

Chapter 11

Section 11-1, page 441

1. $y' = 2x$, critical number 0, local minimum
3. $y' = 4x^3$, critical number 0, local minimum
5. $y' = -\dfrac{4x^{1/3}}{3}$, critical number 0, local maximum
7. $y' = 8x + 8$, critical number -1, local minimum
9. $y' = 3x^2 + 18x + 15$, critical numbers: -5 local maximum, -1 local minimum
11. $y' = 3(x+4)^2$, critical number -4, not a local extremum
13. $y' = e^x(1+x)$, critical number -1, local minimum
15. $y' = 1 + \ln x$, critical number $\dfrac{1}{e}$, local minimum
17. $y' = 3x^2 - 3$, critical numbers: -1 local maximum, 1 local minimum
19. $y' = -10x(1-x^2)^4$, critical numbers: $-1, 1$ not local extrema, 0 local maximum
21. $y' = 5e^{5x}$, no critical numbers
23. $y' = -2x\exp(-x^2)$, critical number 0, local maximum
25. $y' = \dfrac{7(x-7)(x-1)}{3x^{2/3}}$, critical numbers: 0 not local extremum, 1 local maximum, 7 local minimum
27. $y' = (x-4)^7(x+6)^{2/3}(29x+124)/3$, critical numbers: $-\dfrac{124}{29}$ local maximum, -6 not local extremum, 4 local minimum
29. No critical numbers   31. $\left(\dfrac{1}{120}\right)\%$ at $t = 3$
33. $P(40) = \$5120$
35. $\dfrac{N}{2}$ 37. $\dfrac{2}{3}R$
39. $1.5w$ 41. All points on the graph of f
43. All points on the graph of f
45. $y' = 9.44x - 3.25$, critical number: 0.34 local minimum
47. $y' = 9.81x^2 + 10.28x - 9.36$, critical numbers: -1.63 local maximum, 0.58 local minimum

Section 11-2, page 448

1. $y'' = 2$, local minimum at 0
3. $y'' = 12x^2$, no conclusion at 0 since $y''(0) = 0$
5. $y'' = -\dfrac{4x^{-2/3}}{9}$. The second derivative test is not applicable since $y''(0)$ is undefined.
7. $y'' = 8$, local minimum at -1
9. $y'' = 6x + 18$, local maximum at -5 since $y''(-5) = -12 < 0$, local minimum at -1 since $y''(-1) = 12 > 0$
11. $y'' = 6(x+4)$, no conclusion at -4 since $y''(-4) = 0$
13. $y'' = e^x(2+x)$, local minimum at -1 since $y''(-1) = e^{-1} > 0$
15. $y'' = \dfrac{1}{x}$, local minimum at $\dfrac{1}{e}$ since $y''\left(\dfrac{1}{e}\right) = e > 0$
17. $y'' = 6x$, local maximum at -1 since $y''(-1) = -6 < 0$, local minimum at 1 since $y''(1) = 6 > 0$
19. $y'' = -10x \cdot 4(1-x^2)^3 \cdot (-2x) - 10(1-x^2)^4 = 10(1-x^2)^3(9x^2-1)$, no conclusion at -1 and 1, since $y''(-1) = y''(1) = 0$, local maximum at 0 since $y''(0) = -10 < 0$
23. $y'' = -2x\exp(-x^2) \cdot (-2x) - 2\exp(-x^2) = 2\exp(-x^2)(2x^2-1)$, local maximum at 0 since $y''(0) = -2 < 0$
33. $P'' = -0.24q$, local maximum at 40 since $P''(40) = -9.6 < 0$
35. $R''(x) = -2k$, local maximum at $\dfrac{N}{2}$ since $R''\left(\dfrac{N}{2}\right) = -2k < 0$
37. $v'' = k(2R - 6r)$, local maximum at $\dfrac{2}{3}R$ since $v''\left(\dfrac{2}{3}R\right) = -2kR < 0$
45. $y'' = 9.44$, local minimum at 0.34 since $y''(0.34) = 9.44 > 0$
47. $y'' = 19.62x + 10.28$, local maximum at -1.63 since $y''(-1.63) = -21.7 < 0$, local minimum at 0.58 since $y''(0.58) = 21.7 > 0$

Section 11-3, page 457

1. 6 and 6   3. 25×25
5. $5625; sell 75 tickets   7. 56
9. $4 \times 4 \times 2$ 11. $32^{1/3} \times 32^{1/3} \times 32^{1/3}$
13. 128 cu in. $(8 \times 8 \times 2)$ 15. 4800 sq yd (60×80)

17. $360 when $q = 80$ 19. $11.50 when $q = 100$
21. 40   23. 50, 4

Section 11-4, page 463

1. $y_v(v, w) = 3v^2 e^{5w}$, $y_w(v, w) = 5v^3 e^{5w}$, $y_v(2, 0) = 12$, $y_w(2, 0) = 40$

3. $z_x(x, y) = 2x^{-1/2} \ln(3y + 2)$, $z_y(x, y) = \dfrac{12 x^{1/2}}{3y + 2}$, $z_x(9, 0) = \dfrac{2 \ln 2}{3}$, $z_y(9, 0) = 18$

5. $z_v(v, w, x) = 2v \ln x \exp(5w)$, $z_w(v, w, x) = 5v^2 \ln x \exp(5w)$, $z_x(v, w, x) = \dfrac{v^2 \exp(5w)}{x} + 3$, $z_v(2, 1, 6) = 4 \cdot \ln 6 \cdot e^5$, $z_w(2, 1, 6) = 20 \cdot \ln 6 \cdot e^5$, $z_x(2, 1, 6) = \dfrac{4 e^5}{6} + 3$

7. $y_w(w, x) = (x + w)^{1/2} + \dfrac{w(x + w)^{-1/2}}{2}$, $y_x(w, x) = \dfrac{w(x + w)^{-1/2}}{2}$, $y_w(4, 5) = \dfrac{11}{3}$, $y_x(4, 5) = \dfrac{2}{3}$

9. $z_w(w, x, y) = 3 + x^3$, $z_x(w, x, y) = 8x + 3wx^2$, $z_y(w, x, y) = -6$, $z_w(-2, 1, -1) = 4$, $z_x(-2, 1, -1) = 2$, $z_y(-2, 1, -1) = -6$

11. $w_x(x, y, z) = 4x^3 \exp(y^2)$, $w_y(x, y, z) = 2x^4 y \exp(y^2)$, $w_z(x, y, z) = 0$, $w_x(-2, 0, 6) = -32$, $w_y(-2, 0, 6) = 0$, $w_z(-2, 0, 6) = 0$

13. $z_x(x, y) = 3 + 8x$, $z_y(x, y) = -2$, $z_{xx}(x, y) = 8$, $z_{xy}(x, y) = z_{yx}(x, y) = 0$, $z_{yy}(x, y) = 0$

15. $z_x(x, y) = 12 x^2 y^4 - 12 xy^2$, $z_y(x, y) = 16 x^3 y^3 - 12 x^2 y$, $z_{xx}(x, y) = 24 xy^4 - 12 y^2$, $z_{xy}(x, y) = z_{yx}(x, y) = 48 x^2 y^3 - 24 xy$, $z_{yy}(x, y) = 48 x^3 y^2 - 12 x^2$

17. $y_v(v, w) = 16(2v + 3w)^7$, $y_w(v, w) = 24(2v + 3w)^7$, $y_{vv}(v, w) = 224(2v + 3w)^6$, $y_{vw}(v, w) = y_{wv}(v, w) = 336(2v + 3w)^6$, $y_{ww}(v, w) = 504(2v + 3w)^6$

19. $w_x(x, y) = (3x^2 + x^3 y)\exp(xy)$, $w_y(x, y) = x^4 \exp(xy)$, $w_{xx}(x, y) = (6x + 6x^2 y + x^3 y^2)\exp(xy)$, $w_{xy}(x, y) = w_{yx}(x, y) = (4x^3 + x^4 y)\exp(xy)$, $w_{yy}(x, y) = x^5 \exp(xy)$

21. $w_x(x, y, z) = 3 + 3x^2 y^2 \exp(10z)$, $w_y(x, y, z) = 2 + 2x^3 y \exp(10z)$, $w_z(x, y, z) = -4 + 10 x^3 y^2 \exp(10z)$, $w_{xx}(x, y, z) = 6 xy^2 \exp(10z)$, $w_{xy}(x, y, z) = w_{yx}(x, y, z) = 6 x^2 y \exp(10z)$, $w_{xz}(x, y, z) = w_{zx}(x, y, z) = 30 x^2 y^2 \exp(10z)$, $w_{yy}(x, y, z) = 2 x^3 \exp(10z)$, $w_{yz}(x, y, z) = w_{zy}(x, y, z) = 20 x^3 y \exp(10z)$, $w_{zz}(x, y, z) = 100 x^3 y^2 \exp(10z)$

23. $p_v(v, w, x) = 2vw^3$, $p_w(v, w, x) = 3v^2 w^2 - 5x^4$, $p_x(v, w, x) = -20 wx^3$, $p_{vv}(v, w, x) = 2w^3$, $p_{vw}(v, w, x) = p_{wv}(v, w, x) = 6vw^2$, $p_{vx}(v, w, x) = p_{xv}(v, w, x) = 0$, $p_{ww}(v, w, x) = 6v^2 w$, $p_{wx}(v, w, x) = p_{xw}(v, w, x) = -20 x^3$, $p_{xx}(v, w, x) = -60 wx^2$

25. $w_x(x, y) = 3x^2 y^2 + 2xy^3$, $p_y(x, y) = 2x^3 y + 3x^2 y^2$, $w_{xx}(x, y) = 6 xy^2 + 2y^3$, $w_{xy}(x, y) = w_{yx}(x, y) = 6x^2 y + 6xy^2$, $w_{yy}(x, y) = 2x^3 + 6x^2 y$

27. $3.40   29. Increase by $.05
31. a. $4.50   b. $2.00   c. $4.00
33. 103513
35. a. 1550   b. 1700
37. $740,000   39. $535,000
41. $+7.78$ 43. $+4.86$

Section 11-5, page 471

1. $(1, 3)$ saddle point   3. $(4, 1)$ local maximum
5. $(2, -1)$ local minimum   7. $(-1, 1)$ local maximum
9. $(-2, 0)$ local maximum, $(2, 0)$ saddle point
11. $(0, 0)$ saddle point
13. $a = \$1.00$, $c = 50¢$ 15. $x = 3$, $y = 2$

Section 11-6, page 475

1. $16\frac{16}{17}$ at $(48/17, 36/17)$ 3. $32/9$ at $(4/9, -16/9)$
5. 2.5 at $(2.5, 1)$
7. Minimum is -17 at $(-3, -4)$; maximum is 33 at $(3, 4)$
9. $x = \$9$ million, $y = \$8$ million

Section 11-7, page 479

1. $y' = \dfrac{-x^2}{y^2}$ 3. $y' = -\dfrac{2x}{8y + 3}$
5. $y' = -\dfrac{2x}{3 e^{3y}}$ 7. $y' = -3 x^{1/2} y$

9. $y' = \dfrac{4x^3 - 20(4x+3y)^4}{15(4x+3y)^4 - 4y^3}$ 11. $y = -\dfrac{x}{3} + \dfrac{7}{3}$
13. $y = 2x - 6$ 15. 2 ft/hr
17. $625/week   19. $125/week
21. $490/week   23. $60/week

Chapter 11 Review, page 480

1. $(1, 11)$ local maximum
2. $(-3, 0)$ local minimum; $(0, 6561)$ local maximum; $(3, 0)$ local minimum
3. $(-5, 455)$ local maximum; $(4, -274)$ local minimum
4. $\left(1, \dfrac{1}{e}\right)$ local maximum
5. $120.00
6. $230.00
7. $h_x(x, y) = 12x^2 - 6xy + 6y^2$, $h_y(x, y) = -3x^2 + 12xy - 6y^2$
8. $w_x(x, y) = 8(2x - 3y)^3$, $w_y(x, y) = -12(2x - 3y)^3$
9. $z_v = 2v - 2x^3$, $z_w = 6w + x$, $z_x = -6vx^2 + w$
10. $y_p = 2p\exp(2q + 4r)$, $y_q = 2p^2\exp(2q + 4r)$, $y_r = 4p^2\exp(2q + 4r)$
11. $\dfrac{\partial z}{\partial x} = (2x + x^2y^2)\exp(xy^2)$, $\dfrac{\partial z}{\partial y} = 2x^3y\exp(xy^2)$
12. $z_w = \dfrac{2wx^3}{w^2 + 4y}$, $z_x = 3x^2\ln(w^2 + 4y)$, $z_y = \dfrac{4x^3}{w^2 + 4y}$
13. $(5, 2)$ local minimum
14. $(-1, 2)$ saddle point
15. $(0, 1)$ saddle point; $(16, 1)$ local minimum
16. $(0, 0)$ local maximum
17. $\dfrac{216}{5}$ at $\left(\dfrac{18}{5}, \dfrac{12}{5}\right)$ 18. -3.4 at $(0.6, 1.5)$
19. $y' = \dfrac{2x}{3y}$ 20. $y' = \dfrac{4x + 3}{2y}$
21. $y' = \dfrac{xy^2 + 2}{3 - x^2y}$ 22. $y' = \dfrac{5\exp(5x - 4y) - 14}{4\exp(5x - 4y) + 8}$

Chapter 12

Section 12-1, page 487

1. $\dfrac{x^3}{3} + 3x^2 - 3x + c$ 3. $2x^{7/2} + c$

5. $5\ln|x| + c$ 7. $8e^x - \dfrac{x^4}{4} + c$
9. $\dfrac{3x^{8/3}}{8} + \dfrac{3x^{-2/3}}{2} + c$
11. $\dfrac{x^4}{4} + \dfrac{4x^3}{3} - \dfrac{3x^2}{2} + 2x + c$
13. $\dfrac{-4}{x} + \dfrac{5}{2x^2} + c$
15. $\dfrac{5x^4}{4} - 2x^2 + 3\ln|x| + c$
17. $4e^x + 5x + c$ 19. $2\ln|x| - \dfrac{7}{2x^2} + c$
21. $4x + 3\ln|x| + c$ 23. $\dfrac{12x^{5/2}}{5} + \dfrac{4x^3}{3} + c$
25. $C(q) = 0.8q - 0.0002q^2 + 150$
27. $P(q) = 0.6q - 0.0001q^2 - 400$
29. a. $h(t) = -16t^2 + 64t + 96$ b. 160 ft
    c. $t = 2 + \sqrt{10} = 5.16$ sec
31. $\dfrac{x^2}{2} + \ln|x| + c$ 33. $x - 2\ln|x| + c$

Section 12-2, page 494

3. 8   5. 11
7. $I_P = 36$, $C_P = 100$ 9. $I_p = 19.15625$, $C_P = 22.90625$

Section 12-3, page 501

1. 36   3. $\dfrac{64}{3}$ 5. $9\tfrac{1}{3}$
7. $4\ln 6 = 7.17$ 9. $\dfrac{15}{4}$
11. $27 + 2\ln 2 = 28.4$ 13. $e^2 - e = 4.67$
15. $e^2 - 2e = 1.95$ 17. 0
19. -2 21. $-\dfrac{28}{3}$ 23. $3e^4 - 43 = 120.8$
25. $1500 + 40\ln 2.5 = 1537$ minutes
27. a. 3900   b. 5700
29. a. $133   b. $445
31. a. $500{,}000\ln 2$, about $350{,}000$ b. $500{,}000\ln 2$, about $350{,}000$
33. $\dfrac{e^5 - 1}{5} = 29.5$ 35. 14   37. $\dfrac{1}{6}$

Section 12-4, page 511

1. $\displaystyle\int_{-2}^{-1}(x^2 + x)\,dx - \int_{-1}^{0}(x^2 + x)\,dx = 1$

ANSWERS TO ODD-NUMBERED EXERCISES AND REVIEW EXERCISES

3. $-\int_{-2}^{-1}(3x+3)\,dx + \int_{-1}^{2}(3x+3)\,dx = 15$

5. $\int_{0}^{1}(x^2-5x+4)\,dx - \int_{1}^{4}(x^2-5x+4)\,dx = 6\frac{1}{3}$

7. $\int_{0}^{1}(x^2+2x+3)\,dx = 4\frac{1}{3}$

9. $\int_{1}^{2}\left(\frac{4}{x}-2\right)dx - \int_{2}^{8}\left(\frac{4}{x}-2\right)dx = 10 + 4\ln\left(\frac{1}{2}\right)$
$= 7.23$

11. $-\int_{-1}^{0}(e^x-1)\,dx + \int_{0}^{1}(e^x-1)\,dx = e - 2 + \frac{1}{e} = 1.09$

13. $\int_{1}^{2}(e^x-1)\,dx = e^2 - e - 1 = 3.67$

15. $\int_{-4}^{0}(x^2-x)\,dx + \int_{0}^{1}(x-x^2)\,dx = 29.5$

17. $\int_{0}^{4}(x+6-\sqrt{x})\,dx = 26\frac{2}{3}$

19. $\int_{0}^{1}(x-x^2)\,dx = \frac{1}{6}$

21. $\int_{0}^{1}(x^2-x^3)\,dx = \frac{1}{12}$

23. $\int_{2}^{4}(6x-8-x^2)\,dx = 1\frac{1}{3}$

27. $\frac{1}{3}(x^2+1)^{3/2}\Big|_{0}^{1} = \frac{2\sqrt{2}-1}{3} = 0.609$

29. $\frac{-\exp(4-5t)}{5}\Big|_{0}^{1} = \frac{-e^{-1}+e^4}{5} = 10.8$

31. $\frac{1}{8}\ln(4w^2+9) + c$

33. $\frac{-1}{2}\ln|7-t^2| + c$

35. $\frac{1}{2}(7-t^2)^{-1}\Big|_{0}^{2} = \frac{2}{21}$

37. $21.6 million

39. a. 18.5%   b. 26.0%
    c. 32.3%   d. 34.8%
    e. 35.0%   f. 35.0%

41. $10\sqrt{2900} - 10\sqrt{1300} = \$178$

Section 12-5, page 517

1. $\frac{1}{6}(9+4x)^{3/2}\Big|_{0}^{4} = 16\frac{1}{3}$

3. $\frac{1}{2}\ln(x^2+16)\Big|_{0}^{3} = \ln\left(\frac{5}{4}\right) = 0.223$

5. $\frac{(2x-1)^9}{18} + c$

7. $\frac{(3x-2)^{5/3}}{5} + c$

9. $10(2x+3)^{1/2} + c$

11. $\frac{-1}{2}(100-4x)^{1/2}\Big|_{0}^{9} = 1$

13. $\frac{-(5x+10)^{-1}}{5}\Big|_{0}^{2} = \frac{1}{100}$

15. $\frac{-2}{3}(3x+4)^{-0.5} + c$

17. $\frac{-4}{x^2+4} + c$

19. $\frac{1}{3}\ln(x^3+1)\Big|_{1}^{5} = \frac{1}{3}\ln 63 = 1.38$

21. $\frac{1}{2}\exp(x^2-2x+3) + c$

23. $-\exp(3-v) + c$

25. $-\frac{1}{2}\ln(8-2x)\Big|_{0}^{3} = \ln 2 = 0.693$

Chapter 12 Review, page 519

1. $\frac{x^3}{3} - 2x^2 + 7x + c$

2. $3e^x - 4x + c$

3. $-x^{-1} - \ln|x| + c$

4. $\frac{x^4}{4} - \frac{4x^3}{3} - \frac{3x^2}{2} + 8x + c$

5. $\frac{-3}{x^2} - 4\ln|x| + c$

6. $2\sqrt{x} + 3x + c$

7. $\frac{26}{3}$

8. $2\ln 3 = 2.20$

9. 65

10. $2e - 3 = 2.44$

11. $\frac{-2}{9}(7-3x)^{3/2} + c$

12. $\frac{\exp(8t-3)}{8} + c$

13. $0.5\ln\left(\frac{39}{11}\right) = .63$

14. $\frac{2(x^3-4)^6}{3} + c$

15. $\frac{e^5-e^{-5}}{5} = 29.7$

16. $\frac{1}{4}$

17. 24

18. $\frac{e^8-1}{2} = 1490$

19. 2

20. $65\frac{1}{3}$

21. $e^2 - 2e - 1 = 0.95$

22. $9\frac{5}{6}$

23. $7.5 - \ln 4 = 6.11$

24. $\frac{1}{2}$

25. $\frac{32}{3}$

26. $57\frac{1}{6}$

27. $C(q) = 15q - 0.02q^2 + 120$

28. $67,008

29. $2,000,000, $8,000,000

30. 1,000,000 barrels

Chapter 13

Section 13-1, page 526

1. $C(q) = 40 + 10q + \frac{3q^2}{2}$, $R(q) = 40q - q^2$, $P(q) = 30q - 2.5q^2 - 40$, maximum profit is $P(6) = 50$
3. $C(q) = 100q - q^2 + 30$, $R(q) = 110q - 1.5q^2$, $P(q) = 10q - \frac{q^2}{2} - 30$, maximum profit is $P(10) = 20$
5. $C(q) = 50 + 10q + \frac{q^2}{2}$, $R(q) = 60q$, $P(q) = 50q - \frac{q^2}{2} - 50$, maximum profit is $P(50) = 1200$
7. $C(10) = 180 = C(0) + \int_0^{10}(20 - q)\, dq = C(0) + 150$, $C(0) = 30$
9. $C(30) = 2500 = C(0) + \int_0^{30}(80 - 2q - 0.1q^2)\, dq = C(0) + 600$, $C(0) = 1900$
11. 206.25
13. -5
15. $(4, 19)$, C.S. $= 16$, P.S. $= 32$
17. $(8, 34)$, C.S. $= 64$, P.S. $= 96$
19. $(3, 21)$, C.S. $= 13.5$, P.S. $= 18$
21. a. $510 \qquad$ b. $301
23. a. $C(q) = 155q + 1000$
    b. $R(q) = 150q + 500\ln[(q+3)/3]$
    c. $P(q) = 500\ln\left(\frac{q+3}{3}\right) - 1000 - 5q$
    d. $P(97) = \$268$
    e. $\$2.77, 97$

Section 13-2, page 534

1. $\frac{y^3}{3} + \frac{x^3}{3} = c$, $y = (3c - x^3)^{1/3}$
3. $\frac{5y^3}{3} + e^x = c$, $y = \left[\frac{3}{5}(c - e^x)\right]^{1/3}$
5. $2y^2 + 20y - \frac{2}{3}x^{3/2} = c$
7. $\frac{1}{4}\ln|y - 3| + x = c$
9. $\frac{y^2}{2} + 6\ln|x| = c$
11. $y - \frac{4}{x} = c$, $c = 1$, $y = \frac{4}{x} + 1$
13. $|(y+5)x| = c$, $c = 8$, $y = \frac{8}{x} - 5$
15. $\frac{-1}{y} + \ln|x| = c$, $c = 0$, $y = \frac{1}{\ln x}$

17. $\frac{x^3}{3} - \frac{y^3}{3} = c$, $c = 0$, $y = x$
19. $e^y - 2e^{-x} = c$, $c = e - 2$, $y = \ln(e - 2 + 2e^{-x})$
21. $m(t) = 50 \exp\left(\frac{-t^2}{2}\right)$
23. $D = 250{,}000{,}000 \exp\left(\frac{-0.045 t^2}{2}\right)$
25. $\frac{dm}{dt} = -km$

$\frac{dm}{m} = -k\, dt$

$\ln m = -kt + c$

$\ln 50 = -k \cdot 0 + c \quad \text{since } m(0) = 50$

$\ln 25 = -k \cdot 8 + c \quad \text{since } m(8) = \frac{1}{2}m(0) = 25$

$k = \frac{\ln 2}{8}$

$\ln m = \frac{-\ln 2}{8} t + \ln 50 = \ln[50 \cdot 2^{-t/8}]$

$m = 50 \cdot 2^{-t/8}$

27. $\frac{du}{u - 60} = -kt$

$\ln(u - 60) = -kt + c$

$\ln(100 - 60) = \ln 40 = -k \cdot 0 + c = c$

$\ln(90 - 60) = \ln 30 = -3k + c$

$c = \ln 40, \quad k = \frac{1}{3}\ln\left(\frac{4}{3}\right)$

$\ln(u - 60) = \frac{-t}{3}\ln\left(\frac{4}{3}\right) + \ln 40$

$u = 60 + 40\left(\frac{4}{3}\right)^{-t/3}$

$u(6) = 60 + 40\left(\frac{4}{3}\right)^{-2} = 82.5°$

Section 13-3, page 540

1. $26/27$, $\text{EV}(x) = 2.25 \qquad$ 3. $1/3$, $\text{EV}(x) = 7$
5. $10/234$, $\text{EV}(x) = 342/78$
7. $1/2$, $\text{EV}(x) = \frac{3}{\ln 4} = 2.16$

9. $\dfrac{1}{14}$

11. $\dfrac{3}{e^{12}-e^3} = 0.000018$

13. $2e/(e-1) = 3.16$

15. $\dfrac{3}{\ln(32/6)} = 1.79$

17. a. $\dfrac{15}{16}$ b. 8

19. a. $\dfrac{6250}{11250} = 0.56$ b. 100

21. a. $\dfrac{e-\sqrt{e}}{e-1} = 0.62$ b. $\dfrac{e-2}{e-1} = 0.42$

23. a. $\dfrac{28}{108} = 0.26$ b. 7 min

Section 13-4, page 546

1. $x^3 \ln\left(\dfrac{x}{3}\right) - \dfrac{x^3}{9} + c$

3. $\dfrac{3e^4 + 1}{16} = 10.3$

5. $-xe^{-x} - e^{-x} + c$

7. $\dfrac{1}{4} - \dfrac{3}{4e^2} = 0.15$

9. $2x^2 + \dfrac{x^3}{3} + c$

11. $\dfrac{xe^{5x}}{5} - \dfrac{e^{5x}}{25} + c$

13. $\dfrac{98}{3}$

15. $\ln\left(\dfrac{5}{4}\right) = 0.22$

17. 1.00 years

19. $90(1 - e^{-8}) = 89.97$ min

21. $(x^2 - 2x + 2)e^x + c$

23. $e^{3x}(9x^2 - 6x + 2)/27 + c$

Section 13-5, page 549

1. 9.125, 9
3. 20.55, $20\sqrt{10}/3 = 21.08$
5. 4.69, $e^2 - e = 4.67$
7. 5.68, 5.75

Chapter 13 Review, page 550

1. $C(q) = 300q + \dfrac{q^2}{2} + 3000$
2. $R(q) = 500q$
3. $P(q) = 200q - \dfrac{q^2}{2} - 3000$
4. $17,000
5. $250
6. $(4, 33)$, C.S. $= 42\tfrac{2}{3}$, P.S. $= 40$
7. $y = \dfrac{-2}{4x - x^2 + c}$
8. $y = \dfrac{(27 - 12\ln x)^{2/3} - 1}{2}$
9. $R = 10{,}000 - 8000 \exp(1 - t)$
10. $R(5) = 10{,}000 - 8000 \exp(-4) = \$9853$
11. $10,000
12. 23/27, 79/16
13. $\dfrac{2}{\ln 5} = 1.24$
14. 5/12, 242
15. $\dfrac{xe^{8x}}{8} - \dfrac{e^{8x}}{64} + c$
16. $\dfrac{2}{5}x^{5/2}\ln x - \dfrac{4}{25}x^{5/2} + c$
17. 0.26
18. 1.00

Index

Absolute value, 9
Algebraic expression, 45
Amortization, 232
    table, 237–238
Annuity, 220
    due, 225
    ordinary, 221
    present value of, 231
Antiderivative, 483
    formulas for, 484–485, 512–513
    linearity, 485
Area, 489
    between curves, 503–510
    properties of, 489
Augmented matrix, 131

Bayes' Theorem, 282
Binomial probability, 301
Breakeven point, 342

Carbon-14 dating, 380
Carrying cost, 456
Cartesian product, 251
Chain rule, 420
Change in a function, 363
Closed half-plane, 160
Coefficient matrix, 131
Column matrix, 104
Combination, 295
Common logarithm function, 383
Complement, 5, 256, 262
Composite function, 362
Compound interest, 214
Conditional probability, 273, 275
Conformable matrices, 106
Constant function, 348, 400
Constant of integration, 483
Constraint, 168
    nonnegativity, 162
Continuity, 392
Consumers' surplus, 525

Continuously compounded interest, 216, 373
Convex, 168
Coordinates, 88
Cost, 521
    average, 341, 454
    carrying, 456
    fixed, 341, 521
    function, 341
    marginal, 412, 454, 521
    minimum average, 454
    reorder, 455
    variable, 341, 522
Critical number, 432
Critical point, 432, 466

Decibels, 385
Definite integral, 492
    properties of, 501
Demand function, 343
Dependent events, 268
Dependent variable, 338
Derivative, 396, 398
    chain rule, 420, 421
    formulas for, 404, 405, 421
    linearity, 405
    partial, 459
    product rule, 416
    quotient rule, 417
    second, 426
Differentiable, 398
Differential, 484
Differential equation, 528
    separable, 528
Disjoint sets, 4, 255, 264
Domain, 338
Dominated strategy, 332
Dual problem, 200

Economic order quantity, 456
Effective interest rate, 216

Empirical probability, 247
Empty set, 2, 246
Equation, 53
    differential, 528
    general linear, 352
    identity, 56
    inconsistent, 56
    linear, 53
    quadratic, 59
    solution set, 53
Equilibrium point, 343
Equilibrium price, 343
Equilibrium quantity, 343
Event, 245
    dependent, 268
    independent, 268
Expected value, 313, 539
    fair bet, 316
    odds, 316
Exponent, 11–20
Exponential function, 372
    derivative of, 405, 421
    integral of, 485, 513
Extraneous root, 68
Extreme value, 431
    first derivative test, 444
    functions of two variables, 466–471
    original equation test, 437
    second derivative test, 446

Factorial, 290
Factoring, 32–36, 59
Feasible region, 171
First derivative test, 444
Fixed cost, 341, 521
Focus of a parabola, 355
Fractions, arithmetic of, 7
Function, 338
    average cost, 341
    average profit, 342
    average revenue, 341